Natural Products Synthesis

Through Pericyclic Reactions

Natural Products Synthesis
Through Pericyclic Reactions

Giovanni Desimoni

Gianfranco Tacconi

Achille Barco

Gian Piero Pollini

ACS Monograph 180

American Chemical Society
Washington, D.C. 1983

Library of Congress Cataloging in Publication Data

Natural products synthesis through pericyclic reactions.
(ACS Monograph; 180)

Bibliography: p.
Includes index.

1. Natural products. 2. Chemistry, Organic—Synthesis. 3. Cyclic compounds. I. Desimoni, Giovanni, 1936–

QD415.N36 1983 547′.2 83–12303
ISBN 0–8412–0757–7

PRINTED IN THE UNITED STATES OF AMERICA

ACS Monographs

Marjorie C. Caserio, *Series Editor*

FOREWORD

ACS MONOGRAPH SERIES was started by arrangement with the interallied Conference of Pure and Applied Chemistry, which met in London and Brussels in July 1919, when the American Chemical Society undertook the production and publication of Scientific and Technologic Monographs on chemical subjects. At the same time it was agreed that the National Research Council, in cooperation with the American Chemical Society and the American Physical Society, should undertake the production and publication of Critical Tables of Chemical and Physical Constants. The American Chemical Society and the National Research Council mutually agreed to care for these two fields of chemical progress.

The Council of the American Chemical Society, acting through its Committee on National Policy, appointed editors and associates to select authors of competent authority in their respective fields and to consider critically the manuscripts submitted. Since 1944 the Scientific and Technologic Monographs have been combined in the Series. The first Monograph appeared in 1921, and up to 1972, 168 treatises have enriched the Series.

These Monographs are intended to serve two principal purposes: first to make available to chemists a thorough treatment of a selected area in form usable by persons working in more or less unrelated fields to the end that they may correlate their own work with a larger area of physical science; secondly, to stimulate further research in the specific field treated. To implement this purpose the authors of Monographs give extended references to the literature.

ABOUT THE AUTHORS

GIOVANNI DESIMONI studied at the University of Pavia, where he received his doctorate in chemistry. He spent 1968 at the University of East Anglia in Norwich, England, with Prof. A. Kratritzky. In 1975, he became Professor of Heterocyclic Chemistry at the Science Faculty of the University of Pavia, where he is now Dean of the Science Faculty. His research interests lie in the field of heterodiene reactions, in the chemistry of the pyrazole ring, and, recently, in organic syntheses assisted by transition metals.

GIANFRANCO TACCONI earned his doctorate in chemistry from the University of Pavia in 1957. He is currently Associate Professor of Organic Chemistry at the Science Faculty of the University of Pavia. His main scientific interests include the chemistry of the indole ring and heterodiene reactions, particularly those with enamines.

ACHILLE BARCO, after receiving his doctorate in chemistry from the University of Pavia, was appointed Associate Professor of Organic Chemistry at the University of Perugia. He now holds the same position at the Science Faculty of the University of Ferrara. His current research interests are in the synthesis of natural products.

GIAN PIERO POLLINI received his doctorate in chemistry from the University of Pavia; he then taught at the University of Perugia. In 1968, he moved to the University of Ferraro, where he is now Professor of Chemistry of Natural Compounds in the Faculty of Pharmacy. His research interests primarily concern the synthesis of natural products.

CONTENTS

PREFACE

THE METHODOLOGY OF PERICYCLIC REACTIONS has been widely applied to the synthesis of natural products. Notwithstanding these advances, several points need further research: theoretical work on model molecules and experimental determination of HOMO and LUMO energies have to be done to understand several reaction mechanisms and to allow prediction of the regiochemical and stereochemical behavior of several pericyclic reactions.

The field is rapidly growing; a measure of this growth is given by the 50 papers falling within the scope of this book that were published in 1981—in the *Journal of the American Chemical Society* alone.

We hope this review will stimulate research on, or at least interest in, a topic that has resulted in many impressive syntheses.

We wish to thank Remo Gandolfi and Angelo Albini for constructive discussion, and K. N. Houk for useful suggestions and valuable comments.

GIOVANNI DESIMONI
GIANFRANCO TACCONI

Istituto di Chimica
Organica dell'Università
di Pavia
27100 Pavia, Italy

ACHILLE BARCO
GIAN PIERO POLLINI

Istituto Chimico ed
Istituto di Chimica
Farmaceutica dell'Università
di Ferrara
44100 Ferrara, Italy

1

Introduction

Scientific developments are very often the result of human efforts either to mimic Nature or to "create the unnatural," whether in the field of human flight or chemistry. For example, Icarus's attempt to mimic Nature is part of Greek mythology; Wilbur and Orville Wright successfully created the unnatural at Kitty Hawk, N.C., December 17, 1903. In chemistry, along with synthesizing new compounds—creating the unnatural, as it were—an enormous effort has been devoted to synthesizing naturally occurring compounds—copying Nature. For a long time the main difficulty in synthesizing natural compounds was Nature's superior talent for building up structures with many chiral centers, which relatively few scientists were able to synthesize with the correct configurations.

Three short communications in 1965 (*1*), together with a comprehensive review in 1969 (*2*) by Woodward and Hoffmann, opened a new age. Use of molecular orbitals of reactants was found to be a powerful tool for predicting the regiochemistry and stereochemistry, and hence the configuration, of final products.

The original concepts of Woodward and Hoffmann were developed further by the pioneering work of many people, including N. T. Ahn, K. Fukui, W. C. Herndon, K. N. Houk, R. F. Hudson, G. Klopman, L. Salem, and R. Sustmann. The overall field became popular with an excellent book by Ian Fleming (*3*), and terms such as frontier molecular orbital (FMO) and pericyclic reaction are now familiar to students and seasoned chemists alike.

Natural product synthesis gained an enormous advantage from the molecular-orbital approach; in the 1970s many synthetic chemists grasped the opportunity provided by the pericyclic reaction to increase their abilities to assemble complex structures with predictable configurations.

The number of papers dealing with natural product synthesis and featuring a pericyclic reaction as the key step increased greatly as a result of such research. The field has been developing very rapidly and the literature is extensive—these circumstances have motivated the publication of this monograph.

0065-7719/83/0180-0001$06.00/1

Scope and Limitations

This review covers literature published through the end of 1980 dealing with the various types of pericyclic reactions being used as key steps in the synthesis of natural compounds.

It is appropriate to define "key step" and to explain what are considered the natural products.

Singling out one particular reaction within a whole synthetic strategy as the key step is a matter of personal choice. The key step could be, for example, introduction of several chiral centers into a molecule, preparation of a crucial starting material, or finding an elegant route to an important functionality. Naturally, the content of this literature review reflects many such personal choices by the authors.

The goals of the syntheses reported are natural products: materials isolated from plants and animals or metabolites of these materials. Only in exceptional cases are the products intermediates: when the final products are structurally complex molecules, when it is evident that preparation of an intermediate or a simplified molecule is a likely approach to the final goal, and when the goal will be reached along the same pathway.

The main limitations of this review stem from the authors' limited strength. It simply is not possible to survey this broad an area comprehensively. Apologies are due to those whose work is not mentioned.

Nomenclature of Pericyclic Reactions

Woodward and Hoffmann (2) defined pericyclic reactions as those reactions in which all first-order changes in bonding relationships take place in concert on a closed curve, and in which, therefore, the simultaneous redistribution of bonds from reagents to products occurs only through a transition state. The transition state can be of the $[4n+2]$ or $[4n]$ type, and hence have aromatic or antiaromatic character.

The number of electrons involved in the transition state is often substituted for the number of atoms. This nomenclature is especially familiar for cycloadditions and sigmatropic rearrangements, and is generally adopted in this review.

Worthy of further consideration are the electronic states of the reactants, which may be involved in a pericyclic reaction in the ground or in the excited state, and whether the reaction occurs thermally or photochemically.

To illustrate the variety of pericyclic reactions, we discuss examples of thermal six-electron reactions that were shown by Hendrickson (4) to be in the order of thousands. (Table 1.1) These examples involve a six-atom transition state; the familiar nomenclature of direct or inverse reaction is added.

Molecular Orbitals and Pericyclic Reactions

Rationalization of experimental data in terms of molecular-orbital interactions is the excellent result of Woodward and Hoffmann's approach to pericyclic reactions. An introduction to this topic is given in Ref. 3.

Using a perturbation approach, Klopman (*5*) and Salem (*6*) defined the energy (ΔE) gained in a pericyclic reaction as the sum of three terms:

$$\Delta E = -E^{\text{repuls.}}_{\text{occupied/occupied}} + E_{\text{Coulombs}} + E^{\text{attract}}_{\text{occupied/vacant}} \quad (1)$$

Table 1.1

Six-Center Pericyclic Reactions

Type of Pericyclic Reaction (Forward)	*Reagents and Transition States*	*Type of Pericyclic Reaction (Reverse)*
Rearrangement	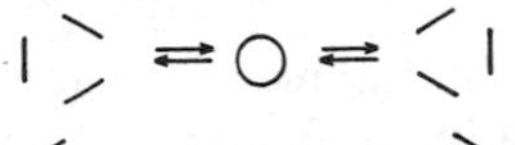	Rearrangement
Addition	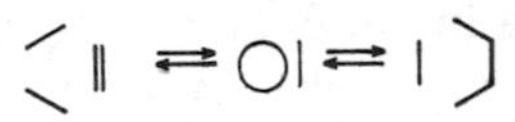	Fragmentation
Rearrangement		Rearrangement
Addition	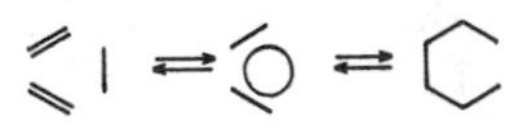	Fragmentation
Transfer		Transfer
[2+2+2] Cycloaddition	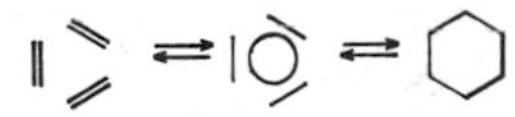	[2+2+2] Cycloreversion
Ene-reaction	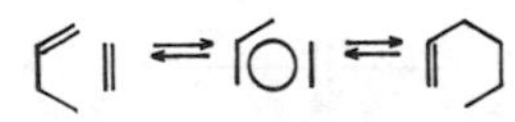	Retro-ene-reaction
[1,4] Addition	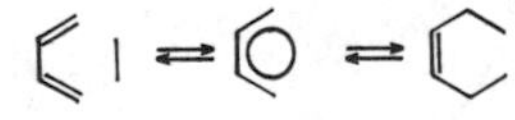	[1,4] Elimination
[3,3] Sigmatropic rearrangement	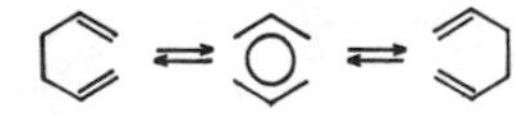	[3,3] Sigmatropic rearrangement
[4+2] Cycloaddition	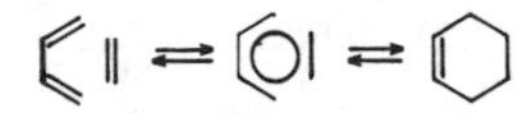	[4+2] Cycloreversion
[1,5] Sigmatropic rearrangement	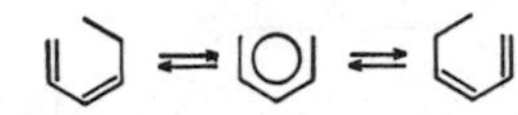	[1,5] Sigmatropic rearrangement
Electrocyclic	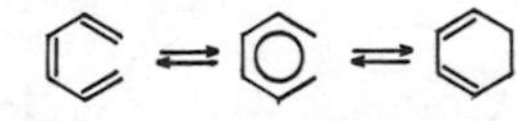	Electrocyclic
Resonance	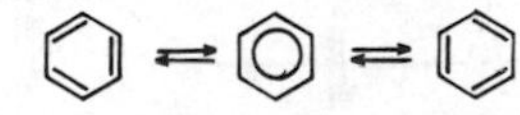	Resonance

The first term is negative, because the interaction between two occupied orbitals always generates an orbital of higher energy than either already existing; hence the energy gained by two electrons in the most "stable" orbital is lost in the less stable one (*7*) (Figure 1.1).

The second term is important only for charged molecules.

The third term is positive, the interaction between an occupied and a vacant orbital yielding a new occupied orbital more stable than the former one; hence there is a net energy gain in the system (Figure 1.2).

Orbital interactions are stronger as they come closer in energy level. The dominant part of the third term is given by the interactions between those molecular orbitals which Fukui (*8*) has named frontier orbitals: the highest occupied molecular orbital (HOMO) of one reagent with the lowest unoccupied molecular orbital (LUMO) of another. With this simplification, which limits the third term only to HOMO/LUMO interactions, Eq. 1 becomes

$$\Delta E = \frac{2\Sigma\,(c_{\mathrm{HO}} \cdot c_{\mathrm{LU}} \cdot \beta)^2}{E_{\mathrm{HO}} - E_{\mathrm{LU}}} \qquad (2)$$

where c is the coefficient at the reacting site,
β is the resonance integral, and
E is the energy of the molecular orbital.

Any condition that increases the nominator or diminishes the denominator of Eq. 2 makes the process easier; this is often the effect of a catalyst.

The importance of the coefficients is emphasized by two further points. An interaction is possible if the symmetry of the interacting coefficients is favorable (Figure 1.3). The reagent can approach most

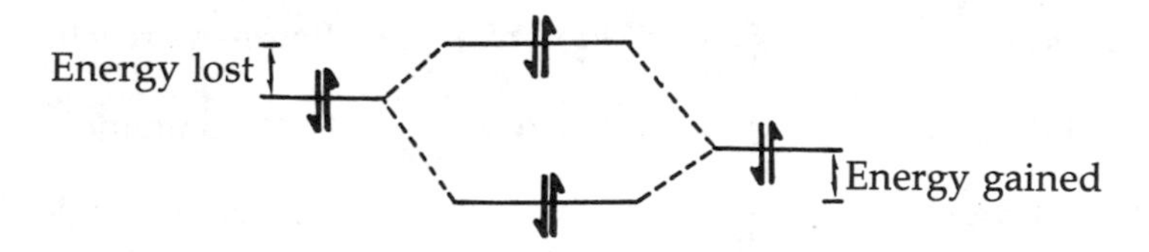

Figure 1.1. Interaction between two occupied orbitals.

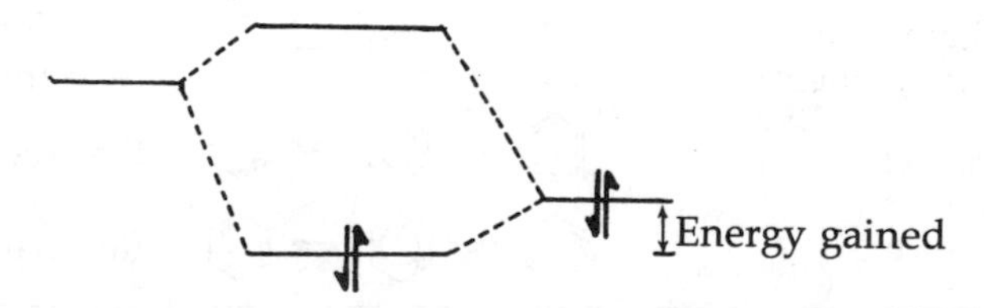

Figure 1.2. Interaction between occupied and vacant orbitals.

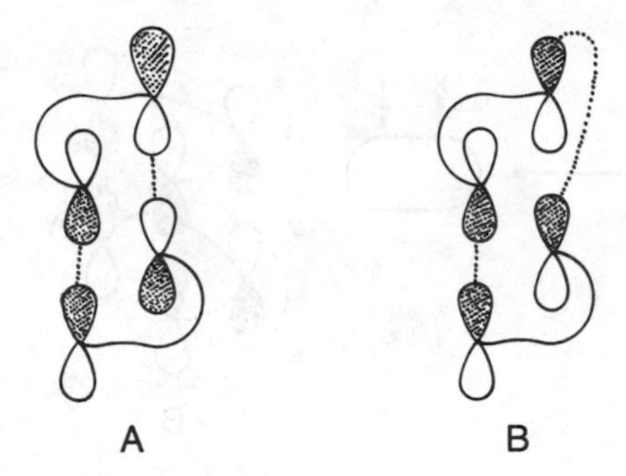

Figure 1.3. Suprafacial (A) and antarafacial (B) cycloaddition.

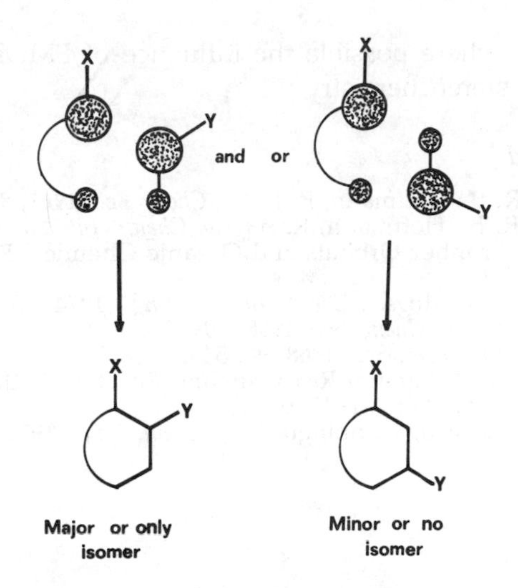

Figure 1.4. Regiochemistry of a reaction from coefficients' amplitude.

favorably along parallel planes (hence suprafacially, as shown in Figure 1.3A); overlap of in-phase lobes, as in Figure 1.3B, requires severe distortions.

Following the principle of maximum overlap, the extent to which the coefficients of the reagents interact with each other (Figure 1.4) will determine the regiochemistry of the reaction.

HOMO/LUMO interactions between centers not directly involved in bond-making often determine the stereochemistry of the product (Figure 1.5).

Keeping these concepts in mind, we turn our attention to the topic of this review—synthesis of natural products based on pericyclic reac-

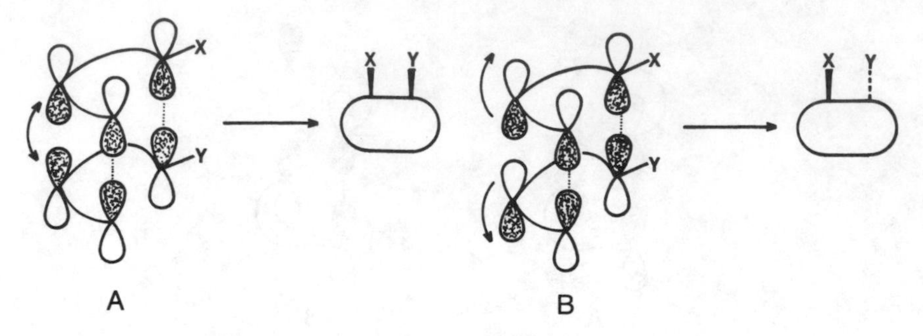

Figure 1.5. Secondary nonbonding interactions favoring an endo *transition state* (cis *adduct formed) (A) or* exotransition state (trans *adduct formed) (B).*

tions, stressing where possible the influence of FMOs on rate, regiochemistry, and stereochemistry.

Literature Cited

1. Woodward, R. B.; Hoffmann, R. *J. Am. Chem. Soc.* **1965**, *87*, 395, 2046, 2511.
2. Woodward, R. B.; Hoffmann, R. *Angew. Chem., Int. Ed. Engl.* **1969**, *8*, 781.
3. Fleming, I. "Frontier Orbitals and Organic Chemical Reactions"; Wiley: London, 1976.
4. Hendrickson, J. B. *Angew. Chem., Int. Ed. Engl.* **1974**, *13*, 47.
5. Klopman, G. *J. Am. Chem. Soc.* **1968**, *90*, 223.
6. Salem, L. *J. Am. Chem. Soc.* **1968**, *90*, 543.
7. Klopman, G. In "Chemical Reactivity and Reaction Paths"; Klopman, G., Ed.; Wiley: New York, 1974; p. 111.
8. Fukui, K.; Yonezawa, T.; Shingu, H. *J. Chem. Phys.* **1952**, *20*, 722.

2

Cheletropic Reactions

Cheletropic reactions are characterized by a component that bears on one atom the two new bonds being formed or broken.

$$\left(>C=C< \right)_n + :X \rightleftharpoons \left(\cdots \right)_{n-1} \; C\text{–}X\text{–}C$$

In the forward sense, therefore, the cheletropic reaction involves cycloaddition of the type $[2n+1]$; in the reverse sense, it involves fragmentation, namely a one-atom extrusion reaction. In the forward sense, the most common cheletropic reaction is the addition of carbenes to an olefin; in the reverse sense a representative reaction is the extrusion of sulfur dioxide from dihydrothiophene dioxide to generate a diene for a Diels–Alder reaction. The latter process will be discussed in Chapter 5, in the section entitled "In Situ Generation of Dienes for Intermolecular Diels–Alder Reactions."

Carbenes can exist in singlet or triplet state (*see* Figures 2.1A and 2.1B, respectively). Only singlet carbenes can undergo pericyclic reactions with olefins. Triplet carbenes require the presence of an intermediate along the reaction pathway that undergoes spin inversion before ring closure.

Carbenes are usually considered to be electrophilic species; however, inspection of FMOs of ethylene, used as a model for olefins, and of three carbenes [$:CCl_2$, $:CH_2$, and $:C(OMe)_2$] (*1, 2, 3*), reveals that this is not always true. The cheletropic reactions might be $HOMO_{olefin}$/$LUMO_{carbene}$-controlled ($:CCl_2$ and electrophilic carbenes), $HOMO_{carbene}$/$LUMO_{olefin}$-controlled [$:C(OMe)_2$ and nucleophilic carbenes], or both HOMO- and LUMO-controlled (ambiphilic carbenes) (Figure 2.2).

Because the most symmetrical linear approach is symmetry-forbidden, a σ-approach or a π-approach (*4, 5*) is preferred respectively for nucleophilic or electrophilic carbenes, where the charge is transferred from the lone pair of the carbene to the π* MO of the olefin, or

0065-7719/83/0180-0007$06.00/1

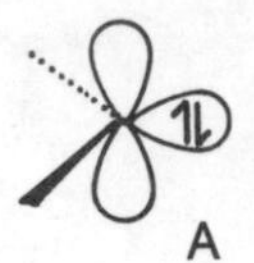

Figure 2.1. Singlet (A) and triplet (B) state carbenes.

from the π MO of the olefin to the empty *p* of the carbene.

The geometries of the two approaches are depicted in Figure 2.3.

Carbenes

Both intermolecular and intramolecular carbene additions have been widely used in the synthesis of natural compounds.

Intermolecular Addition. Naturally occurring cyclopropanes have been obvious candidates for synthesis through cyclopropanation. One of the first papers reporting the synthesis of the monoterpene *trans*-chrysanthemic acid (**2**) (*6*), whose esters are the active principles of Dalmatian pyrethrum flowers (*Chrysanthemum cinerariifolium*), described addition of carbethoxycarbene (from ethyl diazoacetate in the presence of copper bronze) to 2,5-dimethyl-2,4-hexadiene (**1**) as the key step (Scheme 2.1).

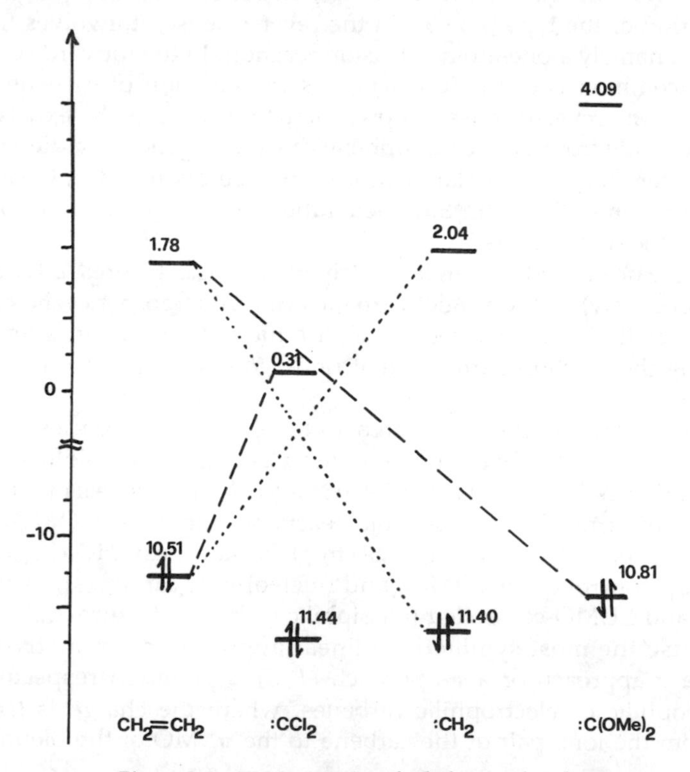

Figure 2.2. FMO energies of ethylene and carbenes.

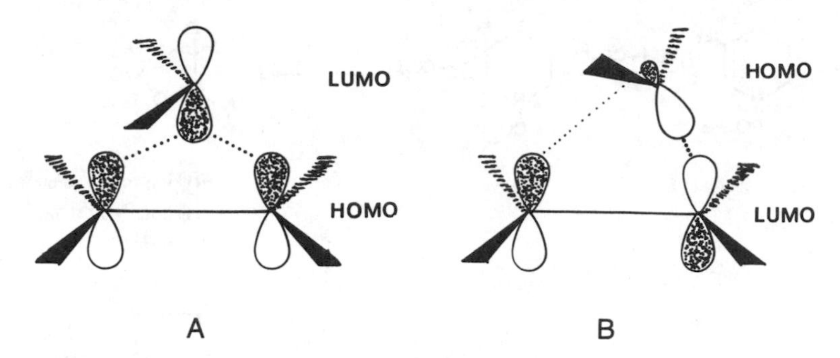

Figure 2.3. π (A) and σ (B) attack for electrophilic and nucleophilic carbenes, respectively.

The copper–bronze-assisted decomposition of diazocarbonyl compounds is still a widely used route to α-carbonyl carbenes, as is illustrated by the synthesis of eburnamonine (*7*) (**3**) (Scheme 2.2), α-cuparenone (*8*) (**4**); and β-vetivone (*8*) (**7**), although the cyclopropane ring is not retained in the final product (Schemes 2.3 and 2.4).

The lower steric hindrance of the endocyclic double bond accounts better for the preferential formation of **5** over **6** (3:1) than perturbation theory does. Thus the HOMO coefficients (CNDO/2) of 1-methoxybutadiene (*9*), used as a model compound, are greater (as a whole) on C1 and C2 than on C3 and C4, and three alkyl groups on C2, C3, and C4 should not significantly change the picture.

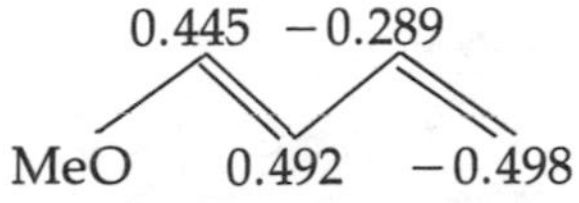

0.445 −0.289
MeO 0.492 −0.498

Alternative methods for generation of carbenes utilize cuprous acetonylacetonate, as in the synthesis of *cis*-jasmone (*10*) (**8**) (Scheme 2.5) in 24% overall yield, or cupric complexes whose eventual chirality (*11, 12*) is responsible for the asymmetric induction in the synthesis of *trans*-chrysanthemic acid (**2**) mentioned earlier. An (*R*) configuration of the catalyst gives dextrorotatory **2** preferentially.

Dihalocarbenes are useful reagents, especially with nucleophilic olefins. Muscone (**10**) was produced in a four-step synthesis starting from 1-ethoxycyclotetradecene (**9**) and dichlorocarbene (*13*)

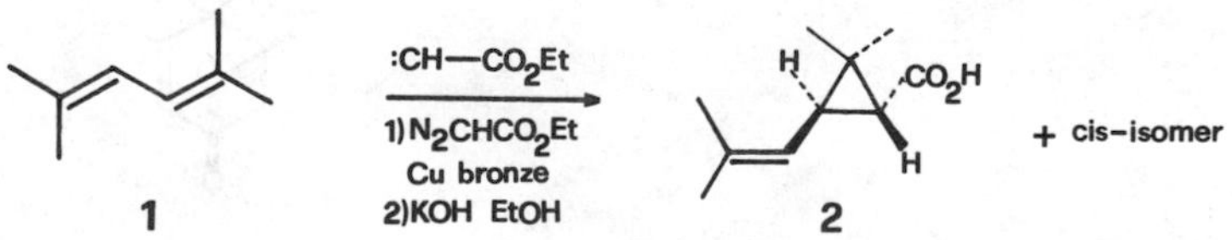

1 → :CH—CO_2Et; 1) N_2CHCO_2Et, Cu bronze; 2) KOH EtOH → **2** (H, CO_2H, H) + cis–isomer

Scheme 2.1

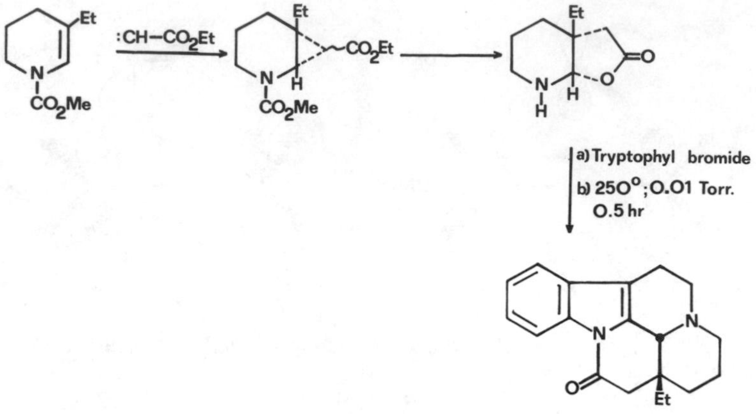
Et
:CH—CO2Et
N
CO2Me
Et
CO2Et
N H
CO2Me
Et
O
O
N H
H
a) Tryptophyl bromide
b) 250°; 0.01 Torr.
0.5 hr
N
N
O
Et

3

Scheme 2.2

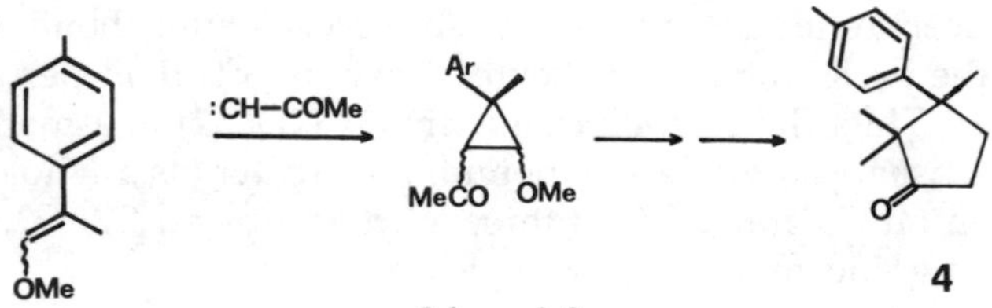
:CH—COMe
Ar
MeCO
OMe
OMe
O

4

Scheme 2.3

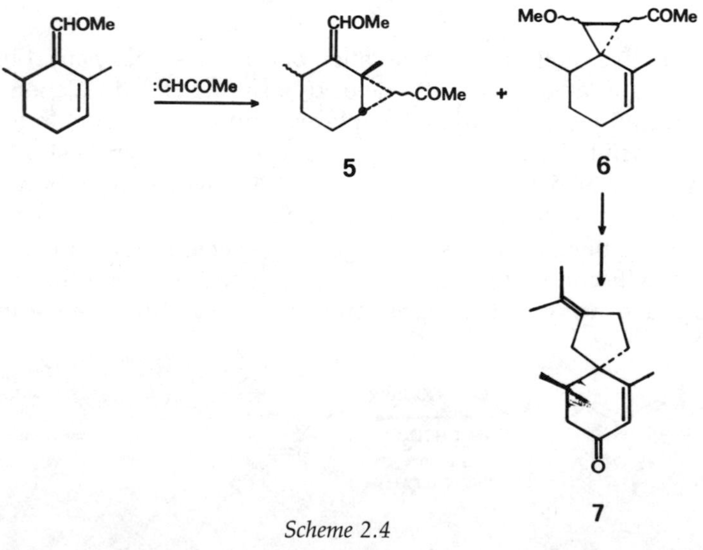
CHOMe
:CHCOMe
CHOMe
COMe
+
MeO
COMe
O

5

6

7

Scheme 2.4

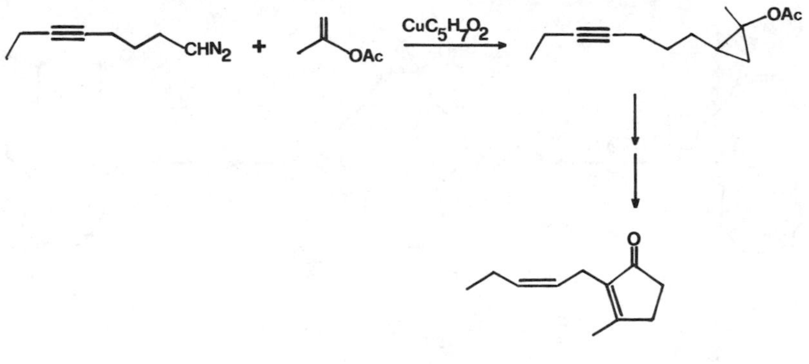

Scheme 2.5

(Scheme 2.6); β, β-cycloeudesmol (**12**) (*14*) was produced from ketal **11** (Scheme 2.7).

The Simmons–Smith reaction (*15*), which involves treatment of olefins with methylene iodide and zinc–copper couple, is still an attractive route to cyclopropanes via methylene addition.

The exact nature of the methylene transfer reagent seems to be $(ICH_2)_2Zn\text{–}Zn\text{–}I_2$ (*16, 17*); the free CH_2 is not involved, as is demonstrated by considerable steric discrimination. A peculiarity of this carbenoid is the high stereoselectivity shown when it is added to substituted olefins bearing an allylic alcohol moiety; this selectivity suggests an oxygen anchimeric assistance through the interaction represented in **13**. This stereoselectivity proved particularly useful in the synthesis of some natural products, as a *cis*-OH-cyclopropane relationship is often required.

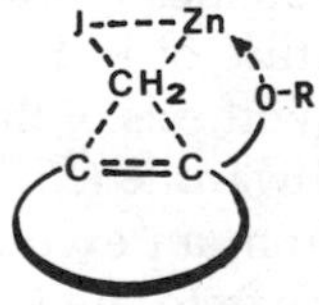

13

The examples that follow clearly demonstrate the assumptions described above. Thus β, α-cycloeudesmol (**15**) was obtained (*14*) from 3-β-octalol (**14**) in 41% yield, 1-valeranone (*18, 19*) (**17**) from **16** in a seven-step synthesis from (+)-carvomenthone, and confertine (**19**) from **18** (*20*) in a highly stereoselective synthesis (Scheme 2.8). Interestingly, both **14** and **18** possess more than one double bond, but the hydroxy group directs the reaction in a regiospecific way.

The $HOMO_{olefin}/LUMO_{carbenoid}$ interaction probably is the origin of the total regiospecificity of the attack on the methoxyketone **20**, which initiates a synthesis of grandisol (**21**) (*19*) (Scheme 2.9).

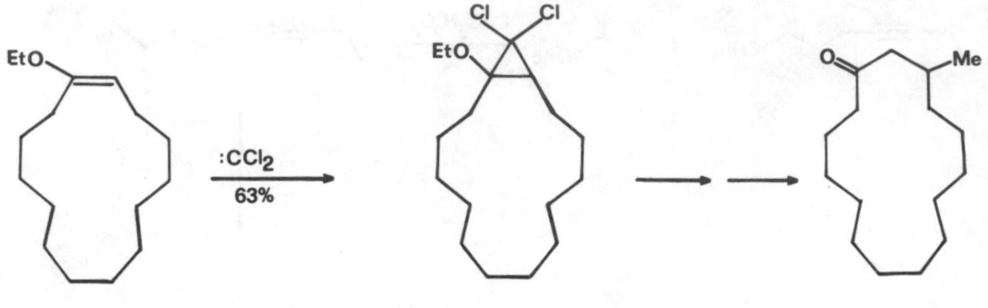

EtO :CCl_2 63% Cl Cl EtO O Me

9 10

Scheme 2.6

A similar species is the carbenoid generated from phenyl(tribromomethyl)mercury, which formally causes addition of dibromomethylene to an olefin. The severe steric requirements of the Seyferth reagent (*21*) render the hydroxyl-directing effect inoperative, although that effect is operative on the same substrate under Simmons–Smith conditions, as is evident in the synthesis of globulol (**23**) (*22*) shown in Scheme 2.10.

While the desmethyl analog of hydroazulene (**22**) gives rise to a cyclopropane ring *cis* to the hydroxy group under Simmons–Smith conditions, Seyferth-like conditions produce a *trans*-dibromo-cyclopropane that is then easily debrominated. Such different behavior provides a useful way to introduce a cyclopropane moiety with different stereochemistry into an organic substrate (Scheme 2.11).

Carbene generation can also be achieved photochemically, as is illustrated in the total synthesis of the fungal metabolite brefeldin A (*23, 24*) (**25**) by irradiation of isopropylidenediazomalonate at 2537 Å. Up to 95% of the mixture of cyclopropanes **24a,b** consists of the desired *trans*-isomer, depending on the nature of R (Scheme 2.12).

Efficient stereoselective synthesis of the racemic modification of the tetracyclic sesquiterpenoid ishwarone (*25, 26*) (**27**) depends on addition of the carbenoid generated from an excess of dimethyl diazomalonate in the presence of copper bronze to the ketal olefin **26** (Scheme 2.13).

Intramolecular Additions. Burke and Grieco (*27*) reviewed the intramolecular reactions of diazocarbonyl compounds, mainly from a preparative point of view.

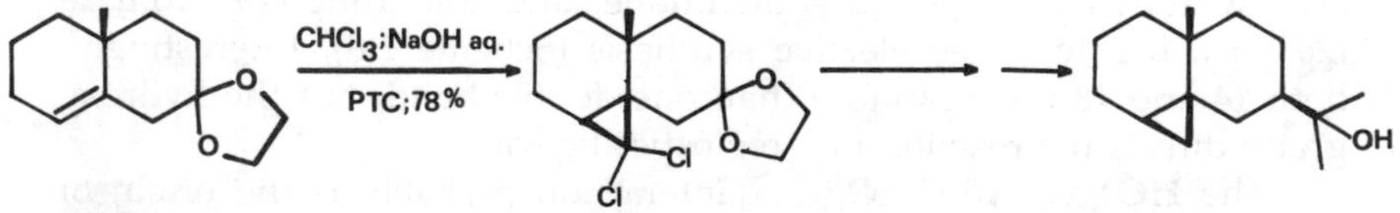

O O $CHCl_3$;NaOH aq. PTC;78% O O Cl Cl OH

11 12

Scheme 2.7

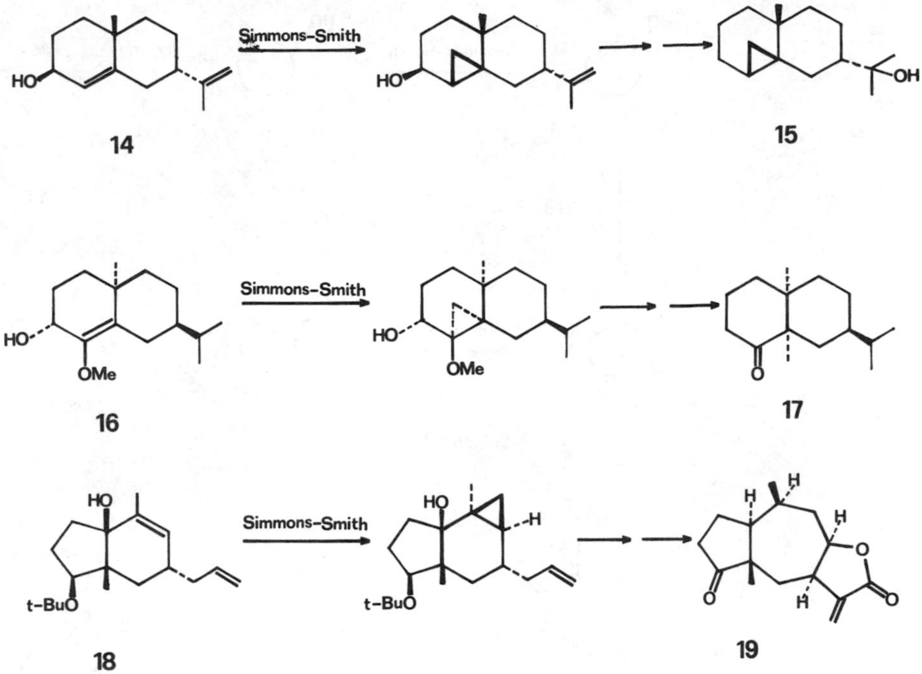

Simmons–Smith

14 15

Simmons–Smith

16 17

Simmons–Smith

18 19

Scheme 2.8

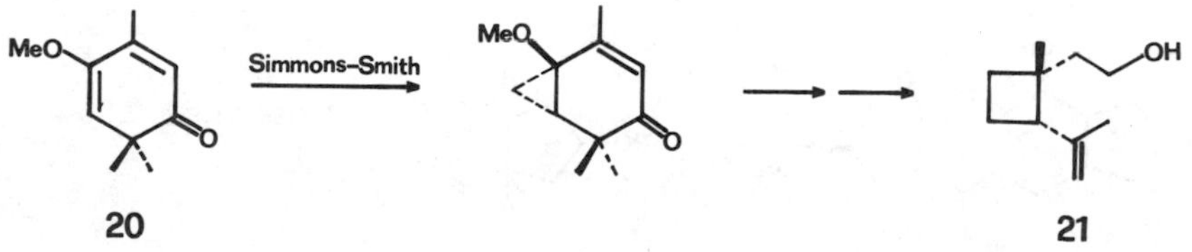

Simmons–Smith

20 21

Scheme 2.9

$PhHgCBr_3$

22 23

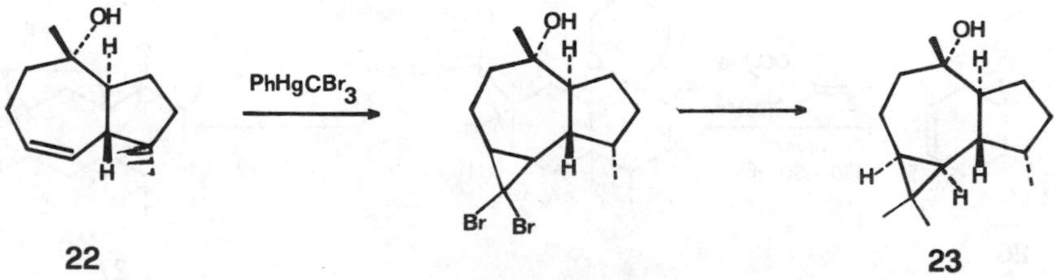

Scheme 2.10

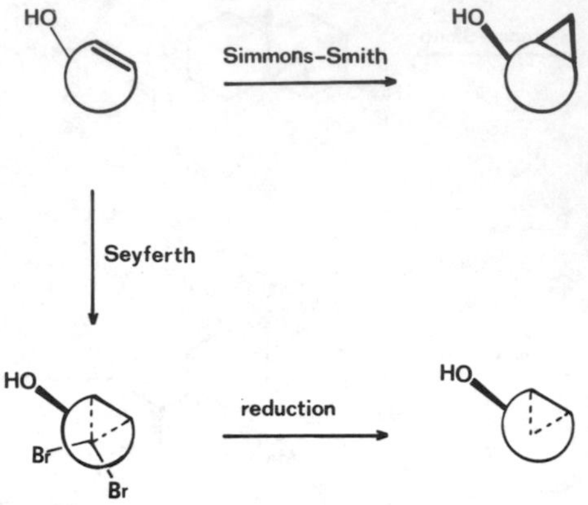

HO
Simmons–Smith
HO
Seyferth
HO
Br
Br
reduction
HO

Scheme 2.11

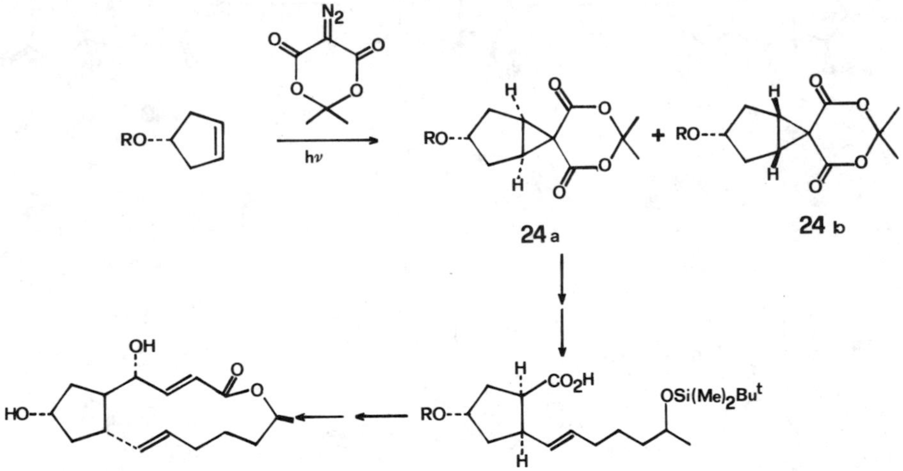

N_2
O O
O O
RO
hν
H
H
O
O
O
O
RO
H
H
O
O
O
O
24 a
24 b
OH
O
O
HO
H
CO_2H
RO
$OSi(Me)_2Bu^t$
H
25

Scheme 2.12

O O
H
26
N_2 CO_2Me CO_2Me
Cu bronze
130–150°; 64%
CO_2Me
CO_2Me
O O
H
H
H
O
H
H
27

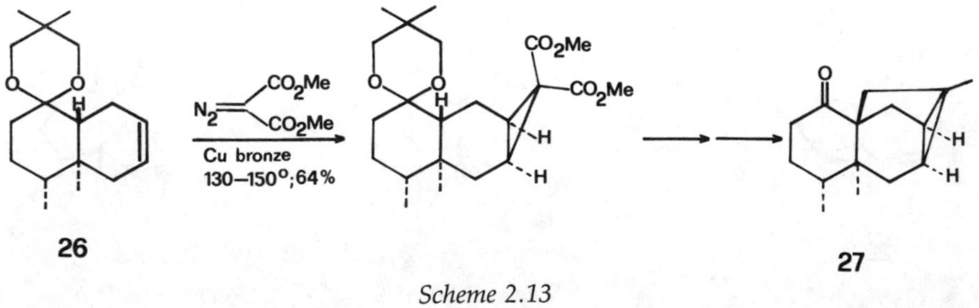

Scheme 2.13

The primary concern in intramolecular additions is what factors affect the stereospecificity of the process. Because orbital interactions are not generally involved in intramolecular additions, minimization of the transition state strain energy is the determining factor.

If the substrate exhibits a single site of addition, the intramolecular addition takes place readily only when an unstrained ring is formed. Thus sabina ketone (**29**) and sabinene (**30**), previously synthesized (*28*) via a Simmons–Smith cyclopropanation, were obtained (*29, 30*) in 40% yield from **28**, and Corey's aldehyde (**32**), a prostaglandins intermediate (*31*), in 50% yield from **31**. In a total synthesis of the natural sesquiterpenes (±)-α-chamigrene (**34**) and (−)-acorenone B (**36**), White and his coworkers (*32–34*) employed as the key stage the intramolecular cycloaddition of diazoketones **33** and **35** in the presence of copper powder, followed by reductive scission of the peripheral bond in the resulting cyclopropanes to construct the necessary spiro-ring junction.

An elegant application of this idea is found in the synthesis of (±)-hinesol (**38**) and (±)-epihinesol (**39**), prepared from diazoketone **37** and followed by fragmentation (*35,36*) (Scheme 2.14).

When the substrate has two or more points of attack, hence at least two double bonds, the strain energy of the transition state plays a role in determining the reaction product; the product occurring through the least strained attack is formed.

A nice example (*14*) is the synthesis of β,α-cycloeudesmol (**15**) from decomposition of the diazoketone **40**, where the ketocarbene **41** faces four different points of attack, two (a and c) downward and two (b and d) upward. Inspection of molecular models reveals that a has the less strained transition state, thus producing **42** as the sole reaction product in 62% yield (Scheme 2.15).

Similar situations have been encountered during the efforts to synthesize (±)-sesquicarene (**46**) and (±)-sirenin (**47**), two compounds that have been obtained by reaction sequences involving intramolecular addition of an α-ketocarbenic end to the slightly different substances **43** (*37–46*), **44** (*47*), and **45** (*48*).

In a similar way **48** was decomposed with copper assistance to give aristolone (**49**) (*49*), with concomitant double bond shift; (*E, E*)-farnesyl diazoacetate (**50**) gave rise to (±)-presqualene alcohol (**51**) (*50*) (Scheme 2.16).

The first step of an intriguing synthesis of (±)-prostaglandin $F_{2\alpha}$ (**53**) (Scheme 2.17) centers on the decomposition of diazocompound **52** in the presence of cupric acetoacetate (*51*).

The synthesis of (±)-majurone (**57**), (±)-thujopsene (**58**), and (±)-thujopsadiene (**59**), combining two different and totally stereospecific intramolecular ketocarbenic additions with a photochemical Wolff rearrangement, and employing in turn three different diazoketones **54–56,** was described (*52*). It must be noted that the last step to **58** had been explored already (*53*) (Scheme 2.18).

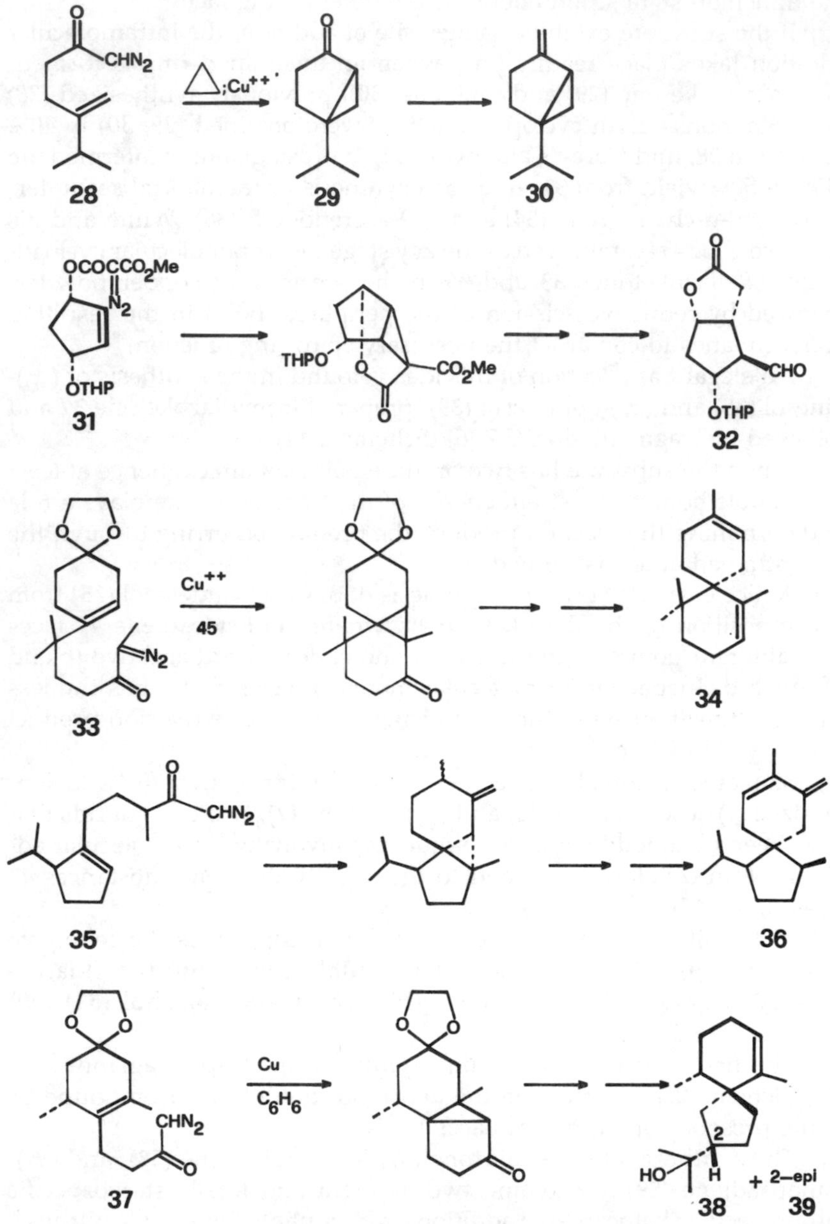
CHN2
;Cu++
28
29
30
OCOCCO2Me
N2
OTHP
31
THPO
CO2Me
CHO
OTHP
32
Cu++
45
N2
33
34
CHN2
35
36
CHN2
37
Cu
C6H6
HO
H
2
+ 2–epi
38
39

Scheme 2.14

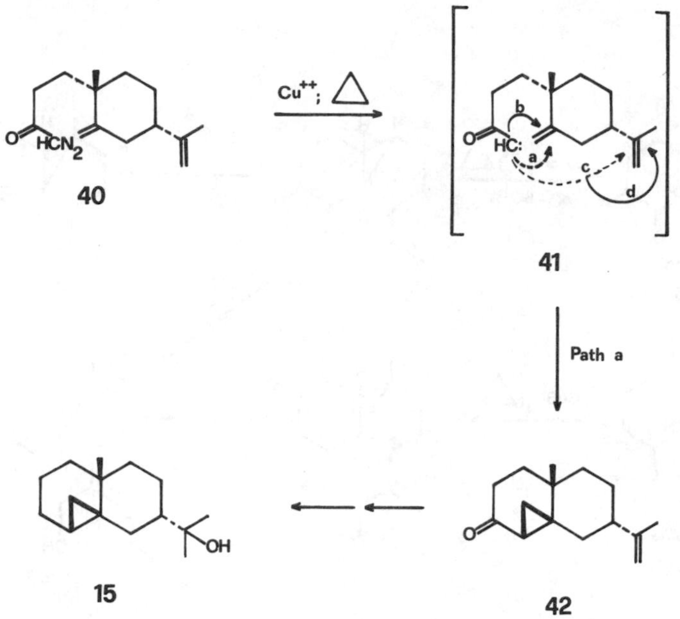

Scheme 2.15

The total synthesis of the tetracyclic sesquiterpene (±)-longicyclene (**61**) isolated from turpentine oil of *Pinus longifolia* also entailed the crucial use of intramolecular carbene addition generated from **60** (*54,55*) (Scheme 2.19).

The synthesis of four necine bases was accomplished in a stereospecific fashion; the key step was an intramolecular α-ketocarbene addition in each synthesis. Isoretronecanol (**63**) (*56*) and trachelanthamidine (**64**) (*56*) were obtained from (*E*)- and (*Z*)-**62**, respectively, whereas dihydroxyheliotridane (**66**) (*57*) and hastanecine (**67**) were obtained from (*E*)- and (*Z*)-**65** (Scheme 2.20).

One of the main topics in the diterpene field has been the synthetic approach to the tetracyclic skeleton belonging to various classes of natural compounds. Wenkert (*58*) suggested this skeleton might arise through cyclization of suitably oriented pimaradienes (**a**). Thus kaurene (**b**), phyllocladene (**c**), and atisine (**d**) might share a common origin with gibbane (**e**) and grayanotoxin (**f**), which could arise by migration of the bonds *i* and *ii* in kaurene (Scheme 2.21).

Along the route to the potent plant-growth stimulator gibberellic acid, some degradation products of gibberellins of general formula **70** were synthesized from tetrahydrofluorenyl diazomethyl ketone **68**. In the presence of activated Cu–CuO, these products gave rise to a single cyclopropyl derivative **69** upon intramolecular addition of the carbene intermediate, paving the way for the synthesis of (±)-gibberone (*59*),

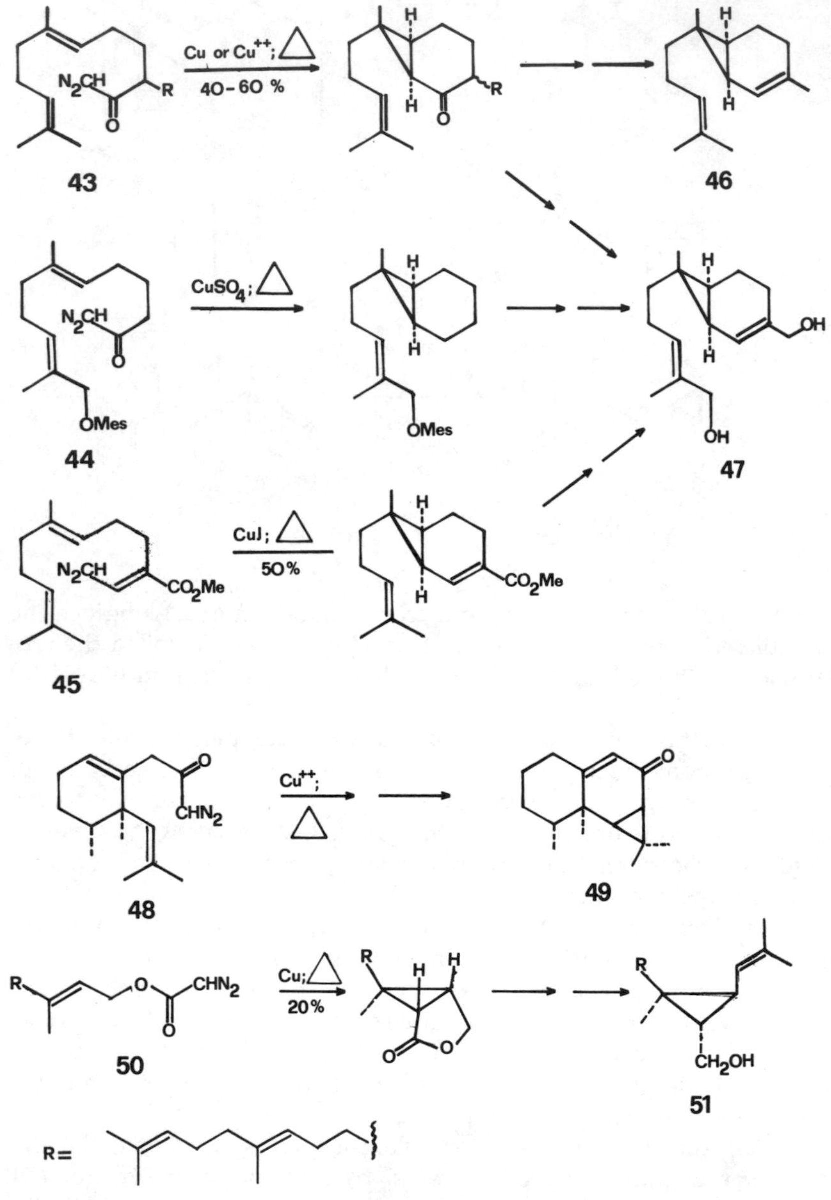

Cu or Cu^{++}; △ 40-60 %

43

46

$CuSO_4$; △

OMes

44

47

CuI; △ 50%

CO_2Me

45

Cu^{++}; △

48

49

Cu; △ 20%

CHN_2

50

CH_2OH

51

R=

Scheme 2.16

Scheme 2.17

Scheme 2.18

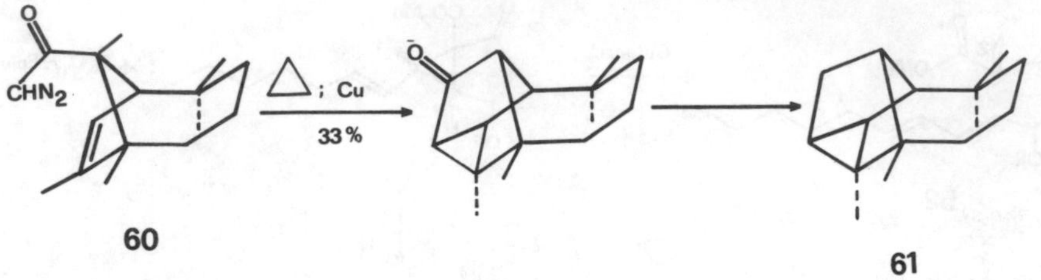

CHN$_2$ △; Cu 33% 60 61

Scheme 2.19

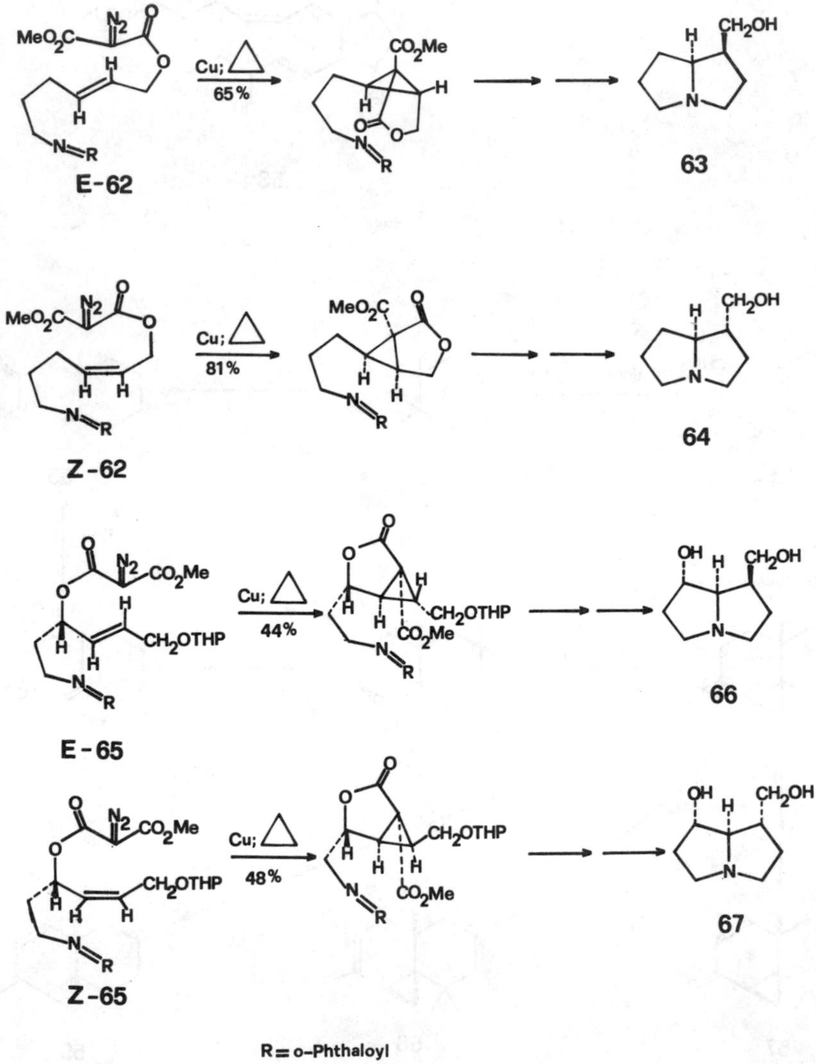

MeO$_2$C N$_2$ Cu; △ 65% CO$_2$Me CH$_2$OH E-62 63

MeO$_2$C N$_2$ Cu; △ 81% CH$_2$OH Z-62 64

N$_2$ CO$_2$Me CH$_2$OTHP Cu; △ 44% OH CH$_2$OH E-65 66

N$_2$ CO$_2$Me CH$_2$OTHP Cu; △ 48% OH CH$_2$OH Z-65 67

R = o-Phthaloyl

Scheme 2.20

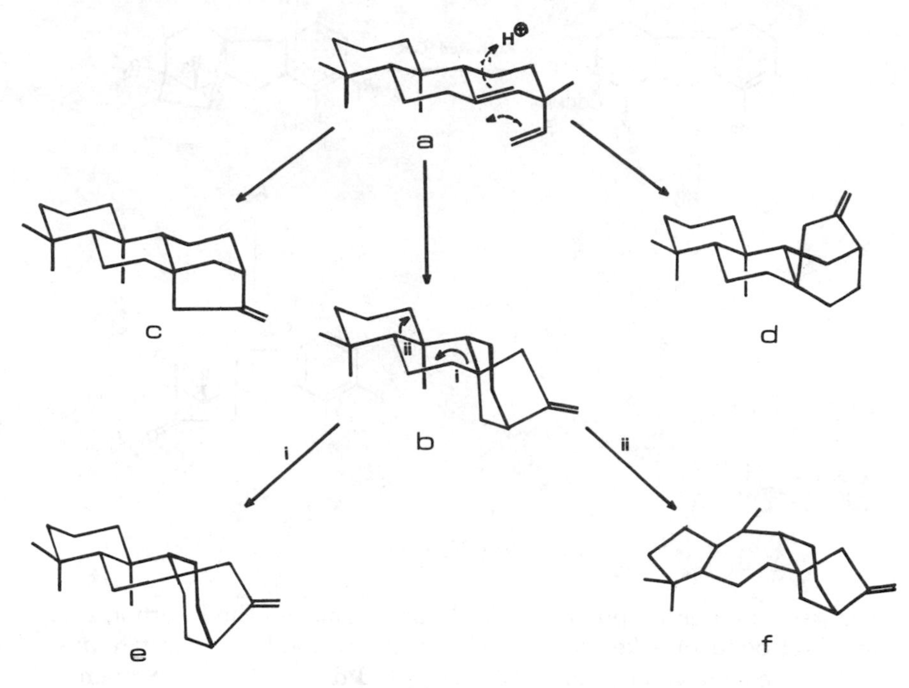

Scheme 2.21

(±)-4-methyl-9β,13α-dihydro-16-oxagibba-1,3,5(10)-triene (*59*) and (±)-2-methoxy-11β-methyl-4b,α,H-gibba-1(10a)-2,4-trien-8-one (*60*) (Scheme 2.22).

Both (+)-phyllocladene (**72**) and (+)-kaurene (**73**) were similarly obtained starting from abietic acid (**71**) (*61, 62*). When the ring A was aromatic, 9α,*H*-phyllocladane derivatives were obtained (*60*) (Scheme 2.23).

Intramolecular Insertion Reactions. Carbenes can undergo intramolecular insertions into acidic X–H bonds, an alternative to Wolff rearrangement, when the substrate is not suitable for addition reactions.

These insertion reactions can be regarded as [2+1] additions where the 2 π component is substituted for a 2 σ component. The considerations applied to addition reactions can be applied to insertion reactions if we keep in mind that σ and σ* molecular orbitals (MOs) are occupied MOs of low energy and vacant MOs of high energy, respectively.

Ketocarbene insertion into a benzylic C–H bond of **74** disclosed a new synthetic route to some intermediates for C_{20} gibberellins and diterpenoid alkaloids like atisine and veatchine. Thus iminopodocarpa-8,11,13-trienes (**75**) were obtained under thermal–photochemical conditions (*63–65*) (Scheme 2.24).

Insertion into a thiazoline C–H bond of **76** featured as the key step an original approach to the penicillin skeleton (**77**) (*66*); the ketocarbene

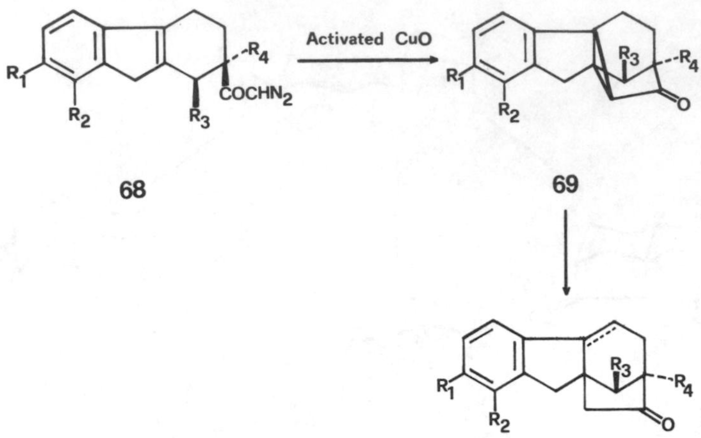

Scheme 2.22

was generated either photochemically or thermally. The insertion into an N–H bond of a ketocarbene produced by rhodium diacetate-promoted decomposition of suitably substituted diazoketoester **78** opened a new entry to thienamycin (**79**) (*67–69*) (Scheme 2.25) in both racemic and optically active forms.

Nitrenes

Nitrenes behave like carbenes, although their chemistry has not been as fully investigated. They are usually obtained by decomposition of azides. In the presence of olefins they afford aziridines, but sometimes the rate of 1,3-dipolar cycloaddition of azide itself is faster than that of decomposition, forming triazolines.

On heating, triazolines may lose nitrogen and yield aziridines, the addition product of nitrenes to olefins; hence this can be the reaction pathway of their formation (Scheme 2.26).

Oxidation of primary amines by means of such oxidants as lead tetraacetate, activated manganese dioxide, yellow mercuric oxide, and *N*-halosuccinimides is a useful alternative preparative method. Nevertheless, in the presence of olefins, aziridines cannot normally be formed from primary amines unless the addition is intramolecular. For large organic molecules, highly strained bridged aziridines **81** are generally formed upon oxidation of δ,ϵ-unsaturated primary amines **80** with 1 equivalent of *N*-chlorosuccinimide (NCS) (*70*) (Scheme 2.27).

Addition of nitrenes has proved particularly useful in natural products synthesis, when the nitrogen atom can be incorporated into the organic framework. Such is the case for alkaloids, where major successes have been achieved with this reaction.

Only one example of intermolecular addition is known to involve the synthesis of an aromatic intermediate to songorine (*71, 72*). Benzenesulfonylnitrene was generated from the corresponding azide, and the adduct **82** was produced in 92% yield. Routine steps transform **82** into **83**. Further stages of the route to songorine (**84**) will be considered in Chapter 3, in the section entitled "Cyclobutane Ring Expanded," where a crucial photochemical pericyclic step occurs (Scheme 2.28).

Two examples of intramolecular addition of nitrene concern the syntheses of two alkaloids: desethylibogamine (**86**) and (±)-serratinine (**89**). The first is centered (*73*) on the aziridine ring opening of **81** (R = H) with indoleacetic anhydride to give **85**, which in a few steps is converted to **86** (Scheme 2.29).

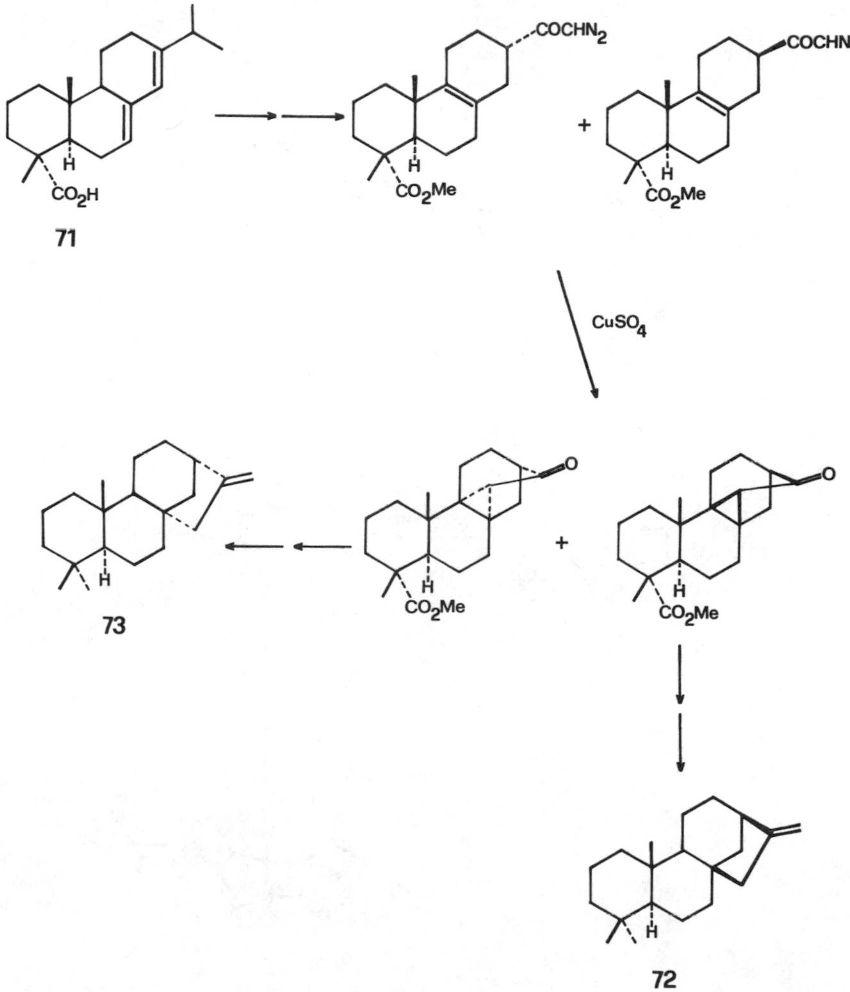

CO_2H **71** $COCHN_2$ CO_2Me + $COCHN_2$ CO_2Me $CuSO_4$ O CO_2Me + O CO_2Me **73** **72**

Scheme 2.23

Scheme 2.24

Scheme 2.25

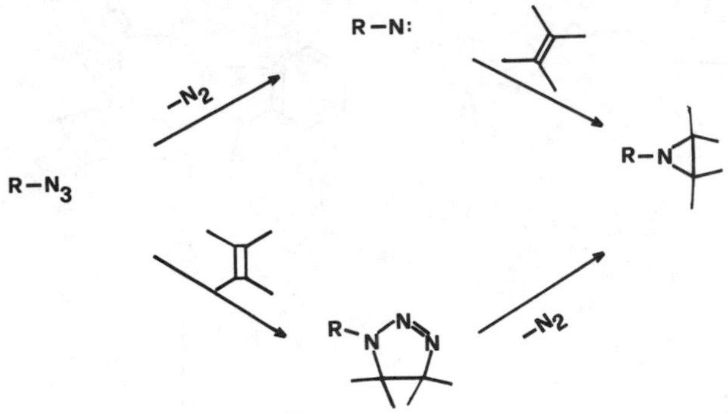

Scheme 2.26

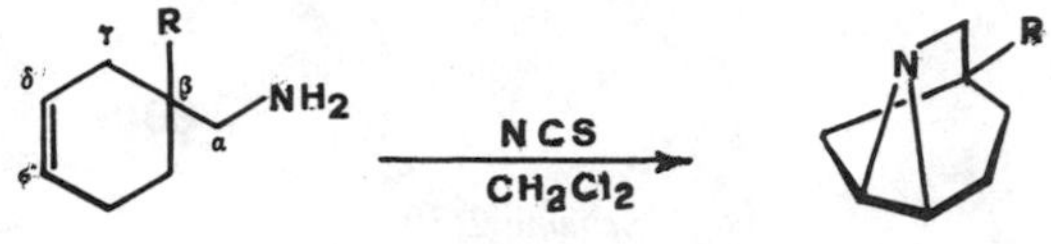

Scheme 2.27

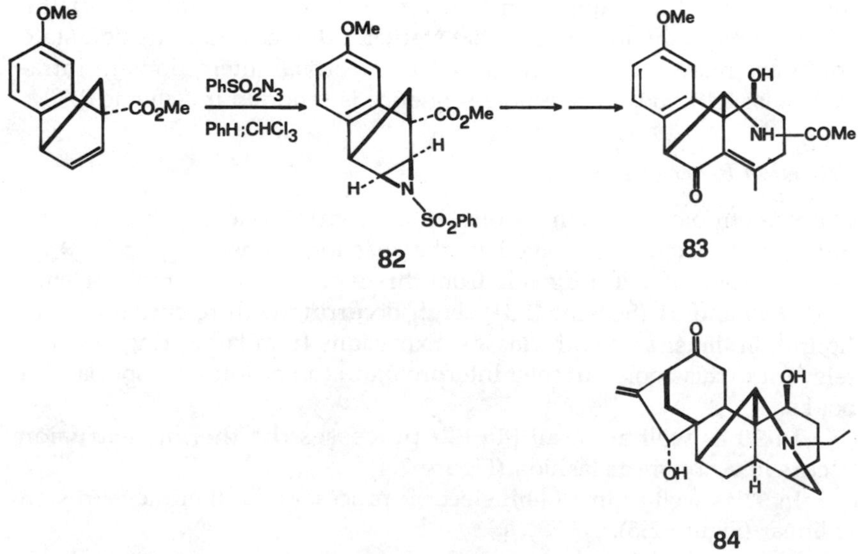

Scheme 2.28

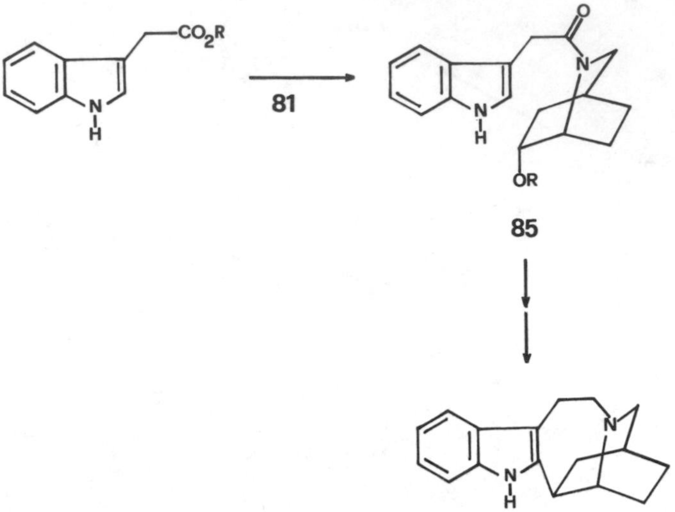

CO_2R 81 O N OR 85 N H 86

Scheme 2.29

The second involves a total synthesis (*74*) of **89**, the major alkaloid of *Lycopodium serratum* Thunb, which relies on an intramolecular nitrene addition to **87** (incidentally prepared by a Diels–Alder reaction). The double bond in **87** was preferentially attacked from the convex face of the molecule, the major isomer being **88** (20%, as against 3% of the minor isomer), demonstrating that steric hindrance and transition state strain are more important than molecular orbital interactions in intramolecular additions. The whole sequence is reported in Scheme 2.30.

Extrusion Reactions

The cheletropic extrusion of one atom from a cycle depends on the number of electrons involved in the reaction, either $[4n]$ or $[4n+2]$. The extrusion of sulfur dioxide from three- or five-membered ring sulfoxides **90** and **91** (Scheme 2.31), both occurring with retention of configuration, illustrates both classes. Extrusions from larger rings are not relevant to this book, but their interpretation can be found in specialized books.

In **90** as well as in all $[4n+2]$ processes the thermic extrusion occurs in a nonlinear fashion (Figure 2.4).

In **91** as well as in all $[4n]$ electron processes the thermic extrusion is linear (Figure 2.5).

Because the latter decomposition generates useful dienes for Diels–Alder reaction, this topic will be considered in Chapter 5, in the section

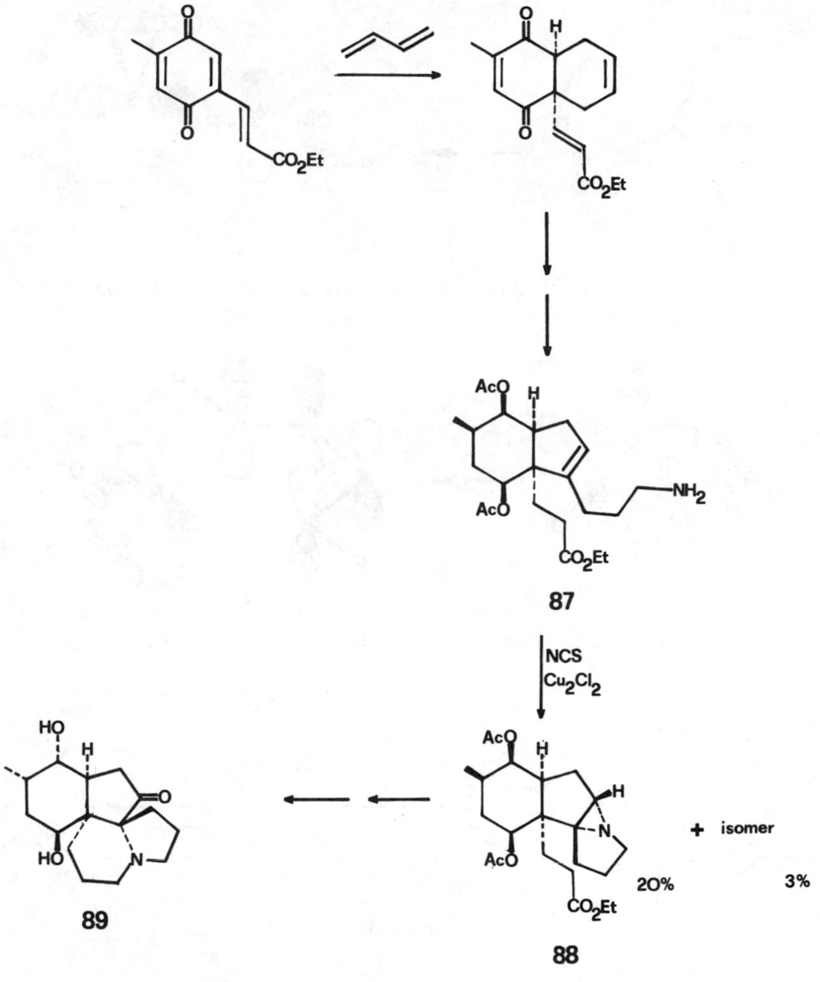

CO_2Et

CO_2Et

AcO H

NH_2

AcO

CO_2Et

87

NCS

Cu_2Cl_2

HO H

O

HO

N

89

AcO H

H

N

AcO

CO_2Et

88

20%

+ isomer

3%

Scheme 2.30

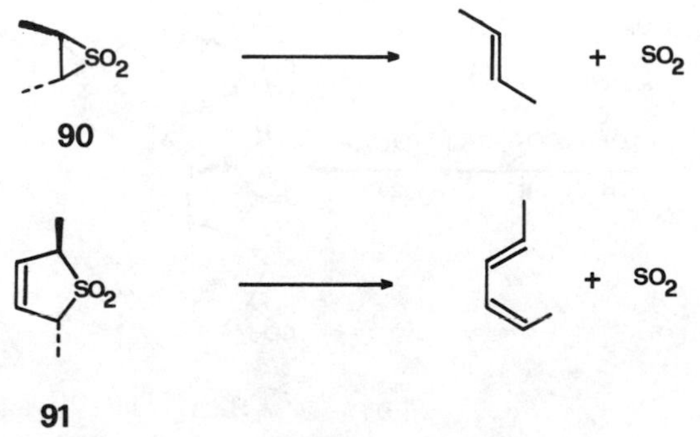

SO_2

90

+ SO_2

SO_2

91

+ SO_2

Scheme 2.31

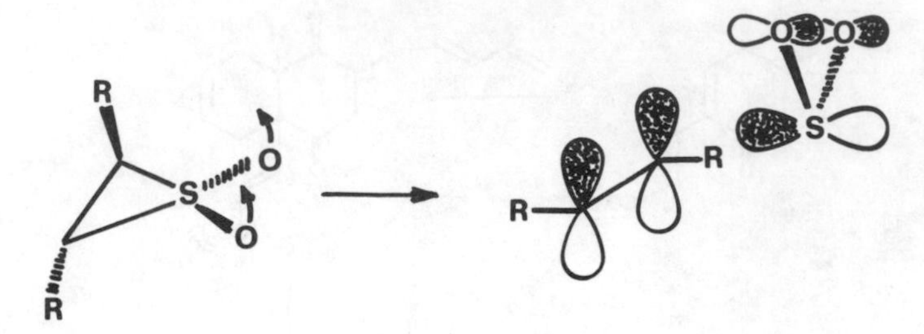

Figure 2.4. Nonlinear cheletropic extrusion of sulfur dioxide.

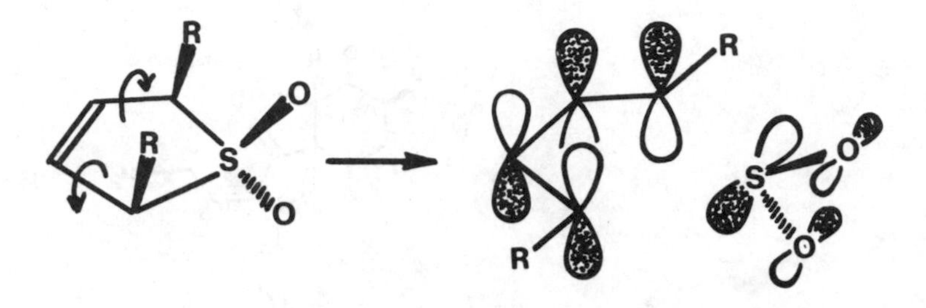

Figure 2.5. Linear cheletropic extrusion of sulfur dioxide.

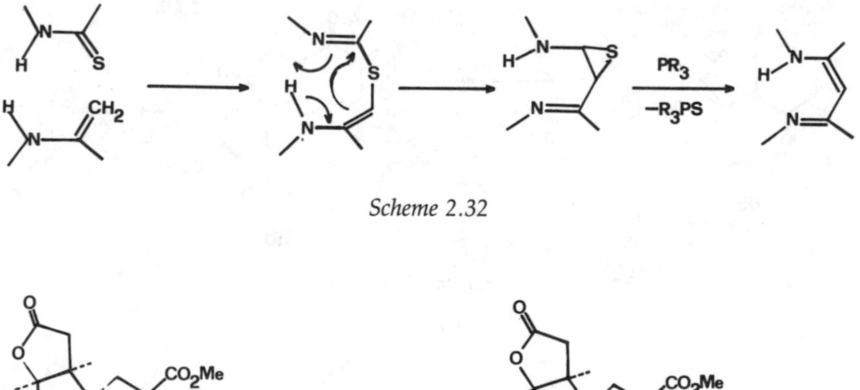

PR_3

$-R_3PS$

Scheme 2.32

1.05 $(PhCO_2)_2$; HCl

CH_2Cl_2 ; RT

$(EtO)_3P$

120°

+ epimer **92b**

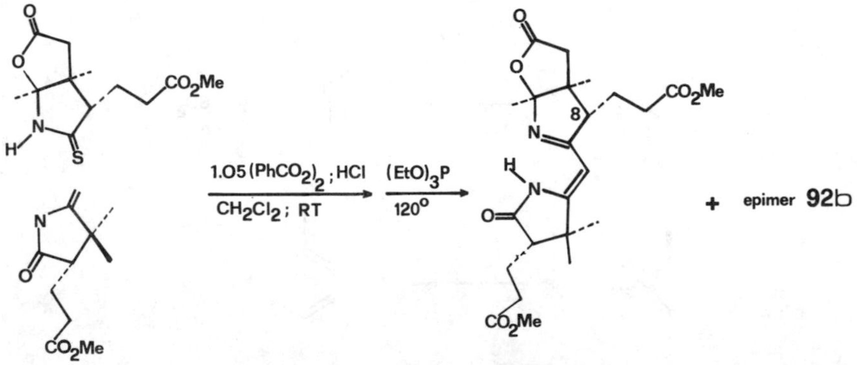

92a

Scheme 2.33

entitled "In Situ Generation of Dienes for Intermolecular Diels–Adler Reactions."

A cheletropic process of the type $[4n+2]$ $(n=0)$ is featured in the Eschenmoser's sulfide coupling (*75*), which requires prior formation of a thiirane ring that reacts subsequently with trivalent phosphorus derivatives to give an olefin and phosphorus sulfide, as is illustrated in Scheme 2.32.

These concepts have been widely applied by Eschenmoser himself to the synthesis of simple corrin models and then extended to the B and C rings of Vitamin B_{12} (Scheme 2.33).

A mixture of the epimers **92a** and **92b**, epimers at C8, was formed in 70% yield, with **92a** being the major isomer. Similarly, a model **93** closely corresponding to the dienone elaeocarpus alkaloids was synthesized (*76*) together with its epimer (Scheme 2.34).

Analogously, the dimethyl ester **96** of the so-called blue pigment isolated by refluxing denatured C-phycocyanin, a photosynthetically active chromoprotein from blue-green algae, has been synthesized (*77*). The key step is the condensation of substituted thiosuccinimide **94** with 2-(2-pyrrolyl)acetic acid derivative **95** in the presence of triphenylphosphine (Scheme 2.35).

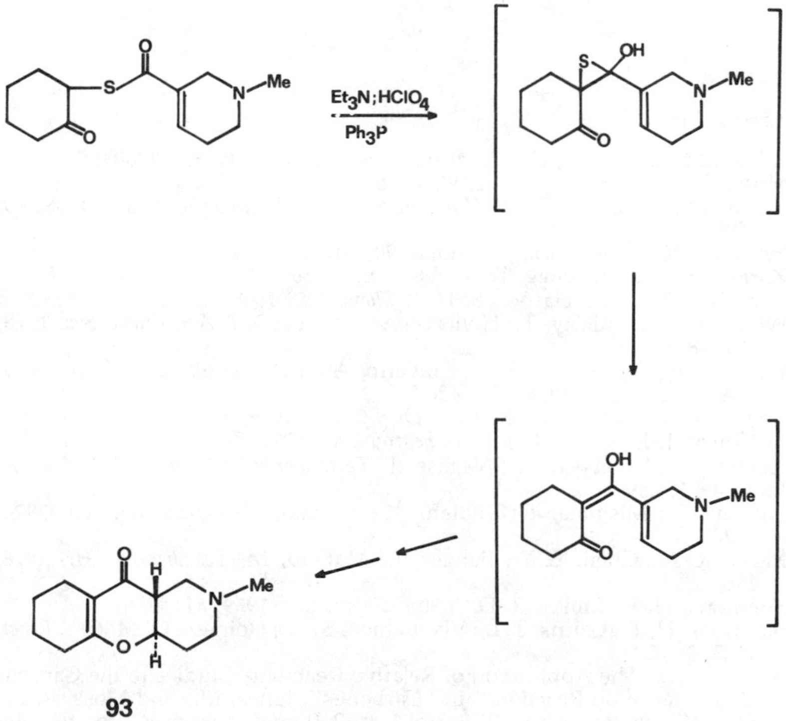

Et_3N;$HClO_4$

Ph_3P

93

Scheme 2.34

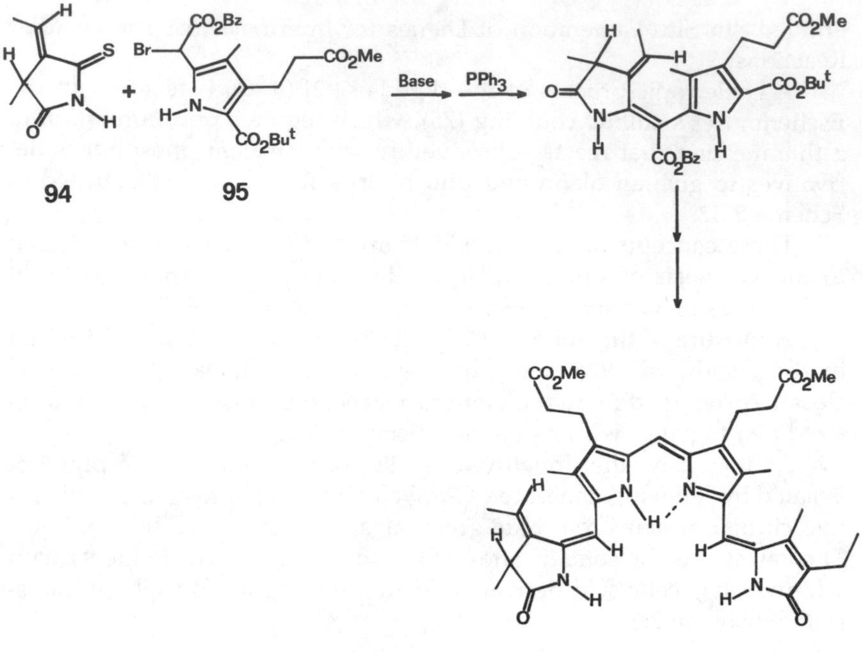

94 + 95 Base PPh$_3$ → 96

Scheme 2.35

Literature Cited

1. Rondan, N. G.; Houk, K. N.; Moss, R. A. *J. Am. Chem. Soc.* **1980**, *102*, 1770.
2. Moss, R. A. *Acc. Chem. Res.* **1980**, *13*, 58.
3. Hoffmann, R.; Zeiss, G. D.; Van Dine, G. W. *J. Am. Chem. Soc.* **1968**, *90*, 1485.
4. Hoffman, R. *J. Am. Chem. Soc.* **1968**, *90*, 1475.
5. Zurawski, B.; Kutzelnigg, W. *J. Am. Chem. Soc.* **1978**, *100*, 2654.
6. Campbell, I. G. M.; Harper, S. H. *J. Chem. Soc.* **1945**, 283.
7. Wenkert, E.; Hudlicky, T.; Hollis Showalter, H. D. *J. Am. Chem. Soc.* **1978**, *100*, 4893.
8. Wenkert, E., Buckwalter, B. L.; Craveiro, A. A.; Sanchez, E. L.; Sathe, S. S. *J. Am. Chem. Soc.* **1978**, *100*, 1267.
9. Alston, P. V.; Ottenbrite, R. M. *J. Org. Chem.* **1975**, *40*, 1111.
10. McMurry, J. E.; Glass, T. E. *Tetrahedron Lett.* **1971**, 2575.
11. Aratani, A.; Yaneyoshi, Y.; Nagase, T. *Tetrahedron Lett.* **1975,** 1707.
12. Ibid., **1977**, 2599.
13. Hiyama, T.; Mishima, T.; Kitatani, K.; Nozaki, H. *Tetrahedron Lett.* **1974**, 3297.
14. Moss, R. A.; Chen, E. Y.; Banger, J.; Matsuo, M. *Tetrahedron Lett.* **1978**, 4365.
15. Simmons, H. E.; Smith, R. D. *J. Am. Chem. Soc.* **1959**, *81*, 4256.
16. Simmons, H. E.; Cairns, T. L.; Vladuchick, S. A.; Hoiness, C. M. *Org. React.* **1973**, *20*, 1.
17. Moss, R. A. "The Application of Relative Reactivity Studies to the Carbene Olefin Addition Reaction." In "Carbenes"; Jones, M., and Moss, R. A., Eds.; Wiley: New York, 1973; Vol **1**, p. 260, and references cited therein.
18. Wenkert, E.; Berges, D. A. *J. Am. Chem. Soc.* **1967**, *89*, 2507.

19. Wenkert, E.; Berges, D. A.; Golob, N. F. *J. Am Chem. Soc.* **1978**, *100*, 1263.
20. Marshall, J. A.; Ellison, R. H. *J. Am. Chem. Soc.* **1976**, *98*, 4312.
21. Seyferth, D. *J. Am Chem. Soc.* **1965**, *87*, 4259.
22. Marshall, J. A.; Ruth, J. A. *J. Org. Chem.* **1974**, *39*, 1971.
23. Livinghouse, T.; Stevens, R. V. *J. Chem. Soc., Chem. Commun.* **1978**, 754.
24. Stevens, R. V. *Pure Appl. Chem.* **1979**, *51*, 1317.
25. Piers, E.; Hall, T. *J. Chem. Soc., Chem. Commun.* **1977**, 881.
26. Piers, E.; Hall, T. *Can. J. Chem.* **1980**, *58*, 2613.
27. Burke, S. D.; Grieco, P. A. *Org. React.* **1979**, *26*, 316.
28. Fanta, W. I.; Erman, W. F. *J. Org. Chem.* **1968**, *33*, 1656.
29. Vig, O. P.; Bhatia, M. S.; Gupta, K. C.; Matta, K. L. *J. Indian Chem. Soc.* **1969**, *46*, 991.
30. Mori, K.; Ohki, M.; Matsui, M. *Tetrahedron* **1970**, *26*, 2821.
31. Corey, E. J.; Fuchs, P. L. *J. Am. Chem. Soc.* **1972**, *94*, 4014.
32. White, J. D.; Torii, S.; Nogami, J. *Tetrahedron Lett.* **1974**, 2879.
33. Ruppert, J. F.; Avery, M. A.; White, J. D. *J. Chem. Soc., Chem. Commun.* **1976**, 978.
34. White, J. D.; Ruppert, J. F.; Avery, M. A.; Torii, S.; Nokami, J. *J. Am. Chem. Soc.* **1981**, *103*, 1813.
35. Mongrain, M.; LaFontaine, J.; Belanger, A.; Deslongchamps, P. *Can. J. Chem.* **1970**, *48*, 3273.
36. LaFontaine, J.; Mongrain, M.; Sergent-Guay, M.; Ruest, L.; Deslongchamps, P. *Can. J. Chem.* **1980**, *58*, 2460.
37. Corey, E. J.; Achiwa, K. *Tetrahedron Lett.* **1969**, 1837.
38. Mori, K.; Matsui, M. *Tetrahedron Lett.* **1969**, 2729.
39. Coates, R. M.; Freidinger, R. M. *J. Chem. Soc., Chem. Commun.* **1969**, 871.
40. Coates, R. M.; Freidinger, R. M. *Tetrahedron* **1970**, *26*, 3487.
41. Vig, O. P.; Chugh, O. P.; Anand, R. C.; Bhatia, M. S. *J. Indian Chem. Soc.* **1970**, *47*, 506.
42. Mori, K.; Matsui, M. *Tetrahedron Lett.* **1969**, 4435.
43. Mori, K.; Matsui, M. *Tetrahedron* **1970**,*26*, 2801.
44. Plattner, J. J.; Bhalerao, U. T.; Rapoport, H. *J. Am. Chem. Soc.* **1969**, *91*, 4935.
45. Bhalerao, U. T.; Plattner, J. J.; Rapoport, H. *J. Am. Chem. Soc.* **1970**, *92*, 3429.
46. Grieco, P. A. *J. Am. Chem. Soc.* **1969**, *91*, 5660
47. Corey, E. J.; Achiwa, K.; Katzenellenbogen, J. A. *J. Am. Chem. Soc.* **1969**, *91*, 4318.
48. Corey, E. J.; Achiwa, K. *Tetrahedron Lett.* **1970**, 2245.
49. Piers, E.; Britton, R. W.; deWaal, W. *Can. J. Chem.* **1969**, *47*, 831.
50. Coates, R. M.; Robinson, W. H. *J. Am. Chem. Soc.* **1971**, *93*, 1785.
51. Kondo, K.; Umemoto, T.; Yako, K.; Tunemoto, D. *Tetrahedron Lett.* **1978**, 3927.
52. Branca, S. J.; Lock, R. L.; Smith, A. B., III *J. Org. Chem.* **1977**, *42*, 3165.
53. Mori, K.; Ohki, M.; Kobayashi, A.; Matsui, M. *Tetrahedron* **1970**, *26*, 2815.
54. Welch, S. C.; Walters, R. L. *Synth. Commun.* **1973**, *3*, 15.
55. Welch, S. C.; Walters, R. L. *J. Org. Chem.* **1974**, *39*, 2667.
56. Danishefsky, S.; McKee, R.; Singh, R. K. *J. Am. Chem. Soc.* **1977**, *99*, 4783.
57. Ibid., 7711.
58. Wenkert, E. *Chem. Ind. (London)* **1955**, 282.
59. Ghatak, U. R.; Chakraborti, P. C. *J. Org. Chem.* **1979**, *44*, 4562.
60. Ghatak, U. R.; Chakraborti, P. C.; Rudra, K. *J. Chem. Soc., Perkin Trans. 1* **1974**, 1957.
61. Tahara, A.; Shimagaki, M.; Ohara, S.; Nakata, T. *Tetrahedron Lett.* **1973**, 1701.
62. Tahara, A.; Shimagaki, M.; Ohara, S.; Tanaka, T.; Nakata, T. *Chem. Pharm. Bull.* **1975**, *23*, 2329.
63. Ghatak, U. R.; Chakrabarty, S. *J. Am. Chem. Soc.* **1972**, *94*, 4756.
64. Chakrabarty, S.; Ray, J. K.; Mukherjee, D.; Ghatak, U. R. *Synth. Commun.* **1975**, *5*, 275.
65. Ghatak, U. R.; Chakrabarty, S. *J. Org. Chem.* **1976**, *41*, 1089.

66. Corey, E. J.; Felix, A. M. *J. Am. Chem. Soc.* **1965**, *87*, 2518.
67. Melillo, O. G.; Shinkai, I.; Ryan, K. M.; Liu, T. M. H.; Sletzinger, M. *Tetrahedron Lett.* **1980**, 2783.
68. Ratcliffe, R. W.; Salzmann, T. N.; Christensen, B. G. *Tetrahedron Lett.* **1980**, 31.
69. Salzmann, T. N.; Ratcliffe, R. W.; Christensen, B. G.; Bouffard, F. A. *J. Am. Chem. Soc.* **1980**, *102*, 6161.
70. Nagata, W. N.; Hirai, S.; Kawata, K.; Aoki, T. *J. Am. Chem. Soc.* **1967**, *89*, 5045.
71. Wiesner, K.; Pak-Tsun Ho; Chang, D.; Blount, J. F. *Experientia* **1972**, *28*, 766.
72. Wiesner, K.; Pak-Tsun Ho; Chang, D.; Kuen-Lam, Y.; Shii Jeou, C.; Yun Ren, W. *Can. J. Chem.* **1973**, *51*, 3978.
73. Nagata, W.; Hirai, S.; Kawata, K.; Okumura, T. *J. Am. Chem. Soc.* **1967**, *89*, 5046.
74. Harayama, T.; Ohtani, M.; Oki, M.; Inubushi, Y. *Chem. Pharm. Bull.* **1975**, *23*, 1511.
75. Eschenmoser, A. *Q. Rev., Chem. Soc.* **1970**, *24*, 366.
76. Jones, T. H.; Kropp, P. J. *Tetrahedron Lett.* **1974**, 3503.
77. Gossauer, A.; Hirsch, W. *Justus Liebigs Ann. Chem.* **1974**, 1496.

3

[2+2] Cycloadditions

Pericyclic [2+2] cycloaddition is a strongly disfavored process (*1*). The interaction between the HOMO of one component and the LUMO of the other cannot occur suprafacially, and an antarafacial approach involves severe steric interactions (Figure 3.1).

Two possibilities are therefore conceivable.

1. The reaction is not pericyclic, but proceeds through a diradical or dipolar intermediate (Scheme 3.1).
2. The reaction occurs throughout HOMO/HOMO or LUMO/LUMO interactions (Figure 3.2). This can be realized only if the orbitals are not both empty or doubly occupied, or if the symmetry of the ethylenic π orbitals of one reagent is reversed by coordination with a metal.

Although the first case falls outside the scope of this survey, it deserves some comment. The intermediate can lose its initial configuration by internal rotation and nonstereospecific reaction results. However, MOs of the reactants determine the regiochemistry of the reaction. Thus the interaction of the greater LUMO coefficient with the greater HOMO coefficient determines the preferential formation of a particular regioisomer. The reaction between a nucleophilic and an electrophilic olefin—for example, dimethylamino isobutene (**97**) and methyl acrylate (**98**) (*2*)—affords a single cyclobutane **100** through the zwitterionic intermediate **99** (Scheme 3.2), despite the recent suggestion by Epiotis (*3*) that, in polar [$2\pi + 2\pi$] cycloaddition, configuration interaction can remove the forbiddenness of a suprafacial approach.

An inspection of frontier orbital energies and coefficients (*4*) suggests that the dominant interaction between the HOMO of **97**, which acts as a donor, and the LUMO of **98**, which behaves as an acceptor (Figure 3.3), determines the observed regiochemistry. In some cases the intermediate is not strictly necessary.

Thermal [2+2] Cycloadditions

Thermal [2+2] cycloadditions become easy when acetylenes or olefins react with a cumulated system. Because the reaction is stereospecific, a concerted pathway (without intermediate) seems to be reasonable (*1*).

0065-7719/83/0180-0033$14.25/1

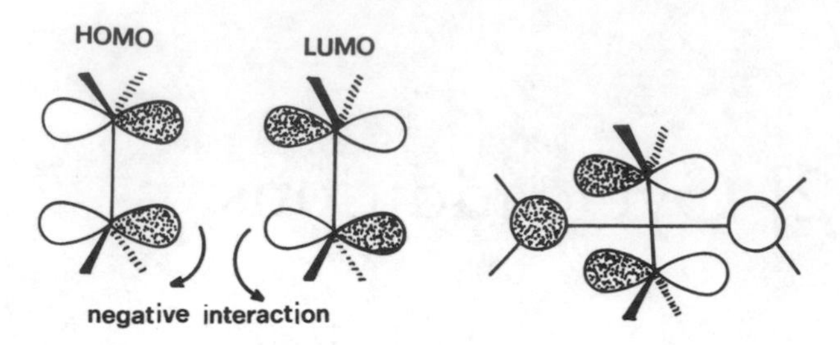

Figure 3.1. HOMO/LUMO interactions in [2 + 2] cycloadditions.

The rate of the reaction is highly dependent on the characteristics of the reagents: one reagent is a donor and the second is an acceptor (ynamines are the best reagents for electrophilic olefins, as are ketenes and isocyanates for nucleophilic olefins). The proximity of the frontier orbitals is the dominant factor, because the closer the HOMO of the donor and the LUMO of the acceptor are, the smaller the denominator of Eq. 2 (p. 4) and the greater the energy gained in the cycloaddition.

The concerted character of the reaction can be explained if we consider that in both cases a second π system is present.

Acetylene Systems. If we examine the cycloaddition of an acetylenic compound to an olefin in the light of the $HOMO_{acetylene}/LUMO_{olefin}$ and $HOMO_{olefin}/LUMO_{acetylene}$ interactions, as is shown in Figure 3.4, we may note that the former interaction (case A) predominates for nucleophilic acetylenes (ynamines). The latter (case B) predominates for electrophilic acetylenes (acetylene carboxylates).

The latter case is well illustrated by an improved synthetic route to the hydroazulenic skeleton of the fungal sesquiterpenoid velleral (*5*) (**104**). Dimethyl acetylenedicarboxylate reacted with the enamine **101**, forming the cyclobutene **102**. Ring opening and hydrogenolytic deamination gave the velleral derivative **103** (Scheme 3.3).

Regiochemistry is determined by the amplitude of coefficients, those of acetylenes paralleling those of the analogous olefins. The results for carboxylate and amine substituents are reported in Figure 3.5.

The following examples apply these concepts. The interaction between the HOMO of the enamine **105** and the LUMO of the acetylene **106** is responsible for the formation of the cyclobutene **107**, from which both velleral (**104**) and vellerolactone (**108**) were obtained (*6*) (Scheme 3.4).

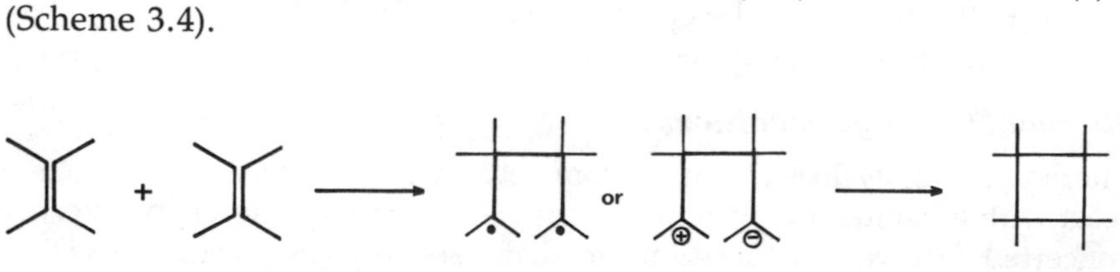

+ or

Scheme 3.1

Figure 3.2. HOMO/HOMO (A) and LUMO/LUMO (B) interactions in [2+2] cycloadditions.

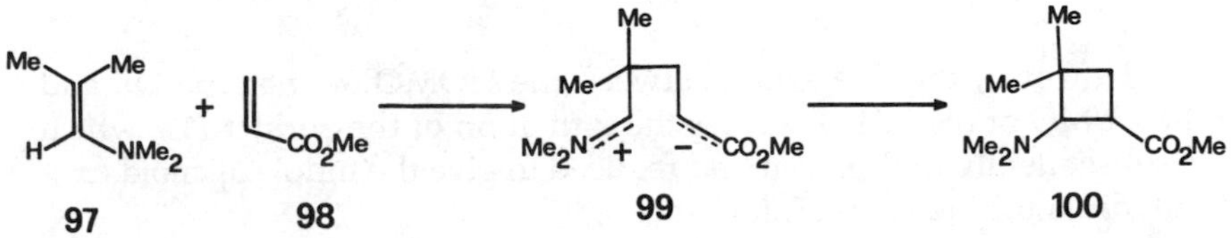

Me Me H NMe_2 **97** + CO_2Me **98** → Me Me Me_2N + − CO_2Me **99** → Me Me Me_2N CO_2Me **100**

Scheme 3.2

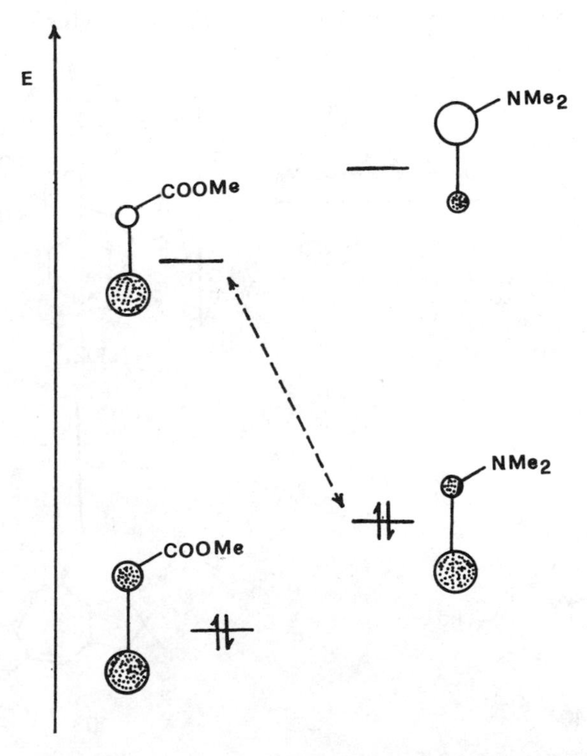

Figure 3.3. Interaction between FMOs of **97** *and* **98.**

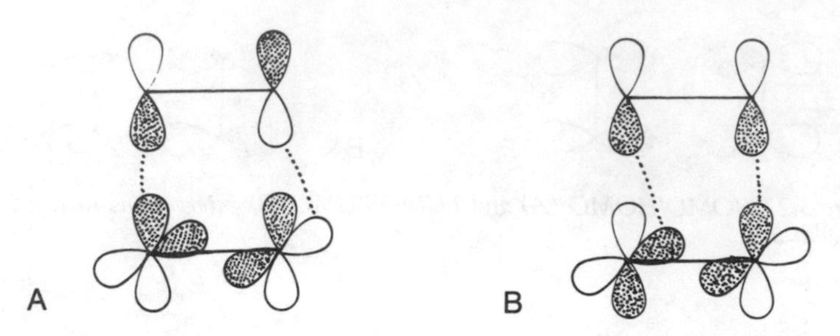

Figure 3.4. HOMO$_{acetylene}$/LUMO$_{olefin}$ (A) and HOMO$_{olefin}$/LUMO$_{acetylene}$ (B) interactions.

However, the interaction between the HOMO of ynamine **110** and the LUMO of olefin **109** secures the formation of the adduct **111**, which is subsequently hydrolyzed and reduced to give the indole alkaloid (±)-dihydroantirhine (**112**) (*7*) (Scheme 3.5).

Cumulated Systems. A concerted pathway for [2+2] cycloaddition with cumulated systems such as ketenes or isocyanates is suggested by the high negative entropy of activation (*8*), the small solvent effect (*8, 9*), and the striking regiospecificity of the reaction (*10, 11*). Although the C=O bond is sometimes involved in cycloaddition to form oxiranes, the most common reaction center is the C=C bond, which behaves as

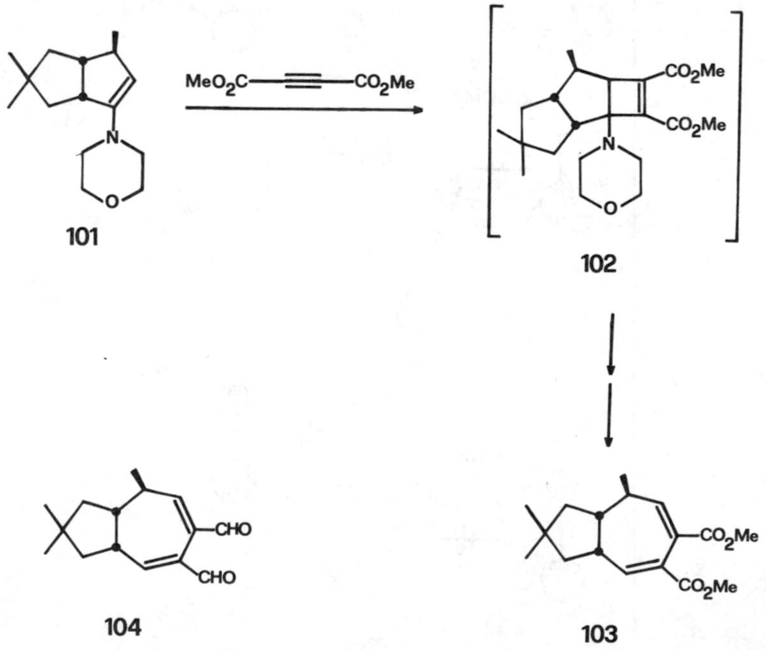

$MeO_2C-C\equiv C-CO_2Me$

101 102 103 104

Scheme 3.3

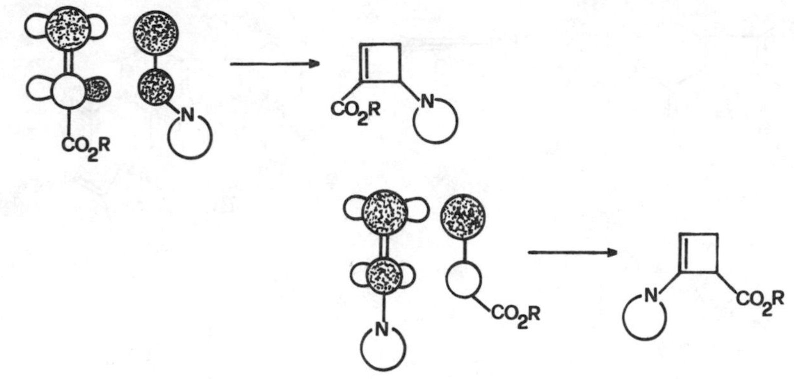

Figure 3.5. FMOs of enamines and ynamines with acetylenic and ethylenic carboxylates.

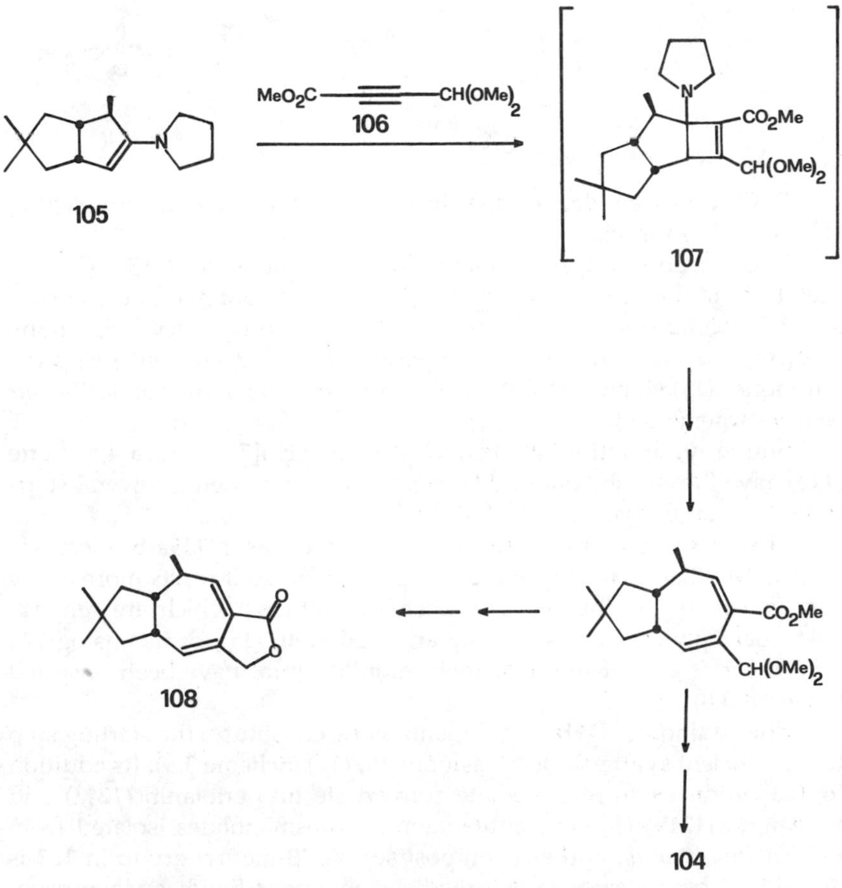

Scheme 3.4

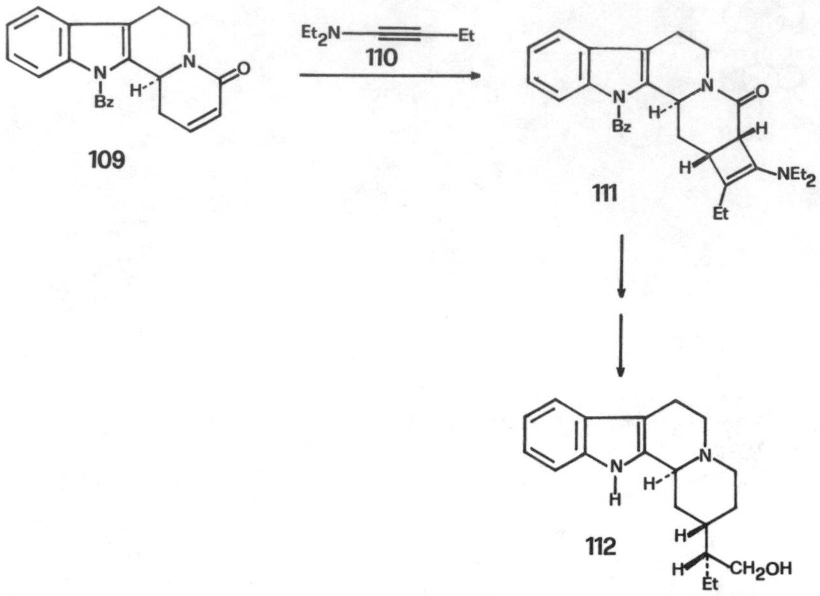

Scheme 3.5

a LUMO-controlled electrophile. In this case the dominant interaction (*12*) is depicted in Figure 3.6.

The reaction between dimethylketene (generated at 130 °C from acetal **113** in the presence of catalytic amounts of potassium carbonate) and 2-(*p*-tolyl)propene, produced 3-*p*-tolyl-2,2,3-trimethylcyclobutanone (**114**) in 31% yield, which was transformed into (±)-α- and (±)-β-cuparenones (*13*) (**4**) and (**115**) isolated from the wood of the conifer *Mayur pankhi* (Scheme 3.6).

Similarly, dimethylketene and *R*(+)-bicyclo[7.1.0]deca-4,5-diene (**116**) gave the cyclobutanone **117**, which was converted in several steps to (+)-isocaryophyllene (*14*) (**118**) (Scheme 3.7).

The presence of one or two chlorine atoms, as in **119a,b,** increases the electrophilic character of the C=C fragment so it reacts more easily with a variety of ethylenic and acetylenic ethers, which are remarkable nucleophiles. In this way squaric acid (**120**) (*15*), or its analog **121** (*16*), the free acid of the mycotoxin moniliformin, have been obtained (Scheme 3.8).

The addition of **119b** to cyclopentadiene constitutes the starting step for an efficient synthesis of *cis*-jasmone (**8**) (*17*) (Scheme 3.9). Its addition to **123** produces an intermediate convertible into eriolanin (**124a**) and eriolangin (**124b**) (*18, 19*), antileukemic eudesmanolides isolated from *Eriophyllum lanatum* Forbes (Compositae). A 4β-methyl group in **123** is introduced by a stereoselective modified Simmons–Smith cyclopropanation of the octalol (**122**) assisted by the presence of the homoallylic al-

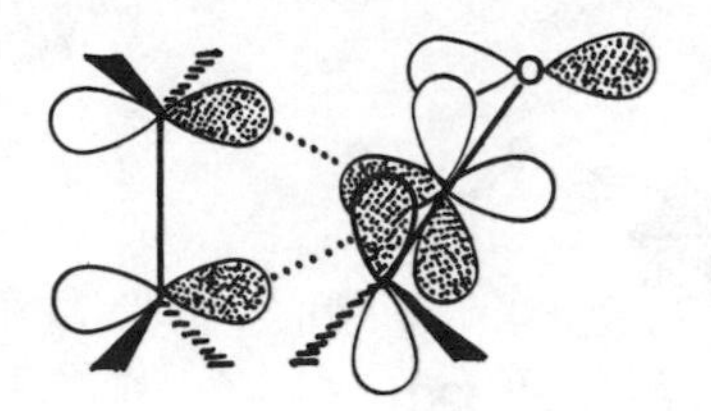

Figure 3.6. HOMO$_{olefin}$/LUMO$_{ketene}$ interaction.

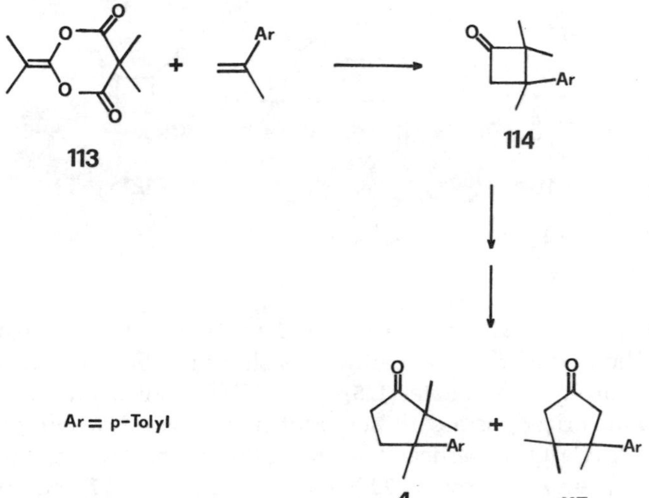

113 + → 114

Ar = p-Tolyl

4 + 115

Scheme 3.6

+ 116 → (80°; 30min. 92 %) 117

118

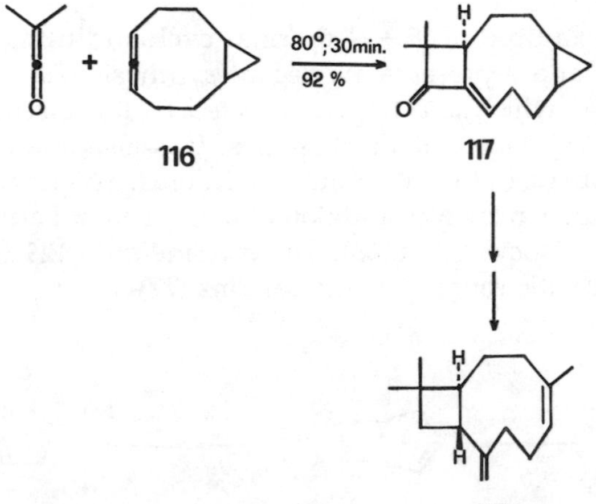

Scheme 3.7

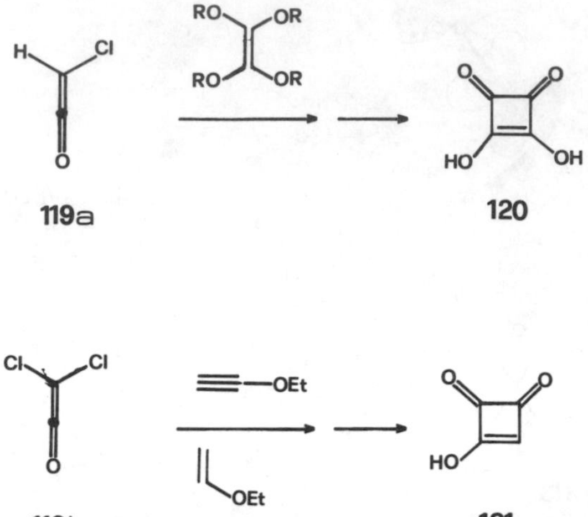

119a

120

119b

121

Scheme 3.8

coholic group (*see* the section in Chapter 2 entitled "Intermolecular Additions"). The overall reaction pathway is shown in Scheme 3.10.

Chlorosulfonyl isocyanate (**125**) is a LUMO-controlled cumulated system, characterized, as are all isocyanates (*20*), by a low-lying vacant MO having the largest coefficient at the carbon atom. Its regiochemical addition to 1-acetoxybutadiene (**126**) and to chromene (**127**) features the starting steps to (±)-thienamycin (**79**) (*21, 22*) and (±)-biotin (**128**) (*23*), respectively (Schemes 3.11 and 3.12). This behavior can be readily understood by looking at the HOMOs of model compounds (*24, 25*) (Figure 3.7), and by comparing the dominant interactions between greater coefficients with reactivity.

A final example of [2+2] thermal cycloadditions is shown in Scheme 3.13, and represents an elegant synthesis (*26*) of the loganin aglycone **(130),** which is a terpene isolated as the glycosyl derivative from several *Strychnos* and *Vinca* species. It is also a key intermediate in the biosynthesis of the major indole alkaloids. The reaction sequence involves the sequential cycloaddition of both dichloroketene **(119b)** and chlorosulfonyl isocyanate **(125).** The intermediate **129** has been also used as a synthetic route to prostaglandins (27).

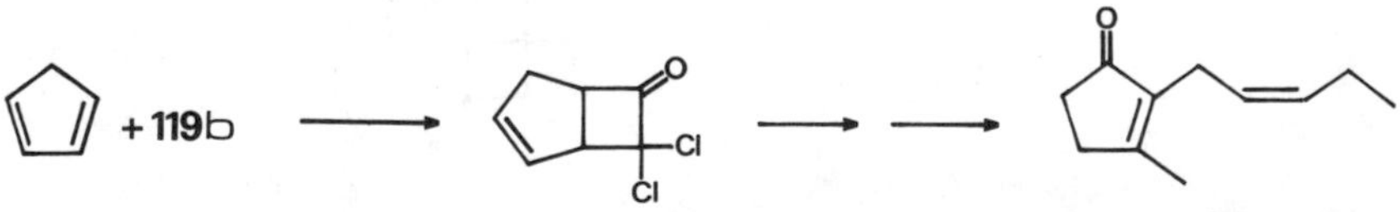

+ 119b

8

Scheme 3.9

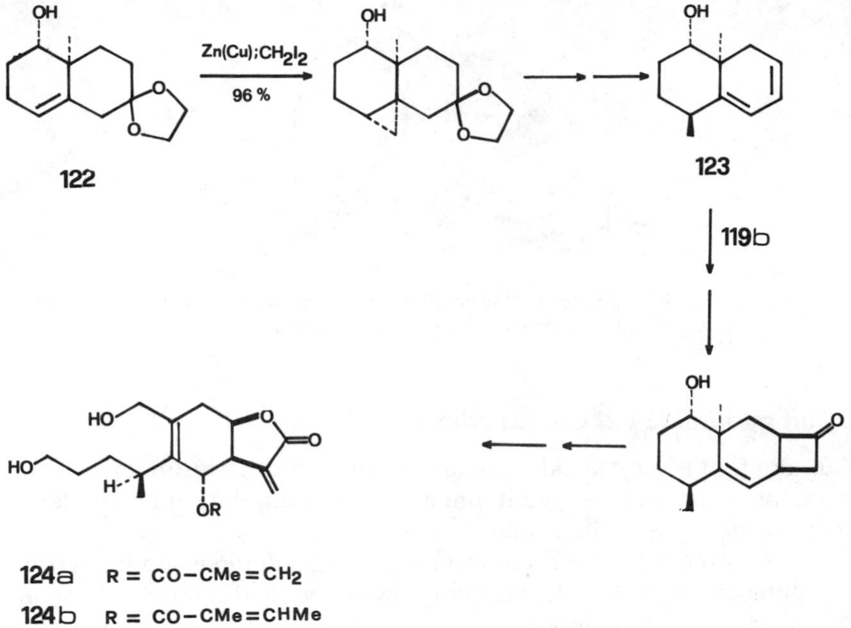

Zn(Cu);CH_2I_2
96 %

122

123

119b

124a R = CO−CMe=CH_2

124b R = CO−CMe=CHMe

Scheme 3.10

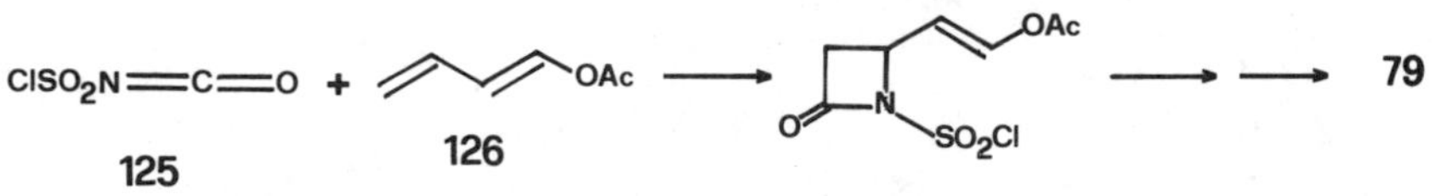

$ClSO_2N=C=O$ + OAc

125 126

OAc

SO_2Cl

79

Scheme 3.11

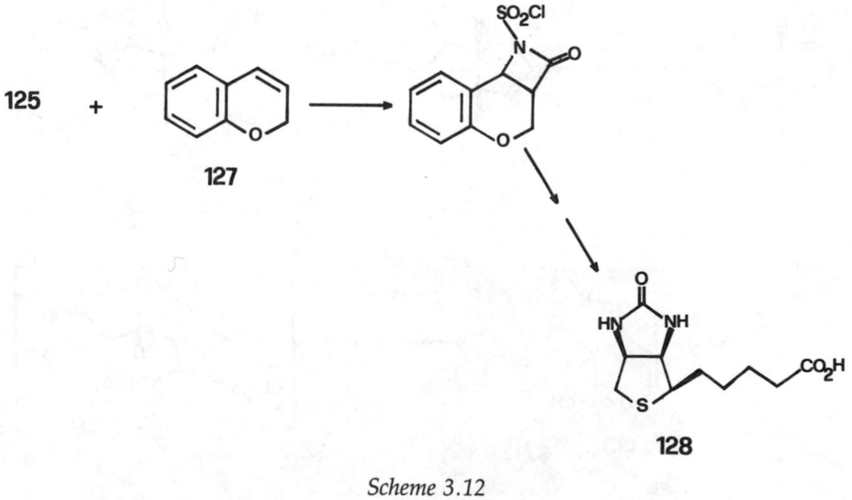

125 +

127

SO_2Cl

128

CO_2H

Scheme 3.12

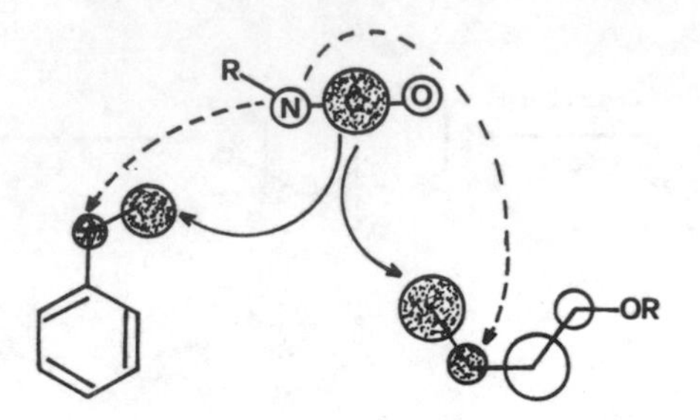

Figure 3.7. Schematic of LUMO coefficients of isocyanates and HOMO coefficients of styrene and alkoxybutadiene with the site of C and N attack in 1,2-cycloadditions.

Photochemical [2+2] Cycloadditions

Although [2+2] cycloadditions are difficult to achieve under thermal conditions, they occur readily under photochemical conditions. Hence they occur in the excited state.

Approaches to [2+2] photochemical cycloadditions are influenced by different scientific interests and they differ significantly if those interests are in photophysics or in preparative organic photochemistry.

Corey (*28*) et al. and De Mayo and coworkers (*29, 30*) investigated

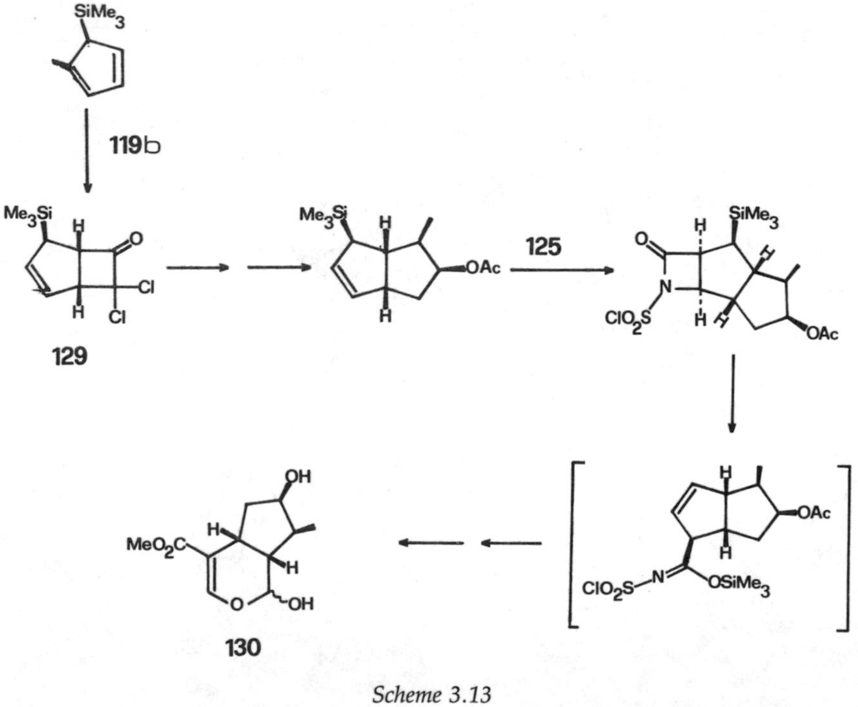

Scheme 3.13

the mechanism of photoanellation of olefins to enones in an attempt to explain the regiochemistry and stereochemistry.

The questions to be answered concern

1. The solvent and the eventual sensitizer effects.
2. The nature of the excited state involved, whether singlet or triplet.
3. The eventual presence of intermediates.
4. The always high regiospecificity of the reactions.
5. The often low stereospecificity of the reactions.

To reach the answers, we may consider as an example the cycloaddition between 2-cyclohexenone (**131**) and 1,1-dimethoxyethene (**132**), which gives rise to a mixture of *trans-* and *cis-*7,7-dimethoxybicyclo[4.2.0]octan-2-one (**133a,b**) in 49 and 21% yield, respectively, and some volatile byproducts in less than 6% yield (*28*) (Scheme 3.14).

The regiospecificity of the reaction is very high but its stereospecificity is low because the original (*Z*) configuration of the olefin **131** is lost in the high-yield adduct **133a**. Furthermore, the experimental conditions suggest that the excited state, initially formed from excitation of **131**, is the $n \rightarrow \pi^*$ singlet; the reacting species, however, is the $n \rightarrow \pi^*$ triplet formed by intersystem crossing (*28*).

The regiospecificity of the reaction has been explained both in terms of a charge-transfer complex, where the negative end of one reagent faces the positive end of the second **134**, and in terms of formation of the more stable biradical intermediate **135a** vs. **135b**, the former being also responsible for the loss of stereospecificity of the reaction (Scheme 3.15).

This explanation fails, however, in several photodimerization reactions. Coumarin (**136**) in the singlet excited state gives a *syn* head-to-head dimer **137**, but in the presence of benzophenone-triplet sensitizer it leads to a mixture of **137** and its *anti* head-to-head isomer **138** (*31*) (Scheme 3.16).

As a personal choice we prefer the Herndon (*32*) approach, which is the application to photochemical cycloaddition of simple perturbation methods. Although the approach has been criticized, it allows an easy semiquantitative insight into the mechanism of the reaction, giving a nice example of the substituents' effect.

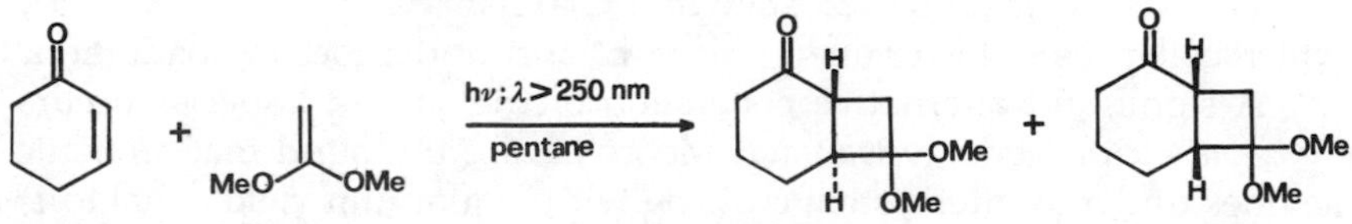

131 + MeO OMe 132 —hν; λ>250 nm, pentane→ 133a + 133b

Scheme 3.14

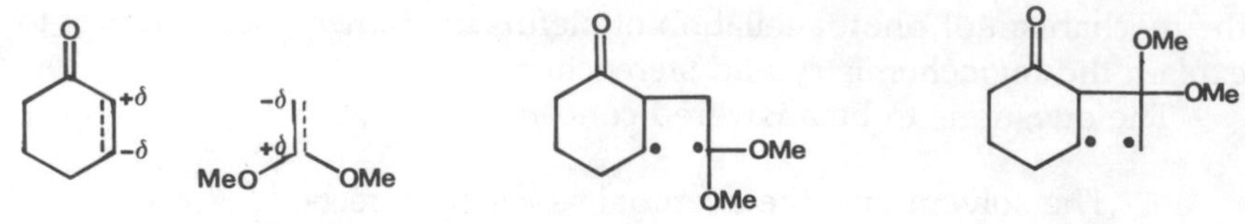

Scheme 3.15

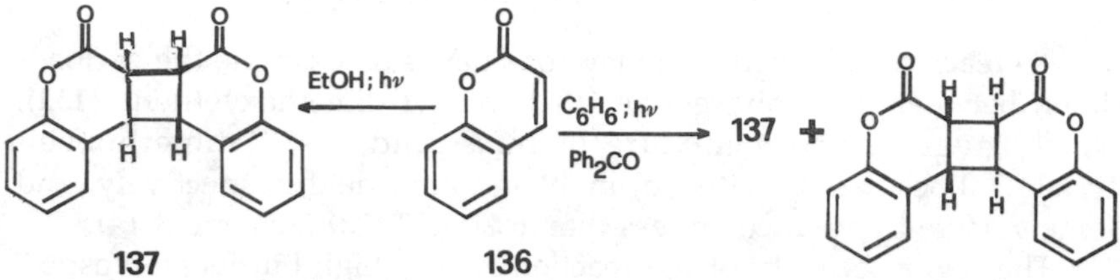

Scheme 3.16

Figure 3.8 is a simple representation of the photochemical process with its stereochemical and regiochemical implications.

When a ground-state molecule $(A=B)^\circ$ is photochemically excited to the singlet $^1(A=B)^*$, it can react with a second ground-state molecule $(C=D)^\circ$ or can be transformed by intersystem crossing to its triplet state $^3(A=B)^*$, which then reacts with $(C=D)^\circ$.

If the cycloaddition occurs between $^1(AB)$ and $(CD)^\circ$, a concerted mechanism is conceivable and the adduct retains the configuration of the reagents. The FMO interactions can be 'LUMO'/LUMO or 'HOMO'/HOMO, giving rise to large energy gain in the perturbation Eq. 1 (in Chapter 1) because of the low orbital energy separation.

If the cycloaddition occurs between $^3(AB)$ and $(CD)^\circ$, a diradical intermediate must be involved and the FMO interactions are relevant only in the regiochemistry of the reaction. Because most intermolecular photocycloadditions occur with α,β-unsaturated carbonyl derivatives, where the obvious cycloaddend undergoes excitation by light, the electron promoted to an antibonding orbital can be in principle a π or an *n* electron.

The first formed excited state in a C=C–C=O system is $^1n \rightarrow \pi^*$, which collapses to lower energy $^3n \rightarrow \pi^*$ and undergoes cycloaddition.

A significant alternative generation of the $^3n \rightarrow \pi^*$ species occurs by triplet–triplet energy transfer. Moore et al. (*33*) found that aromatic ketones undergo intersystem crossing with a quantum yield equal to 1, hence these compounds are widely used as triplet sensitizers (*34*). A schematic process through a sensitizer S becomes: $S \xrightarrow{h\nu} {}^1S^* \pm {}^3S^* \xrightarrow{AB} S$

+ $^3(AB)^*$. Thus one can excite the triplet state of molecules that otherwise would be formed inefficiently.

We usually do not know the amplitude of the MO coefficients of the triplet state. Nevertheless, the energy difference between $^1n \rightarrow \pi^*$ and $^3n \rightarrow \pi^*$ in α,β-unsaturated carbonyls is on the order of 0.2 eV. Therefore it seems reasonable to assume that, apart from its multiplicity, the excited state always has the characteristics of a π^* MO.

In light of this approach we reconsider the reaction between **131** and **132** . If we observe the ionization potentials of **131** (*35*) and **132** (*36*), their electron affinity (*37*), the **132** FMO coefficients (*37*), and the FMO coefficients of acrolein (*38*) (used as a model of **131**), we find that the dominant MO interaction is between 'LUMO' of **131** and LUMO of **132**, the alternatives being larger in energy separation or repulsive forces (Figure 3.9).

The regiochemical implications of this interaction are shown in Figure 3.10, as are those of acrolein dimerization, used as a model of acrolein dimerization of **136** (the MO interactions of photochemical dimerization of α,β-unsaturated carbonyl can be simply obtained by doubling the left part of Figure 3.9).

Bonding interactions (Figure 3.10b) explain the regiochemistry,

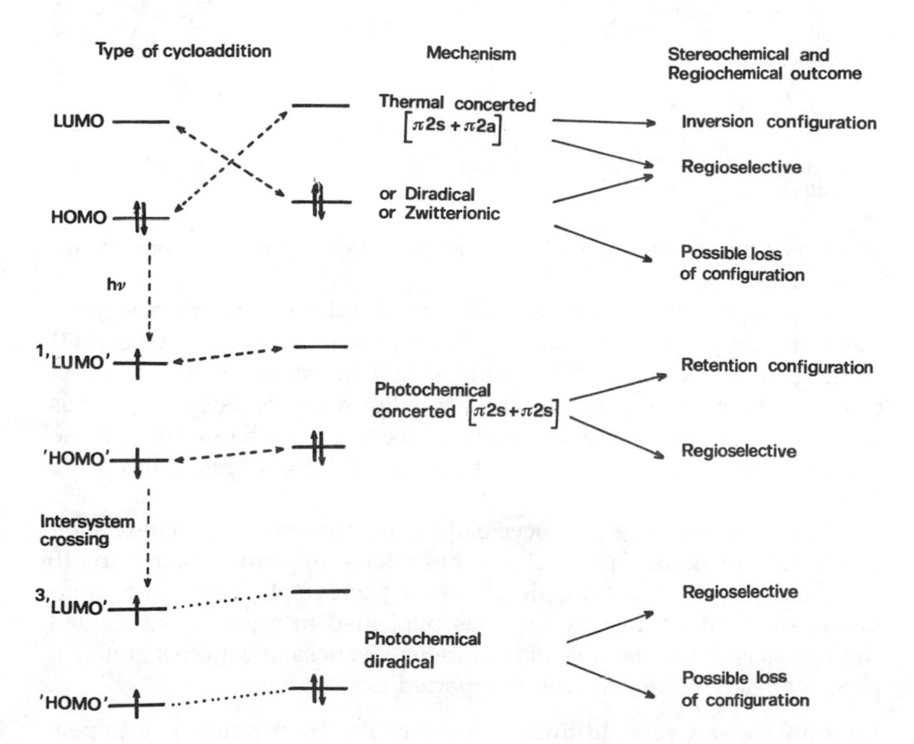

Figure 3.8. [2+2] Cycloadditions: mechanism, regiochemical and stereochemical outcome.

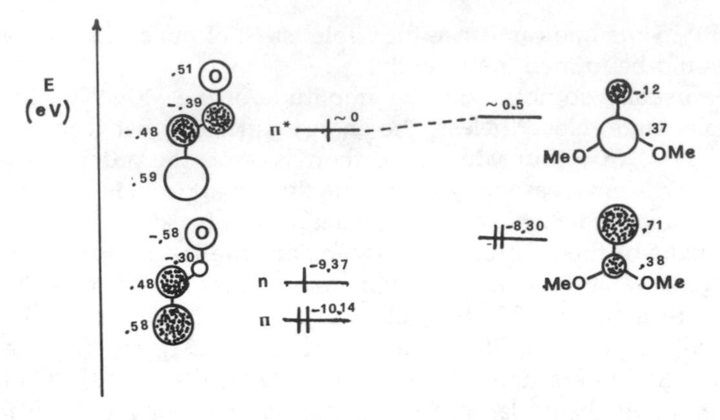

*Figure 3.9. The MO interactions between excited **131** and **132**.*

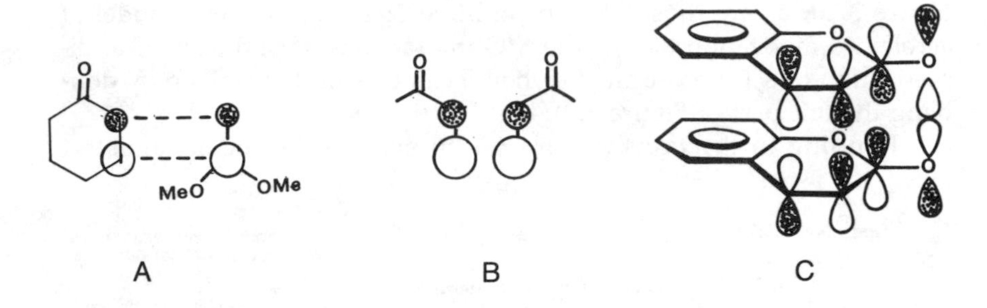

*Figure 3.10. The 'LUMO'/LUMO interaction of **131** and **132** (A) and that of coumarin (**137**) (B and C).*

whereas nonbonding interactions (Figure 3.10c) explain the stereochemistry, of **137**.

This approach can also be applied to complex natural molecules, as exemplified by the Chapman synthesis (*39*) of α-lumicolchicine (**142**) starting from colchicine (**139**). Sunlight first transforms it to β-lumicolchicine (**140**) and its γ-isomer (**141**) through a 4π electrocyclic process (*see* the section on "4π Electrocyclic Processes" in Chapter 9) and the former is further converted by light to its *anti* head-to-head dimer **142** (Scheme 3.17).

Several reviews of photocycloaddition reactions that include their application to natural-product synthesis have appeared. Some are 10 years old (*40–42*); some are simply short surveys (*43, 44*); one is comprehensive and up to date but was published in a journal of limited circulation (*45*). For the sake of comprehensiveness, the more significant papers already reviewed will be reported here as well.

Intermolecular Cycloaddition. In view of the large number of papers on intermolecular cycloaddition to be reviewed and the uncertainty of classifications like symmetry of the excited state, the material is orga-

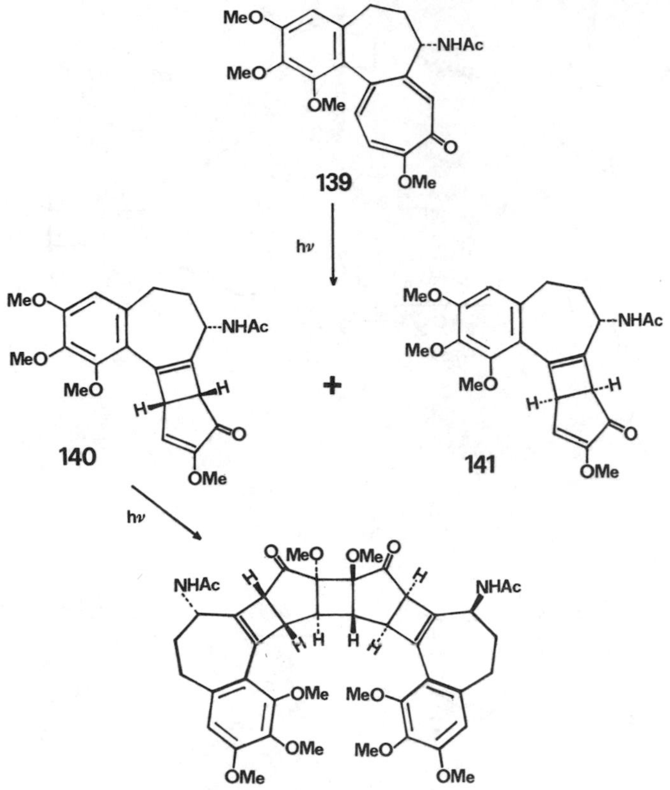

MeO NHAc MeO OMe O 139 OMe

hν

MeO NHAc MeO MeO H H O 140 OMe + MeO NHAc MeO MeO H H O 141 OMe

hν

O MeO OMe O NHAc H H NHAc H H H H OMe MeO OMe MeO OMe OMe

142

Scheme 3.17

nized according to the nature of the adducts and the reactant, an arrangement that parallels the one used by Dilling (*45*).

CYCLOBUTANE RING LEFT INTACT. Grandisol (**21**), the major component of the sex attractant of the male boll weevil, is one of the most popular targets of synthesis.

The first approach (*46*) is centered on photoaddition of methyl vinyl ketone to isoprene to give **143** in about 1% yield. Far better yields were obtained by photoaddition of ethylene to cyclopentenones (*47–49*), cyclohexenone (*50*), or cyclohexenone's lactone analog (*51*) (Scheme 3.18) through intermediates of general formula **144.** Based on this approach, two industrial syntheses of grandisol have been developed by Zoecon Corporation and the U.S. Department of Agriculture.

Fragrantol (**146**), a cyclobutane derivative isomer of grandisol isolated from the roots of *Artemisia fragrans* Willd, was synthesized through the unsymmetrical photodimer **145** of isoprene (*52*) (Scheme 3.19).

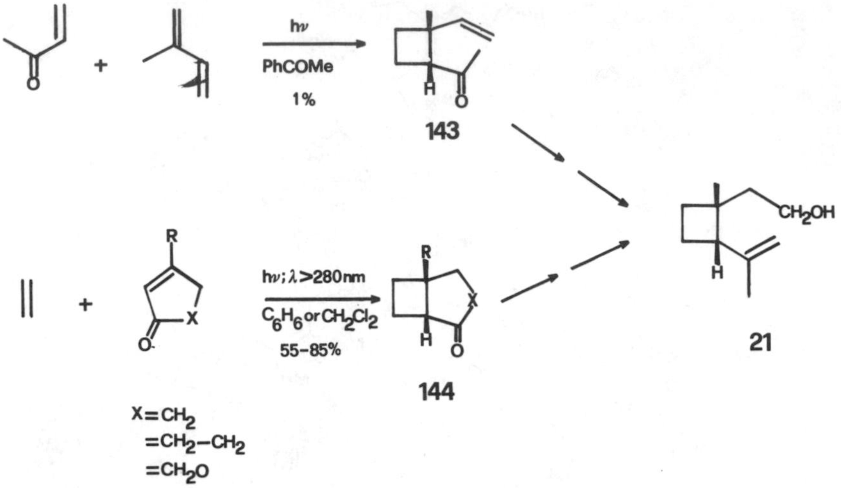

Scheme 3.18

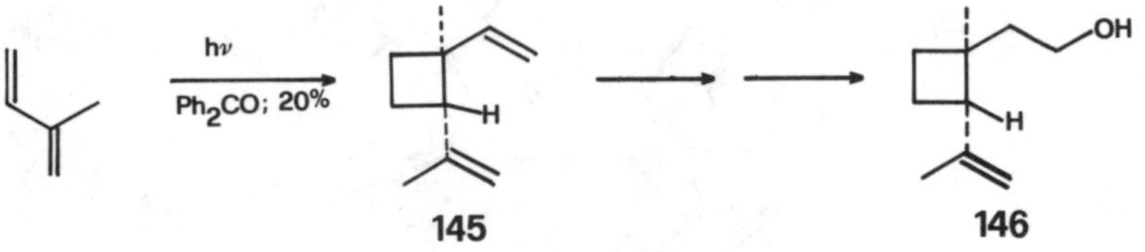

Scheme 3.19

The α,β-unsaturated carbonyl compounds absorb light at much more accessible wavelengths as far as the preparative organic chemist is concerned, and their photocycloaddition to alkenes is probably the most important reaction in photochemical synthesis. Therefore α,β-unsaturated carbonyl compounds are the most used starting material for the photochemical construction of cyclobutanes. However, cyclopentenones are sometimes poorly regiospecific (except when a highly polarized olefin is the second cycloaddend) (*53*). Cyclopentenones were used in the syntheses of *dl*-illudol (**147**) (*53*), a sesquiterpene alcohol isolated from *Clitocybe illudens;* (±)-lineatin (**148**) (*54*), a unique tricyclic pheromone of *Trypodendron lineatum* Olivier; and α- and β-bourbonenes (**149**) and (**150**), two sesquiterpenoid hydrocarbons (*55*) (Scheme 3.20).

A photochemical addition of isobutene to cyclohexenone (*56*) and to *cis,cis*-2,6-cyclononadienone (**151**) (*57*) gave two bicyclic ketones, **152** and **153**, which were further elaborated to (±)-caryophyllene (**154**) and (±)-isocaryophyllene (**118**) (Scheme 3.21). Both reactions are highly regiospecific but not always stereospecific, because **152** represents the low-yield isomer (8%), the *trans* isomer being the high-yield one (35%).

These results may be explained in the same way as those reported in Figure 3.9, because the MO's of isobutene are similar to those of **132**. It is noteworthy that the isolated double bond of **151** is not involved in any cycloaddition, as it cannot be excited under the usual photochemical conditions.

The processes connected with the action of light on human skin are responsible for several changes, from the less important reversible or irreversible processes like reddening or erythemas and aging of the skin tissue to the more dramatic processes degeneration of the skin tissues with formation of cancers.

One of the known effects of light is on the pyrimidine bases of deoxyribonucleic acid (DNA), thymine (**155**), and cytosine, with formation of photodimers by a [2+2] cycloaddition involving the 5,6-double bond of the bases (*58, 59*).

In native DNA the thymine dimerization leads to the *cis-syn* isomer

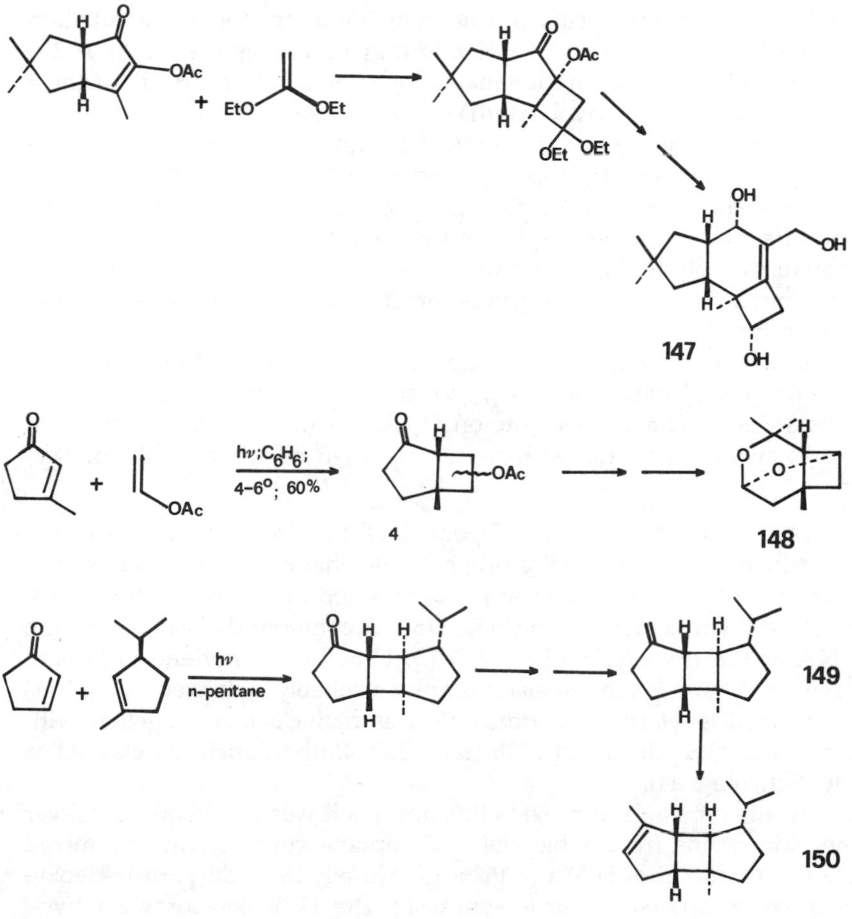

Scheme 3.20

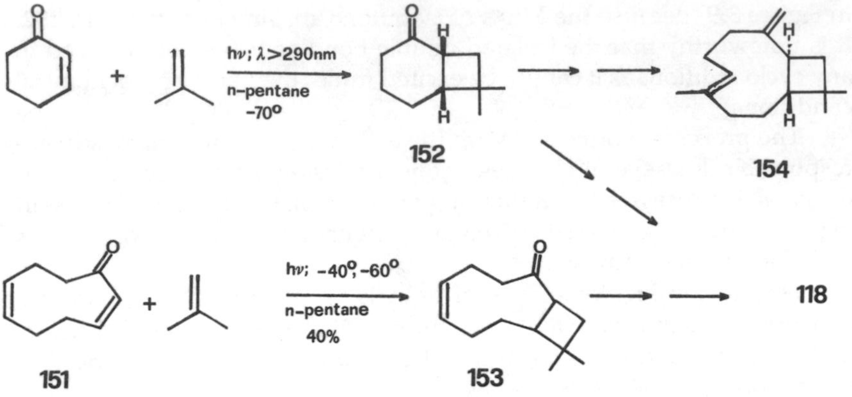

hν; λ>290nm
n-pentane
-70°

152

154

hν; -40°, -60°
n-pentane
40%

118

151

153

Scheme 3.21

156 (*60*) through a reversible process which is a function of the radiation wavelength. At 280 nm the cycloaddition process is preferred; at 239 nm the cycloreversion predominates (Scheme 3.22) as a result of the ε values of the species involved (*61*).

A detailed study of solvent effect on dimerization of dimethyl thymine was performed by Morrison and Kleopfer (*62*), and head-to-head photodimers were found to prevail in high dielectric constant solvents. The effect of singlet vs. triplet excited states was also discussed. The formation of photodimers on skin can be used to advantage in the treatment of psoriasis, a skin degeneration characterized by proliferation of cells in the skin (*63*).

Oral administration of 8-methoxypsoralen (**157**), and more recently of retinoic acid derivatives (*64*), reduces the rate of cell division. The substances presumably operate on DNA, because furocoumarins form mixed cycloadducts **158** with bases as shown in Scheme 3.23 for thymine.

CYCLOBUTANE RING OPENED. Opening of the cyclobutane ring of cycloadducts may be utilized to originate the chains to be introduced into a natural product. A nice example is afforded by the synthesis of (±)-5-(4′,5′-dihydroxypentyl)uracil (*65*) (**160**), a compound isolated from the DNA of *Bacillus subtilis* phage SP–15. Uracil and vinylene carbonate gave, through a preferred *endo* complex probably stabilized by MO interactions, *cis-syn-cis* cycloadduct **159a** as major isomer, together with the *cis-anti-cis* cycloadduct **159b** (ratio 5:2). Both adducts are cleaved to **160** (Scheme 3.24).

In the prostaglandin fields this approach was used both to deliver the side chains from a bicyclo[3.2.0]heptane ring deriving by mixed photocycloaddition of **161** and **162** to give finally 13,14-dihydro-11-deoxyprostaglandin (**163**) (*66*) or to synthesize the 11-deoxy Corey aldehyde (**165**) starting from cyclopentenone and methyl β-acetoxyacrylate (*67*).

The unseparated mixture of regioisomers **164** has been directly worked up to give **165** in 31% yield (Scheme 3.25).

Genepic acid is an antibiotic, isolated from Puerto Rican jagua fruit, to which structure **169** has been assigned. Baldwin and Crimmins (*68*) succeeded in building up this structure by a photochemical cycloaddition of **166** and **167**, followed by ring opening of the cyclobutane ring

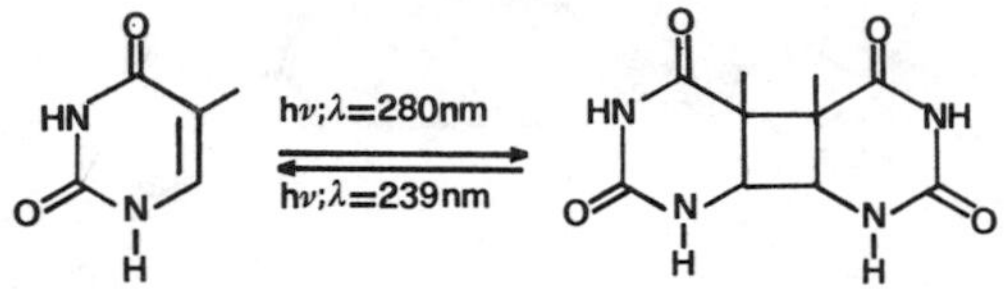

hν;λ=280nm
hν;λ=239nm

155 156

Scheme 3.22

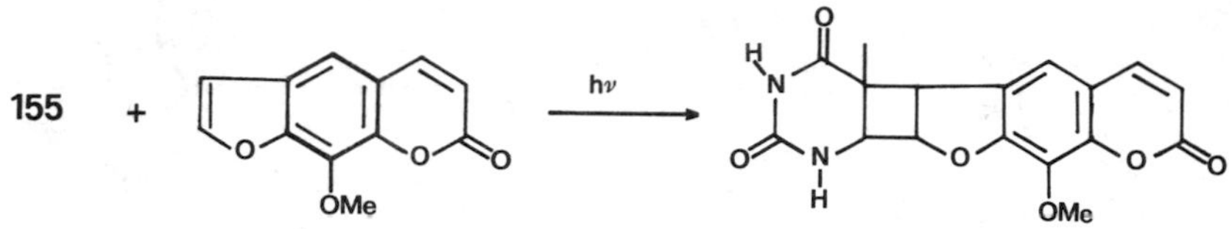

155 + hν

157 158

Scheme 3.23

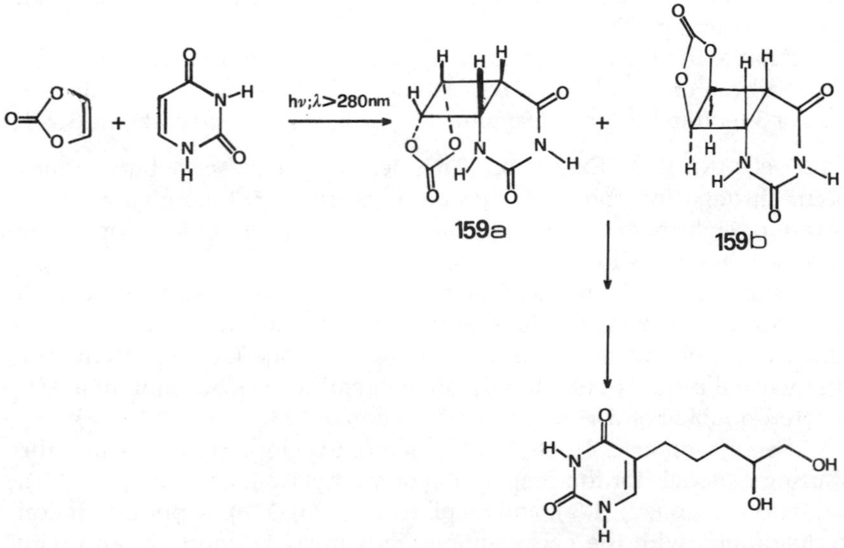

hν;λ>280nm

159a 159b

160

Scheme 3.24

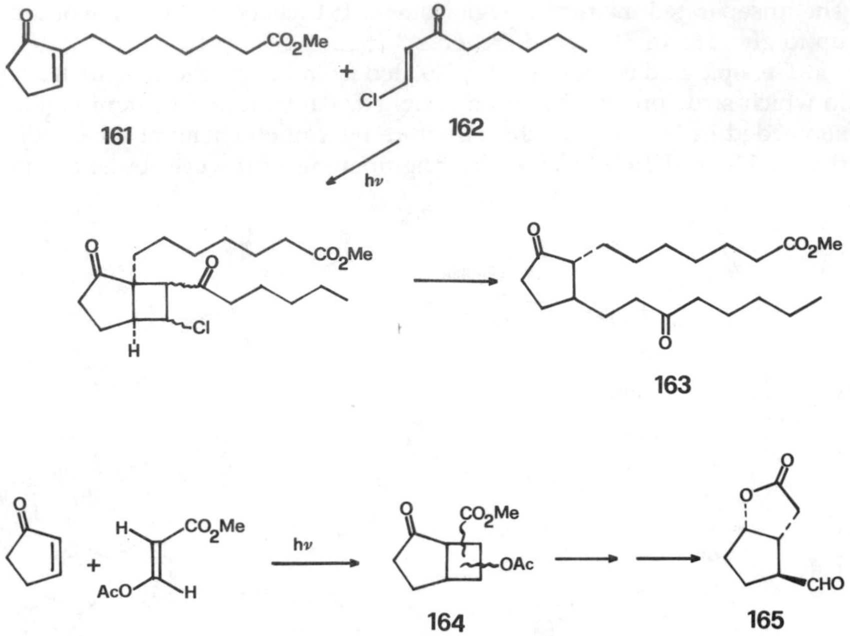

Scheme 3.25

of **168**, to give **169** in several steps (Scheme 3.26). However the product obtained did not match with a sample of genepic acid, which is still a point to clear up.

Finally, the total synthesis of (±)-norketotrichodiene (**171**) was described from one of the two regioisomers obtained by photocycloaddition of 3,6-dimethylcyclohexenone to cyclotene (**170**) (*69*) (Scheme 3.27).

CYCLOBUTANE RING EXPANDED. The cleavage of the cyclobutane may occur through incorporation into an adjacent ring, thus giving rise to seven- or eight-membered rings, depending on whether a cyclopentane or a cyclohexane is involved.

Several natural products have been prepared in accordance with this pathway. The synthesis of α-caryophyllene alcohol (**172**) from 4,4-dimethylcyclopentene and 3-methylcyclohexenone (*70*) represents the first natural product obtained by photochemical cycloaddition of a conjugated double bond with an olefin (Scheme 3.28).

Three isomeric 2-, 3-, and 4-acetoxycyclopentenones are the starting materials for the preparation of methyl isomarasmate (*71*) (**174**), stipitatonic acid (*72*) (**176**), and 5-epikessane (*73*) (**178**) by photochemical cycloaddition with the cyclopentene derivatives **173** and **177** and with dimethylchloromaleate (**175**) (Scheme 3.29).

Piperitone (**179**) reacts with 1-carbomethoxy cyclobutene to give **180**, which in three steps is converted to (±)-10-epijunneol (**181**) (*74*) and reacts with 1,1-dimethoxyethene (**132**) to afford **182**, a useful intermediate to four sesquiterpenes: (±)-sativene **(183)**, (±)-copacamphene **(184)**, (±)-sativenediol **(185)**, and (±)-helminthosporal **(186)** (*75*) (Scheme 3.30).

We wish to emphasize that **180** and **182** are the only regioisomers obtained from piperitone. 'LUMO'/LUMO interactions of an α,β-unsaturated carbonyl system with electron-donating and electron-attracting substituted olefins (*see* Figure 3.10a and 3.10b, respectively) still explain the experimental results.

Ketals of cycloalkenones are useful substrates for [2+2] photocycloaddition, because the protective group of the carbonyl function prevents any Paternò–Büchi reaction (*see* the section entitled "The Paternò–Büchi Reaction" in this chapter), still maintaining the electron-attracting character of the substituent and masking a site suitable for further substitution. Thus (±)-β-himachalene (**190**) was obtained from

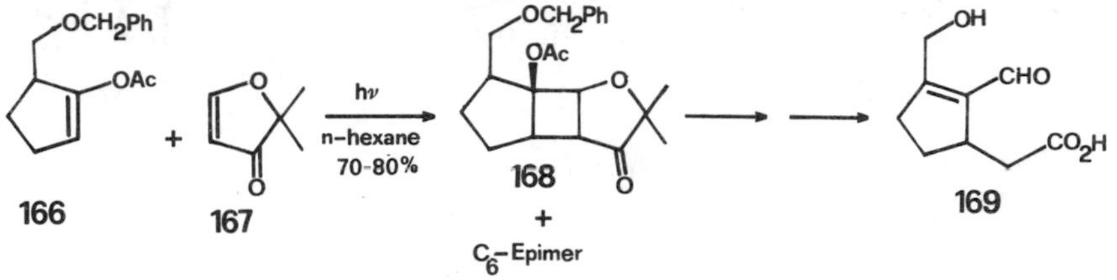

Scheme 3.26

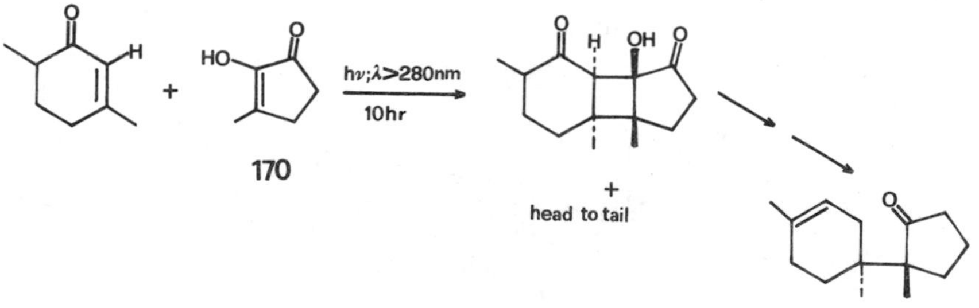

Scheme 3.27

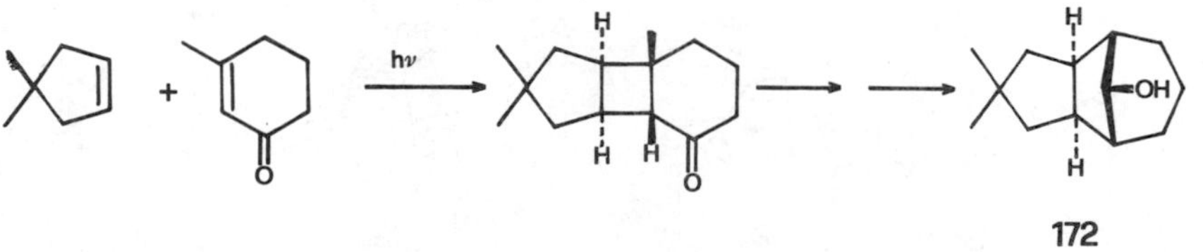

Scheme 3.28

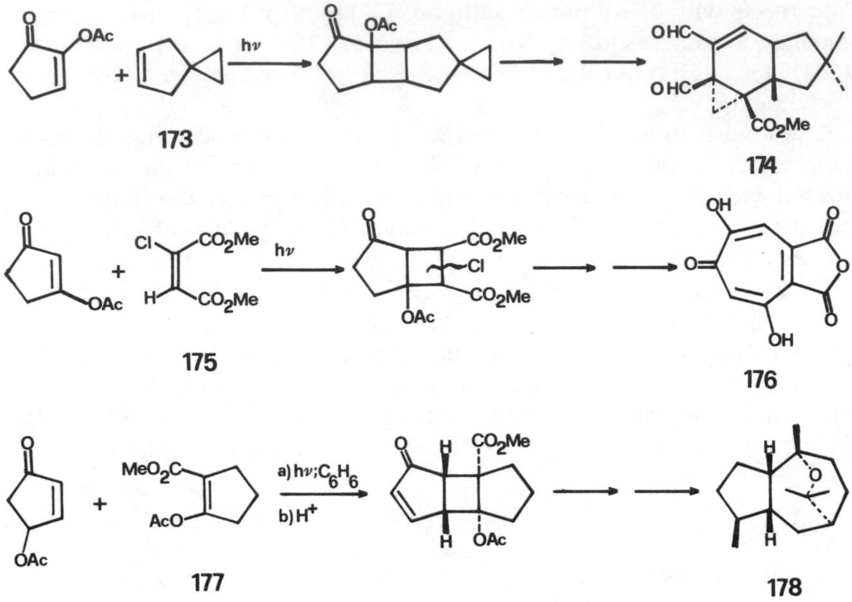

173 174

175 176

177 178

Scheme 3.29

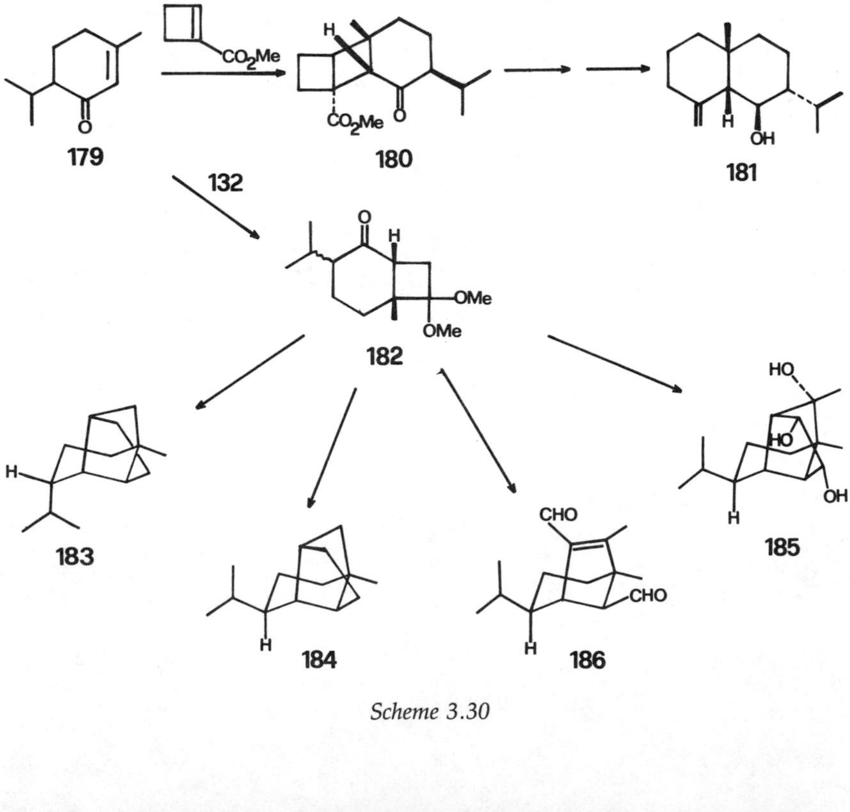

179 180 181

132

182

183 184 186 185

Scheme 3.30

187 (*76, 77*) and (±)-hirsutene (**192**) from **191** (*78*), each having **188** as a counterpart (Scheme 3.31).

The reaction between **187** and **188** has been studied in detail by Challand and De Mayo (*29*) in order to explain the solvent effect of the photocycloaddition on the stereochemistry. Under experimental conditions, **189a** is practically the only adduct, because the regioisomers **189a** and **189b** are formed in the ratio 98:2. In acetonitrile or methanol

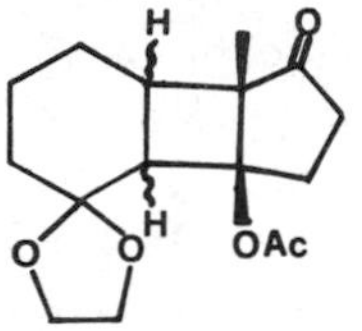

189 b

they are formed in a ratio of 45:55. In spite of the fact that the double bond of **187** has little polar character, determinant dipole interactions are suggested. The formation of **189a** seems to us to be a clear result of the 'LUMO'/LUMO interaction between two olefins carrying electron-attracting substituents. Why polar solvents counterbalance the MO effect remains to be explained. Polar solvents are known to lower MO coefficients through a solvating effect: at the moment their effect on the eventual change of MO coefficients is not known. In keeping with the suggestions of Dauben et al. (*79*) about the effect of solvent polarity changes on the course of photochemical reactions, the specific solvation of the triplet species could provide another explanation.

A [2+2] photochemical cycloaddition of maleic anhydride and 2-

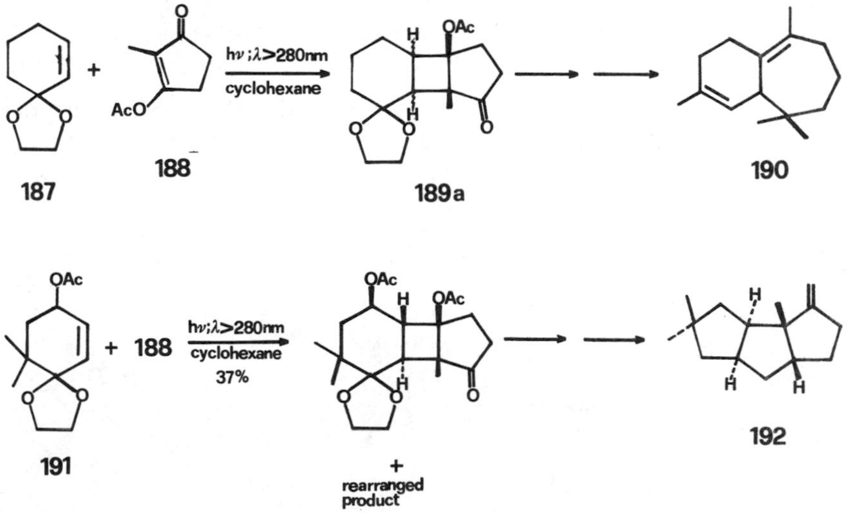

187 + 188 → (hν; λ>280nm, cyclohexane) 189a → → 190

191 + 188 → (hν; λ>280nm, cyclohexane, 37%) + rearranged product → → 192

Scheme 3.31

pentyne is the starting step of a stereospecific total synthesis of (±)-methylenomycin A (**193**), an antibiotic isolated from *Streptomyces violaceoruber* (*80, 81*) (Scheme 3.32).

Two independent routes to loganin (**197a**) and hydroxyloganin (**197b**) have a photochemical cycloaddition as the initial stage. The primary photoproduct **195** was not isolated in either of them because it undergoes retroaldol cleavage and recyclization to the hemiacetal **196**.

Both Büchi (*82*) and Tietze (*83*) started from **194a,** but their strategies diverged at the point of introduction of the 7-substituent; in the former it goes to loganin (**197a**) and in the latter to hydroxyloganin (**197b**). The Partridge et al. (*84*) synthesis begins with the chiral reagent (1*S*,2*R*)-2-methyl-3-cyclopentene-1-ol acetate (**194b**), which therefore has the 7-substituent suitably incorporated to give **197a** (Scheme 3.33).

A final group of papers describes the synthesis of several alkaloids through photocycloaddition of simple olefins to structurally complex polycyclic α,β-unsaturated ketones.

The synthesis of the aromatic intermediate **198** is completed along the route to songorine **(84)** (*85, 86*) through photoaddition of vinyl acetate to **83,** a compound already considered in the section on "Nitrenes" in Chapter 2. This paper seems to be the only one in which the addition to a C=C−C=O system takes place with a regiochemistry opposite to the one predicted by MOs (Scheme 3.34).

The photoaddition of vinyl acetate to **199** gave four isomers with the same "regular" regiochemistry, **200.** The isomers have been used as starting materials for the synthesis of three *Ormosia* alkaloids: ormosanine (**201**), piptanthine (**202**), and panamine (**203**) (*87, 88*) (Scheme 3.35).

Photochemical cycloaddition of ethylene to $\Delta^{8(14)}$-podocarpen-13-one (**204**) (*89*), through **205**, allowed preparation of three hibane derivatives (*90*): hibayl acetate (**206a**), hibaol (**206b**), and hibaone (**206c**), and of hibane itself (**206d**) (Scheme 3.36).

Cumulated Systems. All the syntheses reviewed in this section concern the photocycloaddition of allene to an α,β-unsaturated carbonyl system (or to its ketal derivatives). These reactions are characterized by high regiochemical and stereochemical outcome. Photoaddition of allene **211**, for example, to [3.3.0], [4.3.0], [4.3.0], and [4.4.0] bicyclic enones **207**, **208**, **209**, and **210**, respectively, yields **212**, **213**, **214**, and **215** as the only

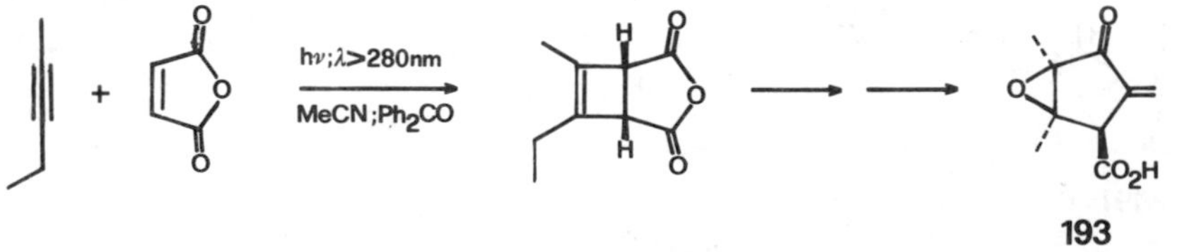

hν;λ>280nm
MeCN;Ph_2CO
CO_2H
193

Scheme 3.32

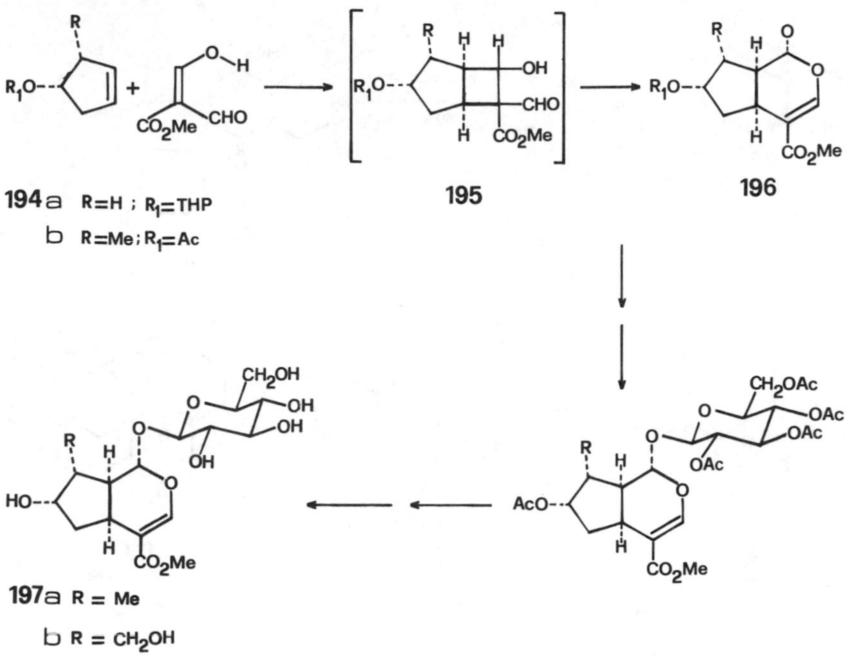

194 a R=H ; R_1=THP
b R=Me; R_1=Ac

195

196

197 a R = Me
b R = CH_2OH

Scheme 3.33

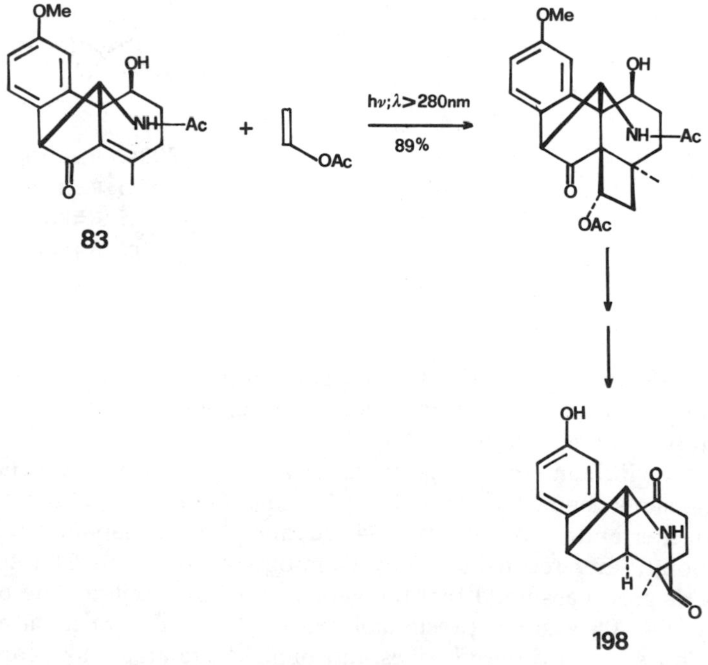

83

hν; λ>280nm
89%

198

Scheme 3.34

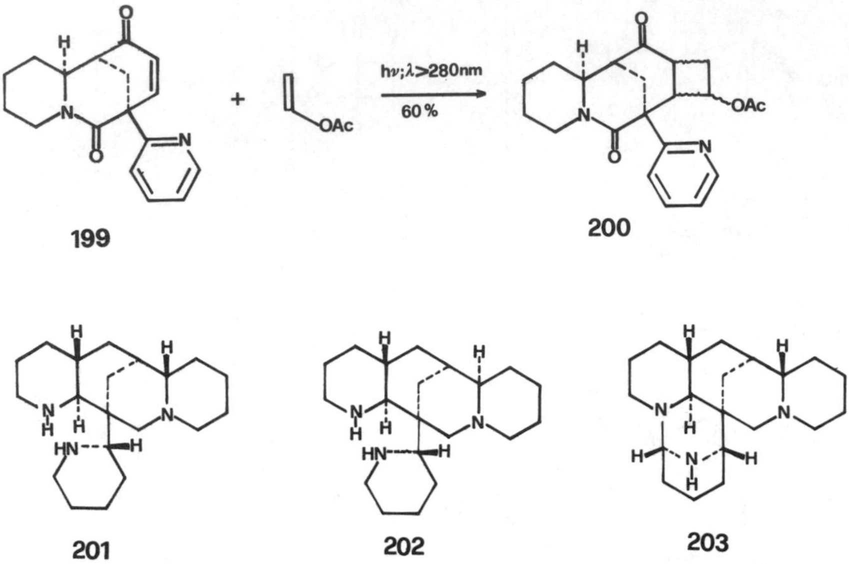

Scheme 3.35

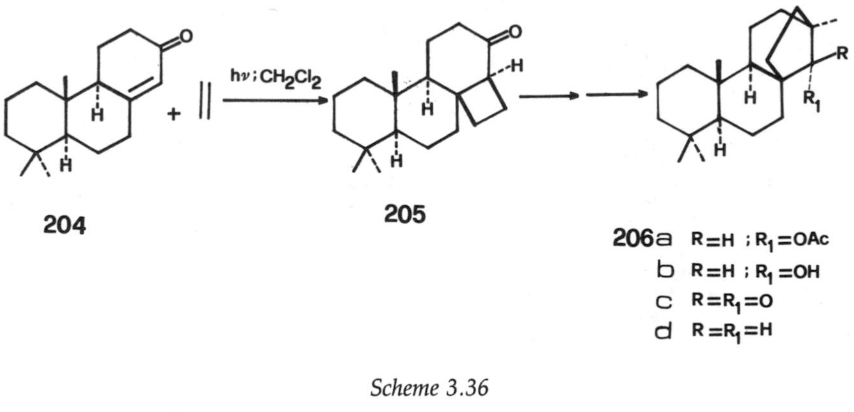

Scheme 3.36

products (*91*) (Scheme 3.37). Two exceptions concern the additions to cyclopentenone and to 3-methylcyclohexenone, where the ratio of head-to-head/head-to-tail adducts is 4:1 (*91, 92*).

The regiochemistry of the attack always gives head adducts (*91, 92*). The stereochemistry that is β for **212** and **213** is α for **214** and **215**.

Wiesner and coworkers (*93, 94*) advanced an explanation for the latter point, later reconsidered by Fleming (*see* Ref. 3 in Chapter 1). Wiesner's group assumed that the geometry of the excited state of the $C_4=C_3-C_2=O_1$ system, presumably $n \rightarrow \pi^*$, has the p orbital at C_4, where lobes are of different sizes and hence have an sp^3-like configuration. Relief of strain in the ring junction results and a more stable configuration is assumed.

It is well known (*95*) that *trans*-decalin is more stable than *cis*-decalin, while *cis*-[3.3.0]-bicyclooctane is more stable than *trans*-[3.3.0]-bicyclooctane. Hydrindanes are borderline, sometimes preferring the *cis*, sometimes the *trans*, isomer, depending on the substituents. Hence the larger lobe will be α for **209** and **210** and β for **207** and **208**, and the overlap with **211** will occur from these sides (Figure 3.11). Hence [4.3.0] systems are expected to give α or β attack depending on the nature of the substituents.

The regiochemical problem was only considered as a matter of fact. A simple consideration of larger lobes interacting in 'LUMO's of unsaturated carbonyl and allene does not explain this behavior, because HOMOs and LUMOs of **211** (*96, 97*) (Figure 3.12) have the larger coefficient in the outer and inner carbon atoms, respectively. Hence 'LUMO'/LUMO should give a regiochemistry opposite to that experi-

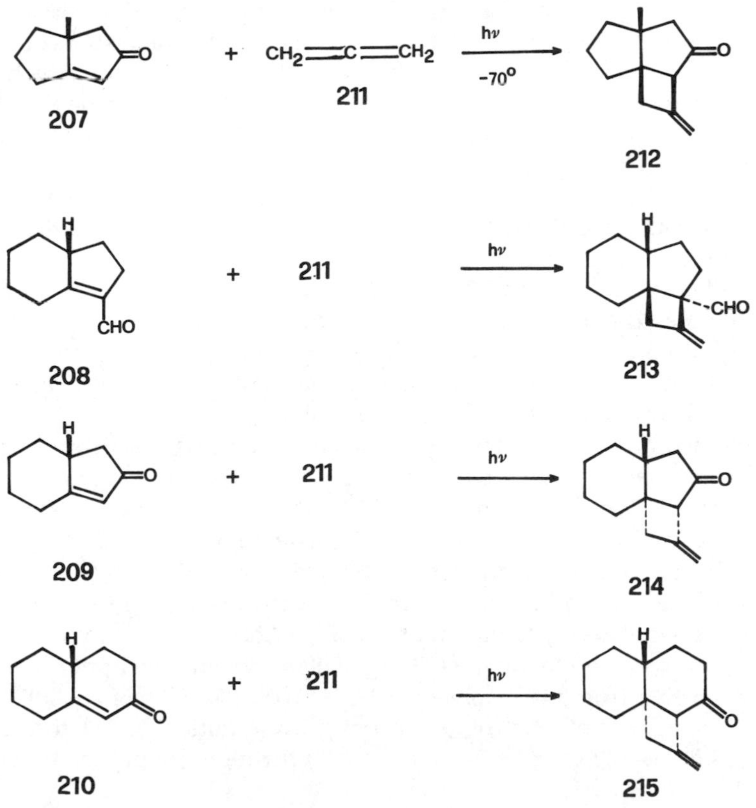

207 + $CH_2{=}C{=}CH_2$ **211** —hν, −70°→ **212**

208 + **211** —hν→ **213**

209 + **211** —hν→ **214**

210 + **211** —hν→ **215**

Scheme 3.37

*Figure 3.11. LUMO configuration of **210** (A) and **207** (B).*

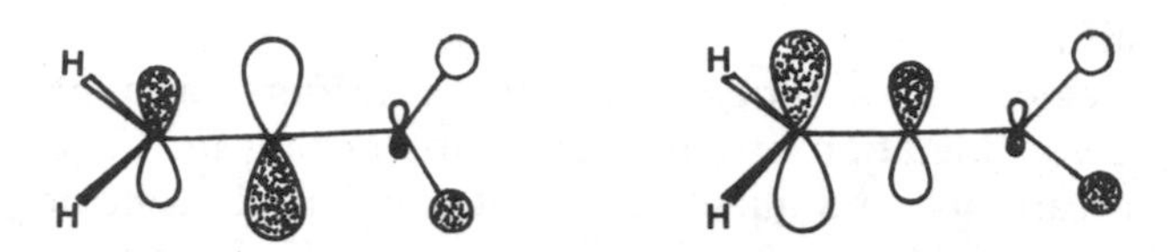

Figure 3.12. LUMOs and HOMOs of allene.

enced, but the 'HOMO'/HOMO interaction, giving rise to the correct regiochemistry, can display an energy gain only for $\pi \rightarrow \pi^*$ excitation.

A possible explanation of the observed regiochemistry could be obtained using the simplest possible Hückel model of allene, which was found to be not much different from more sophisticated models (*98*).

The interaction between 'LUMO' and LUMO of an α,β-unsaturated carbonyl and **211** is represented in Figure 3.13. Nonbonding interactions between the second double bond and the carbonyl group could easily overcome the coefficient effect, which is not large, due to the close proximity of allene coefficients.

Allene (**211**) photoaddition to **216** was the key stage to the alkaloid annotinine (**219**) (*99, 100*), obtained in the optically active form after a long pathway proceeding through its degradation product annatonine (**218**). Both compounds retain the originally formed cyclobutanic structure of the photoadduct **217** (Scheme 3.38).

Ishwarane (**222**), a tetracyclic sesquiterpenoid discovered in the roots of *Aristolochia indica* Linn., has been synthesized by photochemical addition of allene to the octalone **220** (*101*), followed by skeletal rearrangement of the resulting adduct **221**(Scheme 3.39).

The previously mentioned $\Delta^{8(14)}$ -podocarpen-13-one (**204**) undergoes allene photoaddition to give **223**, featuring a novel synthesis of three terpenes: (±)-trachylobane (*102*) (**224**), (+)-isophyllocladene (**225**), and (+)-phyllocladene (**72**), which are obtained in optically active form starting from the resolved (+)-**204** (*103, 104*) (Scheme 3.40).

A similar synthetic route involving the addition of **211** to the tricyclic ketone **226** gave (±)-stemarin (**228**) through the photoadduct **227** (*105*) (Scheme 3.41).

Wiesner's group elaborated many synthetic routes to diterpenic alkaloids characterized by allene photoaddition to tetracyclic structures of general formula **229**, thus obtaining the suitable photoadducts **230a** and

230b. These adducts were further converted to atisine (*106*) (**231**); veatchine (*107*) (**232**); and talatisamine (*108, 109*) (**233**), a highly oxygenated diterpenoid alkaloid isolated from *Aconitum variegatum* (Scheme 3.42).

All the previously mentioned syntheses follow a stereochemical pathway like that of the model octalone **210** with the formation of *trans*-fused cycloadducts. In contrast, the synthesis of (±)-steviol methyl ester (**236**) (*110, 111*) is an example of photochemical addition of allene with

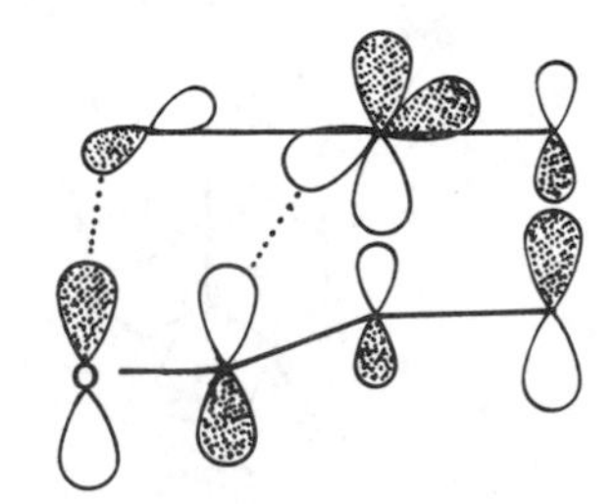

Figure 3.13. 'LUMO'/LUMO interaction of α, β-unsaturated carbonyl system with allene.

Scheme 3.38

Scheme 3.39

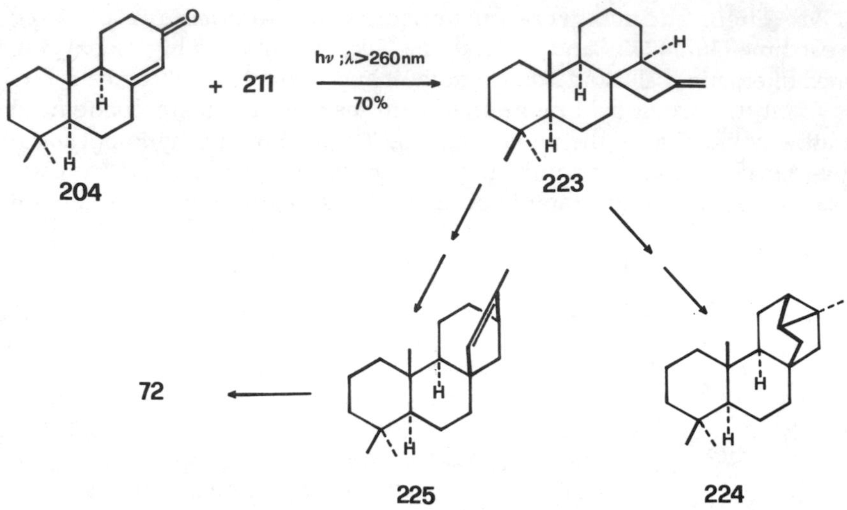

204 + 211 → (hν ; λ>260nm, 70%) 223

223 → 225 → 72

223 → 224

Scheme 3.40

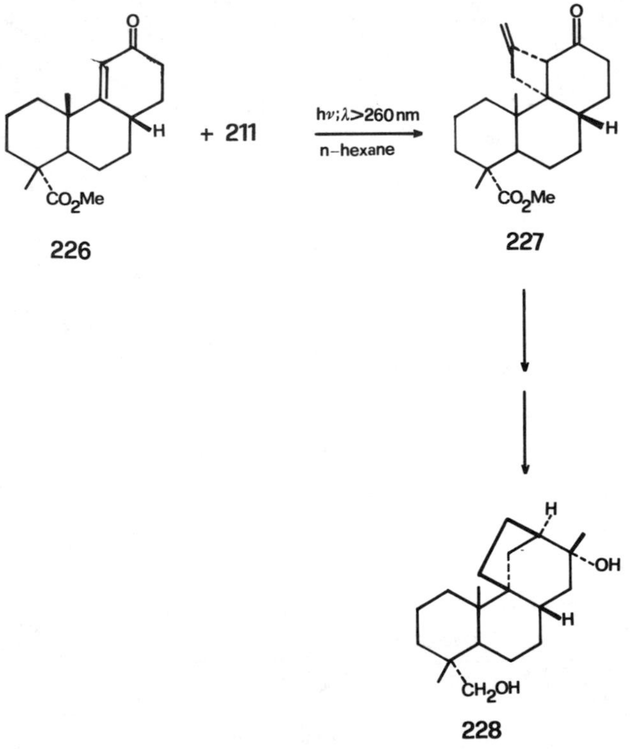

226 + 211 → (hν;λ>260nm, n-hexane) 227 → → 228

Scheme 3.41

a different stereochemical outcome. The *cis*-fused adduct **235** results from the substrate **234**, which behaves similarly to **208** (Scheme 3.43). This confirms that hydrindane-like compounds are in a border region where steric factors can affect the addition to *trans*- or to *cis*-fused cycloadduct. (In addition to a 42% yield of **235**, a second photoadduct, probably a stereoisomer of **235**, was also isolated in 3% yield.)

Intramolecular Cycloadditions. From a theoretical point of view, intramolecular photochemical [2+2] cycloadditions can be treated like the intermolecular ones previously discussed. However, they depend a great deal on the conformational preference of the substrate. Because of possible flexibility, the π system involved may reach a suitable proximity, and a favorable entropy factor can then play a role, one still more relevant in cyclic substrates.

As to the nature of the excited electrons, several examples involve cycloadditions between two pure olefin systems with alkyl groups, as substituents. In this case the excited state must be a $\pi \rightarrow \pi^*$ type. Photochemical excitation of the olefin ground state gives rise to a $^1\pi \rightarrow \pi^*$, which can react as it is or collapse, by intersystem crossing, to $^3\pi \rightarrow \pi^*$, the latter being also formed by triplet–triplet energy transfer from a sensitizer.

An overall scheme like that reported in Figure 3.8 can always be used, with both 'HOMO'/HOMO and 'LUMO'/LUMO stabilizing interactions involved. However, the energy difference between $^1\pi \rightarrow \pi^*$

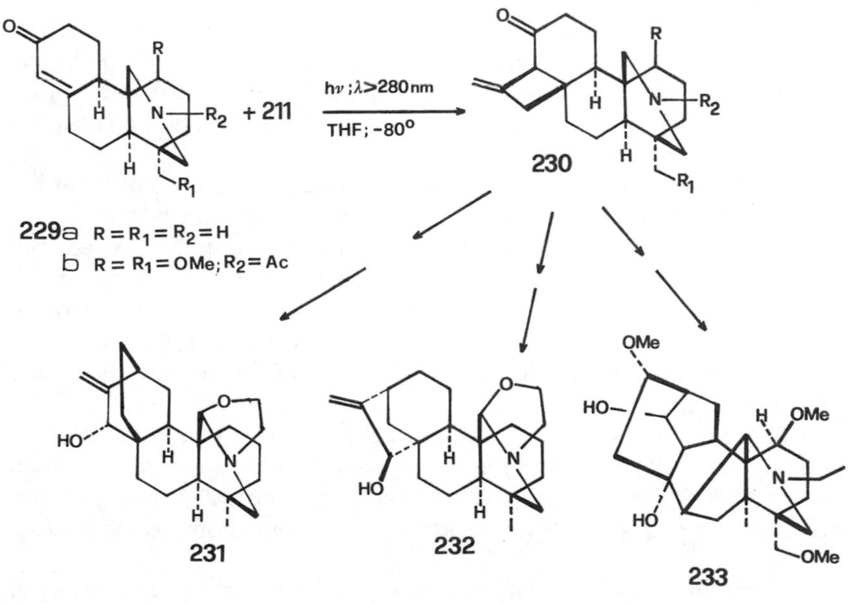

229a R = R1 = R2 = H
b R = R1 = OMe; R2 = Ac

hν; λ>280 nm
THF; -80°

230

231

232

233

Scheme 3.42

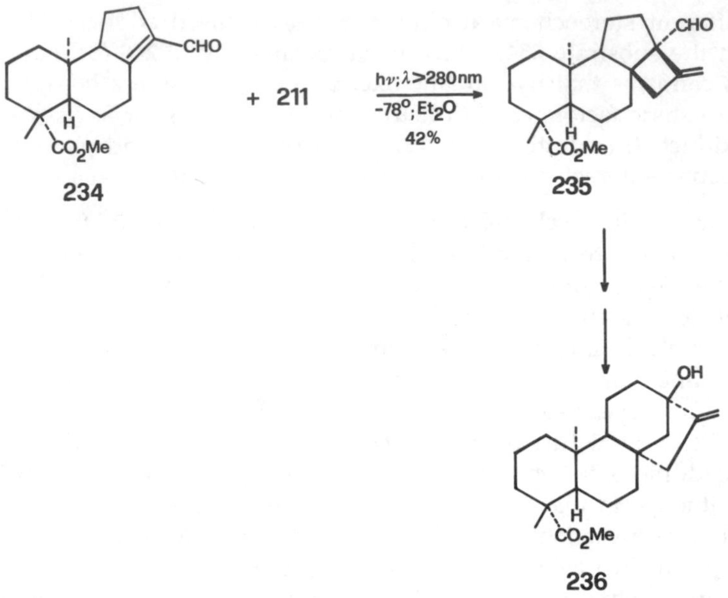

Scheme 3.43

and $^3\pi \rightarrow \pi^*$ is large; furthermore, the excited singlet and triplet states of lowest energy have a twisted geometry (*112*). (For ethylene, *see* Figure 3.14.)

For all these reasons, if a triplet excited state is involved, the parameters of the first antibonding orbital have to be used with great care. Hence, MO interactions are not as necessarily involved in the rationalization of the experimental results reported in this section as they are in those of other sections.

Photocycloaddition involving simple isolated olefins is difficult because these compounds absorb the vacuum–UV region. Cadmium (λ = 229,227,214 nm) or zinc (λ = 214 nm) lamps are sometimes used, but the most popular technique involves the addition of a sensitizer, which produces a triplet excited state by energy transfer. In cyclic nonconjugated dienes, a transannular effect of charge–transfer type lowers the energy required for the excitation; this is useful in the intramolecular photocycloaddition.

These concepts can be applied to the photocycloaddition of (−)-germacrene D (**237**), a methylene-cyclodeca-1,6-diene derivative which has a $\lambda^{\cdot \text{hexane}}$ = 259 nm (ϵ = 4500). When irradiated with a low-pressure mercury lamp it gives high yields of (−)-β-bourbonene (*113*) (**150**) (Scheme 3.44).

A conformation like **238** may easily account for the formation of **150**, probably through a singlet excited state, which also accounts for conversion of myrcene (**239**) (*114, 115*) to β-pinene (**240**). The low yields are simply due to the formation of **241**, coming from $4\pi \rightarrow 2\pi + 2\sigma$ electrocyclic process, because the sensitized triplet process gives **243** (*116*), possibly through diradical species **242a,b** (Scheme 3.45).

Triplets derived by photosensitized energy transfer are involved in the intramolecular cycloadditions of 3-methylene-6-methyl-*trans*-1,5-nonadiene derivative (**244**), which produces α-*trans*- and β-*trans*-bergamotene (*117*) (**246a,b**) through **245** (formed together with its epimer in the ratio 5:3), and of the cyclodeca-1,5-diene derivative (**247**) which gave α and β (±)-longipinenes (**250a,b**) through the diradical **248** and the photoadduct **249** (*118*) (Scheme 3.46).

An interesting route to α- and β-panasinsene (**253a,b**), two sesqui-

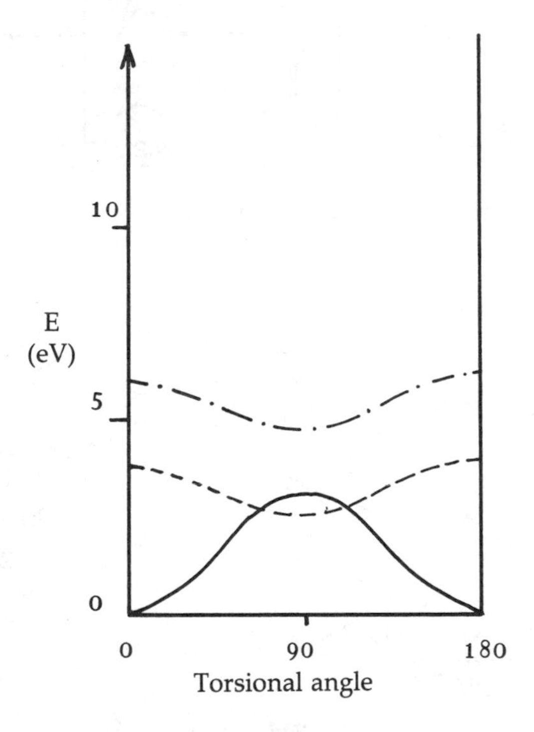

Figure 3.14. Energy diagram for the electronic states of ethylene. Key:—, ground state; - - -, first excited triplet state (π,π); - · - · -, first excited singlet state (π,π*).*

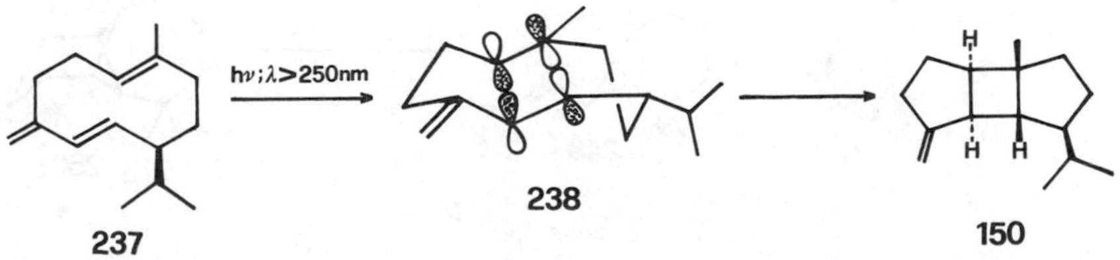

hν; λ>250nm

237 238 150

Scheme 3.44

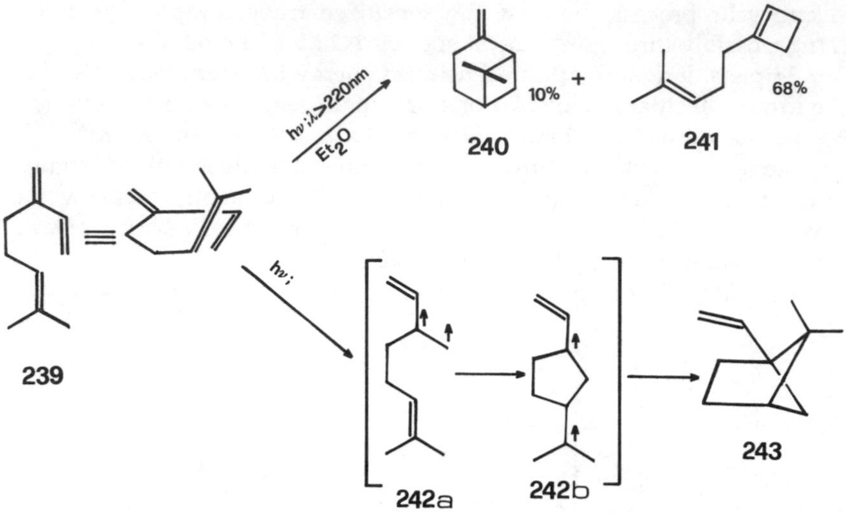

hν; λ>220nm, Et_2O → 240 (10%) + 241 (68%)

239 → hν; → [242a → 242b] → 243

Scheme 3.45

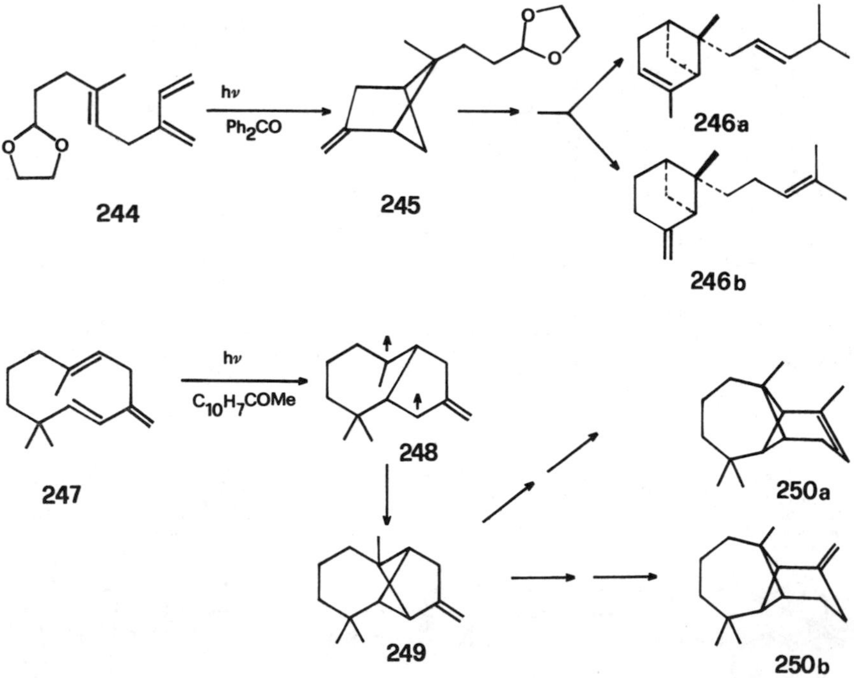

244 → hν, Ph_2CO → 245 → → 246a, 246b

247 → hν, $C_{10}H_7COMe$ → 248 → 249 → → 250a, 250b

Scheme 3.46

terpenes isolated from *Panax ginseng*, was disclosed (*119*). It takes advantage of the cuprous triflate-catalyzed photocyloaddition of an olefin to an allylic alcohol. Thus photolysis of the diene alcohol **251** at 254 nm proceeds smoothly to give the tricyclic product **252** as a mixture of epimeric alcohols, oxidized and elaborated in a straightforward manner to afford a 5:2 mixture of α- and β-panasinsene (**253a,b**) (Scheme 3.47).

All further examples of photochemical intramolecular cycloadditions involve an α,β-unsaturated carbonyl system which becomes the site of excitation through a $n \rightarrow \pi^*$ process. The triplet species, usually indicated as the reacting ones, are sometimes obtained through the action of sensitizers.

Thus cyclodecadienone (**254**) gives **255** (*120*), which is an effective precursor of copaene (**256**); 1,7-dicarbethoxy-3-isopropyl-6-methyl-1,6-heptadiene (**257**) gives 6,7-dicarbethoxy-4-isopropyl-1-methylbicyclo[3.2.0]heptane (**258**), which is converted into the previously described α-bourbonene (**149**) (*121*); and the cyclohexenone derivative **259** gives **260** through a cycloaddition step which produces three contiguous quaternary chiral centers with the stereochemistry necessary for the final conversion into (±)-isocomene (*122*) (**261**) (Scheme 3.48). A somewhat similar process starting from 3-(1,6-dimethylhexene-4-enyl)-cyclohex-2-enone (**262**) gives **263** (*123*), and from 2-(1-methylpent-4-enyl)cyclohex-2-enone (**265**) gives **266** (*124*), the former being an intermediate in a new synthesis of β-cedrene (**264**), the latter in the synthesis of 11-epiprecapnelladiene (**267**) (Scheme 3.48).

Four cyclopentene derivatives, each bearing a double bond in conjugation with a carbonyl group, gave intramolecular cycloaddition to a second double bond of the substrate. Specifically, **268** gave **269** (three chiral centers generated), further converted into the hydroazulenic sesquiterpene β-bulnesene (*125*) (**270**); **271** gave an 82% yield of **272**, whose elaboration provided an expeditious route to a zizaane ring system (*126*) **273**; **274** was photocyclized to **275** (78% yield), which represents the key intermediate for a new and efficient synthesis of (±)-longifolene (*127*) (**276**); dicyclopent-1-enyl-methane (**277**) afforded the hirsutane carbon skeleton (*128, 129*) **279** by addition of methanol to the strained three-membered ring in the presumed intermediate bicyclo[2.1.0]pentane (**278**) (Scheme 3.49).

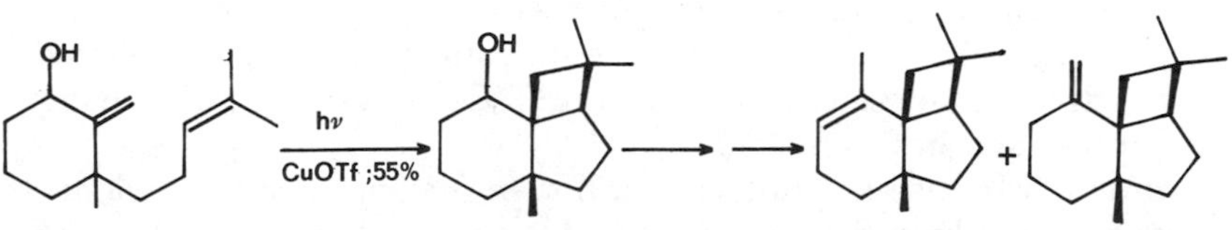

OH hν CuOTf ;55% OH

251 252 253a 253b

Scheme 3.47

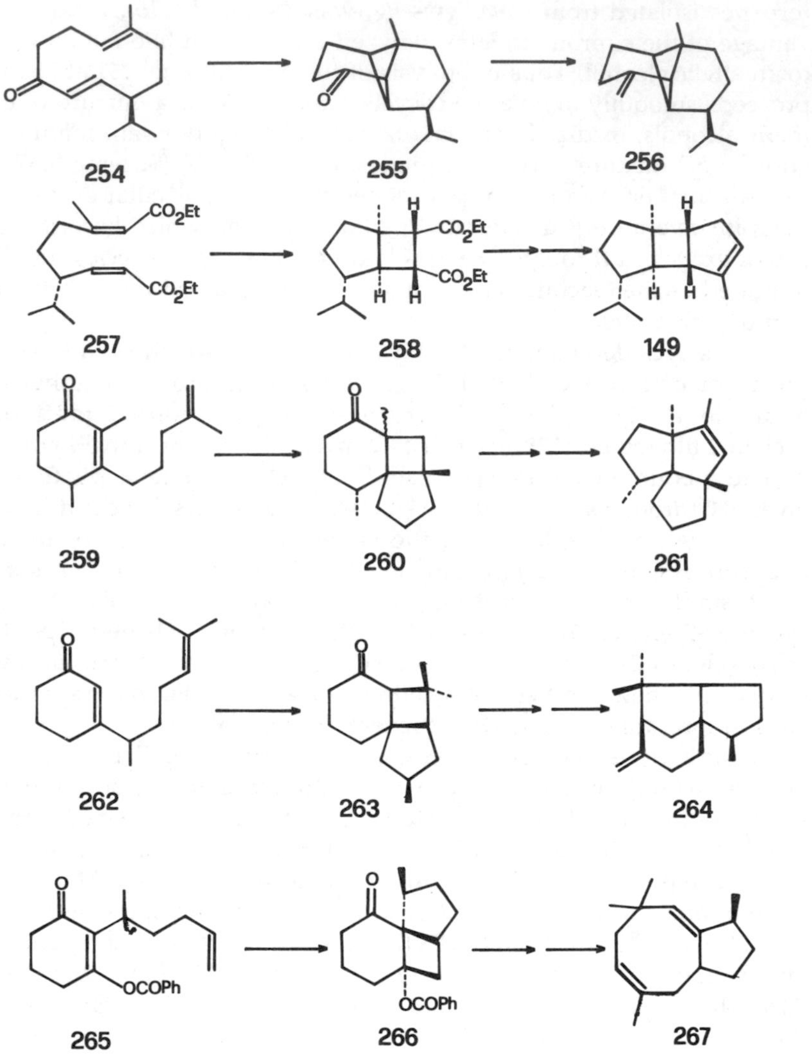
254
255
256
CO2Et
CO2Et
257
H
CO2Et
CO2Et
H H
258
H
H H
149
259
260
261
262
263
264
OCOPh
265
OCOPh
266
267

Scheme 3.48

The [π2 + π2] intramolecular process can compete with the alternative [π2a + π2a + π2a] rearrangement, usually called photochemical di-π-methane rearrangement (*130*); the 'HOMO'/HOMO relationships are represented in Figure 3.15.

Previously, use of photochemical di-π-methane rearrangement in the natural products field was confined to the synthesis of methyl chrysanthemate (**2** R = Me) from the diene **280** (*131*) (Scheme 3.50). Investigations (*132*) by Pattenden and Whybrow extended the scope of this rearrangement through synthesis of the vinylcyclopropane ring system

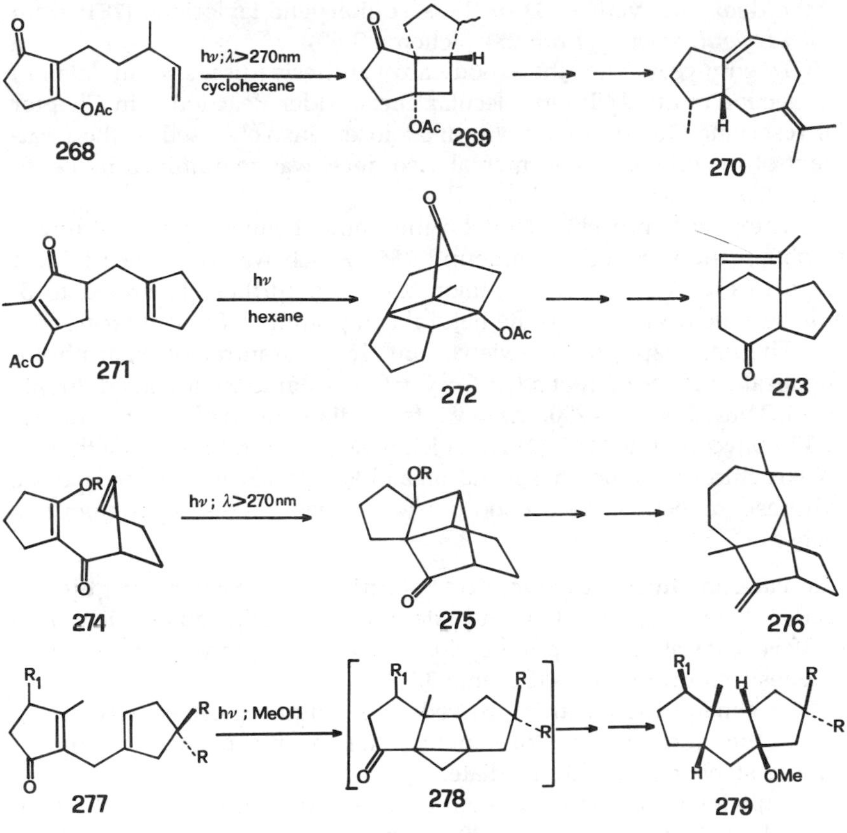

Scheme 3.49

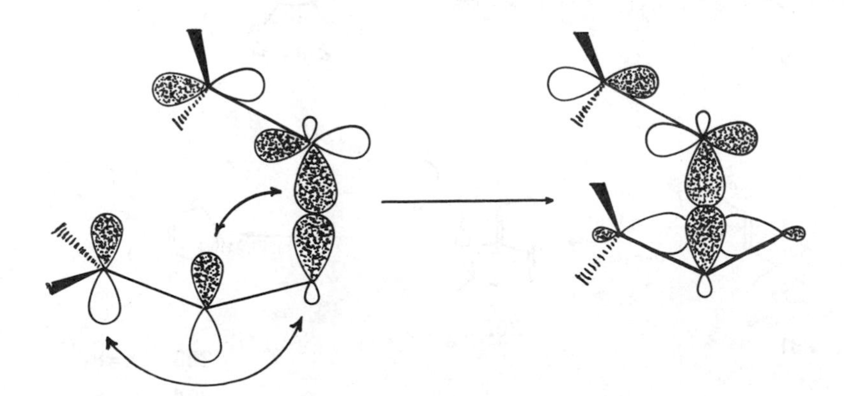

Figure 3.15. 'HOMO'/HOMO relationship in the di-π-methane rearrangement.

of the deoxy-derivative **283** of the sesquiterpene taylorione (**284**) from *Mylia taylorii,* starting from **281** (Scheme 3.50).

Twenty years after the Woodward synthesis (discussed in detail in the section entitled "Intermolecular Diels–Alder Reactions" in Chapter 5), reserpine (**288b**), a *Rauwolfia* alkaloid extensively used in the treatment of hypertension and mental disorders, was reexamined by Pearlman (*133*).

The new approach centers on the photochemical treatment of **285** to obtain the tetracyclic compound **286**, which was converted in two steps to the Woodward precursor **287**, then further elaborated to 3-epireserpine (**288a**) and to the natural compound (**288b**) (Scheme 3.51).

The final paper to be reviewed involves the intramolecular photochemical cycloaddition of a C=C=C=O system to a cumulated double bond. Thus, **289** gives **290**, the key intermediate for an efficient synthesis of 12-epilycopodine (*134*) (**291**), which is not a natural product although its structure is found in several alkaloids. However, a stereospecific synthesis of **291** can be a good entry to these interesting compounds (Scheme 3.52).

The Paternò–Büchi Reaction. The Paternò–Büchi reaction, first discovered by Paternò (*135*) in 1909 and extensively investigated by Büchi (*136*) in 1954, consists of the photocycloaddition of carbonyl compounds to alkenes to form oxetanes (Scheme 3.53).

In general the reaction proceeds with high regioselectivity giving rise to the more hindered oxetane; this has been explained in terms of a more stable radical intermediate.

Only a few papers have reported this approach to the synthesis of natural products. (±)-Apiose (**294**), a pentose first isolated by acid hydrolysis of apiin and subsequently obtained from several plants (*Posi-*

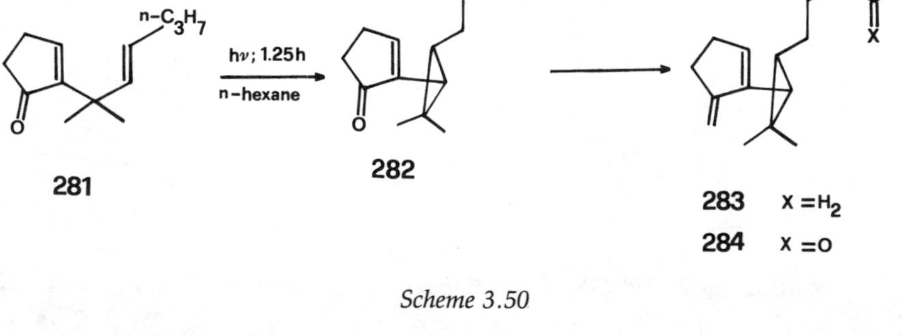

Scheme 3.50

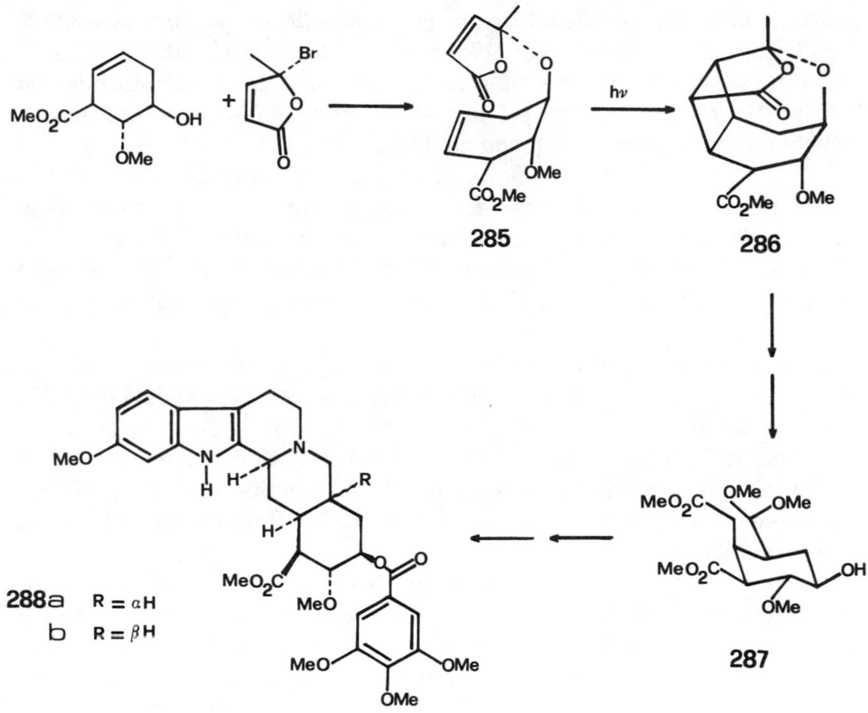

Scheme 3.51

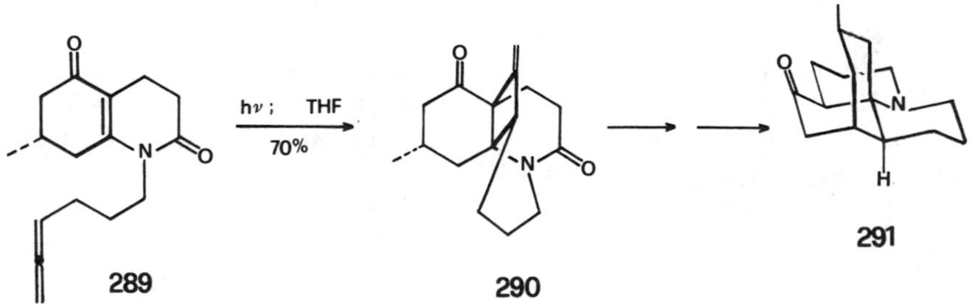

Scheme 3.52

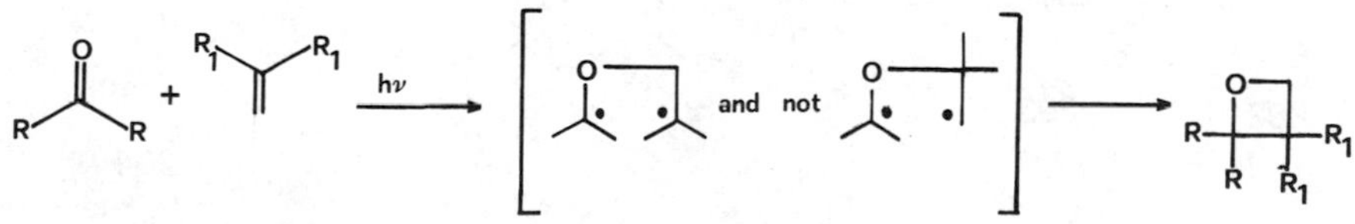

Scheme 3.53

donia australis, *Taraxacum koksaghyz*, *Havea brasiliensis*, etc.), was synthesized from diacetoxyacetone (**292**) and vinylene carbonate (**293**) (*137*) (Scheme 3.54). The oxetane **295** was the intermediate in a photochemical synthesis of optically active 3-oxa-4,5,6-trinor-3,7-inter-*m*-phenylene prostaglandins (**296**) (*138*) (Scheme 3.55).

The synthesis of (*E*)-6-nonen-1-ol (**302**), a component of the sex pheromone of the Mediterranean fruit fly, *Ceratitis capitata*, provides the best application of the Paternò–Büchi reaction (*139*). Thus 1,3-cyclohexadiene (**297**) reacts with propanal (**298**) in acetonitrile to give a single regioisomer **299** in 77% yield (as a mixture of *exo* and *endo* isomers in the ratio 8:2).

Hydrogenation of **299** followed by [2+2] cycloreversion of **300** (*see* the next section) gave (*E*)-6-nonenal (**301**), which was reduced to **302** (Scheme 3.56).

The MO interactions of olefin and carbonyl derivatives were used to explain the regiochemical outcome of this reaction (*140–142*). We reconsidered the interactions between **297** and **298** in order to understand the formation of **299**.

An accurate representation of the MO interactions of olefin and carbonyl derivatives can be derived from ionization potentials and electron affinities of the reagents. The ionization potentials are well known; those of 1,3-cyclohexadiene (*143*) and propanal (*144*) were used. The electron affinities are more difficult to obtain; calculated energies or data from UV transitions are widely used. The recently developed electron transmission spectroscopy (ETS) technique provides accurate data on the electron affinities of **297** (*145*) and of acetaldehyde (*146*), used as a good model for **298**. An accurate representation of MO energies and coefficients is presented in Figure 3.16.

A simple application of Eq. 2 (p. 4) to the π_n **298** /π_2 **297** and $\pi^*_{(CO)}$ **298**/π_3^* **297** interactions using butadiene and formaldehyde coefficients revealed the latter interaction to be the dominant one because of its closer energies and greater overlap.

The $\pi^*_{(CO)}/\pi_3^*$ interaction, however, gives the wrong regiochemical approach (**303**) instead of the correct one (**304**), which is given by the π_n/π_2 interaction. MOs are not useful in this case, therefore, for ratio-

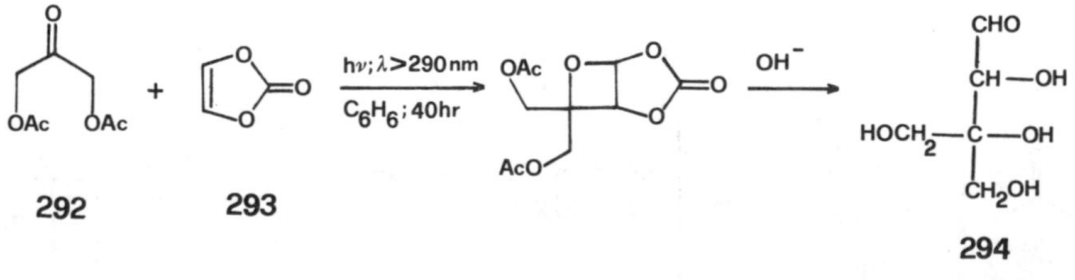

Scheme 3.54

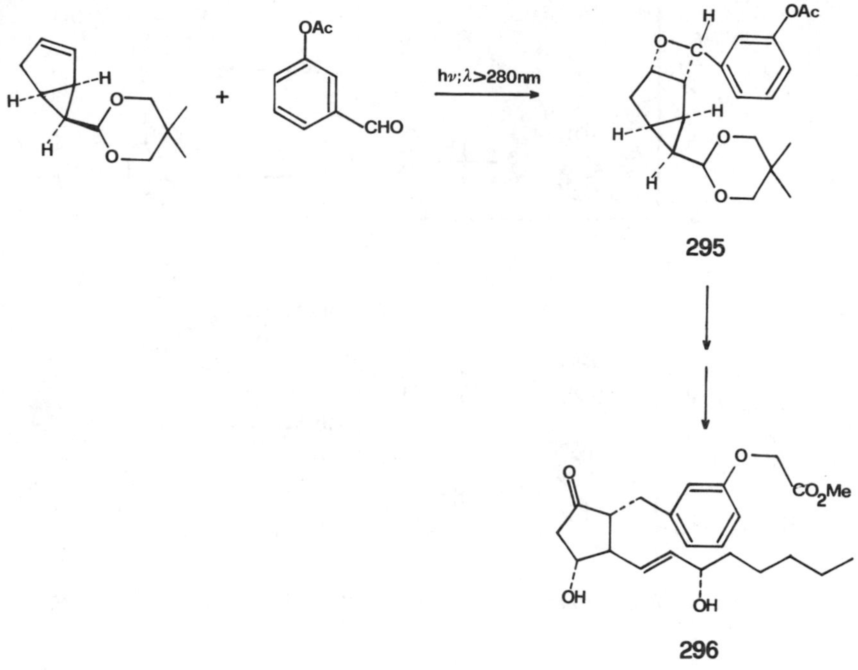

hν; λ>280nm

295

296

Scheme 3.55

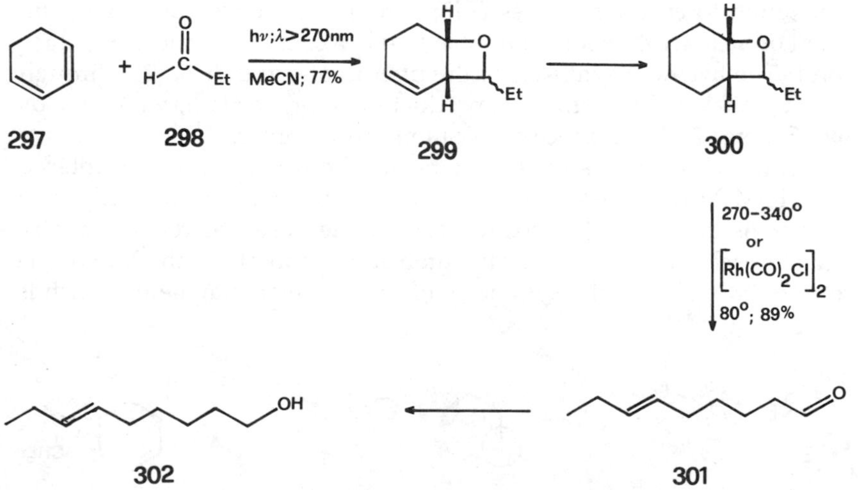

hν; λ>270nm

MeCN; 77%

297

298

299

300

270–340°

or

$[Rh(CO)_2Cl]_2$

80°; 89%

302

301

Scheme 3.56

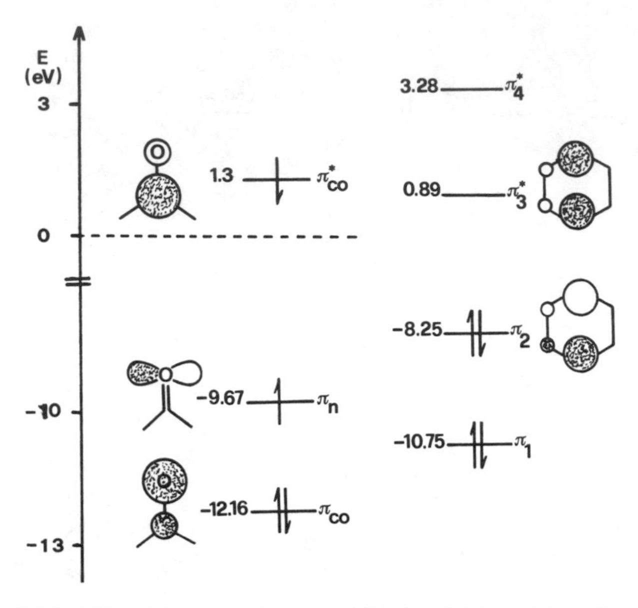

*Figure 3.16. MO interactions of propanal (**298**) and 1,3-cyclohexadiene (**297**).*

nalizing the reaction, which is easily done by taking into account the formation of the most stabilized diradical intermediate **305** (Scheme 3.57).

Metal-Catalyzed [2+2] Cycloadditions

Certain zero-valent nickel complexes can easily catalyze butadiene dimerization to *cis*-cyclobutanes (*147*) through the interaction of vacant metal orbitals with the olefin double bond to form a "covalent metallic" bond. The two-step synthesis of the pheromone grandisol (**21**) through a nickel-catalyzed dimerization product of isoprene (*148*), which yields **306** (Scheme 3.58), is a nice application of this process.

The orbital picture of the interaction between LUMO complexed and HOMO uncomplexed olefins explains not only the unusually mild conditions for a [2+2] cycloaddition but the observed regiochemistry and stereochemistry as well. The interesting property of the MO of the complexed olefins is the symmetry of the ethylenic fragment, which is

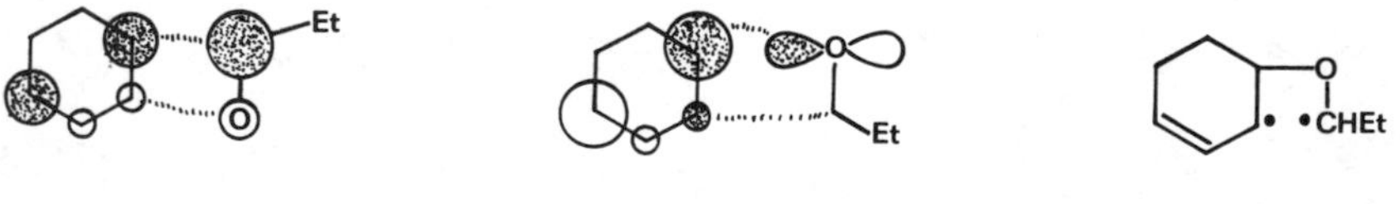

Et
Et
O
• •CHEt

303 **304** **305**

Scheme 3.57

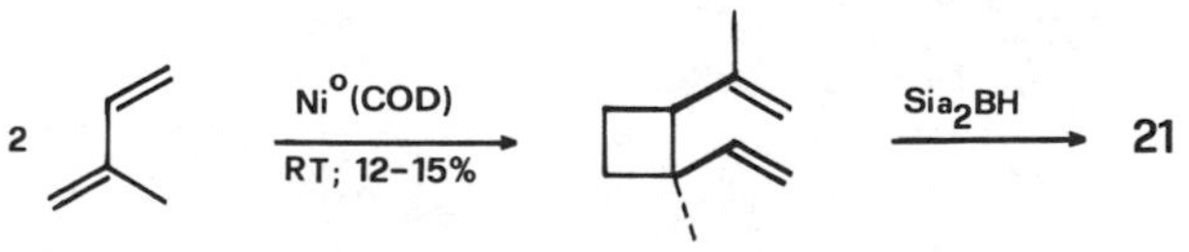

2 Ni°(COD) RT; 12–15% → 306 Sia_2BH → 21

Scheme 3.58

reversed from that of the uncomplexed olefins. This observation is depicted in Figure 3.17 by HOMO (A) and LUMO (B), respectively (*see* Ref. 7 in Chapter 1, p. 161).

It seems reasonable to assume that complexation occurs on the less hindered double bond of isoprene. The dominant interaction concerns the LUMO of the complexed isoprene (both LUMO and HOMO are lowered by complexation) and the HOMO of the uncomplexed isoprene.

Figure 3.18 shows the MO picture under conditions of maximum overlap of the coefficients (*149*), giving a regiochemistry suitable for **306**, with a *cis* relationship of the cyclobutane substituents coming from secondary interactions.

The vinylcyclopropane **307** (R ≠ H) rearranges when heated to cyclopentenes **308** and/or **309**. This transformation can be formally considered a [2 π + 2σ] cycloaddition and represents a tool for cyclopentene anellation, due to the easy synthesis of the starting materials through a [1+2] cycloaddition (*150*) (Scheme 3.59).

By this route a total synthesis of (±)-hirsutene (**192**) was accomplished, which involved [1+2] intramolecular cycloaddition of **310** followed by the rearrangement from **311** to **312** mentioned earlier, where **312** is further converted to **192** (*151*) (Scheme 3.60). The drastic thermal conditions involved severely limited the process, which, being suprafacially forbidden, proceeds in a $[2\pi_s + 2\sigma_a]$ concerted fashion or via a stabilized diradical intermediate.

It was recognized recently that metal-catalyzed rearrangement can take place at room temperature as first exemplified (*152*) on the lithium salt of 2-vinyl-1-cyclopropanol (**307**, R = OLi), which gives rise to 3-cyclopentenol alone (**308**, R = OH).

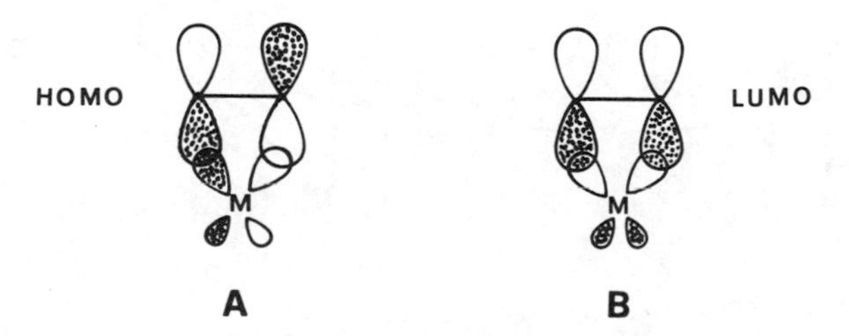

Figure 3.17. HOMO (A) and LUMO (B) of metal-complexed olefin.

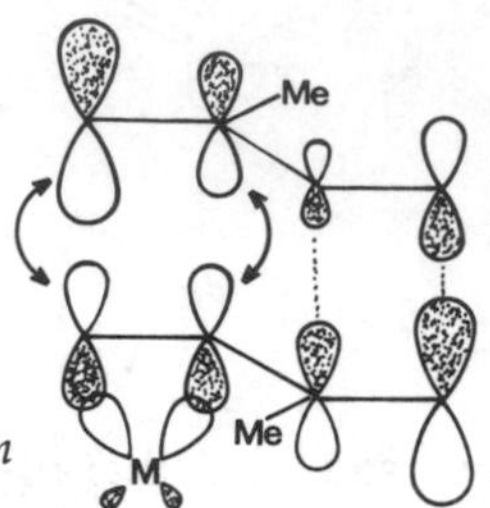

Figure 3.18. $LUMO_{complexed}/HOMO_{uncomplexed}$ interaction in the catalyzed dimerization of isoprene.

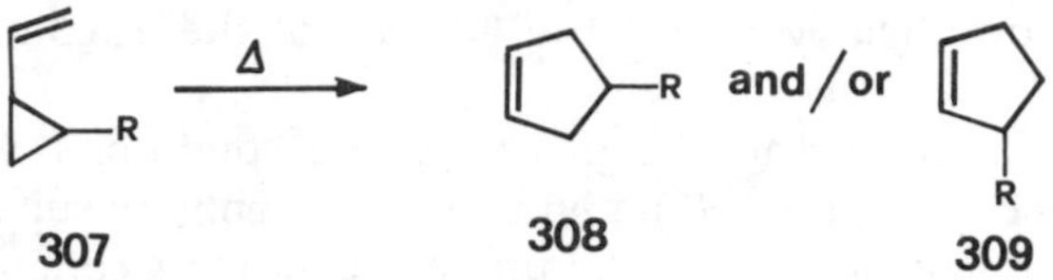

Scheme 3.59

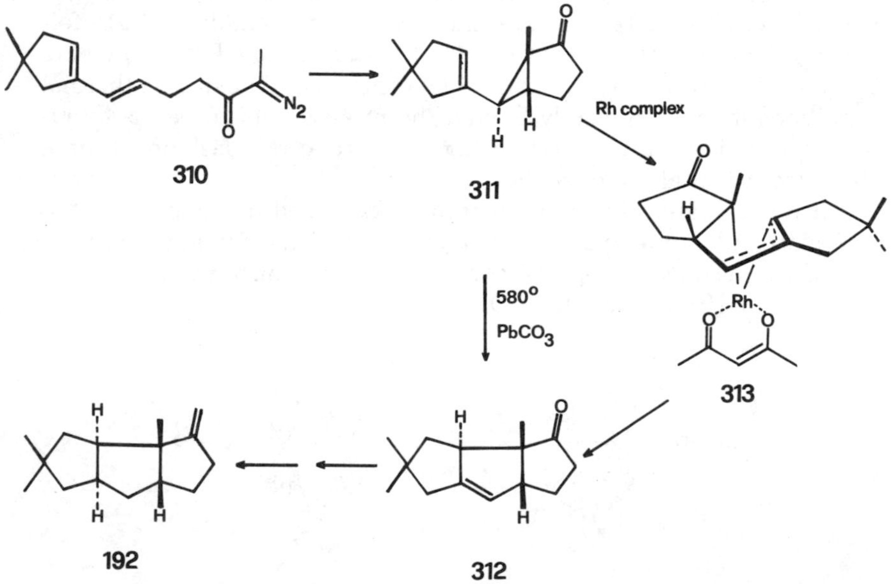

Scheme 3.60

A further example can be found in the transformation of **311** to **312** in the presence of a rhodium catalyst. This process occurs through **313**, which makes the rearrangement a $[2\pi_s+2\pi_s]$ process (*153*), whereby the experimental conditions are significantly improved and stereoselectivity increased.

[2+2] Cycloreversion

Application of orbital symmetry considerations to thermal [2+2] cycloreversion of a cyclobutane into two olefinic fragments requires a $[\sigma 2s+\sigma 2a]$ process (*1*), which gives rise to inversion of the configuration of one of the two double bonds generated (Scheme 3.61).

For example, in the cleavage of bicyclo[2.2.0]hexanes, an all-*endo* hexamethyl derivative **314** produces *erythro*-3,4,5,6-tetramethylhexa-2*Z*,6*E*-diene (**315**) (*154*) together with the all-*exo* isomer **316** (Scheme 3.62). This process supports an early suggestion (*155*) of a two-step process via a diradical intermediate formed by conversion of a σ bond into a π bond, followed by conrotatory opening of the second σ bond.

The photocycloaddition of cyclobutene derivatives with cycloalkenes provided a convenient route to tricyclic intermediates **317** with a bicyclo[2.2.0]hexane fragment. When the photoadduct **317** was heated, a metathetical cycloreversion occurred with formation of medium-sized cycloalkadiene **318** (route a) possessing the sesquiterpenic germacradiene ring skeleton. The alternative [2+2] cycloreversion mode can give *trans*-1,2-divinylcycloalkane **319** (route b) (*156*), known as elemane sesquiterpenes, which also can be formed from germacrenes **318** by a Cope rearrangement (*see* the section in Chapter 7 entitled "The Cope Rearrangement") (Scheme 3.63).

The photocycloaddition–thermal cycloreversion sequence was experienced both on cyclopentene (*156*) and cyclohexene substrates (*157, 158*). When a germacrane skeleton like **320** bears an enophile site, usually an allylic methyl group, and an enic group, usually a carbonyl, an intramolecular ene reaction may take place (*see* the section entitled "Intramolecular Ene Reactions" in Chapter 8), producing the cadinane sesquiterpenic structure (**321**) (*159, 160*) (Scheme 3.64). These routes were used to gain access to sesquiterpenes belonging to all three classes of compounds.

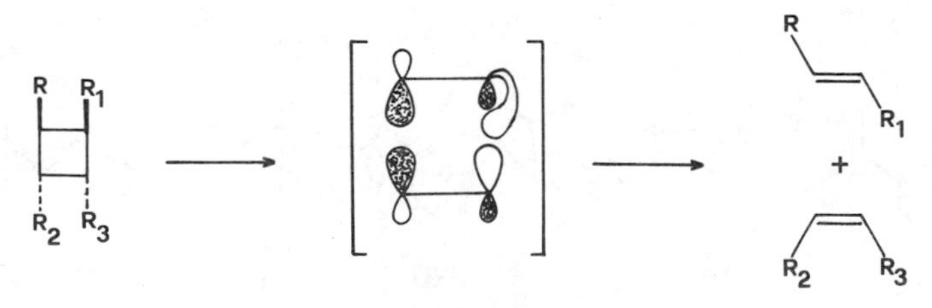

Scheme 3.61

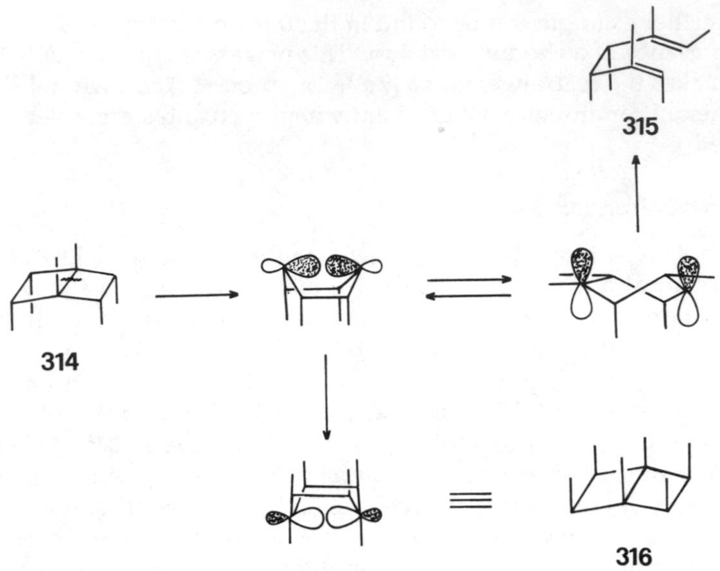

Scheme 3.62

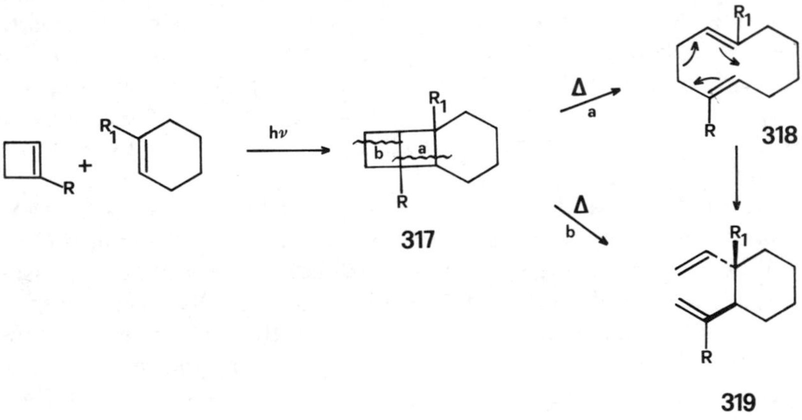

Scheme 3.63

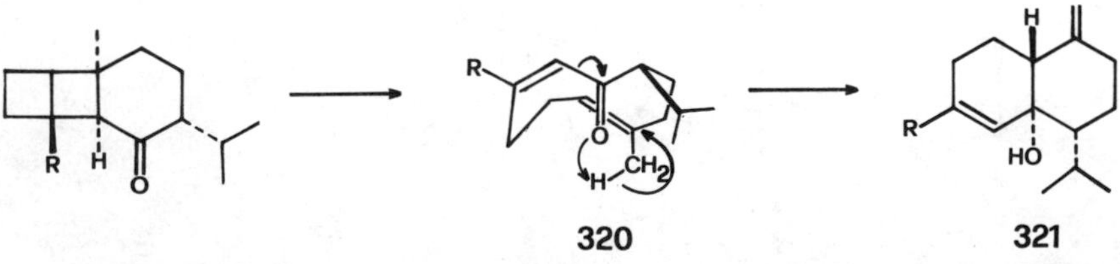

Scheme 3.64

(±)-Isabelin (**325**), a germacranolide dilactone, was obtained through a sequence of 13 operations that started from the photocycloadduct **323**, derived from 3-methyl-2-cyclohexenone and cyclobutenecarboxylate **322**, further converted in several steps to **324**. When the latter was heated at 200 °C for 40 min, the main product, along with its pyro-isomer (*161*), was **325** (Scheme 3.65).

(−)-Shyobunone (**327**), a sesquiterpene of the elemane class, was isolated from *Acorus calamus* L. Its synthesis involves a photochemical [2+2] cycloaddition of methylcyclobutene and L-piperitone (**179**) to give **326**. Flash vacuum pyrolysis gave **327** in 5% yield along with 19% yield of (*R*)-(+)-isoacoragermacrone (**328**), which can be isomerized to (*R*)-acoragermacrone (**329**) (*162, 163*) (Scheme 3.66).

Four cadinanes were obtained from cyclobutenecarboxylate **322** and 3-methylcyclohexenone or piperitone (**179**). Thus the [2+2] cycloreversion of the photoadduct **330** followed by intramolecular ene reaction gave (±)-atractylon (**331**) (*164*), (±)-isoalantolactone (**322**) (*164*), (±)-calameon (**333**) (*165*), and (+)-isocalamendiol (**334**) (*166*) (Scheme 3.67).

Pseudo [σ2+σ2] Cycloreversion

In this section we will consider the elimination of two atoms from a five-membered ring, known as pseudo [σ2+σ2] cycloreversion to give a cyclopropane (Scheme 3.68). From a formal point of view this process can be regarded as a [σ2+σ2] cycloreversion supra/antara under thermal and supra/supra under photochemical conditions.

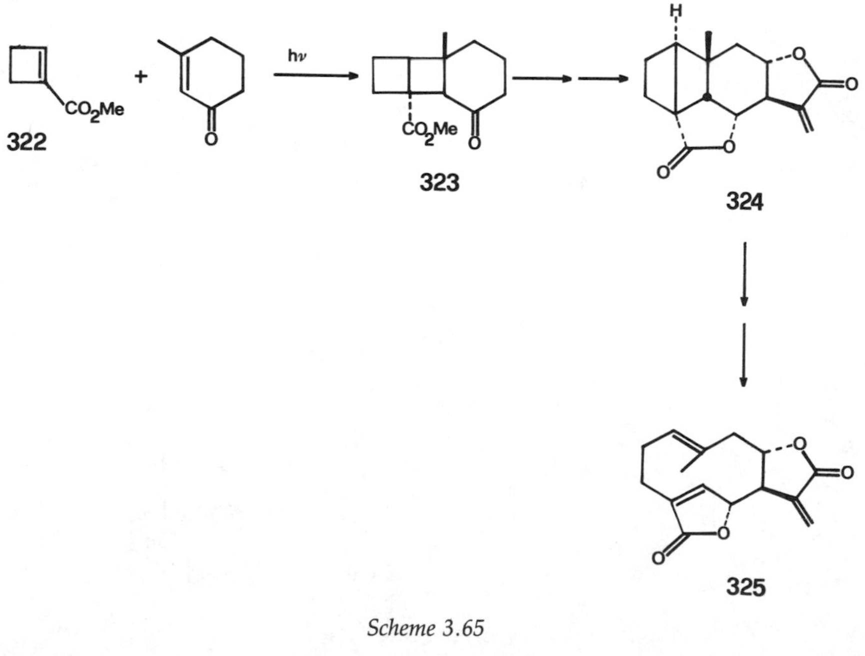

Scheme 3.65

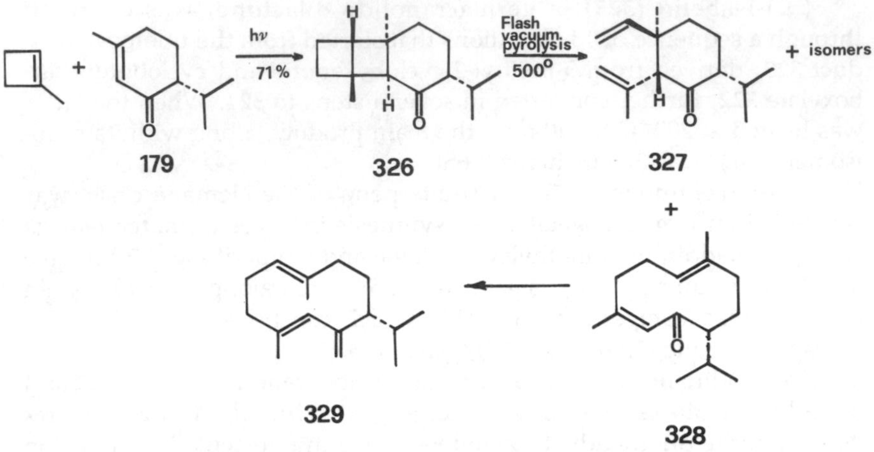

hν 71% Flash vacuum pyrolysis 500° + isomers

179 326 327 + 328 329

Scheme 3.66

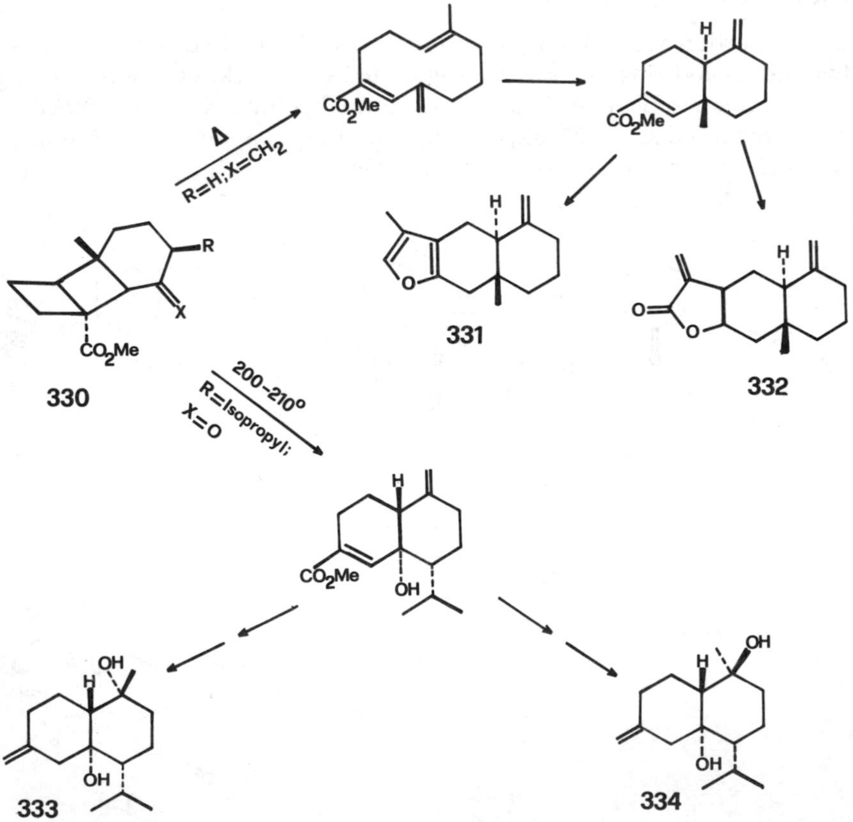

Δ R=H; X=CH_2 CO_2Me R X 330 331 332 200–210° R=Isopropyl; X=O OH 333 334

Scheme 3.67

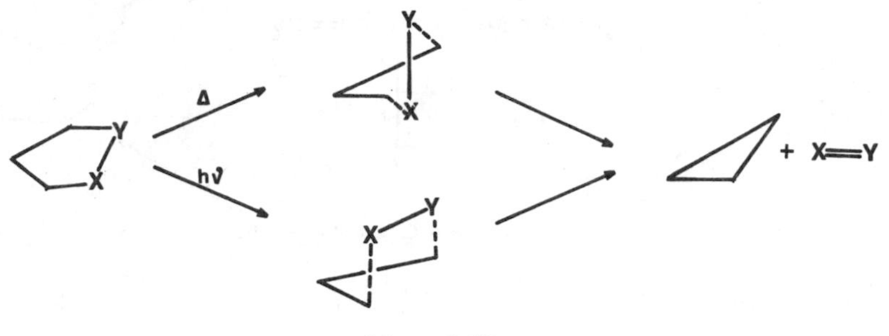

Scheme 3.68

Extrusion of molecular nitrogen from Δ^1-pyrazolines (X = Y = nitrogen) (*167*), both thermally and photochemically, and of carbon dioxide from five-membered lactones under photochemical conditions (*168*), is frequently used in the process.

A detailed investigation of the mechanism of these reactions suggested the following considerations:

1. Thermal extrusion of nitrogen proceeds via a stepwise nitrogen-containing diradical in the *trans* or *gauche* conformations (*169*) (A and B, respectively)

2. Photoextrusion of nitrogen involves both a retro-1,3-dipolar cycloaddition followed by photofragmentation of the derived diazoalkane (Scheme 3.69, path a) or a direct fragmentation (Scheme 3.69, path b). Apart from this, a triplet was observed in ESR spectrum, identical with the phenylcarbene spectrum obtained from phenyldiazomethane irradiated under the same conditions (*170*).
3. Photochemical decarboxylation occurs via a discrete diradical intermediate, supported by the quantum yields of decomposition of model molecules (*171*).

It seems reasonable to regard these reactions as nonpericyclic. Nevertheless their synthetic usefulness for the introduction of a cyclopropane ring into an organic substrate justifies our decision to report several examples.

The sesquiterpene aristolone (**49**) was obtained in good yields (*172*) by photodecomposition of the 1,3-cycloadduct **335** (*see* the section entitled "Diazoalkanes" in Chapter 4) (Scheme 3.70).

During the studies on synthesizing marasmic acid (*173*), isomarasmic acid (**339**) was also obtained by the sequence: Diels-Alder reaction; 1,3-dipolar cycloaddition; nitrogen extrusion to give **336**, **337**,

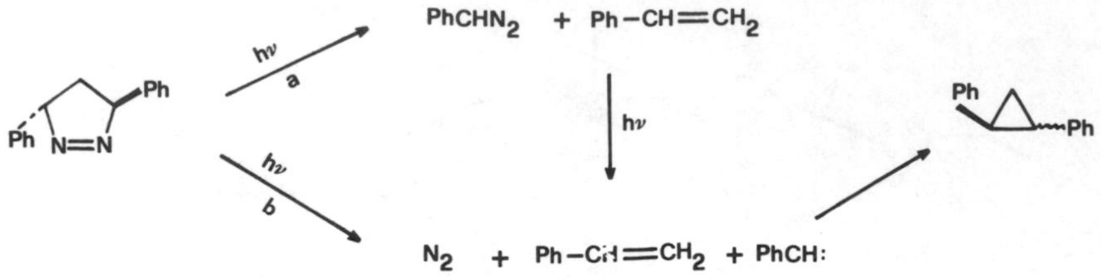

Scheme 3.69

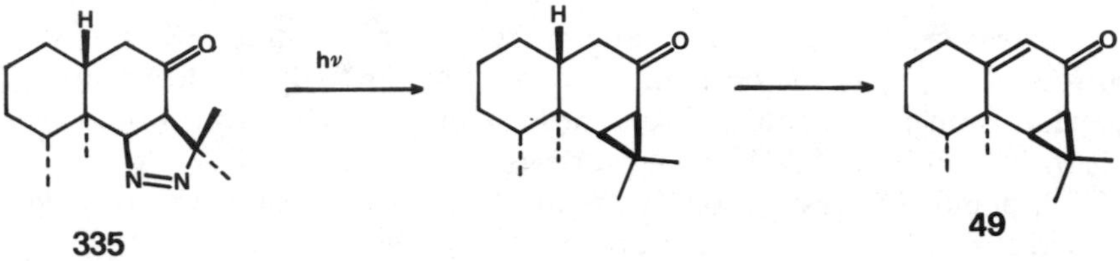

335

49

Scheme 3.70

336

337

338

339

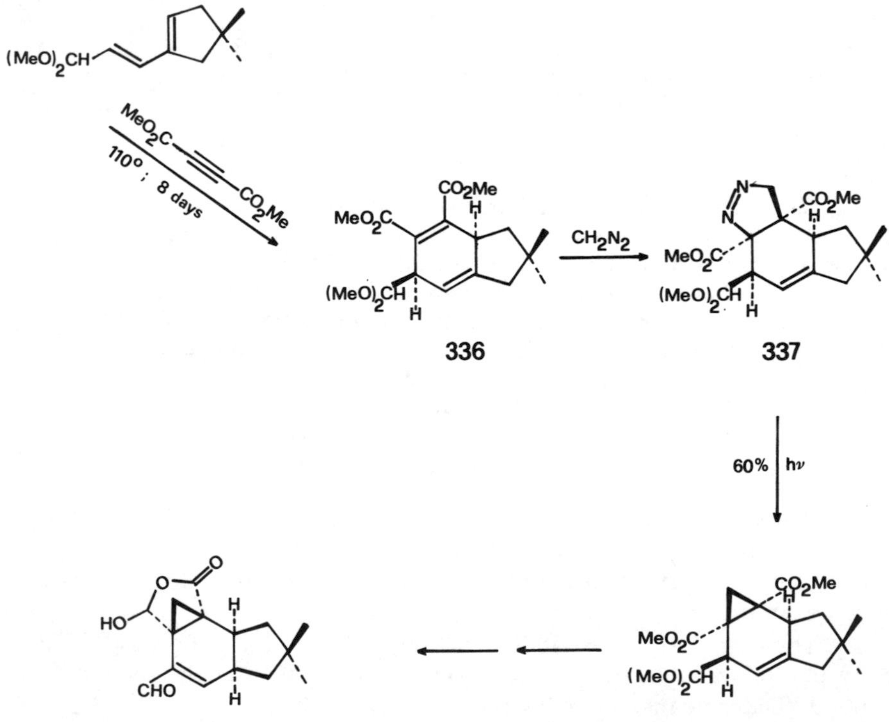

Scheme 3.71

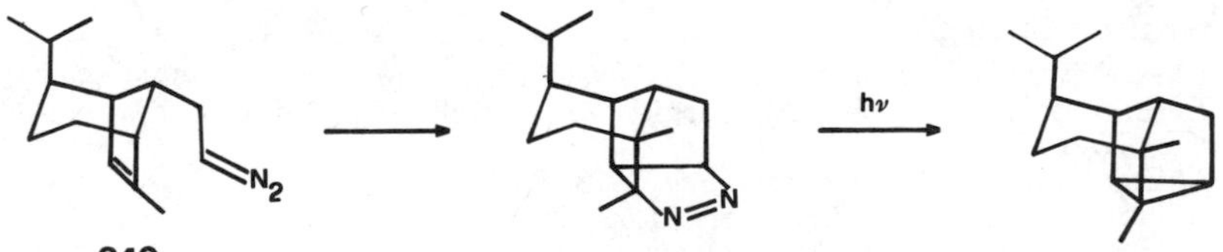

340 341 342

Scheme 3.72

and **338** (Scheme 3.71). This suggests that previous attribution to marasmic acid skeleton (*174*) refers to the *iso* isomer as well.

(−)-Cyclocopacamphene (**342**), a sesquiterpene isolated from vetiver oil, was prepared by nitrogen photoextrusion from the initial adduct **341**, resulting from an intramolecular 1,3-dipolar cycloaddition of the diazoderivative **340** (*175*) (Scheme 3.72).

Two examples of photodecarbonylation are the photolysis of dihydro-1,2-(6β, 11β)-santonine (**343a**) to produce norepimaalienone (*176*) (**344a**) and the synthesis of epimaalienone (**344b**), in the route to α- and β-cyperones (**345**) and (**346**), respectively, from **343b** (*177*) (Scheme 3.73).

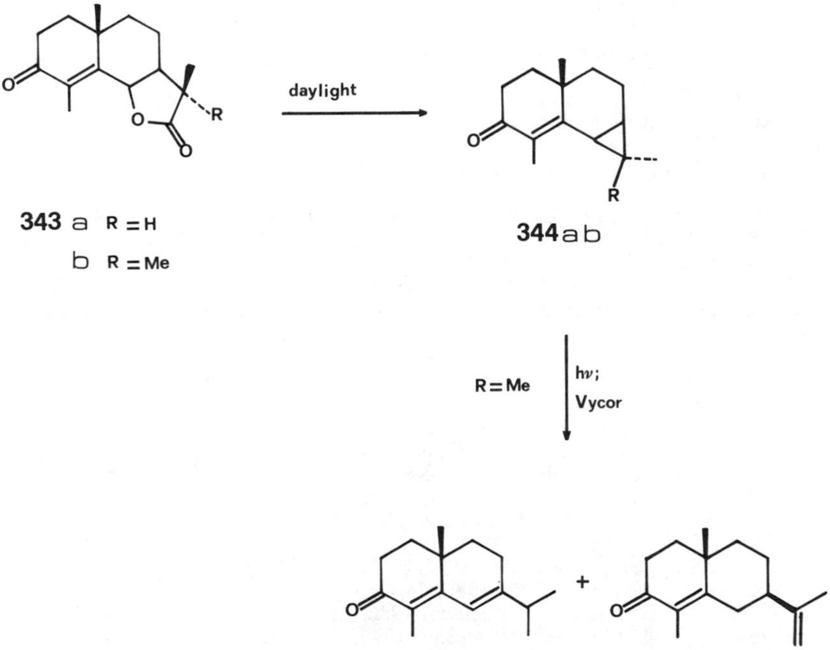

343 a R = H
b R = Me

daylight

344 a b

R = Me hν; Vycor

346 345

Scheme 3.73

Literature Cited

1. Woodward, R. B.; Hoffman, R. *Angew. Chem., Int. Ed. Engl.* **1969**, *8*, 781.
2. Brannock, K. C.; Bell, A.; Burpitt, R. D.; Kelly, C. A. *J. Org. Chem.* **1961**, *26*, 625.
3. Epiotis, N. D. *J. Am. Chem. Soc.* **1973**, *95*, 1191.
4. Houk, K. N. *J. Am. Chem. Soc.* **1973**, *95*, 4092.
5. Fex, T.; Froborg, J.; Magnusson, G.; Thorén, S. *J. Org. Chem.* **1976**, *41*, 3518.
6. Froborg, J.; Magnusson, G. *J. Am. Chem. Soc.* **1978**, *100*, 6728.
7. Ficini, J.; Guigant, A.; D'Angelo, J. *J. Am. Chem. Soc.* **1979**, *101*, 1318.
8. Huisgen, R.; Feiler, L. A.; Otto, P. *Tetrahedron Lett.* **1968**, 4485.
9. Brady, W. T.; O'Neal, H. R. *J. Org. Chem.* **1967**, *32*, 612.
10. Binsch, G.; Feiler, L. A.; Huisgen, R. *Tetrahedron Lett.* **1968**, 4497.
11. Montaigne, R.; Ghosez, L. *Angew. Chem., Int. Ed. Engl.* **1968**, *7*, 221.
12. Sustmann, R.; Ansmann, A.; Vahrenholt, F. *J. Am. Chem. Soc.* **1972**, *94*, 8099.
13. Leriverend, P. *Bull. Soc. Chim. Fr.* **1973**, 3498.
14. Bertrand, M.; Gras, J. L. *Tetrahedron* **1974**, *30*, 793.
15. Bellus, D. *J. Org. Chem.* **1979**, *44*, 1208.
16. Bellus, D.; Fischer, H.; Greuter, H.; Martin, P. *Helv. Chim. Acta* **1978**, *61*, 1785.
17. Grieco, P. A. *J. Org. Chem.* **1972**, *37*, 2363.
18. Grieco, P. A.; Oguri, T.; Gilman, S.; DeTitta, G. T. *J. Am. Chem. Soc.* **1978**, *100*, 1616.
19. Grieco, P. A.; Oguri, T.; Gilman, S. *J. Am. Chem. Soc.* **1980**, *102*, 5886.
20. Houk, K. N.; Strozier, R. W.; Hall, J. A. *Tetrahedron Lett.* **1974**, 897.
21. Johnson, D. B. R.; Schmitt, S. M.; Bouffard, Aileen F.; Christensen, B. G. *J. Am. Chem. Soc.* **1978**, *100*, 313.
22. Bouffard, Aileen F.; Johnston, D. B. R.; Christensen, B. G. *J. Org. Chem.* **1980**, *45*, 1130.
23. Fliri, A.; Hohenlohe-Oehringer, K. *Chem. Ber.* **1980**, *113*, 607.
24. Fleming, I. "Frontier Orbitals and Organic Chemical Reactions"; Wiley: London, 1976.
25. Alston, P. V.; Ottenbrite, R. M. *J. Org. Chem.* **1975**, *40*, 1111.
26. Au-Yeung, B.; Fleming, I. *J. Chem. Soc., Chem. Commun.* **1977**, 81.
27. Ibid., 79.
28. Corey, E. J.; Bass, J. D.; Lamathieu, R.; Mitra, R. B. *J. Am. Chem. Soc.* **1964**, *86*, 5570.
29. Challand, B. P.; De Mayo, P. *J. Chem. Soc., Chem. Commun.* **1968**, 982.
30. Loufty, R. O.; De Mayo, P. *J. Am. Chem. Soc.* **1977**, *99*, 3559.
31. Hoffman, R.; Wells, P.; Morrison, H. *J. Org. Chem.* **1971**, *36*, 102.
32. Herndon, W. C. *Top. Curr. Chem.* **1974**, *40*, 141.
33. Moore, W. M.; Hammond, G. S.; Foss, R. P. *J. Am. Chem. Soc.* **1961**, *83*, 2789.
34. For a review *see* Albini, A. *Synthesis* **1981**, 249.
35. Hentrich, G.; Gunkel, E.; Klessinger, M. *J. Mol. Struct.* **1974**, *21*, 231.
36. Houk, K. N. *Top. Curr. Chem.* **1979**, *79*, 1.
37. Houk, K. N., private communication.
38. Houk, K. N.; Strozier, R. W. *J. Am. Chem. Soc.* **1973**, *95*, 4094.
39. Chapman, O. L.; Smith, H. G.; King, P. W. *J. Am. Chem. Soc.* **1963**, *85*, 806.
40. Sammes, P. G. *Q. Rev., Chem. Soc.* **1970**, *24*, 37.
41. Sammes, P. G. *Synthesis* **1970**, 636.
42. De Mayo, P. *Acc. Chem. Res.* **1971**, *4*, 41.
43. Carless, H. A. J. *Chem. Br.* **1980**, *16*, 456.
44. Pattenden, G. *Chem. Ind. (London)* **1980**, 812.
45. Dilling, W. L. *Photochem. Photobiol.* **1977**, *25*, 605.
46. Tumlison, J. H.; Hardee, D. D.; Gueldner, R. C.; Thompson, A. C.; Hedin, P. A.; Minyard, J. P. *Science* **1969**, *166*, 1010.
47. Cargill, R. L ; Wright, B. W. *J. Org. Chem.* **1975**, *40*, 120.
48. Mori, K. *Tetrahedron* **1978**, *34*, 915.

49. Rosini, G.; Salomoni, A.; Squarcia, F. *Synthesis* **1979**, 942.
50. Zurflüh, R.; Dunham, L. L.; Spain, V. L.; Siddall, J. B. *J. Am. Chem. Soc.* **1970**, *92*, 425.
51. Guelner, R. C.; Thompson, A. C.; Hedin, P. A. *J. Org. Chem.* **1972**, *37*, 1854.
52. Katznellenbogen, J. A. *Science* **1976**, *194*, 139.
53. Matsumoto, T.; Miyano, K.; Kagawa, S.; Yü, S.; Ogawa, J.; Ichihara, A. *Tetrahedron Lett.* **1971**, 3521.
54. Mori, K.; Sasaki, M. *Tetrahedron Lett.* **1979**, 1329.
55. White, J. D.; Gupta, D. N. *J. Am. Chem. Soc.* **1968**, *90*, 6171.
56. Corey, E. J.; Mitra, R. B.; Uda, H. *J. Am. Chem. Soc.* **1964**, *86*, 485.
57. Kumar, A.; Sing, A.; Devaprabhakara, D. *Tetrahedron Lett.* **1976**, 2177.
58. Lamola, A. A. *Pure Appl. Chem.* **1970**, *24*, 599.
59. Ben-Ishai, R.; Ben-Hur, E.; Hornfeld, E. *Isr. J. Chem.* **1968**, *6*, 769.
60. Weinblum, D. *Biochem. Biophys. Res. Commun.* **1967**, *29*, 384.
61. Beukers, R.; Berends, W. *Biochem. Biophys. Acta* **1960**, *41*, 550.
62. Morrison, H.; Kleopfer, R. *J. Am. Chem. Soc.* **1968**, *90*, 5037.
63. Coyle, J. *Chem. Br.* **1980**, *16*, 460.
64. Epstein, J. H. "Photomedicine"; in "The Science of Photobiology"; Smith, K. C., Ed.; Plenum-Rosetta: New York, 1977, p. 201.
65. Bergstrom, D. E.; Agosta, W. C. *Tetrahedron Lett.* **1974**, 1087.
66. Bagli, J. F.; Bogri, T. *J. Org. Chem.* **1972**, *37*, 2132.
67. Ogino, T.; Yamada, K.; Isogai, K. *Tetrahedron Lett.* **1977**, 2445.
68. Baldwin, S. W.; Crimmins, M. T. *J. Am. Chem. Soc.* **1980**, *102*, 1198.
69. Yamakawa, K.; Sakaguchi, R.; Nakamura, T.; Watanabe, K. *Chem. Lett.* **1976**, 991.
70. Corey, E. J.; Nozoe, S. *J. Am. Chem. Soc.* **1964**, *86*, 1652.
71. De Mayo, P.; Helmlinger, D.; Yates, R. B.; Westfelt, L. Abstracts of the International Symposium on "Synthetic Methods and Rearrangements in Alicyclic Chemistry", Oxford, July 1969; The Chemical Society: London, p. 32.
72. Lange, G. L.; De Mayo, P. *J. Chem. Soc., Chem. Commun.* **1967**, 705.
73. Liu, H. J.; Lee, S. P. *Tetrahedron Lett.* **1977**, 3699.
74. Wender, P. A.; Lechleiter, J. C. *J. Am. Chem. Soc.* **1978**, *100*, 4321.
75. Yanagiya, M.; Kaneko, K.; Kaji, T.; Matsumoto, T. *Tetrahedron Lett.* **1979**, 1761.
76. Challand, D. H.; Kornis, G.; Lange, G. L.; De Mayo, P. *J. Chem. Soc., Chem. Commun.* **1967**, 704.
77. Challand, D. B.; Hikino, H.; Kornis, G.; Lange, G.; De Mayo, P. *J. Org. Chem.* **1969**, *34*, 794.
78. Tatsuta, K.; Akimoto, K.; Kinoshita, M. *J. Am. Chem. Soc.* **1979**, *101*, 6116.
79. Dauben, W. G.; Salem, L.; Turro, N. J. *Acc. Chem. Res.* **1975**, *8*, 41.
80. Scarborough, R. M., Jr.; Smith, A. B., III; *J. Am. Chem. Soc.* **1977**, *99*, 7085.
81. Scarborough, R. M., Jr.; Toder, B. H.; Smith, A. B., III; *J. Am. Chem. Soc.* **1980**, *102*, 3904.
82. Büchi, G.; Carlson, J. A.; Powell, J. E., Jr.; Tietze, L. F. *J. Am. Chem. Soc.* **1973**, *95*, 540.
83. Tietze, L. F. *Chem. Ber.* **1974**, *107*, 2499.
84. Partridge, J. J.; Chadha, N. K.; Uskokovic, M. R. *J. Am. Chem. Soc.* **1973**, *95*, 532.
85. Wiesner, K.; Pak-Tsun Ho; Chang, D.; Blount, J. F. *Experientia* **1972**, *28*, 766.
86. Wiesner, K.; Pak-Tsun Ho; Chang, D.; Kuen-Lam, Y.; Shii Jeou, C.; Yun Ren, W. *Can. J. Chem.* **1973**, *51*, 3978.
87. Liu, H. J.; Valenta, Z.; Wilson, J. S.; Yu, T. T. J. *Can J. Chem.* **1969**, *47*, 509.
88. Liu, H. J.; Valenta, Z.; Yu, T. T. J. *J. Chem. Soc., Chem. Commun.* **1970**, 1116.
89. Owley, D. C.; Blomfield, J. J. *J. Chem. Soc. C* **1971**, 3445.
90. Do Khac Manh, D.; Fetizon, M.; Flament, J. P. *Tetrahedron* **1975**, *31*, 1903.
91. Eaton, P. E. *Tetrahedron Lett.* **1964**, 3695.

92. Do Khac Manh, D.; Fetizon, M.; Hanna, I.; Olesker, A.; Pascard, C.; Praugé, T. *J. Chem. Soc., Chem. Commun.* **1980**, 1209.
93. Wiesner, K. *Tetrahedron* **1975**, *31*, 1655.
94. Marini–Bettolo, G.; Sahoo, S. P.; Poulton, G. A.; Tsai, T. Y. R.; Wiesner, K. *Tetrahedron* **1980**, *36*, 719.
95. Eliel, E. L. "Stereochemistry of Carbon Compounds"; McGraw–Hill: New York, 1962, p. 273–80.
96. Pasto, D. J.; Haley, M.; Chipman, D. M. *J. Am. Chem. Soc.* **1978**, *100*, 5272.
97. Domelsmith, L. N.; Houk, K. N.; Piedrahita, C.; Dolbier, W. J., Jr. *J. Am. Chem. Soc.* **1978**, *100*, 6908.
98. Schaad, L. J. *Tetrahedron* **1970**, *26*, 4115.
99. Wiesner, K.; Jirkovsky, I.; Fishman, M.; Williams, C. A. J. *Tetrahedron Lett.* **1967**, 1523.
100. Wiesner, K.; Poon, L.; Jirkovsky, I.; Fishman, M. *Can. J. Chem.* **1969**, *47*, 433.
101. Kelly, R. B.; Zamecnik, J.; Beckett, B. A. *Can. J. Chem.* **1972**, *50*, 3455.
102. Kelly, R. B.; Eber, J.; Hung, H. K. *Can J. Chem.* **1973**, *51*, 2534.
103. Do Khac Manh, D.; Fetizon, M.; Lazare, S. *J. Chem. Soc., Chem. Commun.* **1975**, 282.
104. Do Khac Manh, D.; Fetizon, M.; Lazare, S. *J. Chem. Res. S* **1978**, 22.
105. Kelly, R. B.; Harley, M. L.; Alward, S. J. *Can. J. Chem.* **1980**, *58*, 755.
106. Guthrie, R. W.; Valenta, Z.; Wiesner, K. *Tetrahedron Lett.* **1966**, 4645.
107. Wiesner, K.; Uyeo, S.; Philipp, A.; Valenta, Z. *Tetrahedron Lett.* **1968**, 6279.
108. Wiesner, K. *Pure Appl. Chem.* **1975**, *41*, 93.
109. Wiesner, K. *Chem. Soc. Rev.* **1977**, 413.
110. Ziegler, F. E.; Kloek, J. A. *Tetrahedron Lett.* **1974**, 315.
111. Ziegler, F. E.; Kloek, J. A. *Tetrahedron* **1977**, *33*, 373.
112. Cowan, D. O.; Drisko, R. L. "Elements of Organic Photochemistry"; Plenum: New York, 1976, p. 368.
113. Yoshihara, K.; Ohta, Y.; Sakai, T.; Hirose, Y. *Tetrahedron Lett.* **1969**, 2263.
114. Crowley, J. J. *Proc. Chem. Soc., London* **1962**, 245.
115. Crowley, J. J. *Tetrahedron* **1965**, *21*, 1001.
116. Liu, R. S. H.; Hammond, G. S. *J. Am. Chem. Soc.* **1967**, *89*, 4936.
117. Corey, E. J.; Cane, D. E.; Libit, L. *J. Am. Chem. Soc.* **1971**, *93*, 7016.
118. Miyashita, M.; Yoshikoshi, A. *J. Am. Chem. Soc.* **1974**, *96*, 1917.
119. McMurry, J. E.; Choy, W. *Tetrahedron Lett.* **1980**, 2477.
120. Heathcock, C. H.; Badger, R. M. *J. Chem. Soc., Chem. Commun.* **1968**, 1510.
121. Brown, M. *J. Org. Chem.* **1968**, *33*, 162.
122. Pirrung, M. C. *J. Am. Chem. Soc.* **1979**, *101*, 7130.
123. Fetizon, M.; Lazare, S.; Pascard, C.; Praugé, T. *J. Chem. Soc., Perkin Trans. 1* **1979**, 1407.
124. Birch, A. M.; Pattenden, G. *J. Chem. Soc., Chem. Commun.* **1980**, 1195.
125. Oppolzer, W.; Wylie, R. D. *Helv. Chim. Acta* **1980**, *63*, 1198.
126. Barker, A. J.; Pattenden, G. *Tetrahedron Lett.* **1980**, 3513.
127. Oppolzer, W.; Godel, T. *J. Am. Chem. Soc.* **1978**, *100*, 2583.
128. Kueh, J. S. H.; Mellor, M.; Pattenden, G. *J. Chem. Soc., Chem. Commun.* **1978**, 5.
129. Kueh, J. S. H.; Mellor, M.; Pattenden, G. *J. Chem. Soc., Perkin Trans. 1* **1981**, 1052.
130. Hixson, S. S.; Mariano, P. S.; Zimmerman, H. E. *Chem. Rev.* **1973**, *73*, 531.
131. Bullivant, M. J.; Pattenden, G. *J. Chem. Soc., Perkin Trans. 1* **1976**, 256.
132. Pattenden, G.; Whybrow, D. *J. Chem. Soc., Perkin Trans. 1* **1981**, 1046.
133. Pearlman, B. A. *J. Am. Chem. Soc.* **1979**, *101*, 6404.
134. Wiesner, K.; Musil, V.; Wiesner, K. J. *Tetrahedron Lett.* **1968**, 5643.
135. Paternò, E.; Chieffi, C. *Gazz. Chim. Ital.* **1909**, *39*, 341.
136. Büchi, G.; Inman, C. G.; Lipinsky, E. S. *J. Am. Chem. Soc.* **1954**, *76*, 4327.
137. Araki, Y.; Nagasawa, J. I.; Ishido, Y. *Carbohydr. Res.* **1977**, *58*, 64.
138. Morton, D. R.; Morge, R. A. *J. Org. Chem.* **1978**, *43*, 2093.
139. Jones, G.; Acquadro, M.; Carmody, M. A. *J. Chem. Soc., Chem. Commun.* **1975**, 206.
140. Fleming, I. "Frontier Orbitals and Organic Chemical Reactions"; Wiley: London, 1976.

141. Woodward, R. B.; Hoffmann, R. *Angew. Chem., Int. Ed. Engl.* **1969**, *8*, 781.
142. Herndon, W. C. *Top. Curr. Chem.* **1974**, *40*, 141.
143. Bischof, P.; Heilbronner, E. *Helv. Chim. Acta* **1970**, *53*, 1677.
144. Turner, D. W. "Ionization Potentials"; in "Advances in Physical Organic Chemistry"; Gold, V., Ed.; Academic: New York, 1966; p. 52, and references cited therein.
145. Giordan, J. C.; McMillan, M. C.; Moore, J. H.; Staley, S. W. *J. Am. Chem. Soc.* **1980**, *102*, 4870.
146. Jorda, K. D.; Burrow, P. D. *Acc. Chem. Res.* **1978**, *11*, 341.
147. Heimbach, P.; Brenner, W. *Angew. Chem., Int. Ed. Engl.* **1967**, *6*, 800.
148. Billups, W. E.; Cross, J. H.; Smith, C. V. *J. Am. Chem. Soc.* **1973**, *95*, 3438.
149. Alston, P. V.; Ottenbrite, R. M. *J. Org. Chem.* **1975**, *40*, 1111.
150. Hudlicky, T.; Sheth, Y. P.; Gee, V.; Barnvos, D. *Tetrahedron Lett.* **1979**, 4889.
151. Hudlicky, T.; Kutchan, T. M.; Wilson, S. R.; Mao, D. T. *J. Am. Chem. Soc.* **1980**, *102*, 6351.
152. Danheiser, R. L.; Martinez-Davila, C.; Morin, J. M., Jr. *J. Org. Chem.* **1980**, *45*, 1340.
153. Hudlicky, T.; Koszyk, F. Y.; Kutchan, T. M.; Sheth, Y. P. *J. Org. Chem.* **1980**, *45*, 5020.
154. Sinnema, A.; Rantwijk, F. V.; De Koning, A. J.; Van Wijk, A. M.; Van Bekkum, H. *J. Chem. Soc., Chem. Commun.* **1973**, 364.
155. Paquette, L. A.; Schwartz, J. A. *J. Am. Chem. Soc.* **1970**, *92*, 3215.
156. Wender, P. A.; Lechleiter, J. C. *J. Am. Chem. Soc.* **1977**, *99*, 267.
157. Lange, G. L.; Huggins, M. A.; Neidert, E. *Tetrahedron Lett.* **1976**, 4409.
158. Wilson, S. R.; Phillips, L. R.; Pelister, Y.; Huffman, J. C. *J. Am. Chem. Soc.* **1979**, *101*, 7373.
159. Lange, G. L.; McCarthy, F. C. *Tetrahedron Lett.* **1978**, 4749.
160. Williams, J. R.; Callahan, J. F. *J. Chem. Soc., Chem. Commun.* **1979**, 405.
161. Wender, P. A.; Lechleiter, J. C. *J. Am. Chem. Soc.* **1980**, *102*, 6340.
162. Williams, J. R.; Callahan, J. F. *J. Chem. Soc., Chem. Commun.* **1979**, 404.
163. Williams, J. R.; Callahan, J. F. *J. Org. Chem.* **1980**, *45*, 4475.
164. Wender, P. A.; Letendre, L. J. *J. Org. Chem.* **1980**, *45*, 367.
165. Wender, P. A.; Hubbs, J. C. *J. Org. Chem.* **1980**, *45*, 365.
166. Williams, J. R.; Callahan, J. F. *J. Org. Chem.* **1980**, *45*, 4479.
167. For a review on this topic, *see* Engel, P. S. *Chem Rev.* **1980**, *80*, 99.
168. Simonaitis, R.; Pitts, J. N., Jr.; *J. Am. Chem. Soc.* **1969**, *91*, 108.
169. Hiberty, P. C.; Jean, Y. *J. Am. Chem. Soc.* **1979**, *101*, 2538.
170. Buchwalter, S. L.; Closs, G. L *J. Org. Chem.* **1975**, *40*, 2549.
171. Givens, R. S.; Oettle, W. F. *J. Am. Chem. Soc.* **1971**, *93*, 3301.
172. Berger, C.; Frank-Neumann, M.; Ourisson, G. *Tetrahedron Lett.* **1968**, 3451.
173. Greenlee, W. J.; Woodward, R. B. *J. Am. Chem. Soc.* **1976**, *98*, 6075.
174. Wilson, S. R.; Turner, R. B. *J. Org. Chem.* **1973**, *38*, 2870.
175. Piers, E.; Britton, R. W.; Keziere, R. J.; Smillie, R. D. *Can. J. Chem.* **1971**, *49*, 2623.
176. Perold, G. W.; Ourisson, G. *Tetrahedron Lett.* **1969**, 3871.
177. Greene, A. E.; Muller, J. C.; Ourisson, G. *Tetrahedron Lett.* **1971**, 4147.

4

[3+2] Cycloadditions

The most popular class of [3+2] cycloadditions involves a three-atom fragment (a−b−c) having a positive and negative charge at the ends in a resonance structure. Interaction of the fragment with a multiple bond system (x−y) leads to an uncharged five-membered ring, through a pericyclic reaction known as 1,3-dipolar cycloaddition (*1*).

These 1,3-dipoles can be represented as a resonance between structures having a sextet or an octet; the sextet either does or does not contain a double bond.

sextet	octet	
⊕ − ⊖ a=b−c	⊕ ⊖ a=$\overline{\text{b−c}}$	with double bond (b̲=N)
⊕ − ⊖ a−b−c	⊕ ⊖ a=b−c	without double bond (b̲=N−R,O)

Among 1,3-dipoles that have a double bond (or that are of the propargyl-allenyl–anion type) we find (a,b,c) nitrile ylides (C,N,C), nitrilimines (C,N,N), nitrile oxides (C,N,O), diazoalkanes (N,N,C), and azides (N,N,N). Among those that have no double bond (or that are of the allyl–anion type) we find azomethine ylides (C,N,C), azomethine imines (C,N,N), nitrones (C,N,O), and ozone (O,O,O).

This classical approach was revised in the early seventies by Sustmann and Houk. Sustmann (*2*) proposed a simplification of Eq. 2 for a homogeneous series of substrates, thus deriving Eq. 3, whose graphic

$$\Delta E = A\,\beta^2 \left[\frac{1}{E_{HO}^{a\,c} - E_{LU}^{xy}} + \frac{1}{E_{HO}^{x\,y} - E_{LU}^{a\,c}} \right] \qquad (3)$$

0065-7719/83/0180-0089$08.25/1

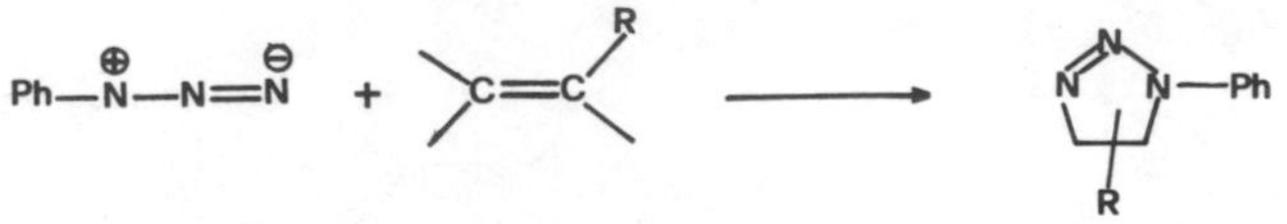

Scheme 4.1

representation is two branches of hyperbola that form a parabola when superimposed. A correlation of ln K_2 for cycloaddition of phenyl-azide with olefins against the HOMO energies of the olefins has a parabolic shape (*2,3*), where *R* is both electron attracting and electron donating (Scheme 4.1) increase the rate.

The 1,3-dipolar cycloadditions therefore generally can be regarded as pericyclic reactions under the control of both HOMO and LUMO of the 1,3-dipole. The former predominates with electron-attracting substituted olefins, the latter with olefins bearing electron-donating substituents (*4*).

The MO representation of 1,3-dipoles (*5*) (Figure 4.1) may also be evoked to explain the regioselectivity of the cycloaddition.

Because the greater coefficient of the LUMO of 1,3-dipole is on the a atom and that of the HOMO on the c atom, a reversed regiochemical outcome can be expected when azides are allowed to react with electron-rich and electron-deficient dipolarophiles. The predominance of the regioisomer **347** or **348** respectively (*6*) accounts for the different MO interactions involved (Figure 4.2).

Both the classical and the MO approaches consider the 1,3-dipolar cycloaddition to be a pericyclic reaction—even calculations by methods including overlap (ab initio STO–3G and EHT) suggest that the eventual distortion of the transition state is remarkably small. With these simple concepts in mind, let us turn back to our topic.

Intermolecular 1,3-Dipolar Cycloadditions

In the last decade intermolecular 1,3-dipolar cycloaddition proved to be one of the most powerful tools to synthesize natural products because

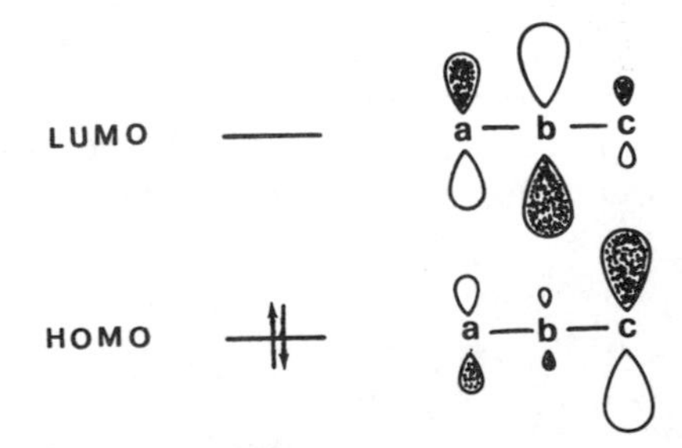

Figure 4.1. FMOs of 1,3-dipoles.

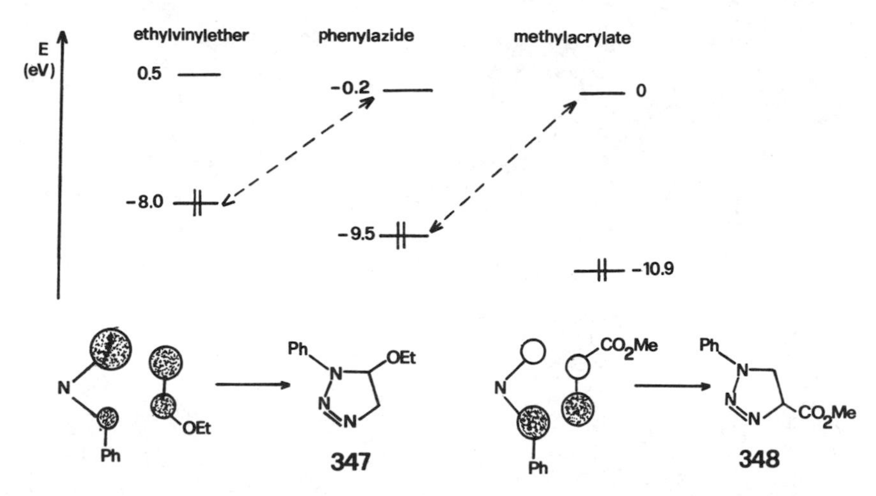

Figure 4.2. FMO interactions of phenyl azide with ethyl vinyl ether and methyl acrylate, showing the predominant regioisomers.

of the mild experimental conditions involved and the high level of stereochemical and regiochemical control. Several 1,3-dipoles were employed in this approach.

Nitrile Oxides. Efforts directed toward the synthesis of vitamin B_{12} (**349**) led to the development of methods of synthesizing fascinating macrocyclic ligands such as corrin (**350**) and corphin (**351**) (Scheme 4.2).

The Woodward–Eschenmoser approach, the first route to vitamin B_{12}, represented the main achievement of modern synthetic organic chemistry. A fundamentally different approach to the synthesis of corrinoid substances was disclosed by Stevens (*7*). The method depends on the use of the isoxazole nucleus, readily obtainable by 1,3-dipolar cycloaddition of a nitrile oxide on terminal acetylenes, as a means of elaborating the crucial ring-bridging vinylogous amidines chromophores (Scheme 4.3).

The utility of the isoxazole nucleus for the synthesis of vitamin B_{12} was originally suggested by Conforth (*8*) in various lectures and first applied to simple models by an Italian group (*9*).

The synthesis of the semicorrin **355**, already incorporated into corrin systems, became the initial goal and was reported independently and almost simultaneously by the Stevens (*10–12*) and Traverso (*13, 14*) groups. The key step was the 1,3-dipolar cycloaddition of the nitrile oxide **352** to the acetylene **353** to afford the isoxazole scaffold **354**, further converted to **355**, as outlined in Scheme 4.4.

Further work by Stevens and his associates developed the strategy (*15*) and realized the synthesis (*16*) of the southern half of the vitamin in the latent form **358**, obtained through cycloaddition of **356**, probably the most complicated 1,3-dipole ever synthesized, to the acetylene **357** (Scheme 4.5).

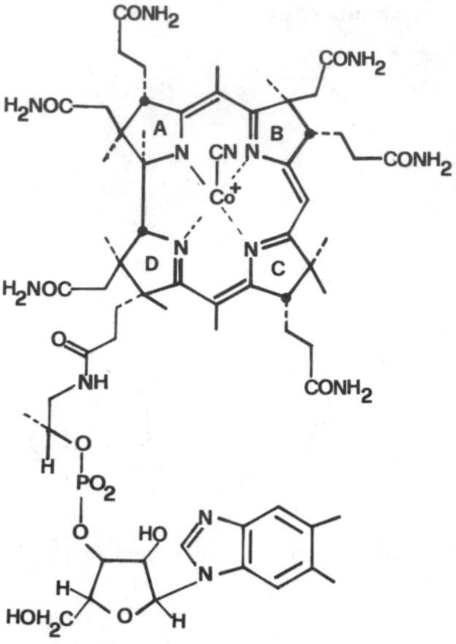

349

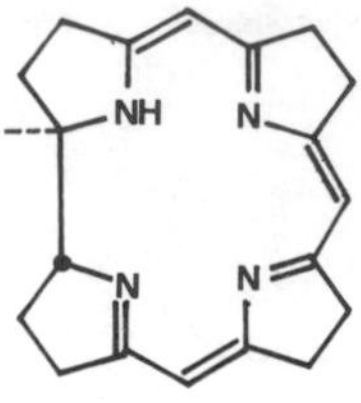

350

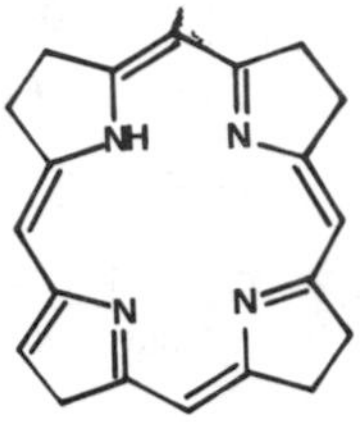

351

Scheme 4.2

The Stevens group synthesized the corphin ring (**351**) using a similar approach, which involved a spectacular sequence of three 1,3-dipolar cycloadditions of nitrile oxides to acetylenes to construct the appropriately substituted triisoxazole **359**. The latter was then transformed into the nickel precorphin complex **360**, finally converted into the nickel(II) octamethylcorphin (**361**) (*17*) (Scheme 4.6).

Although the monumental work of Stevens and coworkers seems to be the most spectacular demonstration of the synthetic flexibility and usefulness of the 1,3-dipolar cycloaddition of nitrile oxides, other papers in the field are worthy of mention. In 1967 Stork and coworkers (*18–20*) realized one of the most original approaches to the steroid field. They described the total synthesis of (±)-progesterone (**368**) via 1,3-dipolar cycloaddition of the nitrile oxide **362** (generated from the corresponding nitro derivative with $POCl_3$) on the enamino ester **363**. The loss of pyr-

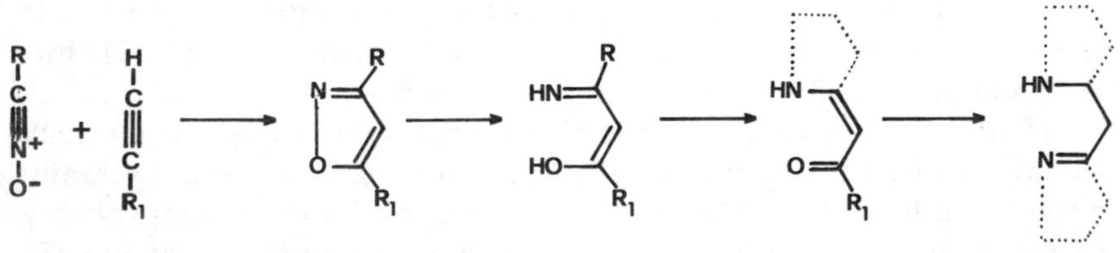

Scheme 4.3

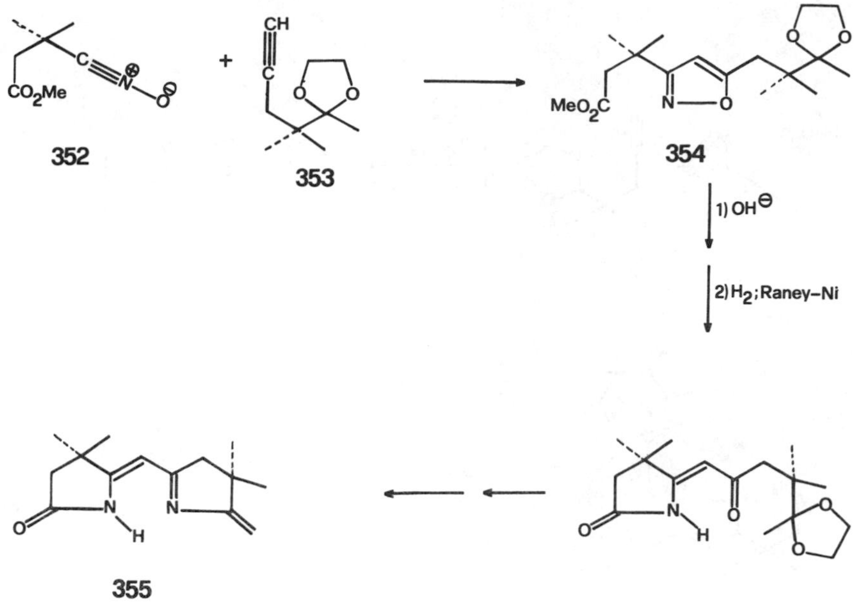

Scheme 4.4

rolidine gave the isoxazole **364**, the latent precursor of the rings A and B, further elaborated into the corresponding chloromethyl derivative **365**, which reacts with the enolate of 10-methyl-$\Delta^{1,9}$-octalin-1,5-dione (**366**) to give the key intermediate **367** (Scheme 4.7).

The regiochemical control leading to the crucial 4-carboxylic ester **364** was secured by the presence of the amino residue of **363**, which gives a definite polarization of the coefficients determining the nature of the HOMO dipole–LUMO dipolarophile transition state **369a**. The alternative LUMO dipole–HOMO dipolarophile interactions, as represented in **369b**, should play a significant role only if the nitrogen residue polarizes the double bond more strongly than the carboxylate does (Figure 4.3). But this has to be demonstrated.

It was recently discovered that *Streptomyces sviceus* produces

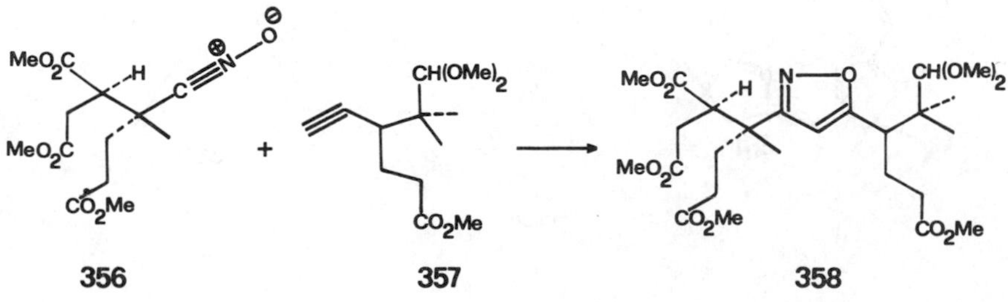

Scheme 4.5

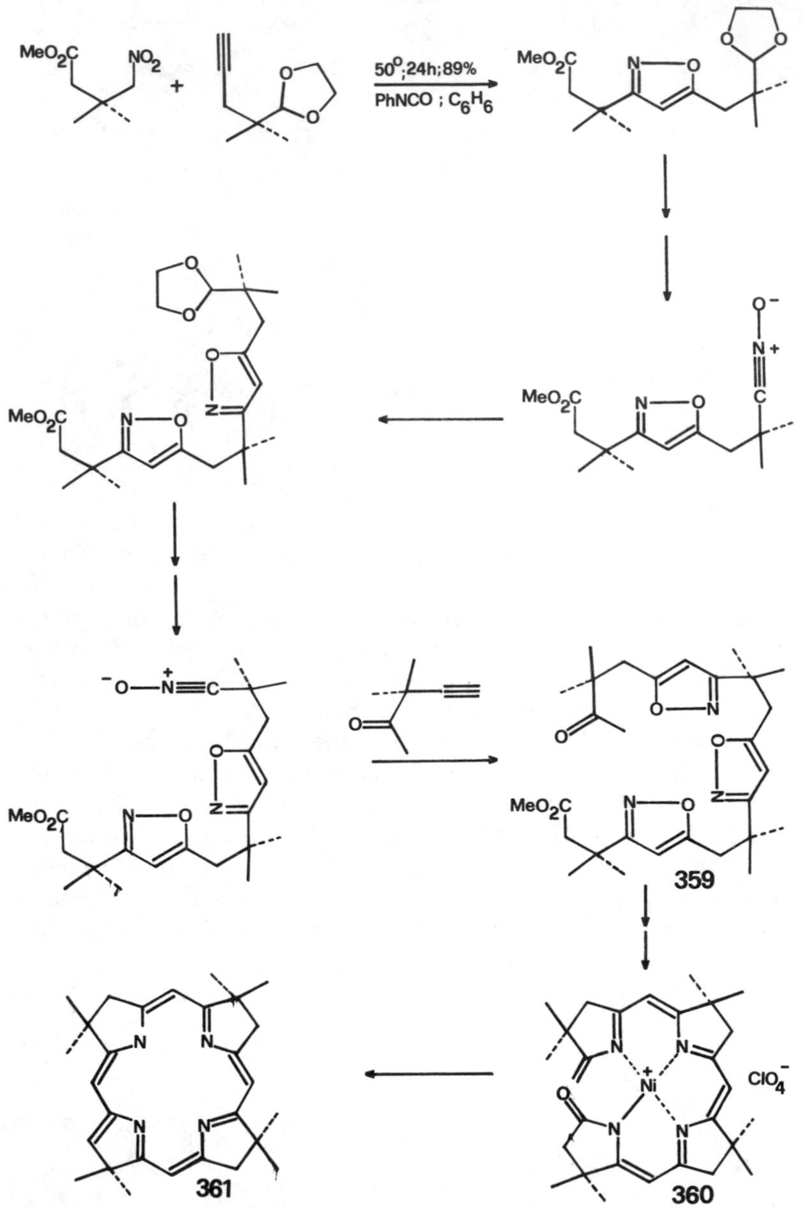

MeO_2C NO_2 +

50°;24h;89%
PhNCO ; C_6H_6

MeO_2C N—O

359

ClO_4^-

360

361

Scheme 4.6

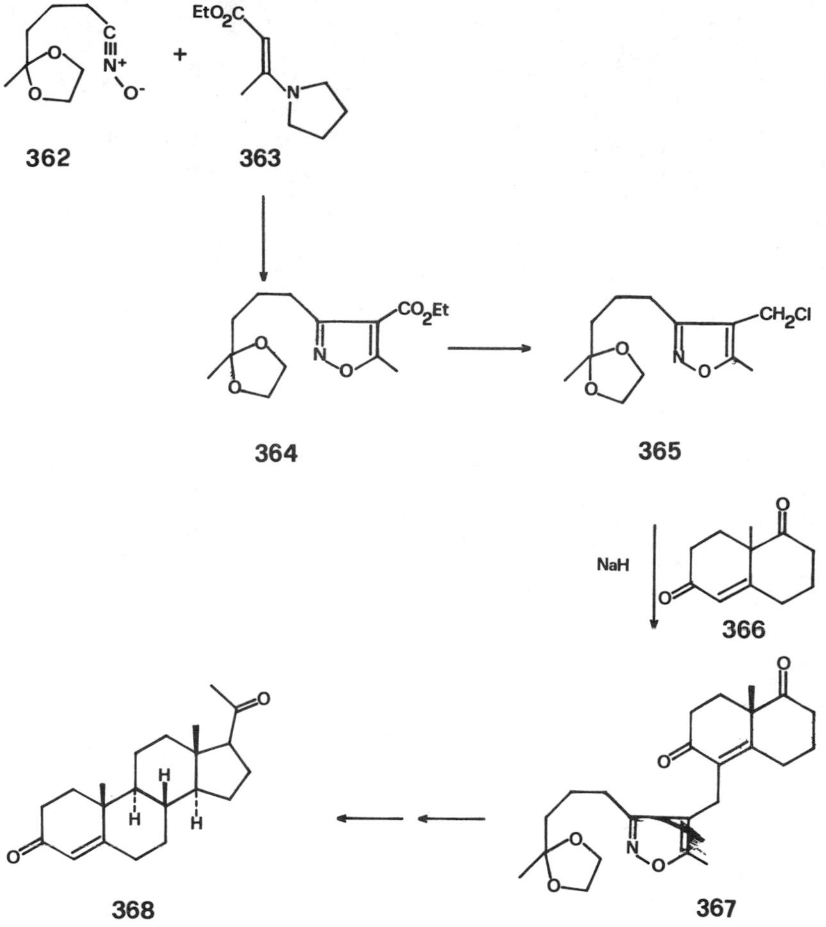

362 363

364 365

NaH

366

368 367

Scheme 4.7

(αS,5S)-α-amino-3-chloro-4,5-dihydroisoxazole-5-acetic acid (**372**, R =Cl), which exhibits antitumoral properties both in vitro and in vivo. Chloronitrile oxide **370**, (R =Cl) was initially added to the salt of α-nitrocrotonic ester to produce nonstereospecifically a methylated analog (*21*). A stereorandom synthesis of the bromo analog **372**, (R = Br) was achieved by direct 1,3-dipolar cycloaddition of bromonitrile oxide **370** (R = Br), on racemic vinylglycine (**371**) (*22*) (Scheme 4.8).

Several syntheses involved ring-opening products of isoxazole cycloadducts. Thus hentriacontane-14,16-dione (**374**), a naturally occurring β-diketone, was synthesized starting from the isoxazole **373** (*23*) (Scheme 4.9).

The antibiotic (−)-vermiculine (**378**) isolated from the fermentation broth of *Penicillum vermiculatum* Dangeard, was synthesized by [1,3]

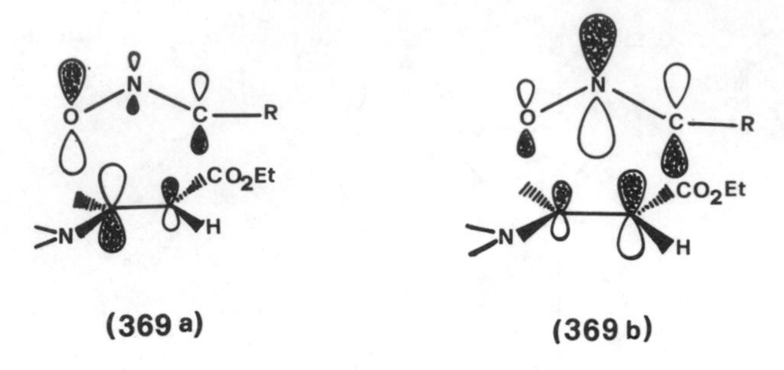

Figure 4.3. FMO interactions of nitrile oxides and enamino esters.

Scheme 4.8

Scheme 4.9

dipolar cycloaddition of acetonitrile oxide on **375** to give the isoxazoline **376**, which was reductively opened to **377** (*24*) (Scheme 4.10).

By a similar approach thienamycin (**79**), the antibiotic isolated from the fermentation broth of the soil microorganism *Streptomyces cattleya*, was obtained from **380**, the reduction product of the isoxazoline **379** (*25*) (Scheme 4.11).

A novel approach to the synthesis of the 11-deoxy derivative **383** of a popular prostaglandin synthon featured a 1,3-dipolar cycloaddition of the nitrile oxide **381** on 1-heptyne to furnish the isoxazole **382** containing the C13–C20 framework destined to become the ω chain of prostanoids (*26*) (Scheme 4.12).

Diazoalkanes. The use of 1,3-dipolar cycloaddition of dimethyldiazomethane and diazomethane was described in the section of Chapter 3 entitled "Pseudo [σ2+σ2] Cycloreversion" during the synthesis of aristolone (*27*) (**49**) as well as in the synthesis of isomarasmic acid (**339**) (*28,29*).

Vinyldiazoalkane (**384**) and (*E,Z*)-1,3,5-octatriene (**385**) gave a 55% yield of pyrazoline **386**, which was photolyzed to give dictyopterene B (**387**), isolated from a Pacific brown algae in the waters off Hawaii together with **388**. The latter was converted by a Cope rearrangement (*see* the section on "The Cope Rearrangement" in Chapter 7) to ectocarpene

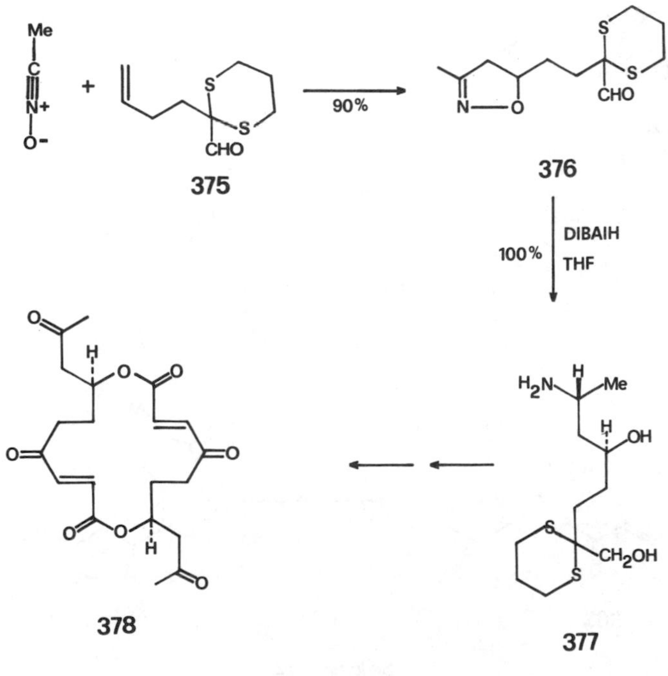

Scheme 4.10

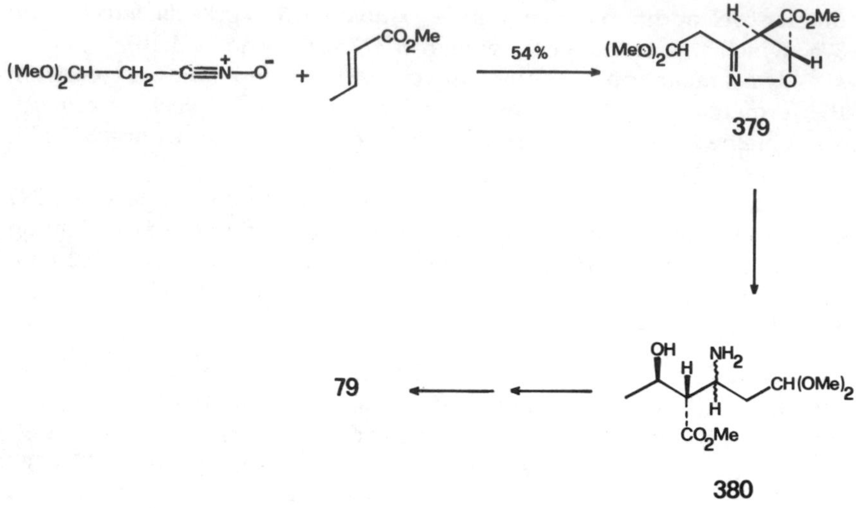

(MeO)$_2$CH—CH$_2$—C≡N$^+$—O$^-$ + CO$_2$Me 54% 379

380 79

Scheme 4.11

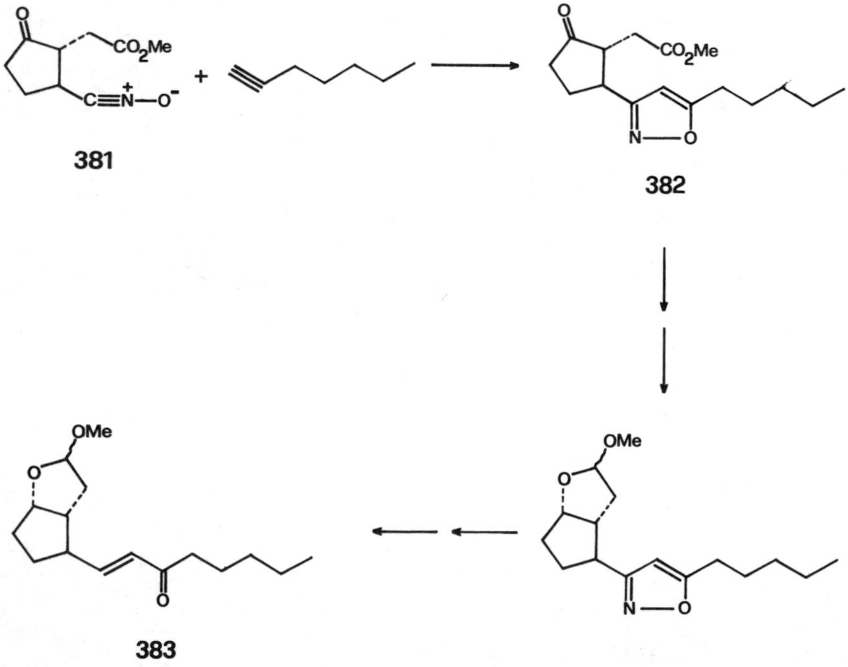

381 + 382 383

Scheme 4.12

(**389**), produced by the female gametes of *Ectocarpus siliculosus* (*30*) (Scheme 4.13).

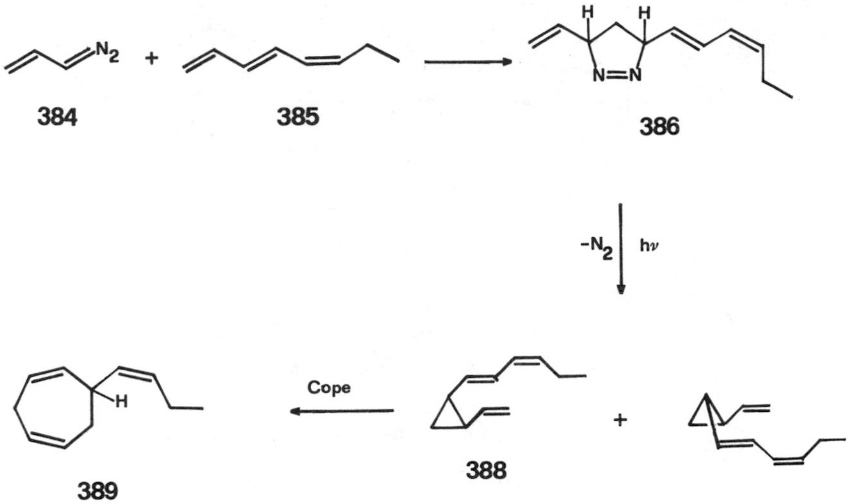

384 385 386 -N2 hν 389 Cope 388 387

Scheme 4.13

A noteworthy paper (*31*) concerns the synthesis of β-thujaplicin (**390**) and β-dolabrin (**391**), two natural tropolones, where the approach was quite difficult in spite of their simple structures.

In principle three steps should provide the structure: addition of dimethyldiazomethane to tropone (**392**), decomposition of the cycloadduct with transformation of the dimethyl-Δ^1-pyrazoline ring into the isopropylidene group, and, finally, conversion of tropone to tropolone.

Difficulties arose immediately during the 1,3-dipolar cycloaddition stage, because **392** reacted with dimethyldiazomethane in the "wrong" way and produced the useless regioisomer **393** (*32*) (Scheme 4.14).

O OH (390) O OH (391)

This behavior is explained by considering the FMOs of diazomethane and tropone (**392**) (*33*), whose coefficients were calculated by a self-consistent field (SCF) method (*34*) (Figure 4.4).

Because the dominant interaction occurs between the LUMO of tropone and the HOMO of diazoalkane, where the principle of maximum overlap is taken into account, a transition state such as **394** furnishes **393**.

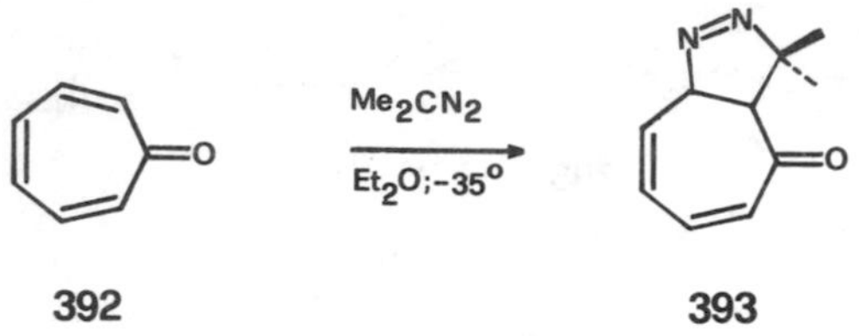

Me_2CN_2

Et_2O;-35°

392

393

Scheme 4.14

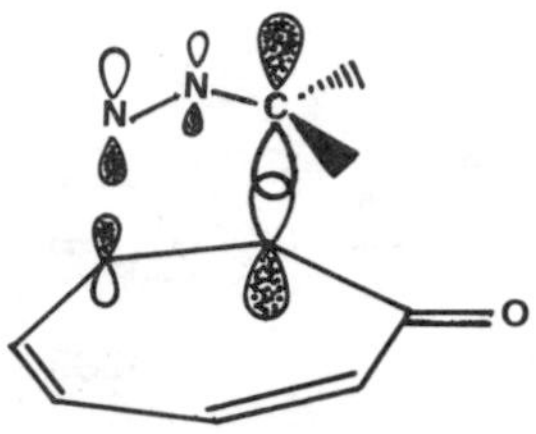

(394)

Conjugation of the α,β-unsaturated carbonyl system determines that the α coefficient is higher than the β, in contrast, for example, with acrolein (*35*). In an attempt to promote a reversal of regiochemistry by lowering this effect, tropone was complexed with iron tricarbonyl. The results were unsuccessful and the regiochemistry remained unchanged (*36*), with the conjugation lowered but not suppressed (Scheme 4.15).

Inversion of regiochemistry was achieved by placing on the tropone ring both the $Fe(CO)_3$ group and an electron-attracting substituent in the α position of the reacting C=C−C=O system (*31*), thus performing the cycloaddition on **395**. Consideration of the substituent effects indicates that the electron-attracting group both lowers the LUMO (increasing the rate) and enhances the β coefficient (inverting the regiochemistry), thus giving **396**. The results are reported in Scheme 4.16.

Azomethine Ylides. Azomethine ylides are 1,3-dipoles without an orthogonal double bond; hence they are of the allyl anion type. They began to be studied in the sixties by Huisgen and his group, who discovered the most representative class of these 1,3-dipoles, the oxazolium-5-oxides (**397**), and clarified the mechanism of their cycloaddition to alkenes to produce Δ^2-pyrrolines (*37*) (**398**) (Scheme 4.17). When a triple bond is the counterpart as dipolarophile, a pyrrole nucleus is derived.

Application of azomethine ylides to the natural product field opened a straightforward route to the pyrrolizidine skeleton, which is present in several widely distributed alkaloids.

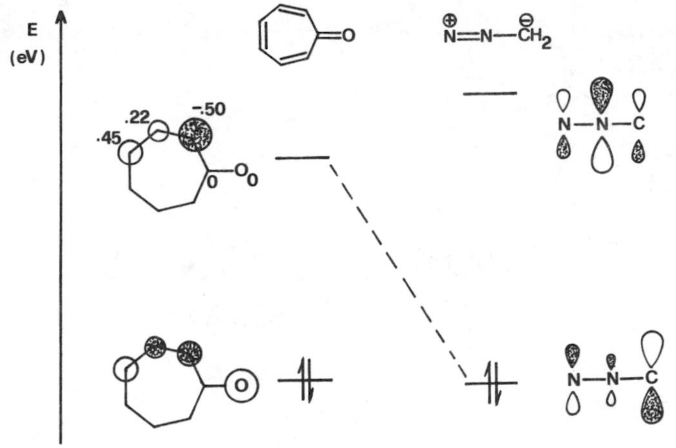

Figure 4.4. FMO interaction of tropone and diazomethane.

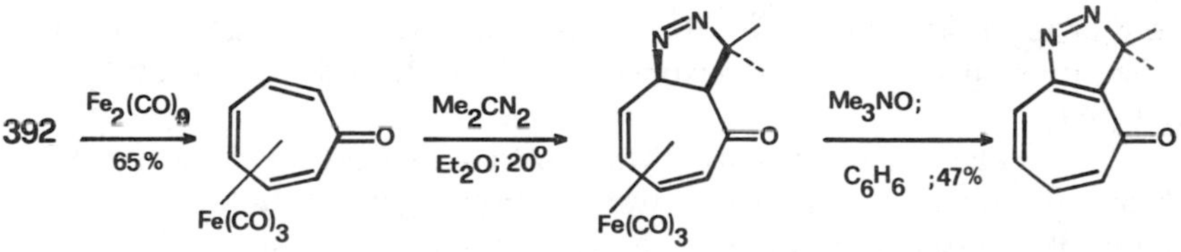

Scheme 4.15

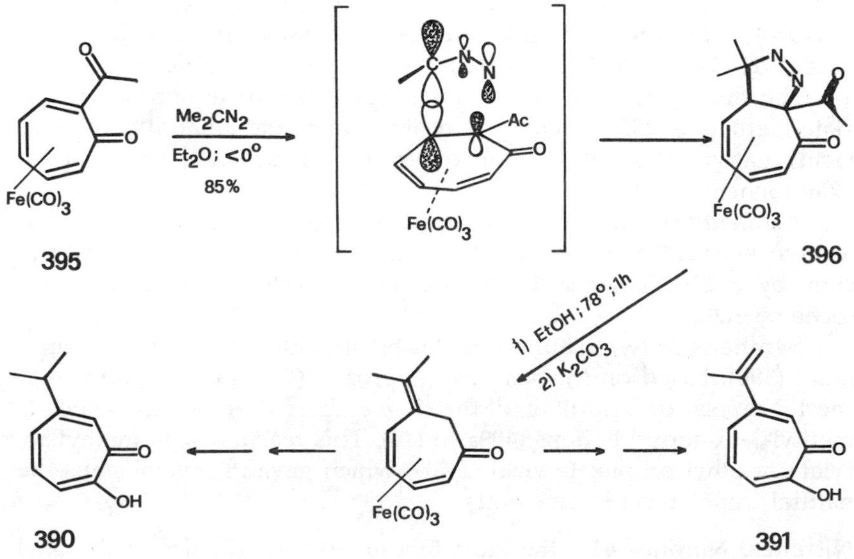

Scheme 4.16

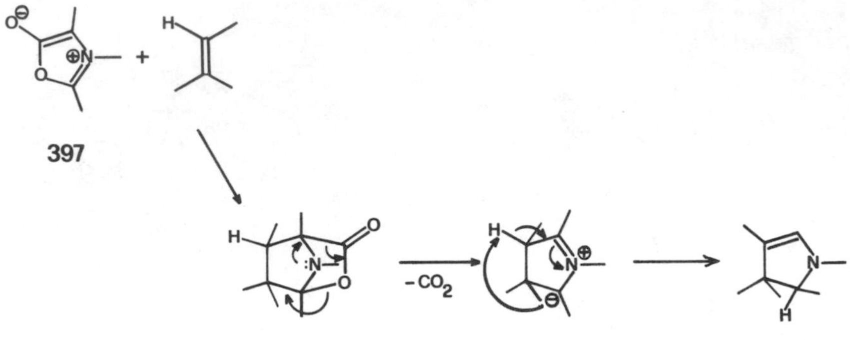

397

$-CO_2$

398

Scheme 4.17

Regiospecific cycloaddition (*38*) of ethyl propiolate to the postulated azomethine ylide generated by loss of carbon dioxide from **400a**, formed by heating *N*-formyl-L-proline (**399a**) with acetic anhydride, gave the dehydropyrrolizidine ester **401a**, easily converted to 1-substituted pyrrolizidines such as (±)-isoretronecanolate (**402**) and (±)-isoretronecanol (**63**).

In an elegant extension of this approach, starting from the NO-diformyl derivative of (−)-4-hydroxy-L-proline (**399b**), Robins and Sakdarat (*39*) were able to obtain the chiral hydroxy ester **401b**. This compound was further converted into isoretronecanol (**63**), trachelanthamidine (+)-laburnine (**64**), and (+)-supinidine (**403**).

We can further report (*40*) that cycloaddition of **400a** using proparglyic aldehyde as a dipolarophile led to the synthesis of 1-formyl-6,7-dihydro-5*H*-pyrrolizidine (**404**), a pheromone isolated from *Danaus affinis affinis* (Fab.) and *Utetheisa lotix* (Cram.) (Scheme 4.18).

The first application (*41*) of a 1,3-dipolar cycloaddition of the imidate methylide (**405**) to methyl acrylate gave rise to an adduct (in 55% yield) that was the key for a stereospecific synthesis of (±)-retronecine (**406**) (Scheme 4.19).

Azomethine ylides were also employed in the synthetic approach to mitosenes (**407**), the chemical dehydration products of the mitomycins, by a 1,3-dipolar cycloaddition with acetylenedicarboxylate (*42*) (Scheme 4.20).

Synthesis of two natural tropolones, stipitatic acid (**411**) and hinokitiol (**390**), based on the Katritzky approach (*43*), entailed the use of the 1,3-dipolar cycloaddition of the six-membered azomethine ylides 1-methyl-3-oxidopyridiniums (**408a,b**) (*44*). This reaction with methyl acrylate or ethyl propiolate yielded **409**, which gave in several steps the natural tropolones (Scheme 4.21).

Nitrones. Nitrones **412**, the most investigated 1,3-dipoles of the allyl anion type, react with double or triple bonds to give Δ^4-isoxazolines **413** or isoxazolidines **414** (Scheme 4.22).

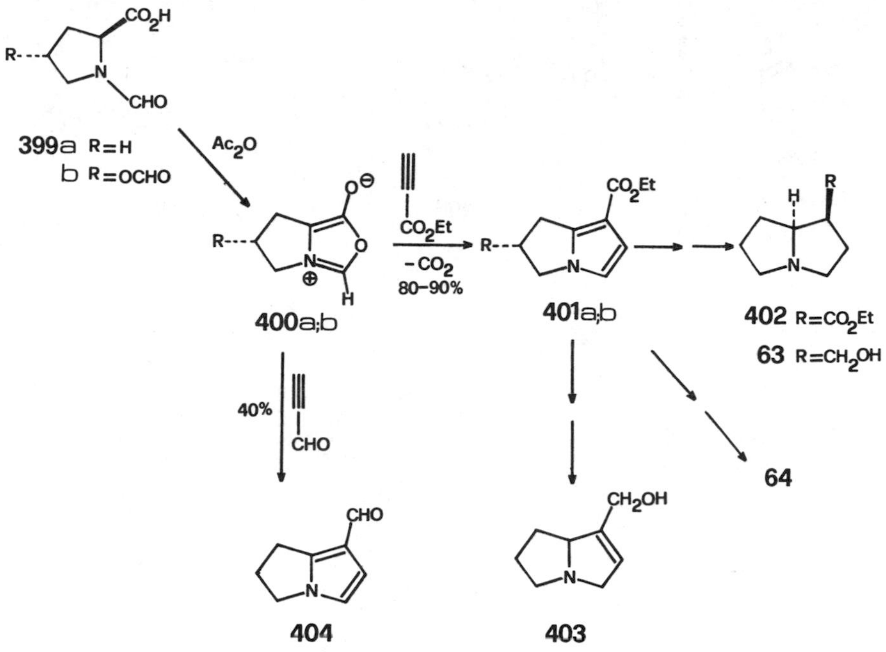

Scheme 4.18

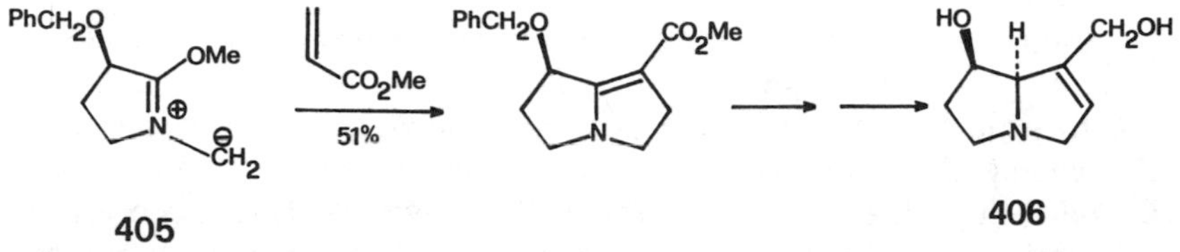

Scheme 4.19

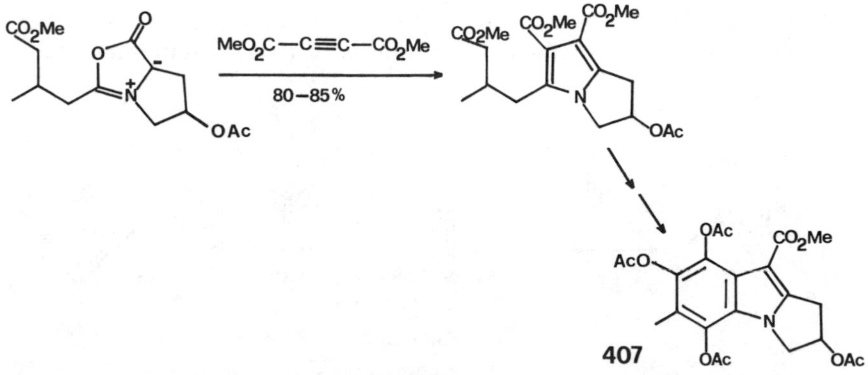

Scheme 4.20

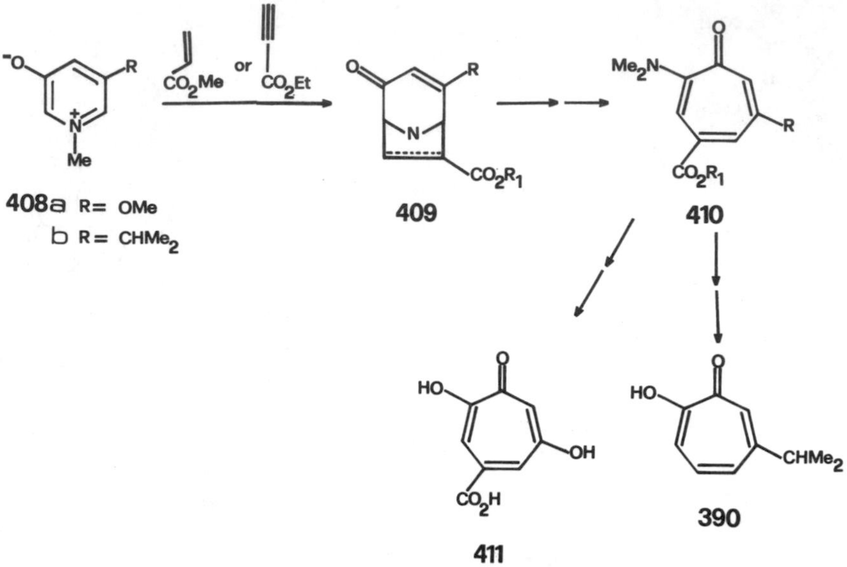

408a R= OMe
b R= CHMe2

Scheme 4.21

Frontier orbital energies indicate that this cycloaddition is $HOMO_{nitrone}/LUMO_{dipolarophile}$-controlled (*4,45*), except when dipolarophiles are electron-donating substituted and the nitrones bear an electron-attracting substituent (*46*) (Figure 4.5).

The isoxazolidine adducts are useful precursors in the synthesis of alkaloids. A recent review by Tufariello, a leader in the field (*47*), covered this topic. Therefore, we report only the structural formula of the alkaloids synthesized throughout this route: supinidine (**403**), retronecine (**406**), lupinine (**415**), epilupinine (**416**), myrtine (**417**), elaeocarpine (**418**), isoelaeocarpine (**419**), elaeokanine A (**420**), cocaine (**421**), and hydroxynicotine (**422**) (Scheme 4.23). The detailed synthetic routes are reported in the review or in the references quoted herein.

New and interesting papers appeared after publication of Tufariello's authoritative review, and we report them in more detail here.

Tufariello (*48*) described the synthesis of (±)-allosedamine (**429**),

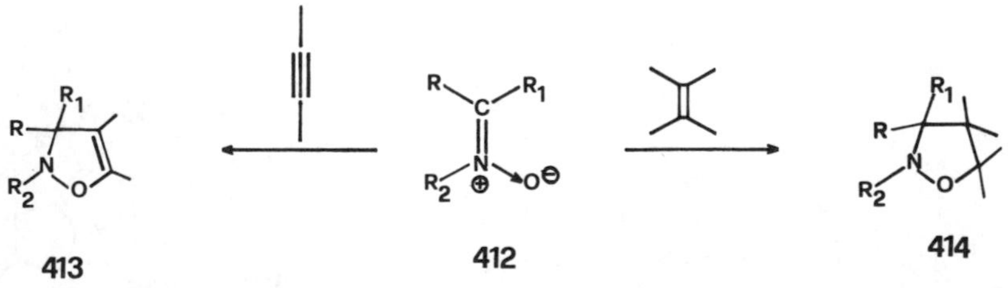

Scheme 4.22

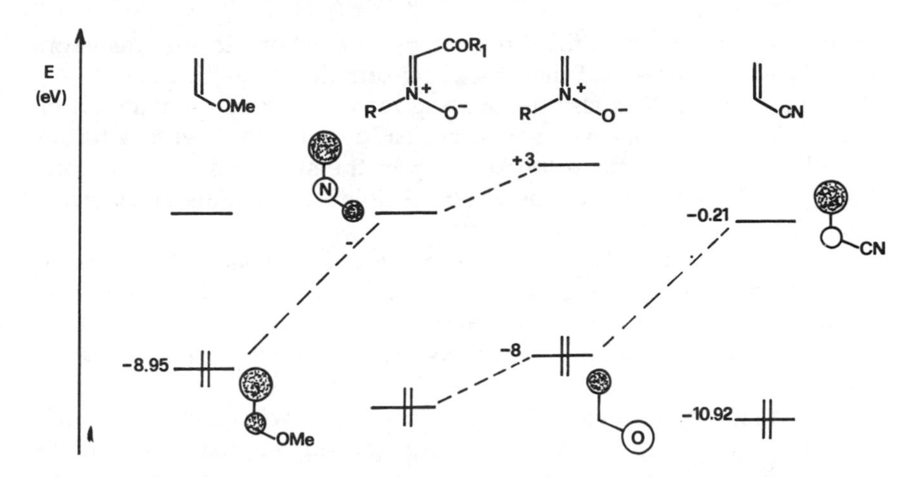

Figure 4.5. FMO interactions of nitrones with methyl vinyl ether and acrylonitrile.

(±)-sedamine (**430**), and (±)-sedridine (**428**), starting from the six-membered ring nitrone **423**, which gave 1,3-dipolar cycloaddition with propene and styrene yielding two diastereoisomers, **425** and **427**. The former resulted from an *exo*-oriented transition state (**424**) and the latter from an *endo*-oriented transition state **426**.

Activated alkenes capable of secondary nonbonding interactions (styrene and methyl acrylate as well) give a mixture of *exo*- and *endo*-adducts. The *exo*-adducts are of high yield; unactivated alkenes (e.g.,

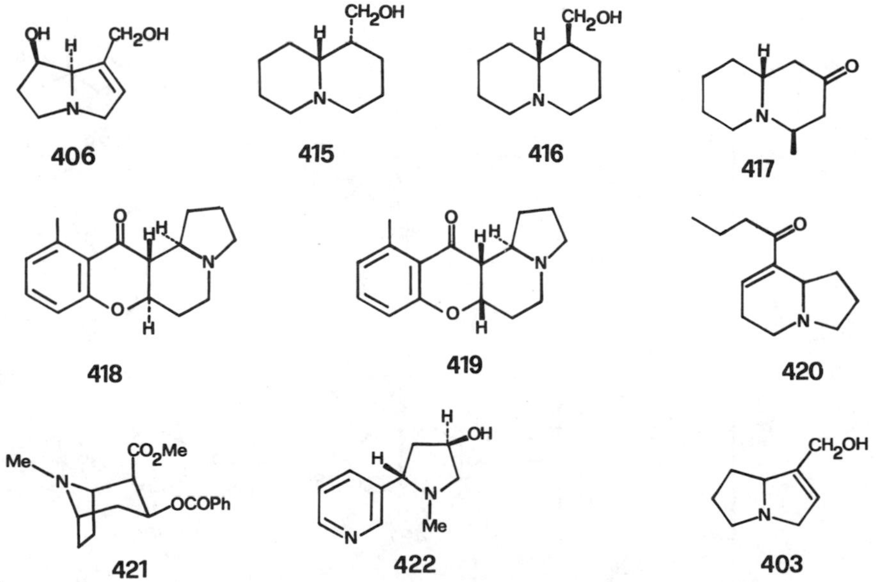

Scheme 4.23

propene) give an *exo*-adduct only. Steric interactions in the transition state explain this result. Scheme 4.24 reports the synthetic path.

The nitrone **423** is also the starting material for the preparation of a new silylated nitrone **431**, whose cycloaddition with 1-pentene to the *exo*-adduct **432** constituted the key step in the synthesis of (±)-porantheridin (**433**) (*49*), one of the main alkaloids of the Australian plant *Poranthera corymbosa* (Scheme 4.25).

Another alkaloid obtained via nitrone was (±)-septicine (**436**), an indolizine alkaloid isolated from *Ficus septica* and synthesized (*50*) by a Japanese group through a cycloaddition reaction of 1-pyrrolidine-1-oxide (**434**) with 2,3-bis(3,4-dimethoxyphenyl)butadiene (**435**) (Scheme 4.26).

A straightforward route to thienamycin (**79**) was exploited via nitrone (*51*). Thienamycin is a popular target already approached via both [2+2] cycloaddition and nitrile oxide (*see* the section in Chapter 3 entitled "Cumulated Systems" and the section in Chapter 4 entitled "Nitrile Oxides"). The isoxazolidine ring of **416** gave rise to the 1-hydroxyethyl substituent of the antibiotic ring **438** (Scheme 4.27).

Intramolecular 1,3-Dipolar Cycloadditions

Few examples of the use of intramolecular 1,3-dipolar cycloadditions to synthesize natural products have been reported in the Tufariello review (*47*) mentioned earlier or in reviews by Padwa (*52*) and Oppolzer (*53*). The latter is mainly devoted to intramolecular [4+2] cycloadditions. The favorable entropy factor promised by this approach has obviously fas-

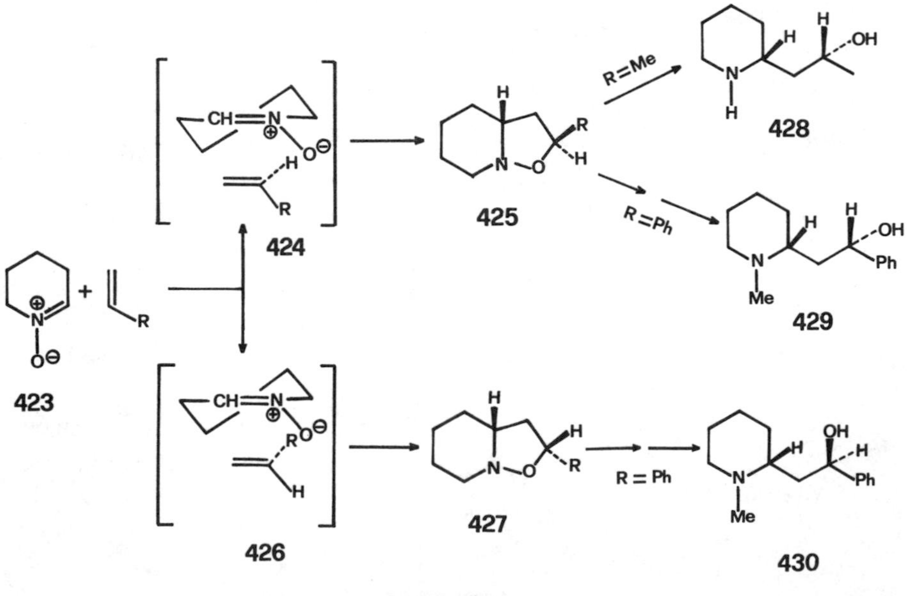

Scheme 4.24

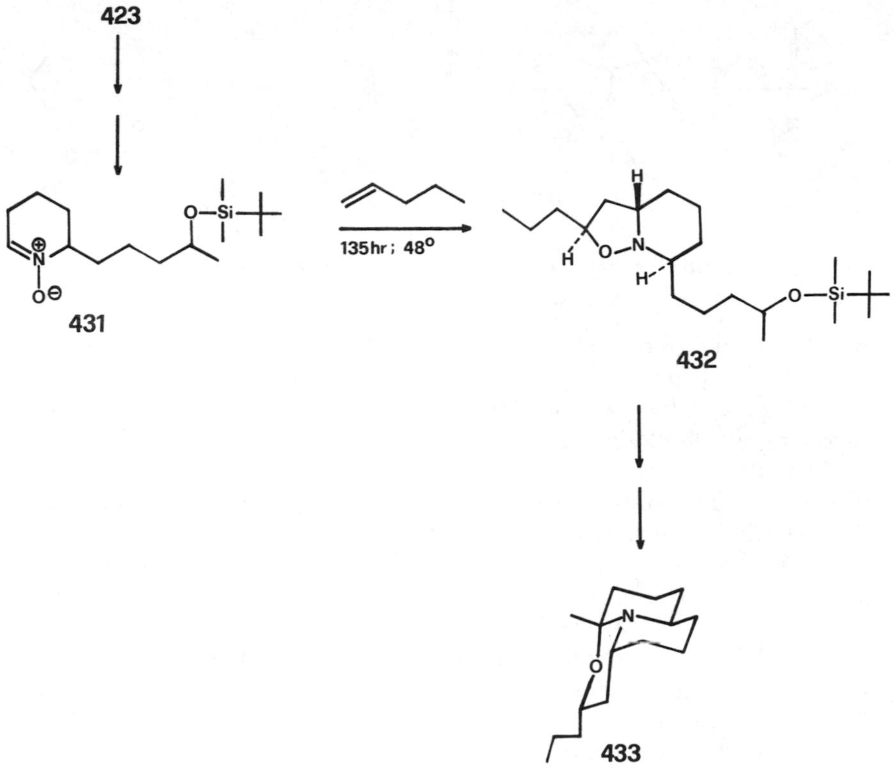

423

431

135hr ; 48°

432

433

Scheme 4.25

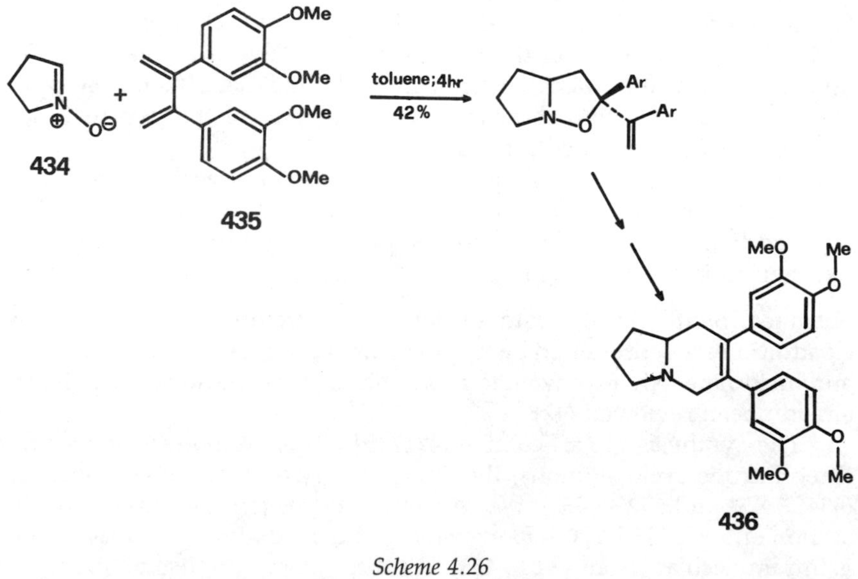

434

435

toluene;4hr
42%

436

Scheme 4.26

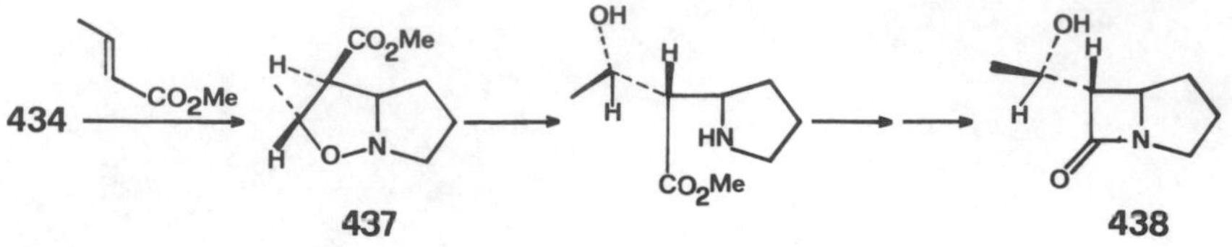

Scheme 4.27

cinated many researchers, however, as is demonstrated by its intensive application in recent years.

Nitrile Oxides. Biotin (**128**), a member of the B vitamin complex important in nutrition as a growth factor but relatively unavailable from natural sources, has been obtained through two stereospecific syntheses involving nitrile oxide cycloadditions as the key step.

The first approach (*54*) builds up the bicyclic intermediate **441** from the acyclic precursor **439** under dehydrating conditions, probably through the dinitrile oxide **440**, which undergoes dimerization to the furoxane. Four further steps give **128** in 30% yield from **441**. The second approach (*55,56*) offers some advantages: the starting material (cycloheptene) is readily available and three chiral centers develop in the cycloaddition stage from **442** to **444** (Scheme 4.28), the nitrile oxide **443** being generated from the nitro group using phenyl isocyanate at room temperature in the presence of a trace of triethylamine, following Mukaiyama and Hoshino's procedure (*57*). By using the same procedure, **445** was converted to **446** in the key step of a synthesis of the ergot alkaloid chanoclavine I (*58*) (**447**) (Scheme 4.29).

Diazoalkanes. We have described, in the section on "Pseudo [$\sigma 2 + \sigma 2$] Cycloreversion" in Chapter 3, the synthesis of (−)-cyclocopacamphene (**342**) (*59*) by photolysis of the pyrazoline **341**. The intramolecular 1,3-dipolar cycloaddition of the intermediate olefinic diazoalkane **340** proceeded easily, probably because of the proximity of the two reacting groups in a fairly strained system.

The synthesis of the sesquiterpene (+)-cyclosativene **450** was accomplished by using a very similar route (*60*) that involved intramolecular addition of the diazoalkane **448** followed by photoelimination of nitrogen from the pyrazoline **449** (Scheme 4.30).

Nitrones. Synthetic routes to alkaloids via intramolecular nitrone cycloaddition are represented by two such outstanding examples—cocaine and luciduline—that we want to resummarize them although they have already been reviewed (*47*).

The synthesis of (±)-cocaine (**421**) (*61,62*) is a unique sequence of three nitrone cycloadditions, the first two of which are intermolecular (**434** → **451** and **452** → **453**). We will return to this point in this chapter's section entitled "[3+2] Cycloreversion." The third nitrone cycloaddition is intramolecular (from **454** to **455**). All three stages are useful and flex-

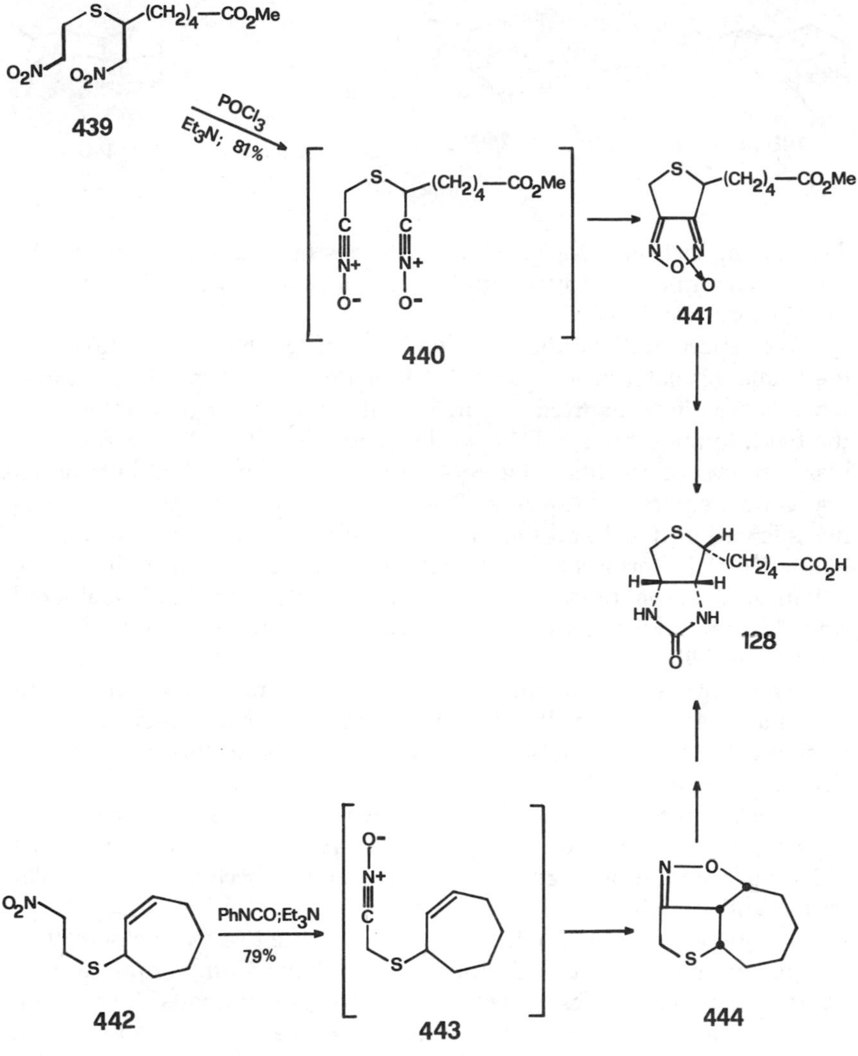

439

$POCl_3$

Et_3N; 81%

440

441

128

442

$PhNCO; Et_3N$

79%

443

444

Scheme 4.28

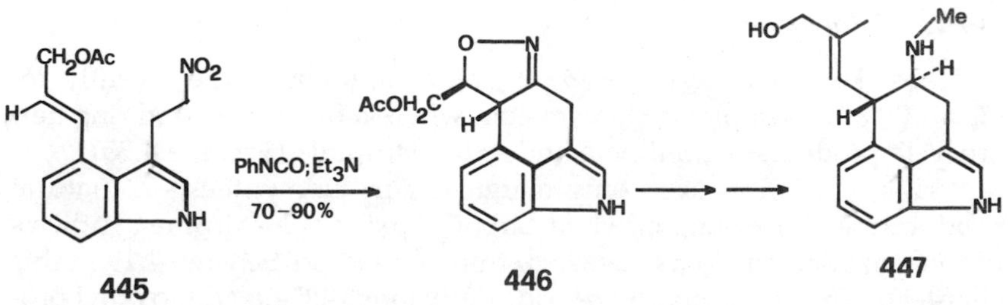

445

$PhNCO; Et_3N$

70–90%

446

447

Scheme 4.29

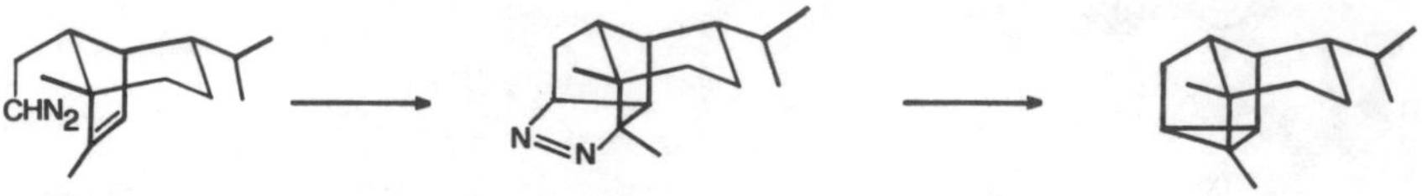

448 449 450

Scheme 4.30

ible. In the last stage, a substrate with a single chiral center can be transformed into an adduct with four chiral centers strategically placed for **421** (Scheme 4.31).

We report in detail the total synthesis of (±)-luciduline (**461**) (*63*), the alkaloid isolated from *Lycopodium lucidulum,* in Scheme 4.32 because when it was first reported (*64*) in 1976 it aroused our own interest in the field. In the catalyzed Diels–Alder sequence **456** → **457**, intramolecular nitrone cycloaddition **458** → **460a** generates an adduct bearing all five chiral centers of luciduline. The nature of the transition state was investigated in detail and the predominance of **459a** over **459b** was explained (*65*) in terms of the formation of the C–C bond being more advanced than that of the C–O bond. Hence **460a** offers a six-membered ring closure entropically favored over seven-membered ring formation **460b** from **459b**.

The syntheses of chanoclavine I (**447**), isochanoclavine I (**465**), and α-bisabolol (**469**) are novelties in the field. (The synthesis via nitrile oxide of chanoclavine is mentioned in the earlier section of this chapter entitled "Nitrile Oxides.")

The Oppolzer synthesis (*66*) of the two alkaloids isolated from *Claviceps purpurea* involves a key step of **462** to **464** via the transient nitrone **463**, which undergoes a regiospecific and stereospecific intramolecular cycloaddition (Scheme 4.33).

Japanese workers (*67*) described a new route to α-bisabolol (**469**), a sesquiterpene present in various plants of the Compositae family, starting from (6Z)-farnesal (**466**). The nitrone **467**, prepared by conventional methods, undergoes cycloaddition to give the isoxazolidine **468**, which is then transformed into the desired target in the correct configuration (Scheme 4.34).

[3+2] Cycloreversion

The best known class of [5→3+2] fragmentations is the recently reviewed (*68*) 1,3-dipolar cycloreversion, where a five-membered ring heterocycle produces a 1,3-dipole and a dipolarophile (Scheme 4.35).

This reaction is used only marginally in the synthesis of natural products, as, for example, when carbon dioxide is lost in the synthesis of pyrrolizidine alkaloids via azomethine ylides (*see* Scheme 4.19 in this chapter). When oxazolinium-5-oxide derivatives **400a,b** add to ethyl propiolate, the tricyclic adduct **470a,b** undergoes 1,3-dipolar cyclore-

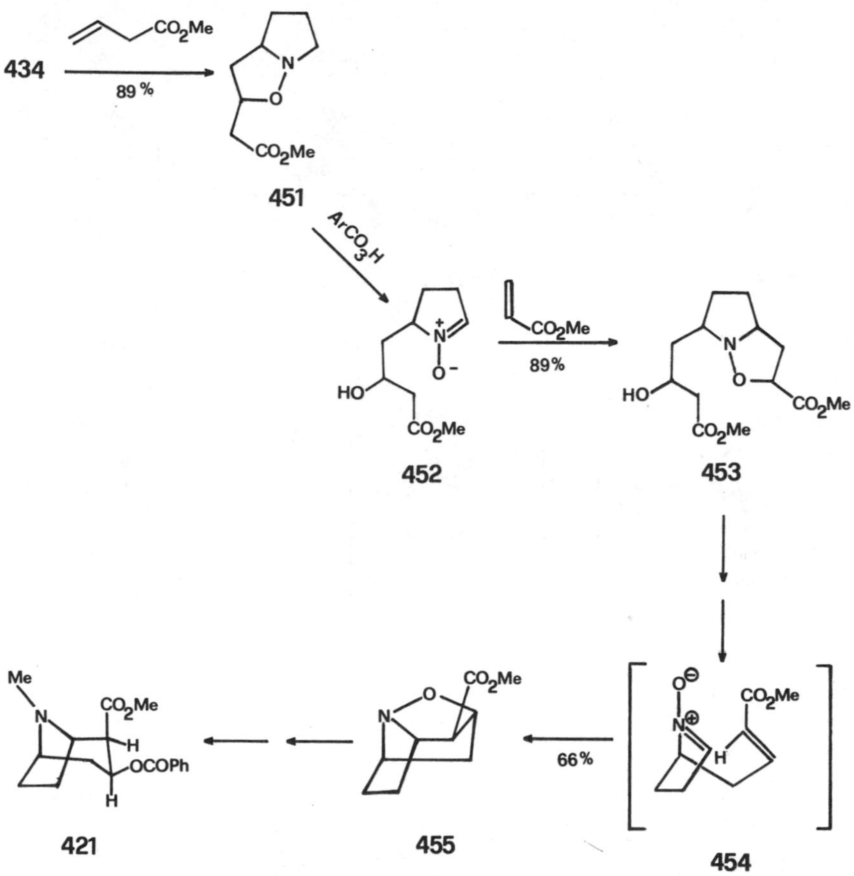

Scheme 4.31

version; carbon dioxide is the dipolarophile and the azomethine ylide system **471a,b** gives **401a,b** through electron reorganization (Scheme 4.36).

The ozonolysis of double bonds also involves a 1,3-dipolar cycloaddition to produce 1,2,3-trioxolane (**472**). In accordance with the modified Criegee mechanism (*69*), the initial cycloadduct suffers 1,3-dipolar cycloreversion, yielding a carbonyl oxide (**473**) and a carbonyl compound whose 1,3-dipolar cycloaddition finally produces the 1,2,4-trioxolane **474** (Scheme 4.37).

The best application of 1,3-dipolar cycloreversion to natural products synthesis is in Tufariello's synthesis of (±)-cocaine (**421**) (*61,62*) (*see* Scheme 4.31 in the section of this chapter entitled "Nitrones"). We have already mentioned the sequence of three nitrone cycloadditions that features the synthetic strategy. The second, which involves adding methyl acrylate to the hydroxy ester nitrone **452** to give **453**, protects the nitrone moiety during the subsequent reaction with methanesulfonyl

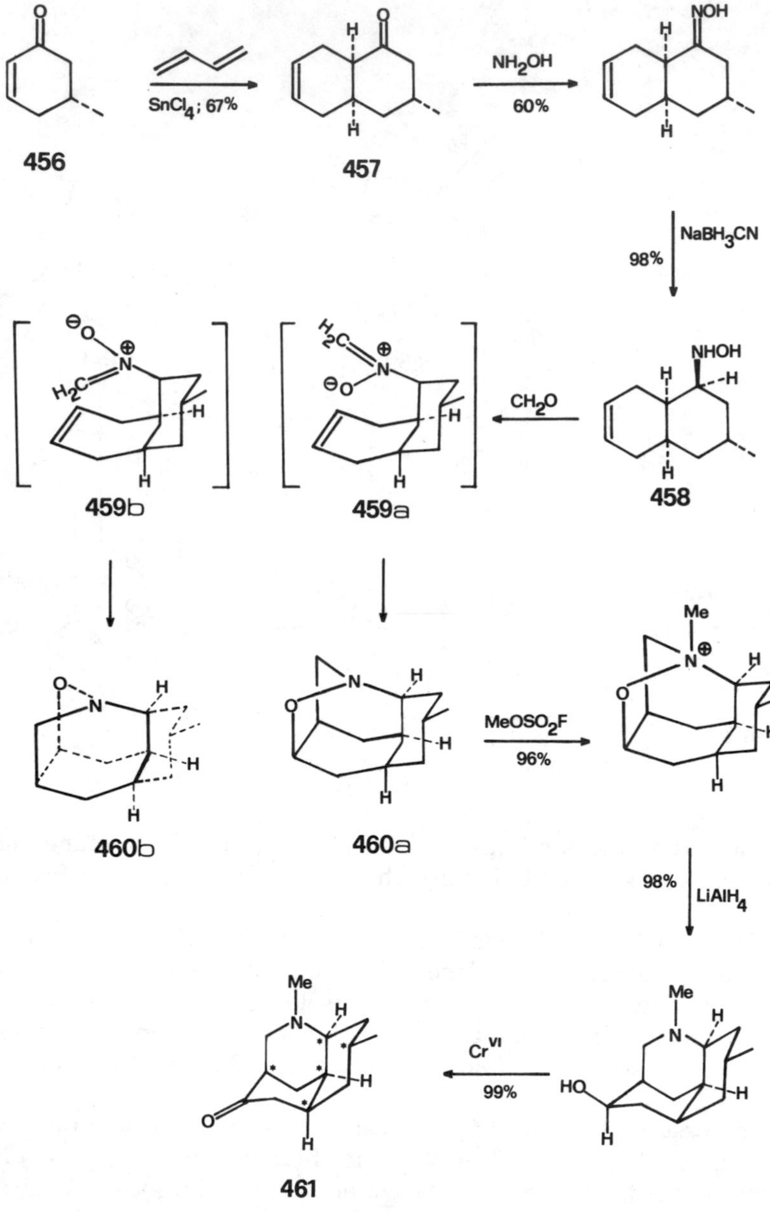
456
SnCl4; 67%
457
NH2OH
60%
NOH
98%
NaBH3CN
NHOH
458
CH2O
459b
459a
460b
460a
MeOSO2F
96%
Me
98%
LiAlH4
Cr^VI
99%
HO
461

Scheme 4.32

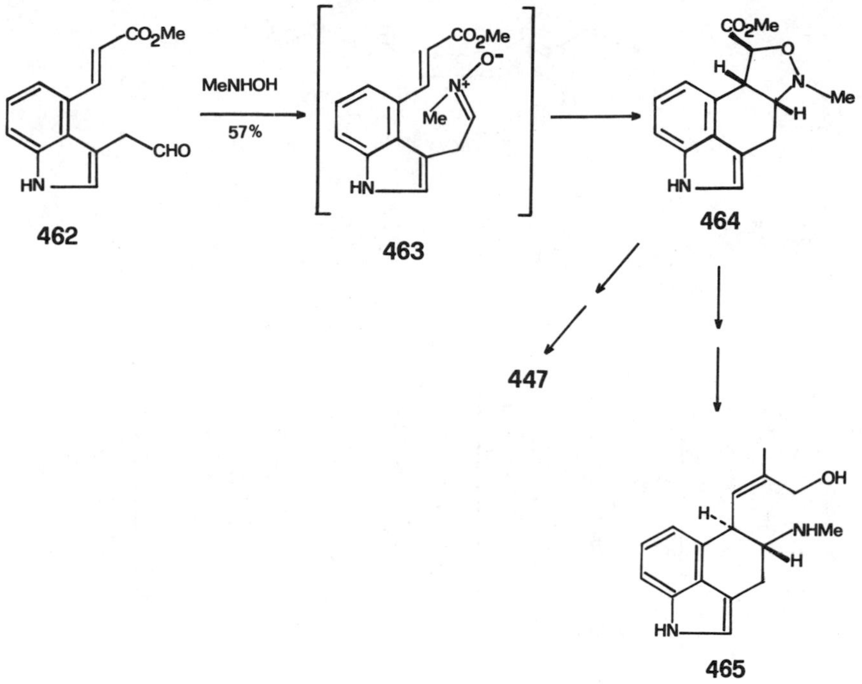

MeNHOH
57%
462
463
464
447
465

Scheme 4.33

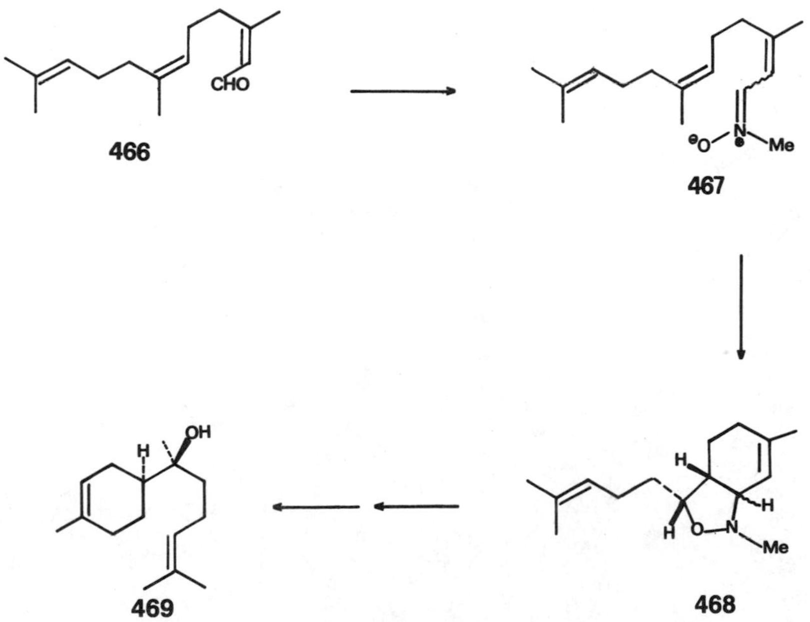

466
467
468
469

Scheme 4.34

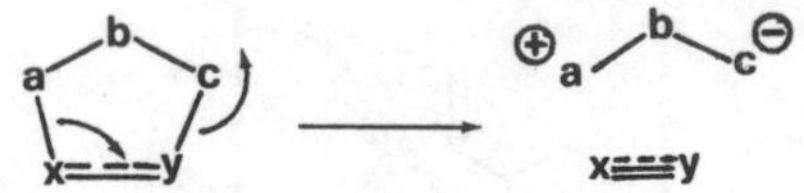

Scheme 4.35

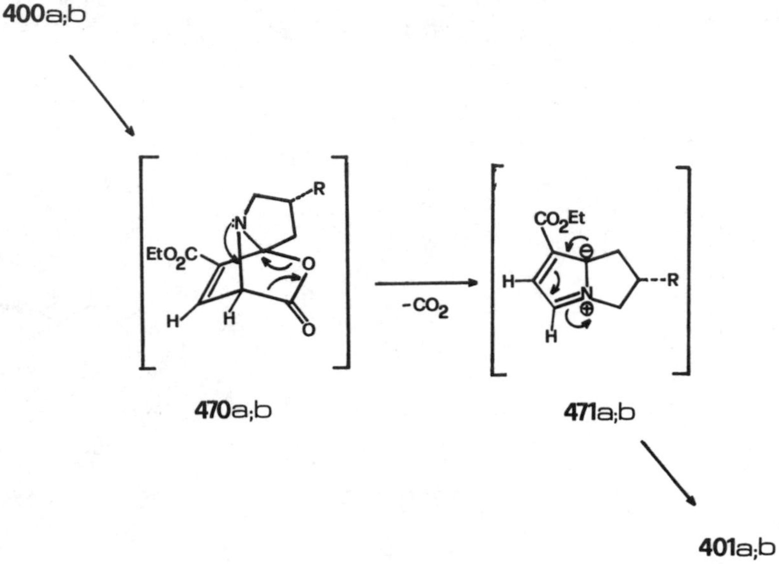

400a;b

470a;b

$-CO_2$

471a;b

401a;b

Scheme 4.36

472

473

474

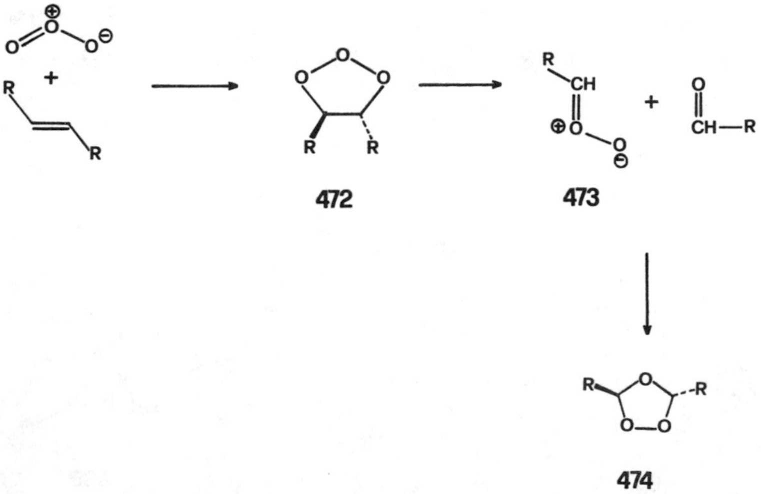

Scheme 4.37

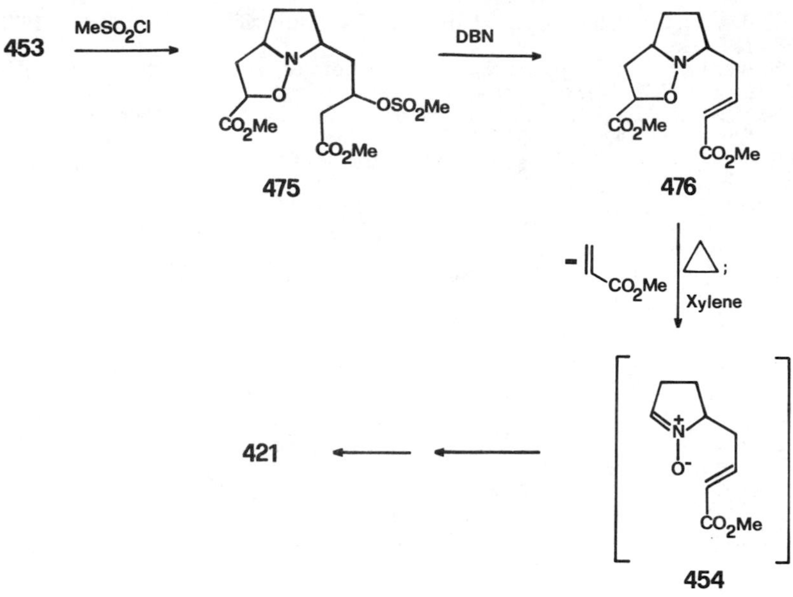

Scheme 4.38

chloride and production of the corresponding methane sulfonate **475**. Upon β-elimination, the methane sulfonate affords the α,β-unsaturated ester isoxazolidine **476** that, refluxed in xylene, undergoes 1,3-dipolar cycloreversion to methyl acrylate and the nitrone **454** (Scheme 4.38), a sequence that provides an unusual example of protection and regeneration of a 1,3-dipole.

Literature Cited

1. Huisgen, R. *Angew. Chem., Int. Ed. Engl.* **1963**, *2*, 565.
2. Sustmann, R.; Trill, H. *Angew. Chem., Int. Ed. Engl.* **1962**, *11*, 838.
3. Sustmann, R. *Pure Appl. Chem.* **1974**, *40*, 569.
4. Houk, K. N.; Sims, J.; Watts, C. R.; Luskus, L. J. *J. Am. Chem. Soc.* **1973**, *95*, 7301.
5. Houk, K. N. "Theory of Reactive Intermediates and Reaction Mechanisms," in "Reactive Intermediates"; Jones, M., Jr.; Moss, R. A., Eds.; Wiley-Interscience: New York, 1978, p. 279.
6. Houk, K. N. *Top. Curr. Chem.* **1979**, *79*, 1.
7. Stevens, R. V. *Tetrahedron* **1976**, *32*, 1599.
8. Cornforth, J. W. Unpublished results reported in Jackson, A. H.; Smith, K. M. "The Total Synthesis of Natural Products"; Vol. 1; Apsimon, J., Ed.; Wiley-Interscience: New York, 1973; p. 143.
9. Traverso, G.; Barco, A.; Pollini, G. P.; Anastasia, M.; Pirillo, D. *Il Farmaco Ed. Sci.* **1969**, *24*, 946.
10. Stevens, R. V.; Kaplan, M. *J. Chem. Soc., Chem. Commun.* **1970**, 822.
11. Stevens, R. V.; Christiansen, C. G.; Edmonson, W. L.; Kaplan, M.; Reid, E. B.; Wentland, M. P.; *J. Am. Chem. Soc.* **1971**, *93*, 6629.
12. Stevens, R. V.; Dupree, L. E., Jr.; Edmonson, W. L.; Magid, L. L.; Wentland, M. P. *J. Am. Chem. Soc.* **1971**, *93*, 6637.
13. Traverso, G.; Barco, A.; Pollini, G. P. *J. Chem. Soc., Chem. Commun.* **1971**, 926.

14. Traverso, G.; Pollini, G. P.; Barco, A.; De Giuli, G. *Gazz. Chim. Ital.* **1972**, *102*, 243.
15. Stevens, R. V.; Fitzpatrick, J. M.; Germeraad, P. B.; Harrison, B. L.; La Palme, R. *J. Am. Chem. Soc.* **1976**, *98*, 6313.
16. Stevens, R. V.; Cherpeck, R. E.; Harrison, B. L.; Lai, J.; La Palme, R. *J. Am. Chem. Soc.* **1976**, *98*, 6317.
17. Stevens, R. V.; Christensen, C. G.; Cory, R. M.; Thorsett, E. *J. Am. Chem. Soc.* **1975**, *97*, 5940.
18. Stork, G.; Danishefsky, S.; Ohashi, M. *J. Am. Chem. Soc.* **1967**, *89*, 5459.
19. Ohashi, M.; Kamachi, H.; Kakisawa, H.; Stork, G. *J. Am. Chem. Soc.* **1967**, *89*, 5460.
20. Stork, G.; McMurry, J. E. *J. Am. Chem. Soc.* **1967**, *89*, 5461, 5463, 5464.
21. Baldwin, J. E.; Hoskins, C.; Kruse, L. *J. Chem. Soc., Chem. Commun.* **1976**, 795.
22. Hagedorn, A. A., III; Miller, B. J.; Nagy, J. O. *Tetrahedron Lett.* **1980**, 229.
23. Bianchi, G.; De Amici, M. *J. Chem. Soc., Chem. Commun.* **1978**, 968.
24. Burri, K. F.; Cardone, R. A.; Chen, W. Y.; Rosen, P. *J. Am. Chem. Soc.* **1978**, *100*, 7069.
25. Kametani, T.; Huang, S. P.; Yokohama, S.; Suzuki, Y.; Ihara, M. *J. Am. Chem. Soc.* **1980**, *102*, 2060.
26. Barco, A.; Benetti, S.; Pollini, G. P.; Baraldi, P. G.; Simoni, D.; Guarneri, M.; Gandolfi, G. *J. Org. Chem.* **1980**, *45*, 3141.
27. Berger, C.; Franck-Neumann, M.; Ourisson, G. *Tetrahedron Lett.* **1968**, 3451.
28. Greenlee, W. J.; Woodward, R. B. *J. Am. Chem. Soc.* **1976**, *98*, 6075.
29. Wilson, S. R.; Turner, R. B. *J. Org. Chem.* **1973**, *38*, 2870.
30. Schneider, M. P.; Goldbach, M. *J. Am. Chem. Soc.* **1980**, *102*, 6114.
31. Franck-Neumann, M.; Brion, F.; Martina, D. *Tetrahedron Lett.* **1978**, 5033.
32. Franck-Neumann, M.; Martina, D. *Tetrahedron Lett.* **1975**, 1755.
33. Mukherjee, D.; Watts, C. R.; Houk, K. N. *J. Org. Chem.* **1978**, *43*, 817.
34. Kuroda, H.; Kunii, T. *Theor. Chim. Acta* **1977**, *7*, 220.
35. Houk, K. N.; Strozier, R. W. *J. Am. Chem. Soc.* **1973**, *95*, 4094.
36. Franck-Neumann, M.; Martina, D. *Tetrahedron Lett.* **1975**, 1759.
37. Knorr, R.; Huisgen, R.; Staudinger, G. K. *Chem. Ber.* **1970**, *103*, 2639.
38. Pizzorno, M. T.; Albonico, S. M. *J. Org. Chem.* **1974**, *39*, 731.
39. Robins, D. J.; Sakdarat, S. *J. Chem. Soc., Chem. Commun.* **1979**, 1181.
40. Pizzorno, M. T.; Albonico, S. M. *Chem. Ind. (London)* **1978**, 349.
41. Vedejs, E.; Martinez, G. R. *J. Am. Chem. Soc.* **1980**, *102*, 7994.
42. Rebek, J., Jr.; Gehret, J. C. E. *Tetrahedron Lett.* **1977**, 3027.
43. Katritzky, A. R.; Takeuchi, Y. *J. Am. Chem. Soc.* **1970**, *92*, 4134.
44. Tamura, Y.; Saito, T.; Kiyokawa, H.; Chen, L.; Ishibashi, H. *Tetrahedron Lett.* **1977**, 4075.
45. Houk, K. N. *Acc. Chem. Res.* **1975**, *8*, 361.
46. Commisso, R.; Desimoni, G.; Righetti, P. P.; Tacconi, G.; Torresani, M. *Abst. Congr. Heterocycl. Chem., 0th*, **1981**, 301.
47. Tufariello, J. J. *Acc. Chem. Res.* **1979**, *12*, 396.
48. Tufariello, J. J.; Asrof Ali, S. *Tetrahedron Lett.* **1978**, 4647.
49. Gössinger, E. *Tetrahedron Lett.* **1980**, 2229.
50. Iwashita, T.; Suzuki, M.; Kusumi, T.; Kakisawa, H. *Chem. Lett.* **1980**, 383.
51. Tufariello, J. J.; Lee, G. E.; Senaratne, P. A.; Al-Nuri, M. *Tetrahedron Lett.* **1979**, 4359.
52. Padwa, A. *Angew. Chem., Int. Ed. Engl.* **1976**, *15*, 123.
53. Oppolzer, W. *Angew. Chem., Int. Ed. Engl.* **1977**, *16*, 10.
54. Marx, M.; Marti, F.; Reisdorff, J.; Sandmeier, R.; Clark, S. *J. Am. Chem. Soc.* **1977**, *99*, 6754.
55. Confalone, P. N.; Lollar, E. D.; Pizzolato, G.; Uskokovic, M. R. *J. Am. Chem. Soc.* **1978**, *100*, 6291.
56. Confalone, P. N.; Pizzolato, G.; Confalone, D. L.; Uskokovic, M. R. *J. Am. Chem. Soc.* **1980**, *102*, 1954.
57. Mukaiyama, T.; Hoshino, T. *J. Am. Chem. Soc.* **1960**, *82*, 5339.
58. Kozikowski, A. P.; Ishida, H. *J. Am. Chem. Soc.* **1980**, *102*, 4265.
59. Piers, E.; Britton, R. W.; Keziere, R. J.; Smillie, R. D. *Can. J. Chem.* **1971**, *49*, 2623.

60. Piers, E.; Geraghty, M. B.; Soucy, M. *Synth. Commun.* **1973**, *3*, 401.
61. Tufariello, J. J.; Mullen, G. B. *J. Am. Chem. Soc.* **1978**, *100*, 3638.
62. Tufariello, J. J.; Mullen, G. B.; Tegeler, J. J.; Trybulski, E. J.; Chun Wong, S.; Asrof Ali, S. *J. Am. Chem. Soc.* **1979**, *101*, 2435.
63. Oppolzer, W.; Petrzilka, M. *Helv. Chim. Acta* **1978**, *61*, 2755.
64. Oppolzer, W.; Petrzilka, M. *J. Am. Chem. Soc.* **1976**, *98*, 6722.
65. Oppolzer, W.; Siles, S.; Snowden, R. L.; Bakker, B. H.; Petrzilka, M. *Tetrahedron Lett.* **1979**, 4391.
66. Oppolzer, W.; Gayson, J. I. *Helv. Chim. Acta* **1980**, *63*, 1706.
67. Iwashita, T.; Kusumi, T.; Kakisawa, H. *Chem. Lett.* **1979**, 947.
68. Bianchi, G.; De Micheli, C.; Gandolfi, R. *Angew. Chem., Int. Ed. Engl.* **1979**, *18*, 721.
69. Criegee, R. *Angew. Chem., Int. Ed. Engl.* **1975**, *14*, 745.

5

[4+2] Cycloadditions

The [4+2] cycloaddition is the best way to generate the six-membered rings so common in natural products. As a consequence, there are many papers in the field to review. The most popular reaction, the Diels–Alder reaction, will be the focus of our discussion.

Intermolecular Diels–Alder Reaction

Because some of the legendary syntheses of the 20th century have involved strategic approaches based on the Diels–Alder reaction, they provide the best introduction to the topic. Much of the philosophy of organic synthesis was born in papers written about these syntheses.

R. B. Woodward first realized that "the synthesis of substances occurring in Nature provides a measure of the condition and the power of the science," (*1*) and his synthetic plans are prophetic.

In 1952 Woodward approached the task of synthesizing naturally occurring steroids (*2*). The first step was a Diels–Alder reaction between butadiene and 2-methoxy-5-methylbenzoquinone to form the suitably substituted A and B rings. From **477**, 19 steps gave **478**, a crucial intermediate to progesterone; desoxycorticosterone; testosterone; androsterone; cortisone (**479**), the key member of the class; and cholesterol (**480**), a target with eight chiral centers (Scheme 5.1).

In the same field Stork (*3*) used a Diels–Alder approach in the synthesis of the key compound **481**, the cyclization of which leads to the 8-methyl-1-hydrindanone system **482** with the correct steroid stereochemistry of rings C and D (Scheme 5.2).

In 1956 the first total synthesis of morphine (**486**) was completed by Gates and Tschudi (*4*). The formation of ring III was accomplished by a Diels–Alder reaction between the 1,2-naphthoquinone derivative **483** and butadiene to give **484**. This latter compound contains a double bond—a convenient site of operations and substituents—suitably placed to give in a single step the correct configuration for the piperidine ring of **485** (Scheme 5.3).

In 1958 reserpine (**288b**) was synthesized by the Woodward group

0065-7719/83/0180-0119$31.00/1

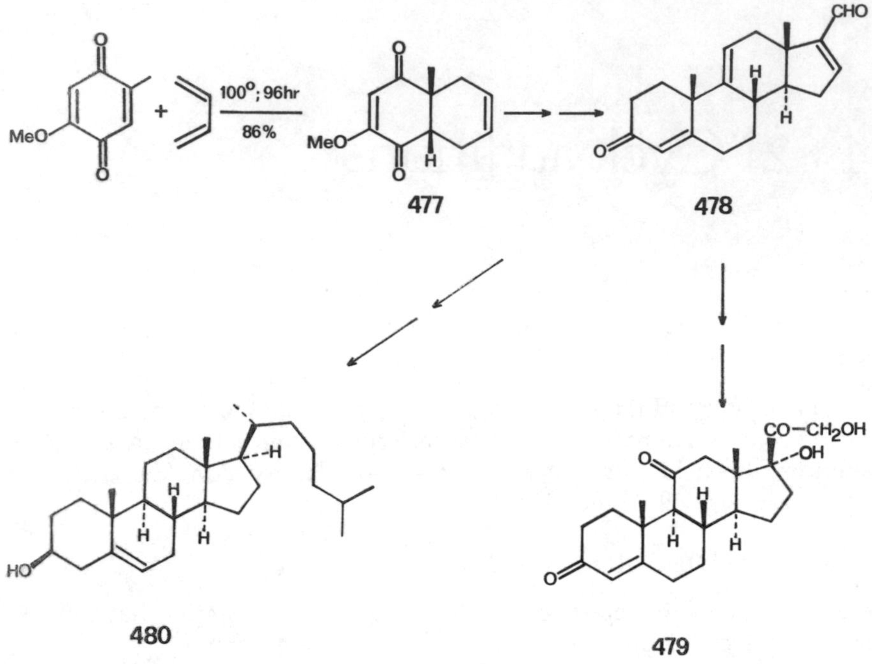

MeO
100°; 96hr
86%
MeO
H
477
CHO
H
H
O
478
H
H
H
HO
480
CO–CH2OH
OH
O
H
H
H
O
479

Scheme 5.1

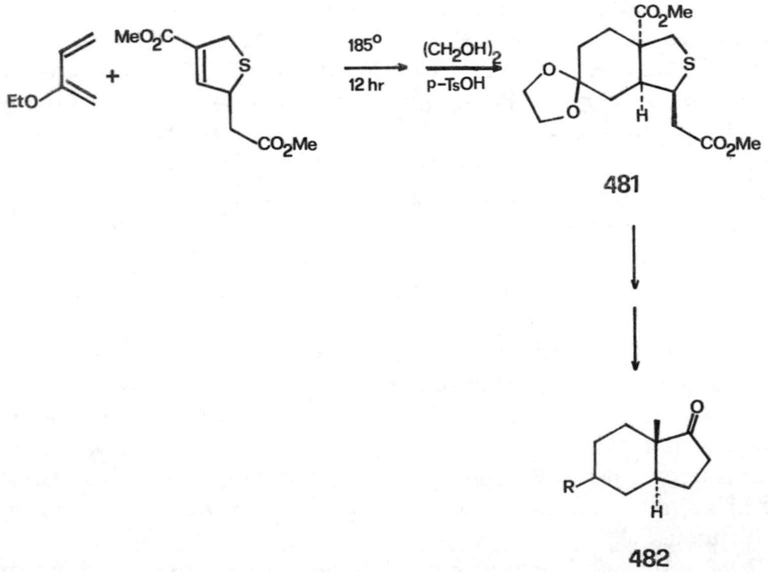

MeO2C
S
EtO
CO2Me
185°
12 hr
(CH2OH)2
p-TsOH
CO2Me
S
O
O
H
CO2Me
481
O
R
H
482

Scheme 5.2

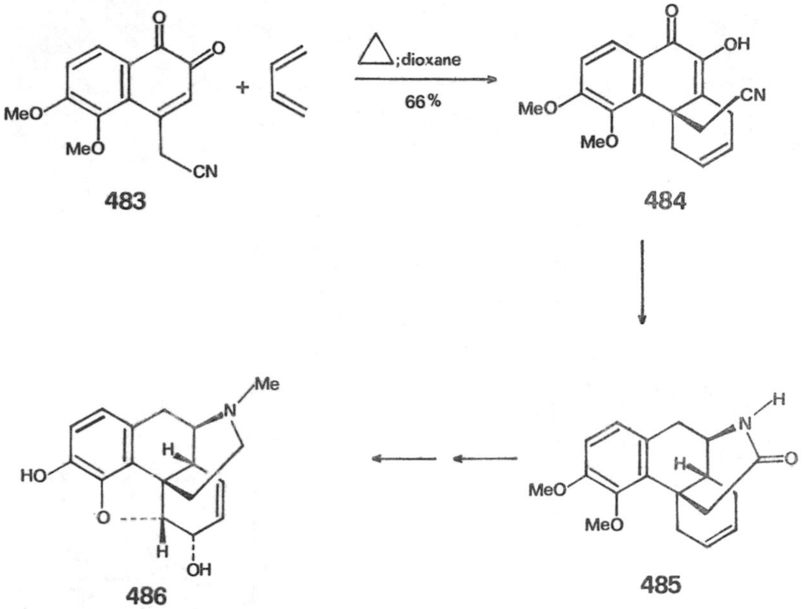

Scheme 5.3

at Harvard (*5*). The well-known stereospecificity of the Diels–Alder reaction suggested that the first step be the reaction of vinylacrylic acid with quinone, resulting in the adduct **487**. The "six-membered ring . . . was destined to become ring E of reserpine. . . . [I]t contained already a felicitously placed carbonyl group, a double bond of good augury for the introduction of oxygen atoms at appropriate positions, and three asymmetric carbon atoms properly oriented"—just what was needed for the key intermediate **488**. (Scheme 5.4).

The interest in reserpine is still high. In 1980, 22 years after Woodward's synthesis and 1 year after Pearlman's (*6*), a third route involving a Diels–Alder approach was explored (*7*). Rings D and E of **288b** were secured by the reaction of 2-acetoxyacrylate with methyl-1,2-dihydropyridine-1-carboxylate. The major stereoisomer **490** was converted to the *cis*-hydroisoquinoline **491**, which played the same role as **488** in the Woodward approach. Its reaction with 6-methoxytryptophyl bromide gave rise to an intermediate similar to **489**, which was then easily transformed into **288b** (Scheme 5.5).

We have already mentioned colchicine (**139**) in the section in Chapter 3 on "Photochemical [2+2] Cycloadditions." This alkaloid of *Colchicum autunnale* L. is used in the treatment of gout and as a mitosis inhibitor because of its photoconversion to α-lumicolchicine. The first synthesis of colchicine was performed in 1961 by the Eschenmoser group (*8*); the key step was the Diels–Alder addition of chloromethylmaleic anhydride (**493**) to the α-pyrone **492**. The resulting product **494**

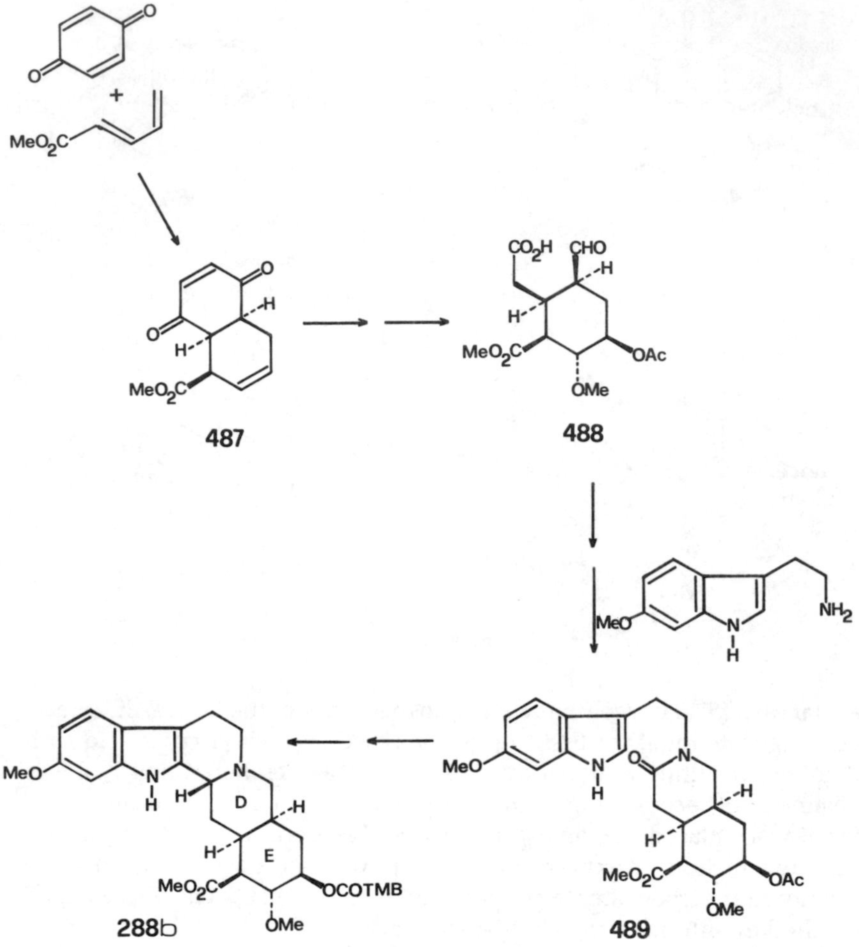

MeO$_2$C

487

488

289

489

288b

Scheme 5.4

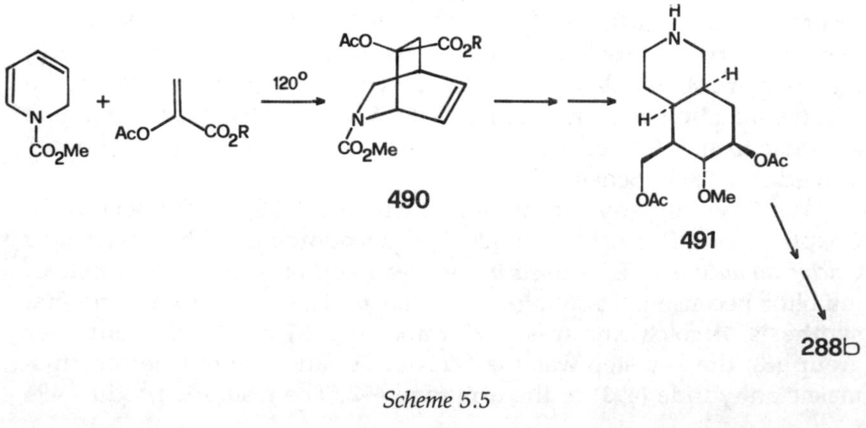

120°

490

491

288b

Scheme 5.5

derives from a [4+2] cycloreversion (*see* the section on "Retro-Diels–Alder Reactions" in Chapter 5) of the original adduct through a spontaneous loss of carbon dioxide, which can be regarded as the dienophile expelled in a retro-Diels–Alder reaction. In the further steps, a third pericyclic reaction occurs, namely an electrocyclic ring opening of the norcaradiene derivative **495** (*see* the section on "6π Electrocyclic Reactions" in Chapter 9), to give the cycloheptatriene dimethyl ester **496**. This compound can be converted in a few steps to colchicine (**139**). (Scheme 5.6).

Five years before the Woodward–Hoffmann rules were revealed, a synthesis involving three pericyclic steps demonstrated how much knowledge was at that time ready to be developed.

The Diels–Alder reaction is not only a suitable tool for multiple-step synthesis of complex structures, but also an elegant way to synthesize uncomplicated products otherwise obtainable only through long and complicated efforts. The recent simple and efficient total synthesis of cantharidin (**499**) (*9*), the active principle of *Cantharis vesicatoria*, performed in two steps from furan and 2,5-dihydrothiophene-3,4-dicarboxylic anhydride (**497**) in 85% yield, amply demonstrates this state-

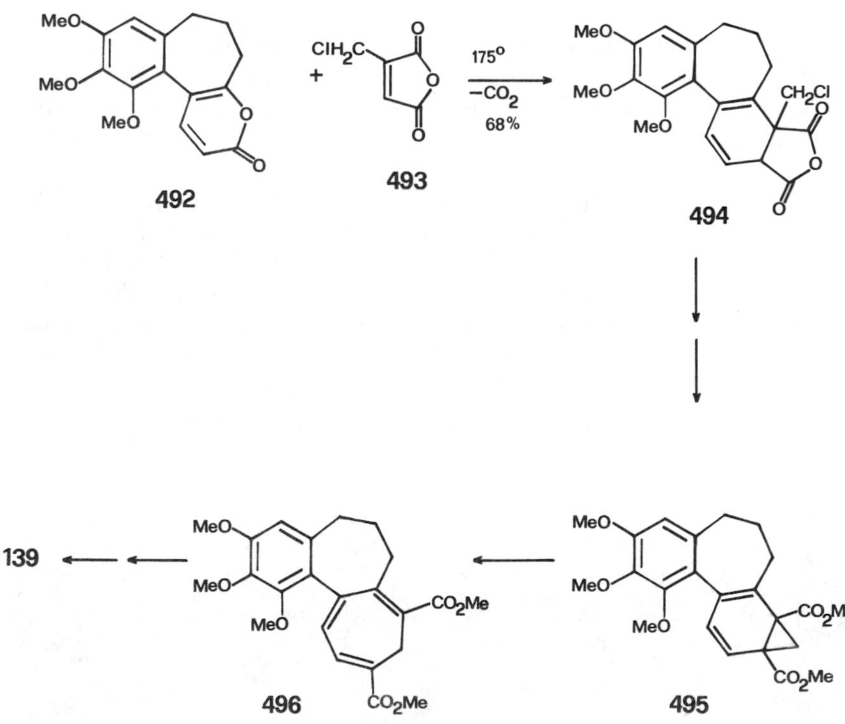

Scheme 5.6

ment. The *exo* adduct **498** is generally preferred in the high-pressure Diels–Alder reaction (Scheme 5.7).

Two recent syntheses of the bicyclic intermediate **502**, a key compound in the route to prostaglandin $F_{2\alpha}$ (see the section entitled "Intramolecular Additions" in Chapter 2) are noteworthy. Prostaglandin $F_{2\alpha}$ was one of the significant targets of the seventies. The 6-acetoxyfulvene (**500**) underwent Diels–Alder reaction with 2-chloroacrylonitrile to give 7-acetoxymethylene-2-chlorobicyclo[2.2.1]hept-5-ene-2-carbonitrile (**501**), which was further converted to **502** (R = CHO) (*10*). The cyclopentadienyl derivative **503** reacted with dihaloketene to give **504**, which is easily transformed to **502** (R = CH_2OMe) (*11*) (Scheme 5.8).

Diels–Alder Cycloadducts as Starting Materials. Several syntheses of natural products involve in the early stages a Diels–Alder reaction addressed to the construction of the starting materials only.

We report these references for the sake of completeness, but because the beauty of the syntheses lies in their further stages, we have not shown them schematically. Only the structures of the natural compounds synthesized are reported, with the parts having their origins in the Diels–Alder reaction shown in the drawings.

We have assembled the syntheses according to the nature of the dienophile, affording a simple key to understanding the synthetic pathways.

All the Diels–Alder reactions reviewed in this section occur with normal electron demand (*12, 13*), and the dienophiles bear electron-attracting substituents. Several terpenes such as α-atlantone (**505**) (*14*), stachenone (**506**) (*15*), nootkatone (**507**) (*16*), and δ-damascone (**508**) (*17*) were synthesized from acyclic olefins (Scheme 5.9).

Fulvoplumierin (**509**) was synthesized from the cumulated olefin penta-2,3-dienoate (*18*). Dimethyl acetylenedicarboxylate was the dienophile used in the route to fujenoic acid (**510**) (*19, 20*) and in the syntheses (*21*) of three arylnaphthalide lignins: justicidin B (**511a**), taiwanin C (**511b**), and dehydroanhydropicropodophyllin (**511c**), in addition to the stereospecific total synthesis of warburganal (**512**) (*22*) and isodeoxypodophyllotoxin (**513**) (*23*) (Scheme 5.10).

Cyclopentenone was the dienophile in the synthesis of (±)-coronafacic acid (**514**) (*24*), whereas lactones or lactams were used in the

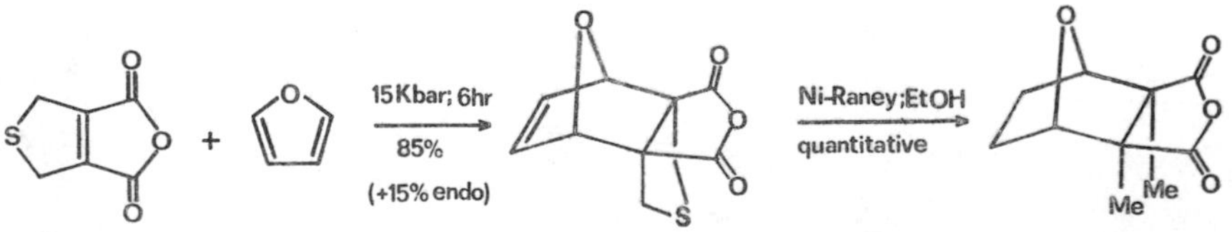

497 + furan → (15 Kbar; 6 hr, 85%, (+15% endo)) → 498 → (Ni-Raney; EtOH, quantitative) → 499

Scheme 5.7

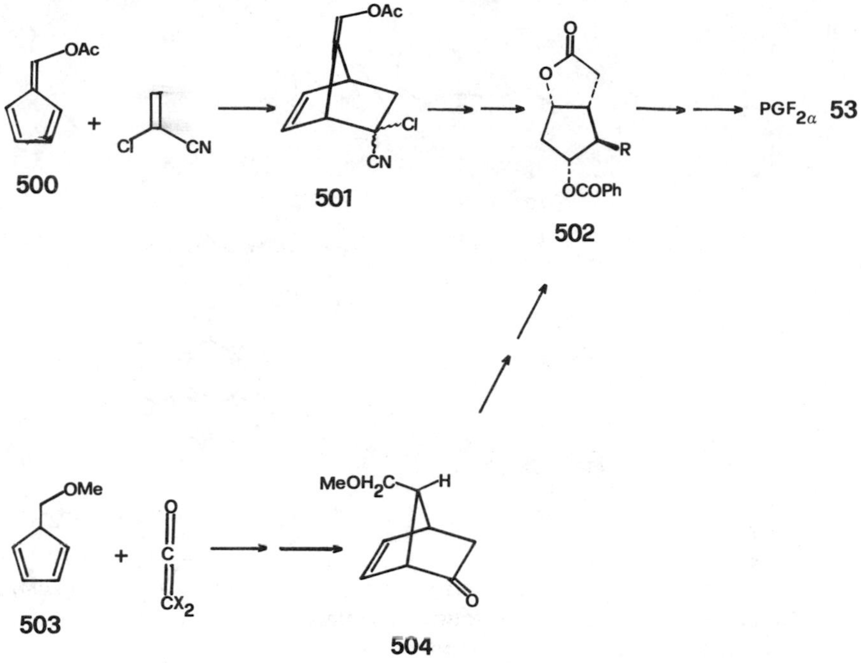

OAc
+
Cl CN
500
OAc
Cl
CN
501
O
O
R
OCOPh
502
$PGF_{2\alpha}$ 53
OMe
+
O
C
CX_2
503
$MeOH_2C$ H
O
504

Scheme 5.8

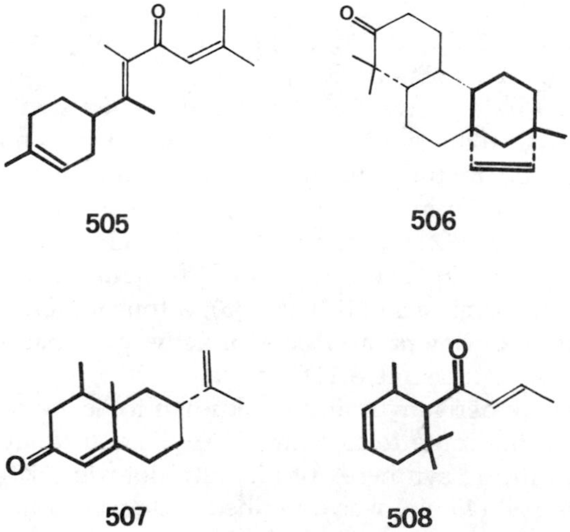

O
505
O
506
O
507
O
508

Scheme 5.9

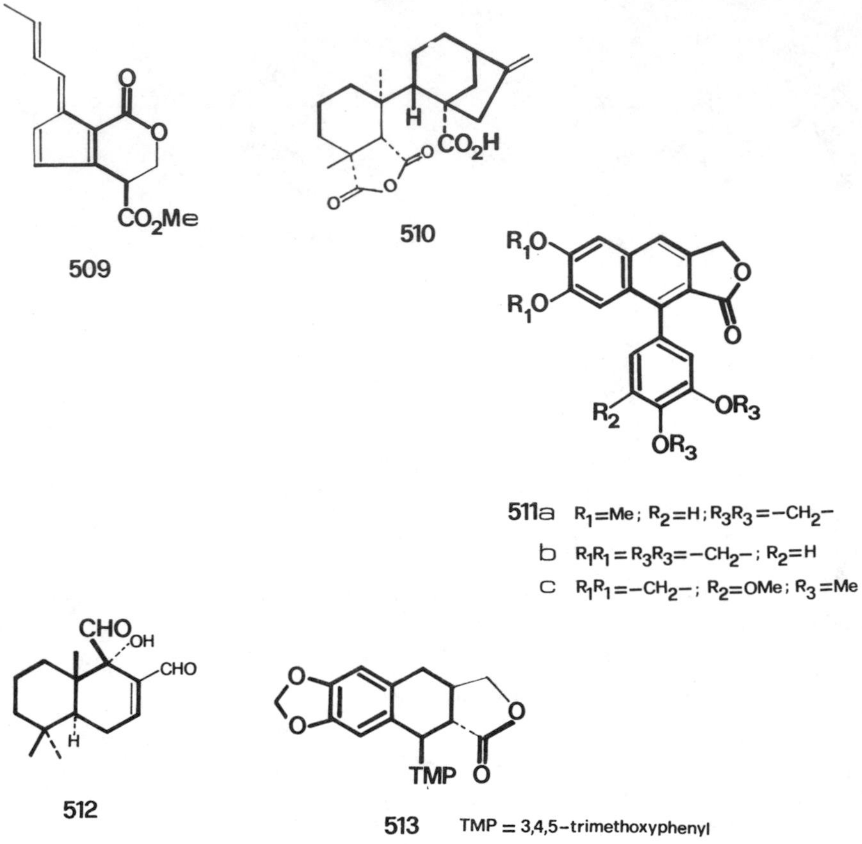

509

510

511a R_1=Me; R_2=H; R_3R_3=$-CH_2-$
b R_1R_1 = R_3R_3=$-CH_2-$; R_2=H
c R_1R_1=$-CH_2-$; R_2=OMe; R_3=Me

512

513 TMP = 3,4,5-trimethoxyphenyl

Scheme 5.10

route to verrucarol (**515**) (*25, 26*), (±)-menthol (**516**) (*27*), and haemanthamine (**517**) (*28, 29*) (Scheme 5.11).

Maleic anhydride derivatives are known to be useful dienophiles, and their wide use in the synthesis of natural products is not surprising. They are used to clarify the structure of albene (**518**) (*30, 31*); in the route to clerodane (**519**) (*32*); in the total synthesis of dihydrolycorine (**520**) (*33*), haemanthidine (**521**) (*34*), tazettine (**522**) (*34*), methyl elenolate (**523**) (*35*), and palitantin (**524**) (*36*); and in the synthesis of isomarasmic acid (**339**) and marasmic acid (**525**) (*37, 38*), a fungal metabolite isolated from basidiomycetes having antibacterial activity, in particular against *Staphylococcus aureus* (Scheme 5.12).

Benzoquinone derivatives also are known to be good dienophiles, and their use in the route to siccanin (**526**) (*39*) and to ajugarin I (**527**) (*40*), as well as in the syntheses of (±)-tetrodotoxin (**528**) (*41–44*) and (±)-serratinine (**89**) (*45, 46*) was described (Scheme 5.13).

In the following sections the Diels–Alder reaction plays a more relevant role in the synthetic design than we have seen thus far. We report

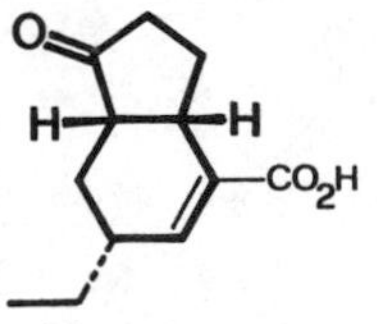

514 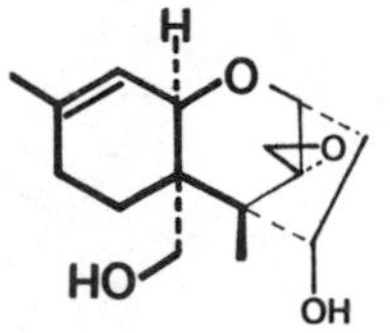 515

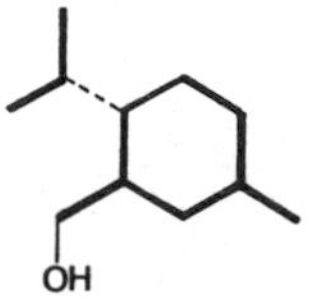

516 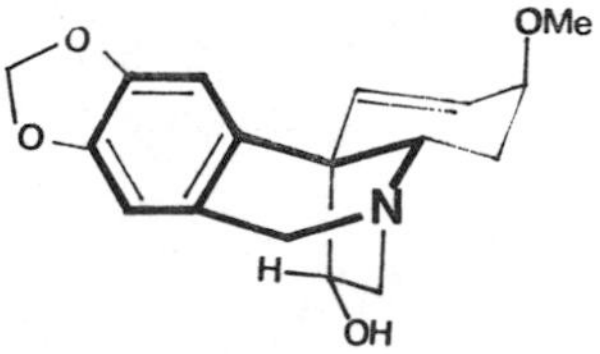 517

Scheme 5.11

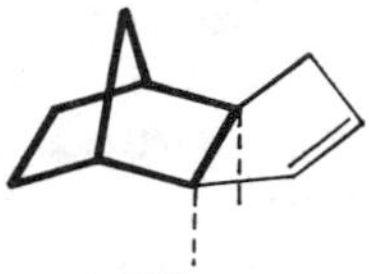

518 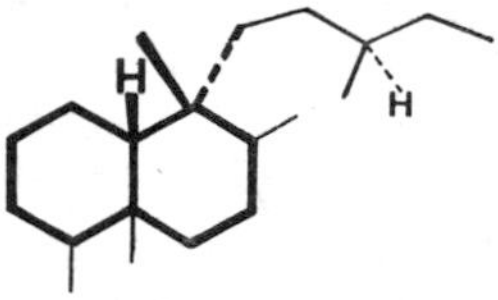 519 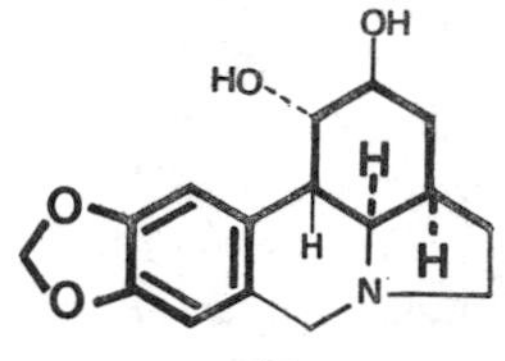 520

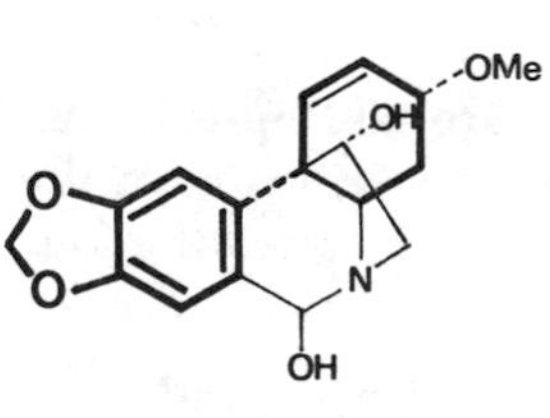

521 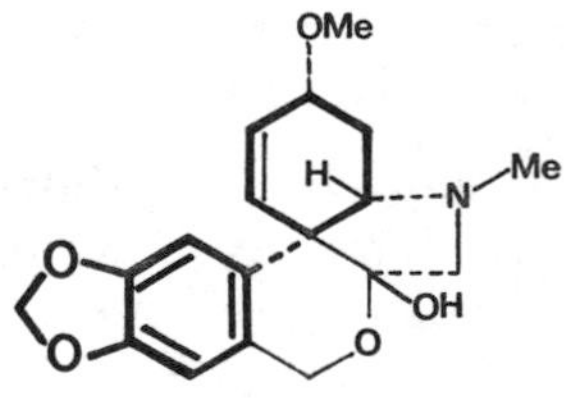 522 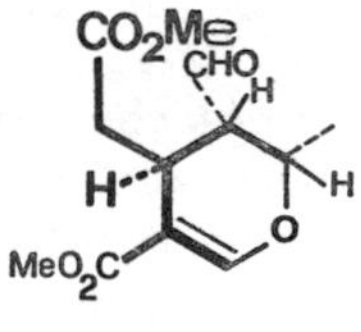 523

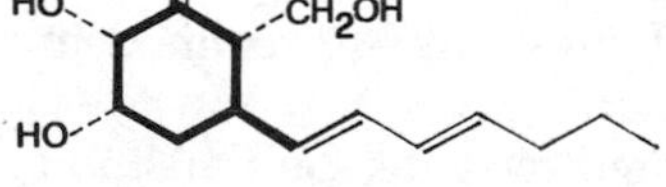

524 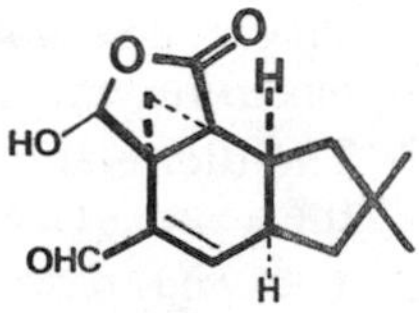 525

Scheme 5.12

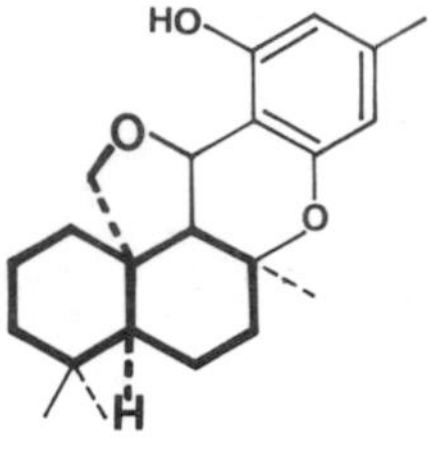

Scheme 5.13

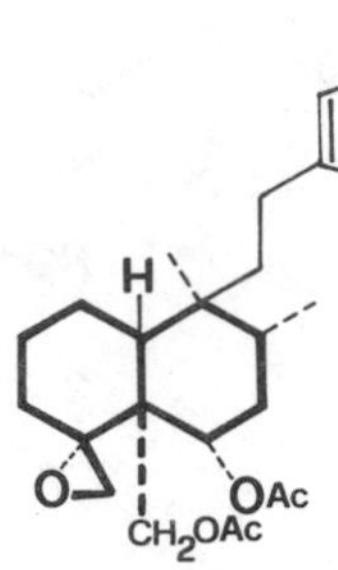

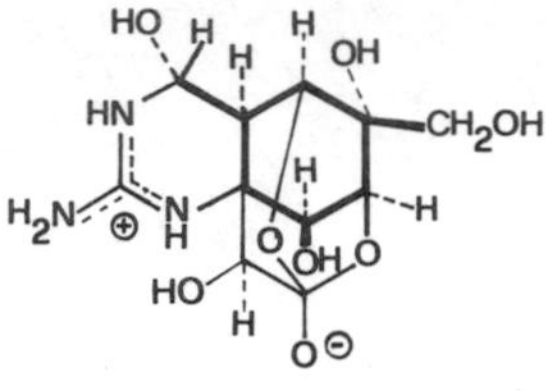

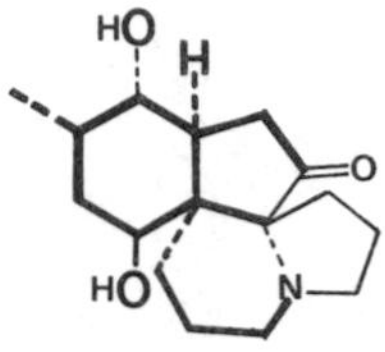

two short preparations as an introduction. One preparation involves the synthesis of solanoquinone (**531**) (*47*), a constituent of tobacco, from neophytadiene (**529**) and 2,3-dimethylbenzoquinone (Scheme 5.14). The other involves synthesis of the aglycone of the antibiotic chartreusin (**535**) (*48*) from juglone (**532**) and a styrene derivative **533** (Scheme 5.15). Both adducts **530** and **534** have the comprehensive framework of the natural products. In particular, the second synthesis provides a good example of the way fantasy and a good strategic view of the synthesis can make a difficult route easy (Scheme 5.15).

Regiochemistry, Stereochemistry, and Molecular-Orbital Implications. One of the best applications of Woodward–Hoffmann rules is in the field of the Diels–Alder reaction. MO interactions have clarified the four well-known rules (*12, 13*).

1. Electron-attracting substituents (Z) on dienophiles and electron-donating ones (X) on dienes increase the rate of the reaction (Diels–Alder with direct electron demand). The reverse substituent effects (Diels–Alder with inverse electron demand) likewise increase the rate.
2. The diene and dienophile configurations are retained in the adduct (the *cis* principle).
3. The *endo* transition state is favored over the *exo* transition state (the *endo* rule).
4. Z-Substituted dienophiles react with 1-substituted butadienes in Diels–Alder with normal electron demand to

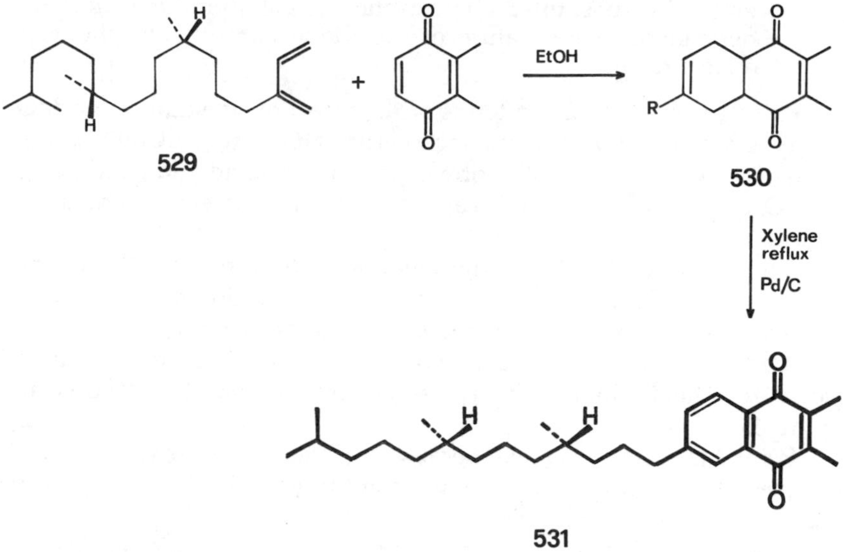

529 + EtOH → 530 (R) → Xylene reflux, Pd/C → 531

Scheme 5.14

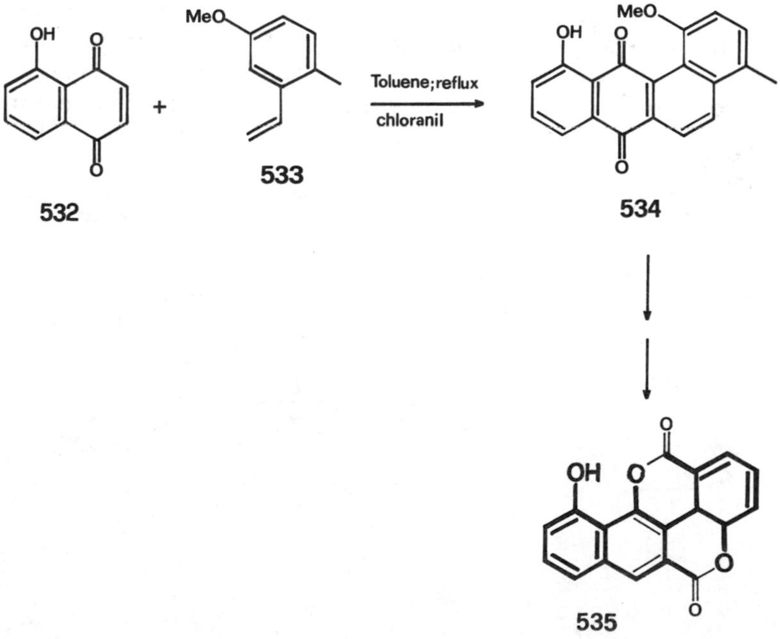

532 + 533 (MeO) → Toluene; reflux, chloranil → 534 (OH, MeO) → → 535 (OH)

Scheme 5.15

give 3,4-disubstituted cyclohexenes as main products, independent of the nature of the diene substituents (the *ortho* effect).

Point 1 is rationalized in terms of the effect of substituents on MO energy levels. The dominant interactions are $HOMO_{diene}/LUMO_{dienophile}$ in Diels–Alder reactions with normal electron demand and $LUMO_{diene}/HOMO_{dienophile}$ in Diels–Alder reactions with inverse electron demand (Figure 5.1).

The lower the LUMO of the dienophile (lowered by Z) and the higher the HOMO of the diene (raised by X), the closer are the interacting levels and the smaller is the denominator of Eq. 2, which gives a higher energy gain in the cycloaddition (*49–51*). The same situation occurs for Diels–Alder with inverse electron demand, as shown in Figure 4.4.

Point 2 is easily explained: Because the Diels–Alder reaction occurs in a [$4\pi s$ + $2\pi s$] way under thermal conditions, there is no reason to lose or invert the configuration.

Point 3 is rationalized in terms of secondary nonbonding interactions if the dienophile has additional orbitals that can be involved. For example, furan reacts with maleimide in ether at 30 °C to give a 95% yield of the *endo*-adduct **537** (*52*), whose transition state **536** is stabilized by the secondary overlap of the bonding type that occurs between the HOMO of the furan and the LUMO of maleimide (Scheme 5.16).

Point 4 was first successfully explained simply in terms of "hard and soft" centers in diene and dienophile (*53, 54*). The larger terminal coefficients on each cycloaddend become bonded preferentially in the transition state (*55*); this is true for either electron-attracting or electron-

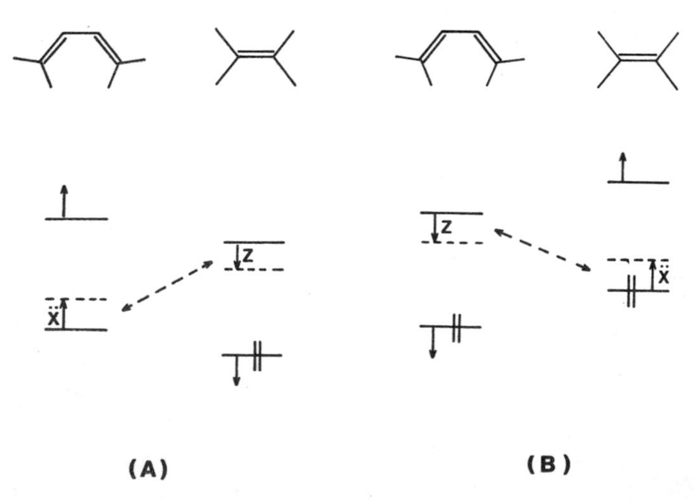

Figure 5.1. MO interactions in normal (A) and inverse (B) Diels–Alder reactions with the substituent effect. Key: Z, electron acceptors; and X, electron donors.

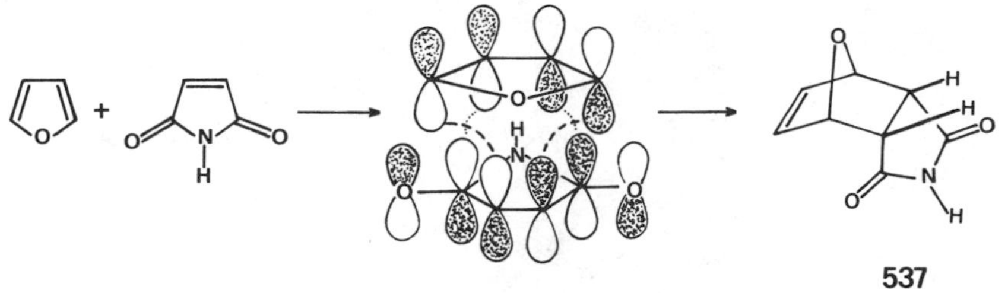

536

537

Scheme 5.16

donating substituents (*56*). The predicted preferred regioisomer in *Z*- or *X*-monosubstituted dienes and dienophiles is reported in Scheme 5.17 (largely taken from Houk's works). The experimental results are consistent with prediction (*53, 54, 56*). Tegmo-Larsson et al. (*57*) recently suggested a new explanation for the regiochemistry of Diels–Alder reaction between unsymmetrical electron-rich dienes and electron-rich alkenes; this explanation involves interactions among substituents.

By using these simple models, the regiochemical and stereochemical behavior of several Diels–Alder cycloadditions along the route to natural products can be easily understood. The synthesis of the two stereoisomeric β- and α-lycoranes (*58*) (**551** and **552**, respectively) begins with the Diels–Alder reaction of β-nitrostyrene (**546**) and the 1-substituted butadienes **126** and **547** (Scheme 5.18). The stereochemistry and the regiochemistry of the resulting adducts **548** and **549** are comparable to **538** in Scheme 5.18.

Similarly, *trans*-1-*N*-acylamino-1,3-butadiene (*59*) (**553**) reacted with crotonaldehyde to afford **554**, which is converted in two steps into (±)-pumiliotoxin (**555**) (*60*), or with ethyl atropate to produce **556**, easily transformed in the analgesic tilidine (**557**) (*61*) (Scheme 5.19). Researchers have (*62*) described the synthesis of the alkaloid perhydrogephyrotoxin (**558**) starting from **553** (R = OCH_2Ph) and replacing crotonaldehyde with *trans*-4-benzyloxy-2-butenal. Because an interesting rearrangement is involved, this topic is detailed in Chapter 7 in the section entitled "Polar [3,3] Rearrangements."

The triene **559** can be regarded analogously because its HOMO has the same symmetry and coefficients' amplitude at C4 as an electron-donating substituted butadiene. It reacts with α-bromoacrolein to give **560**, transformed in two steps into (±)-fumagillin (**561**) (*63*) (Scheme 5.20).

A similar explanation can be proposed for the addition of nitroethylene to **562**, which leads to **563** in the route to (−)-aspidofractinine (**564**) (*64*). The latter compound is a very important alkaloid, constituting the fundamental skeleton of more than 20 members of the pleiocarpine series (Scheme 5.21).

The reaction between 1-methoxycyclohexa-1,3-diene and *trans*-6-

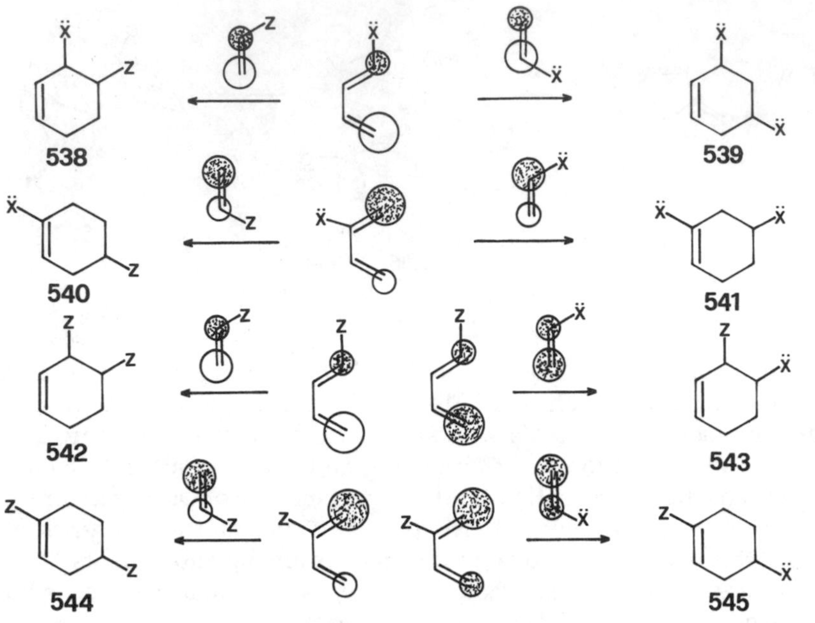

538 539 540 541 542 543 544 545

Scheme 5.17

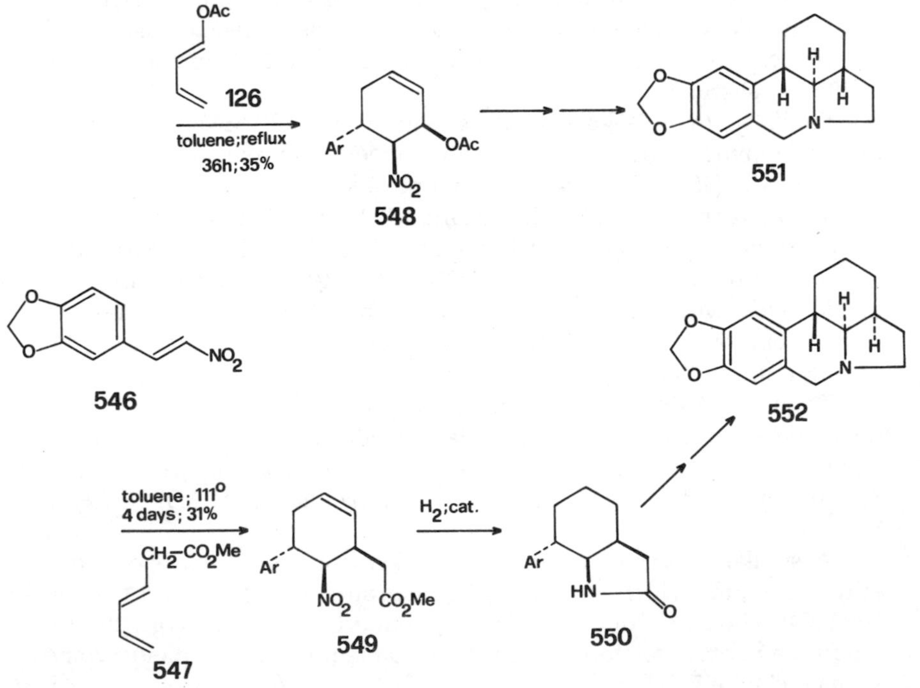

126

toluene;reflux 36h;35%

548

551

546

toluene; 111° 4 days; 31%

CH_2-CO_2Me

547

549

H_2;cat.

550

552

Scheme 5.18

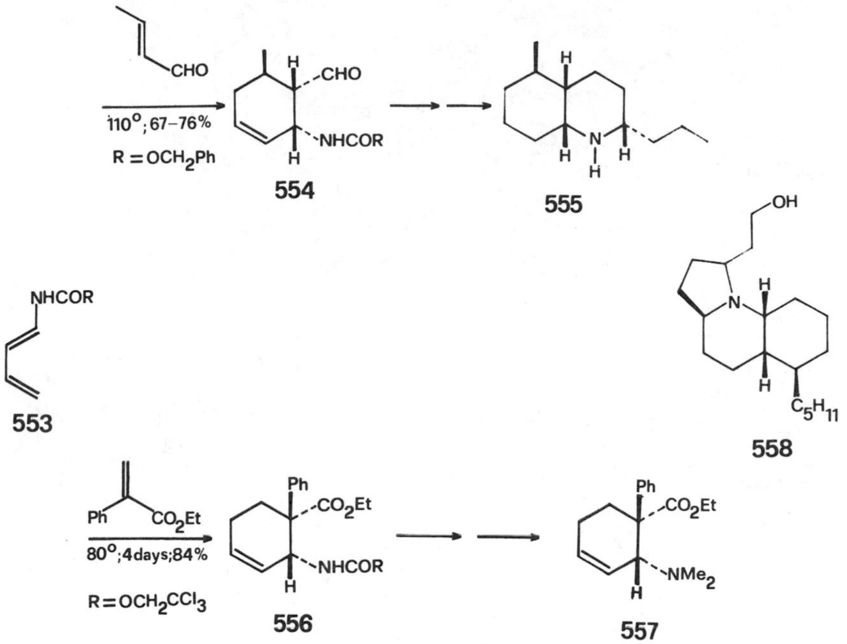

CHO
110°; 67–76%
R = OCH_2Ph
H
CHO
NHCOR
H
554
H
N
H H
H
555
NHCOR
553
OH
H
N
H
C_5H_{11}
558
Ph CO_2Et
80°;4days;84%
R=OCH_2CCl_3
Ph
CO_2Et
NHCOR
H
556
Ph
CO_2Et
NMe_2
H
557

Scheme 5.19

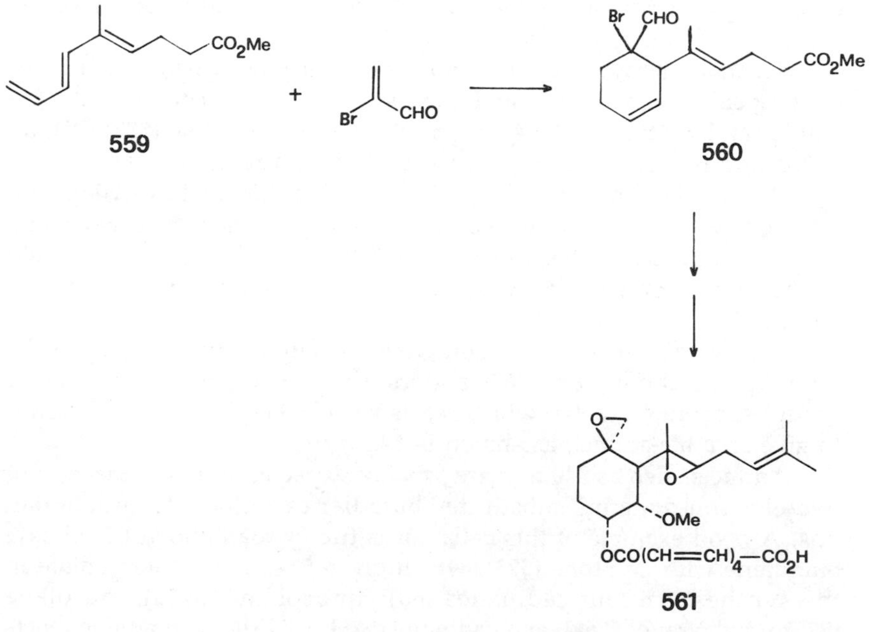

CO_2Me
559
+
Br CHO
Br CHO
CO_2Me
560
O
O
OMe
OCO(CH=CH)$_4$—CO_2H
561

Scheme 5.20

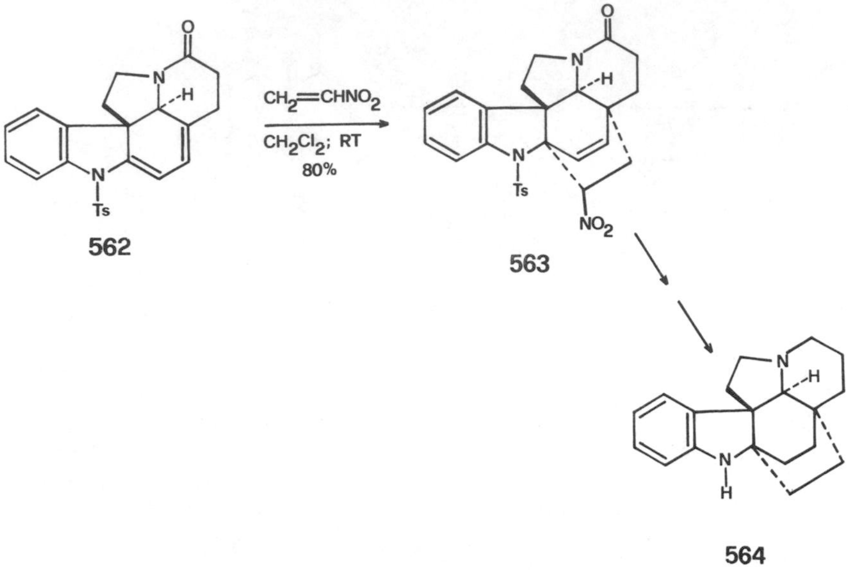

562 → $CH_2{=}CHNO_2$, CH_2Cl_2; RT, 80% → 563 → 564

Scheme 5.21

methylhept-2-en-4-one (*65, 66*) yields two adducts, **565** and **566**, in 1:1 ratio with a total yield of 80%. The loss of *endo*-acyl stereospecificity is disappointing, because only **565** is a useful intermediate to the sesquiterpene (±)-juvabione (**567**) (Scheme 5.22). Nevertheless the reaction is totally regiospecific—both *endo*-adduct **565** and *exo*-adduct **566** have the *ortho* regiochemistry as predicted for **538**.

Reactions between dienes and dienophiles both with electron-attracting substituents (such as the model **542**) are less common. An example can be found in the synthetic studies of ryanodine (**572**) (*67*), an insecticide isolated from *Ryana speciosa* Vahl. Ryanodine is an ester of pyrrole-2-carboxylic acid with ryanodol, and is classified as a 20-carbon diterpene with 11 asymmetric centers. Heating the diene **568** in benzene with a slight excess of the vinyl ketone **569** afforded a quantitative yield of the adduct **570**, which was successively converted to anhydroryanodol (**571**) (Scheme 5.23).

The overall balance of substituent effects in the diene moiety of the monoketal orthoquinone **568** renders it a Z-type (electron-attracting group) substituted diene, which reacts with **569** (a Z-substituted olefin) to give an *ortho*-adduct, as shown in **542**.

Adducts such as **540** that are *para* are expected from the reaction of two-electron-donating substituted butadienes with electron-poor olefins. A good example of this behavior is the cycloaddition of 2-ethoxybutadiene with cryptone (**573**), which gives **574**, a key intermediate in the synthesis of four cadinanes (*68*): amorphane (**575a**), murolane (**575b**), bulgarane (**576a**), and cadinane (**576b**). 2-Ethoxybutadiene reacts in the same way with 3,3-dimethyl-2-methylenecyclohexanone (**577**) to

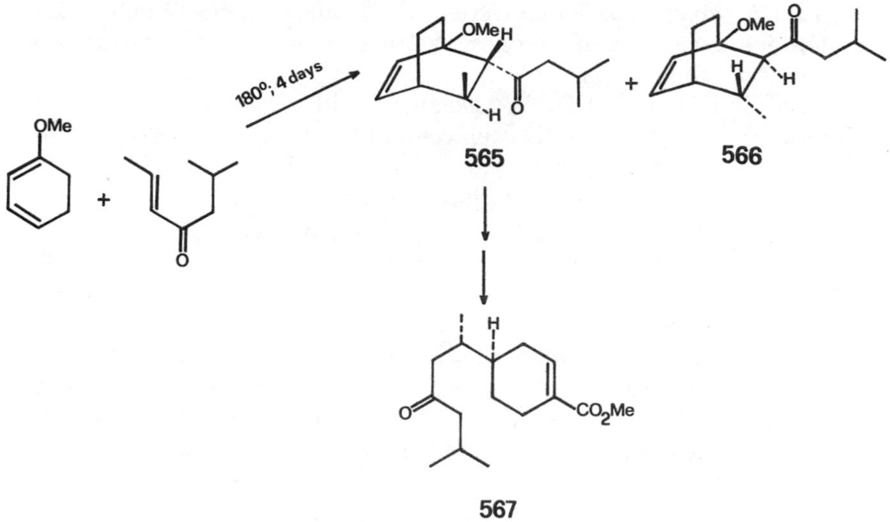

180°; 4 days

565

566

567

Scheme 5.22

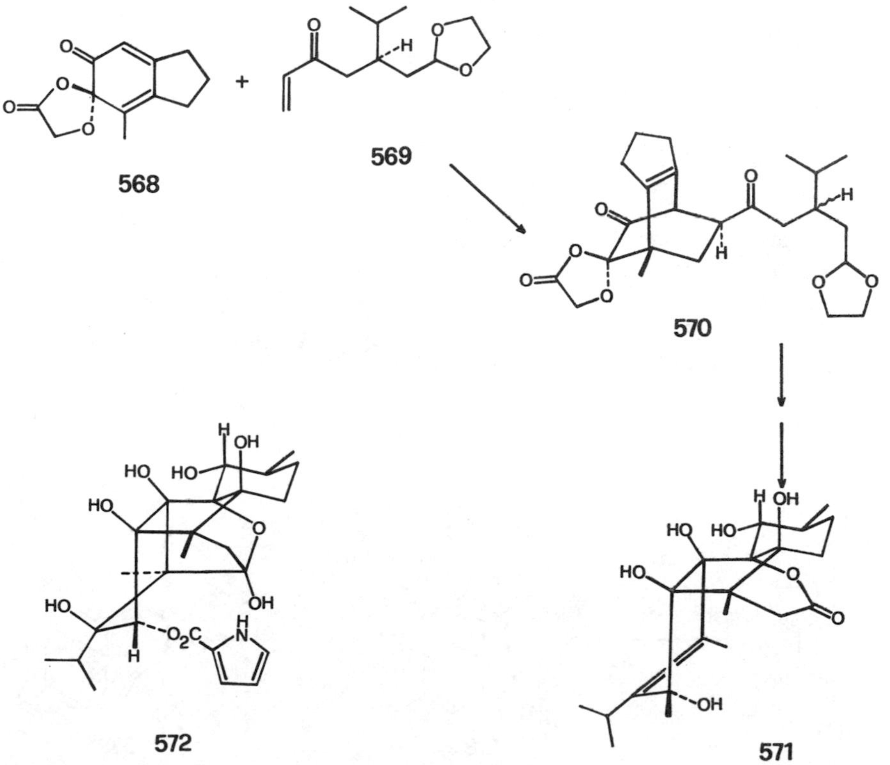

568

569

570

571

572

Scheme 5.23

afford **578**, which is easily converted to β-chamigrene (**579**) (*69*), a sesquiterpene isolated from the essential oils of leaves of *Chamaecyparis taiwanensis* (Scheme 5.24).

Both INDO and CNDO/2 calculations with 2-methyl-1,3-butadiene (*70*) indicate that the HOMO coefficient at C1 (0.614 and 0.621, respectively) is sufficiently greater than that at C4 (0.506 and 0.498, respectively) to determine the same regiochemistry in the adduct. This feature was important in Corey's synthesis of prostaglandin E_1 (**582**) (*71*), where 2-alkyl-substituted butadiene **580** reacts with *trans*-1-nitro-8-cyano-1-octene to give **581** as the major product, along with small amounts of the isomeric position adduct (Scheme 5.25).

The reactivity of 1-alkylbutadienes, which preferentially afford *ortho*-adducts (*72*) (Scheme 5.26), is much more difficult to understand. The HOMO coefficients at C1 and C4 are nearly identical (*70, 73*) with those of 2-methyl-1,3-butadiene, but the two compounds react differently.

The question was resolved, in spite of some controversy (*74*), by invoking secondary nonbonding orbital interactions (*70, 73*). The *ortho*-

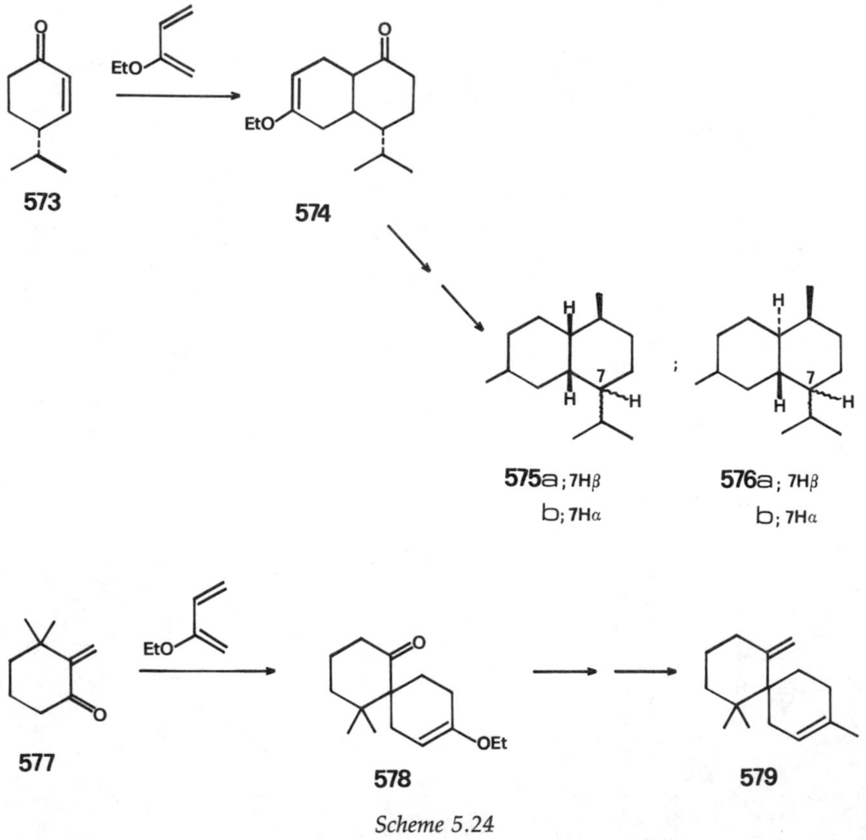

573 574 575a; 7Hβ b; 7Hα 576a; 7Hβ b; 7Hα 577 578 579

Scheme 5.24

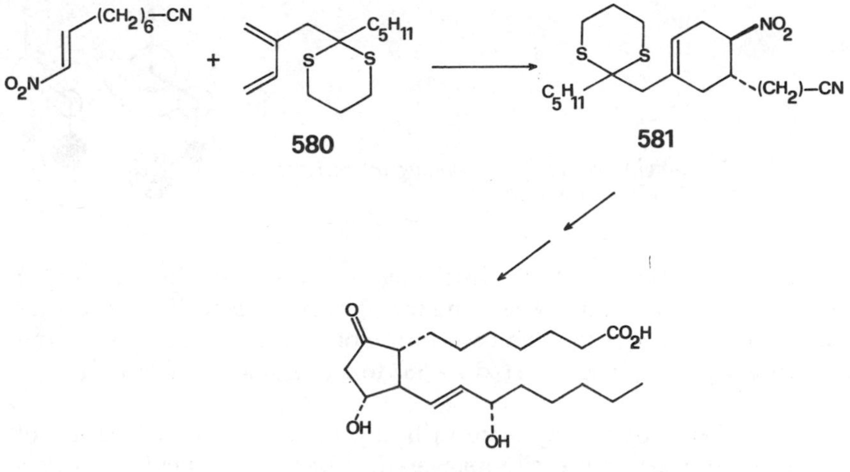

580 581 582

Scheme 5.25

isomer is preferred because the nonbonding interactions are better between the coefficient of the dienophile on C3 and that of the diene on C2. The latter coefficient is greater than that at C3 (*70*) (Figure 5.2). These conditions may occur only in the *endo* transition state. A nice application of these concepts can be found in the synthesis developed by Corey and coworkers (*75, 76*) of gibberellic acid (**589**), a plant growth-promoting substance. Gibberellic acid was previously produced only by the fungus *Gibberella fujikuroi;* the previous syntheses of Mori et al. (*77*) and Nagata et al. (*78*) were limited to the more simple gibberellins A_4 and A_{15}.

The starting point for this synthesis was the Diels–Alder reaction between *trans*-2,4-pentadien-1-ol (**583**) and 2-methoxy-6-(2′-benzyloxy)-

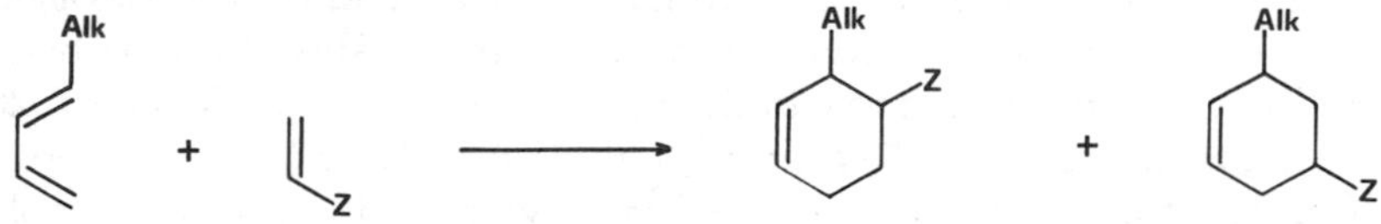

Alk	Z	ratio "orto" / "meta"
= Me	= CHO	= 8
= Me	= CN	= 10
= Me	= CO_2Me	= 6.8
= i-C_3H_7	= CO_2Me	= 5
= n-C_4H_9	= CO_2Me	= 5.1
= t-C_4H_9	= CO_2Me	= 4.1

Scheme 5.26

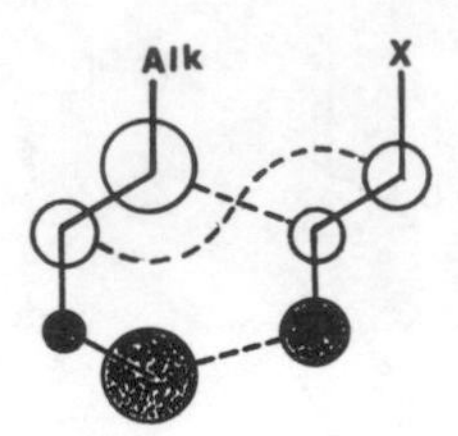

Figure 5.2. FMOs with secondary nonbonding interactions that favor formation of the ortho *isomers.*

ethyl-1,4-benzoquinone (**584**), which gave a single crystalline adduct **585** in 91% yield upon reflux in benzene for 30 h. A few steps later a further Diels–Alder addition (in this case intramolecular) produced **587** from **586**, which was finally converted via **588** to gibberellic acid (**589**) (Scheme 5.27).

Several questions may arise in trying to explain the formation of **585**; some of these are easily answered, others are difficult. It seems reasonable that the reaction occurs on the dienophile double bond with lower electron density (i.e., not at C2–C3), through an *endo* transition state obviously involving nonbonding interactions between C2–C3 of the diene and C1–C4 of the dienophile.

However, because the HOMO coefficients at the two ends of the diene are nearly equal and the coefficient at C2 is greater than that at C3 (*70*), **585** should be obtained. The values of LUMO coefficients of **584** at C1 are greater than those at C4. A simple explanation can be advanced by considering the LUMO of **584** as the mixing of empty orbitals of the unpolarized O=C−C=C−C=O with the polarized C=C−X system. Figure 5.3 represents the interactions involved and the resulting polarization with the coefficients at C1 greater than those at C4 (*79*). The transition state leading to **585** can be pictured as in Figure 5.4.

An alternative to the steric hindrance we suggested can be put forward to explain the reaction occurring between myrcene (**239**) and hemigossypolone (**590**) to produce heliocide H_2 (**591**), a substance isolated in the subepidermal pigment glands of Upland cotton (*Gossypium hirsutum* L.). This compound reduces the growth rates of the bollworm (*Heliothis zea*) and the tobacco budworm (*H. virescens*) (*80*) (Scheme 5.28).

Thus the LUMO of **590** can be derived by mixing the empty orbital of an unpolarized O=C−C=C−C=O system with a strongly polarized benzene bearing an electron-attracting group *para* to an electron-donating one. The coefficient at C4 should be increased more than that at C1. Therefore, the regioisomer **591** is preferred if an *endo* transition state does occur (as we suggested).

Secondary nonbonding orbital interactions are very useful in explaining the behavior of disubstituted butadienes (*81*). For example, 1,2-dimethylbutadiene (**592**) reacts with acrylonitrile and methyl methacrylate to give 2,3,4-trisubstituted cyclohexenes **593** as the major isomers in a ratio of 6:1 over the 2,3,5-trisubstituted ones **594** (*82*) (Scheme 5.29).

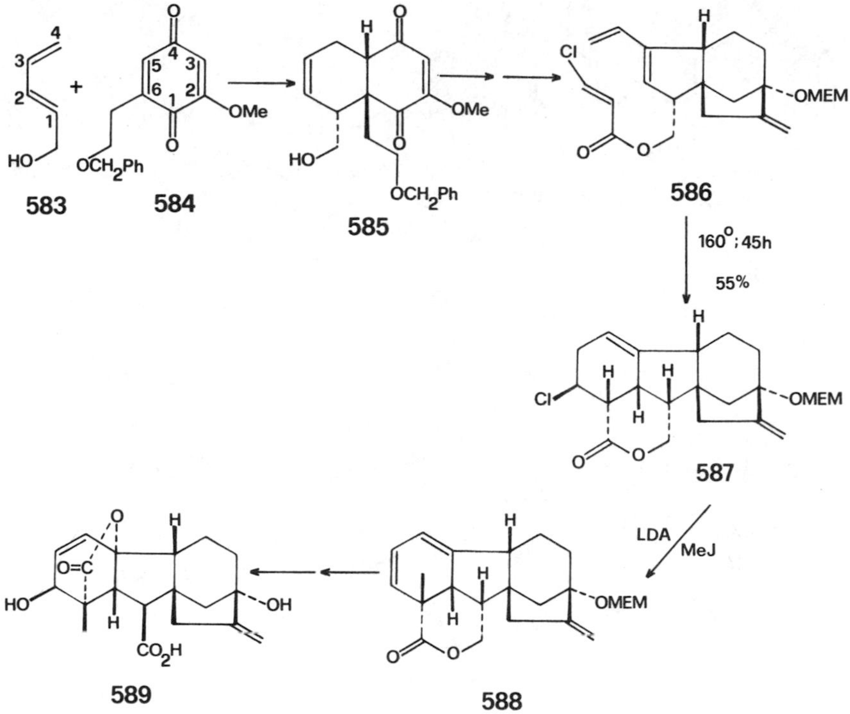

Scheme 5.27

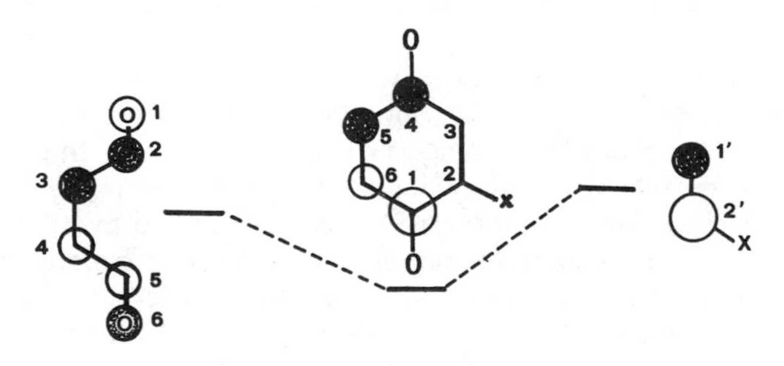

*Figure 5.3. Mixing of empty MOs to give the LUMO of **583**.*

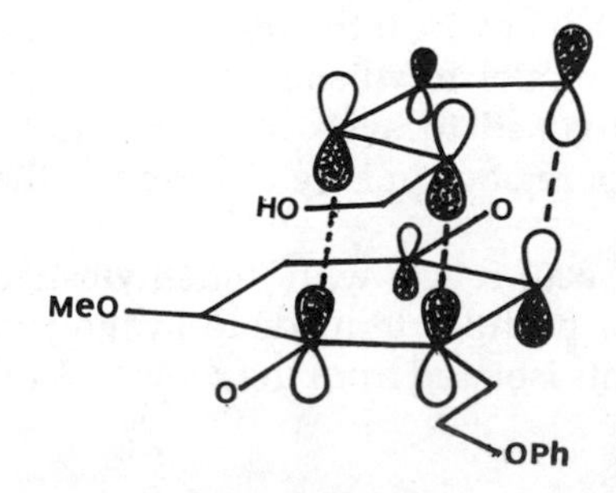

*Figure 5.4. FMO interactions that yield **585**.*

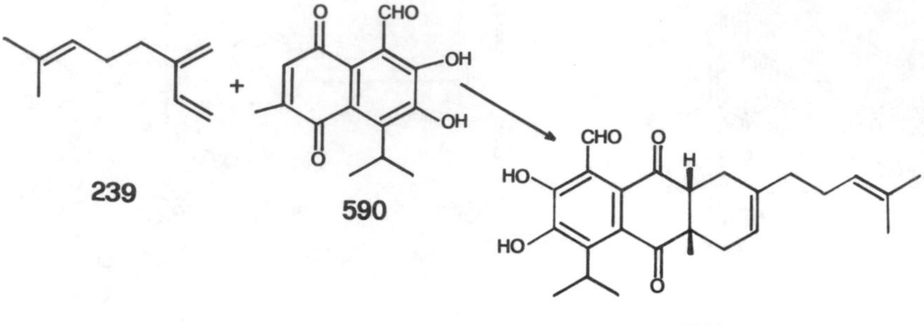

Scheme 5.28

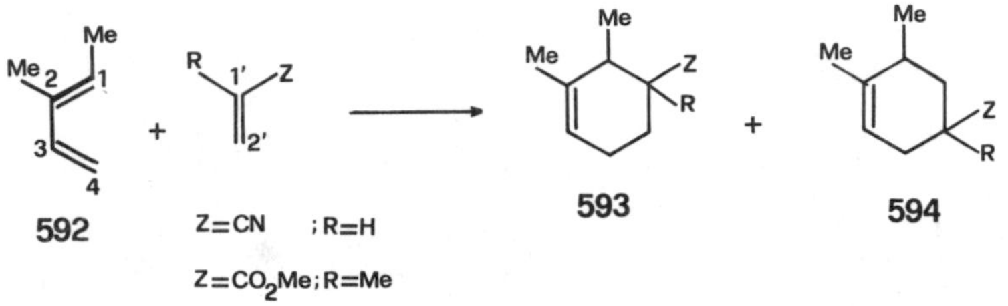

Z=CN ; R=H

Z=CO_2Me; R=Me

Scheme 5.29

The values of HOMO coefficients of **592**, calculated through different methods and reported in Table 5.1 (*81*), show in all cases C1 > C4 and C2 > C3.

Taking into account primary orbital interactions only, the formation of **594** is favored by 1.8 (acrylonitrile) or 3.0 (methyl acrylate) kcal/mol, the greater coefficients being at C1 in **592** and at C2′ in the dienophile. However, when secondary nonbonding interactions (C2 > C3) are also considered, the regioisomer **593** becomes favored by 1.7 (acrylonitrile) or 2.0 (methyl acrylate) kcal/mol (*81*), as is found experimentally.

Along the route to quasimarin (**599**), a tricyclic intermediate **598** was synthesized (*83*). The BC unit was accomplished by using a Diels–Alder reaction of diene **595** with quinone **596** (prepared in situ) to afford **597** (Scheme 5.30).

Obviously the LUMO of 2-carbomethoxy-1,4-benzoquinone (**596**) has the coefficient at C3′ greater than that at C2′. If the reaction is under MO control and if **592** is taken as a model for diene **595**, secondary nonbonding interactions should be invoked to overcome the primary interactions (Figure 5.5), which favor a regiochemistry opposed to that of **597**.

2-Vinyl-3,3-dimethylcyclohexene (**600**) reacts with 3-methylbenzofuran-4,7-quinone (**601**) to give **602**, a useful intermediate in the synthesis of four tanshinones (*84*) (pigments isolated from the roots of *Salvia* *miltiorrhiza*): isotanshinone II (**603**); isocryptotanshinone (**604**); and, via **605**, cryptotanshinone (**606**) and tanshinone II (**607**) (Scheme 5.31).

Many difficulties are encountered in explaining, in terms of simple MO concepts, the behavior of the complex molecules used in the synthesis of natural products. Only a small part of the mystery is ever cleared up. These statements are exemplified by the synthesis of (±)-ligularone (**611**) a furanoeremophilane family sesquiterpenoid isolated from *Ligularia sibirica* Cass. It was easily synthesized from the Diels–Alder adduct **610** of 1-methyl-2-ethoxybutadiene (**608**) and 3,5-dimethylbenzofuran-4,7-quinone (**609**) (*85*) (Scheme 5.32).

Table 5.1

HOMO Coefficients of 592

MO Method	P_z Coefficient			
	C1	C2	C3	C4
CNDO/2	0.561	0.456	−0.284	−0.468
INDO	0.562	0.453	−0.296	−0.475
CNDO/S	0.588	0.469	−0.322	−0.514
Hückel	0.592	0.444	−0.316	−0.546

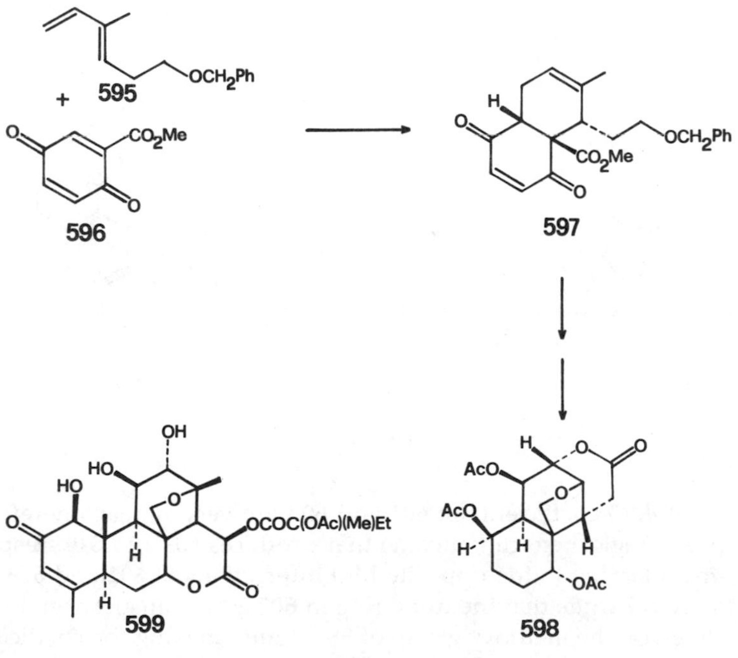

Scheme 5.30

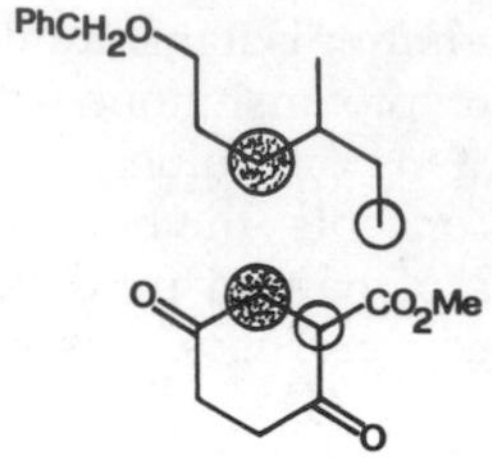

PhCH$_2$O
CO$_2$Me

Figure 5.5. FMO interactions that yield 597.

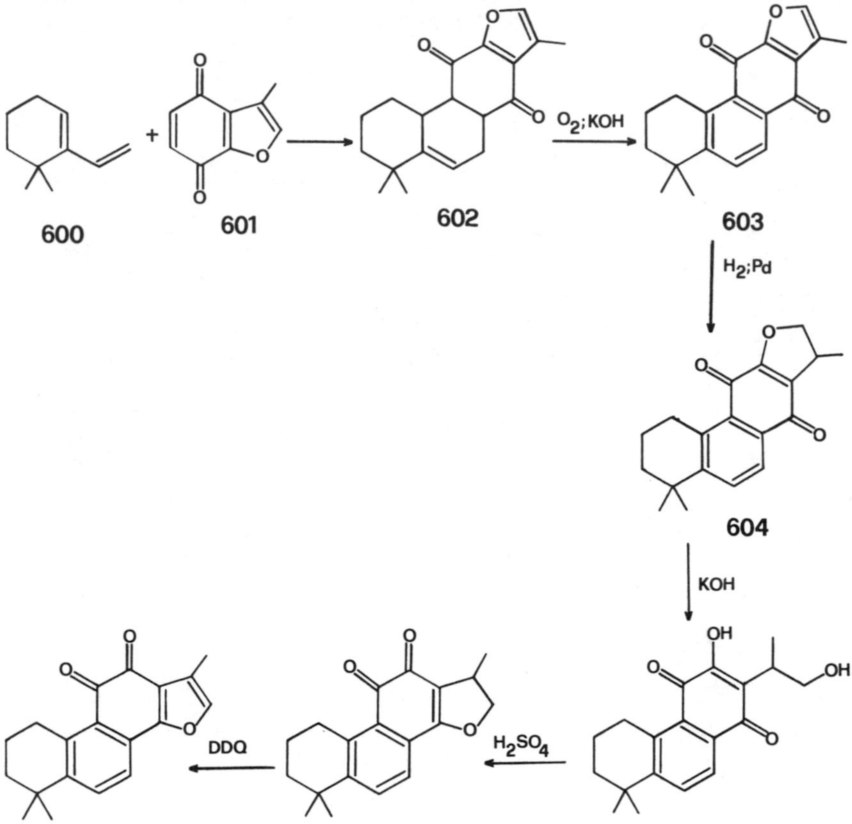

O$_2$; KOH
H$_2$; Pd
KOH
H$_2$SO$_4$
DDQ
600 601 602 603 604 605 606 607

Scheme 5.31

The HOMO coefficients of **601** and **608** are very similar; therefore, the opposite regiochemical outcome that produces **603** and **610**, respectively, must be due to the opposite MO interactions of **602** and **609**.

One could argue that the furan ring in **602** acts as an electron donor group, likewise the methoxy group of **583**, hence making the coefficient at C7 greater than that at C4; the methyl group of **609** makes the coefficient at C4 greater than that at C7. However, this argument is only conjecture; other possible explanations and suitable calculations are needed.

Silyloxysubstituted Dienes. In 1974 Danishefsky and Kitahara (*86*) reported the synthesis of a new diene: *trans*-1-methoxy-3-trimethylsilyloxy-1,3-butadiene (**613**), which was easily obtained from *trans*-4-methoxybuten-2-one (**612**). Its reaction with a variety of dienophiles provided new and effective routes to benzene and naphthalene derivatives as well as cyclohexenones (*87, 88*) (Scheme 5.33).

At the outset **613** seemed to be one of the many useful dienes of the chemical literature, but 3 years later it made its triumphal entry into natural products syntheses, becoming the protagonist of a new phase of the Diels–Alder reaction.

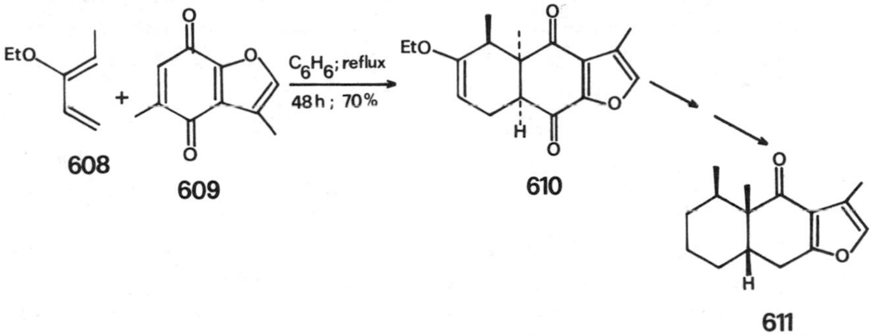

EtO + 608 609 C_6H_6; reflux 48 h; 70% EtO 610 611

Scheme 5.32

OMe 612 Et_3N; $ZnCl_2$ Me_3SiCl OMe Me_3SiO 613 DMAD CO_2Me Me_3SiO CO_2Me

OMe Me_3SiO H_2O H^+ OAc AcO OAc Ac_2O OMe Me_3SiO

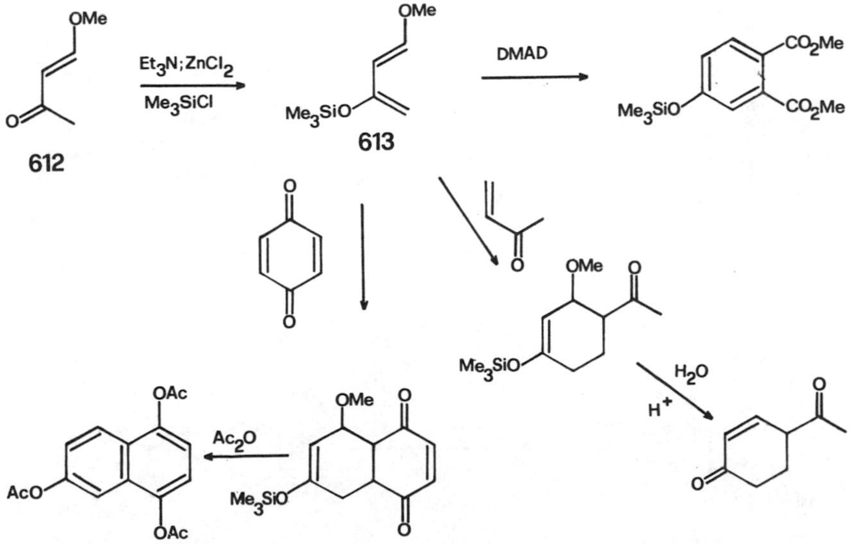

Scheme 5.33

The first synthesis of natural products concerned the disodium salt of the rather elusive prefenic acid (**617**), biogenetically arising from shikimic acid by way of chorismic acid. This salt is an important precursor of the aromatic amino acids tyrosine and phenylalanine.

The diene reacts with the β-phenylsulfinyl-α,β-unsaturated lactone (**614**) to give an adduct **615** that loses the silyl, methoxy, and phenylsulfoxide groups to afford the dienone **616**, which was easily converted to **617** (together with its epimer) (*89–92*) (Scheme 5.34).

After the first successful approach, several natural products were synthesized by the Danishefsky group. In 1978 the sesquiterpene antibiotic tumor-inhibitory agent pentalenolactone (**621**) was obtained from **613** and the unusual dienophile **618** through the adducts **619** and **620** (*93, 94*) (Scheme 5.35).

In 1980 the total synthesis (*95*) of the alkaloid tazettine (**522**) was achieved from the reaction of **613** and sulfone **622**. A mixture of epimers **623** was produced, and they were easily cyclized to the isooxindolic ring **624** and then transformed into **522** (Scheme 5.36).

However, among the most fascinating achievements was the synthesis (*96*) of the tumor inhibitors (±)-vernolepin (**629**) and (±)-vernomenin (**630**) realized in early 1977. The starting steps were two consecutive Diels–Alder reactions, the first to build up the dienophile **625**, which reacts with **613** in the second one to give the bicyclic *cis*-fused cyclohexenone **626**, further converted to the lactone **627**. The latter compound was then elaborated to the δ-lactone **628**, the key intermediate for **629** and **630** (Scheme 5.37).

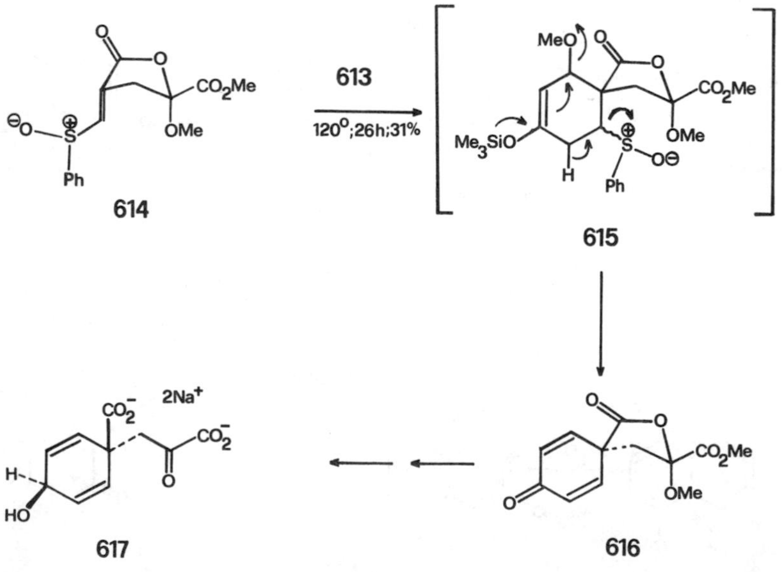

614 → 613, 120°;26h;31% → [615] → 616 → → 617

Scheme 5.34

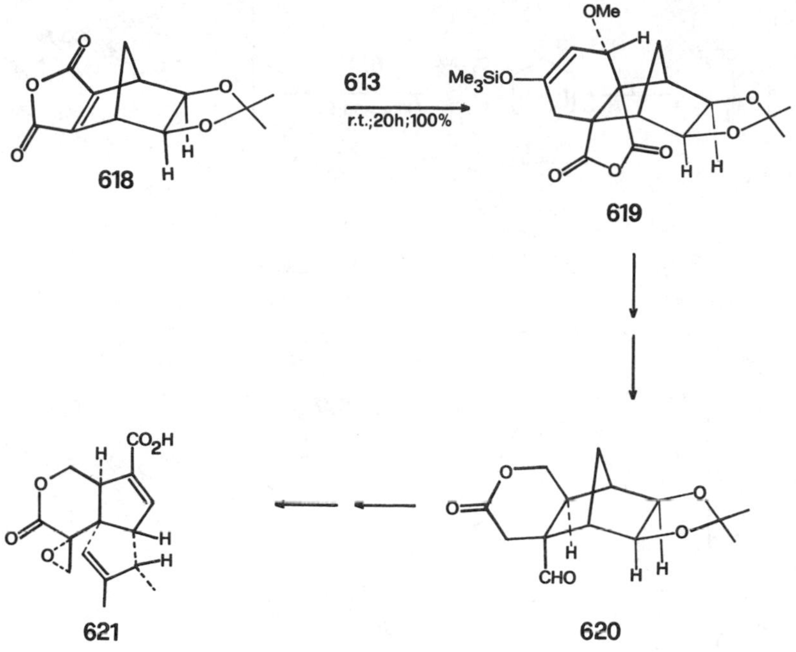

618

613

r.t.;20h;100%

OMe

Me_3SiO

619

620

CHO

CO_2H

621

Scheme 5.35

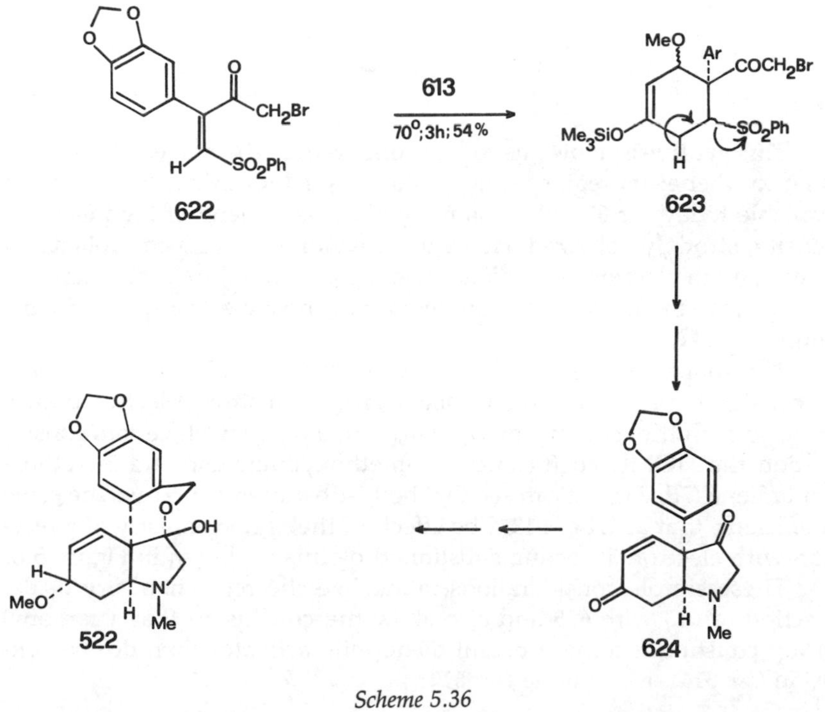

622

CH_2Br

SO_2Ph

613

70°;3h;54%

MeO

Ar

$COCH_2Br$

Me_3SiO

SO_2Ph

623

624

Me

522

OH

MeO

Me

Scheme 5.36

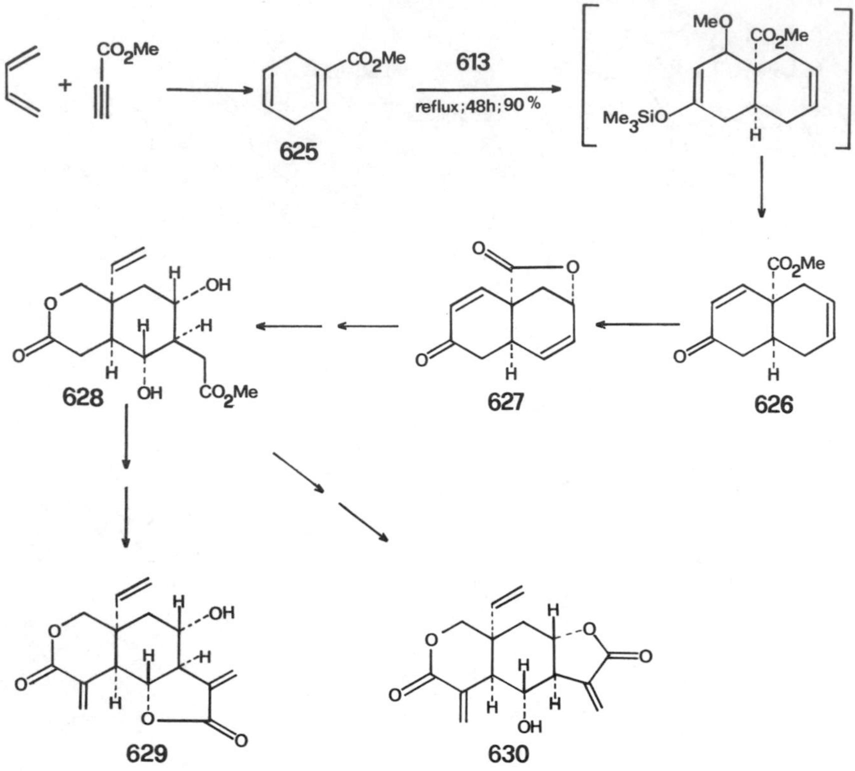

613
reflux;48h;90 %

625 626 627 628 629 630

Scheme 5.37

This synthesis allows us to infer one of the advantages of the Danishefsky diene: its regiospecificity. Although MO calculations are not available today for **613**, there is no doubt it is a diene of high electron density, strongly polarized. Hence its behavior is HOMO-controlled and it reacts with electron-poor dienophiles as illustrated by **625**, where the attack site is the more highly hindered, but most electron-poor, double bond.

The shape of the HOMO coefficients of **613** can be simply derived by considering that the trimethylsiloxy group is a strong electron donor, more so than the methoxy group if $\sigma_{Si} \cong \sigma_C$ (*97*). If we take into consideration the HOMO coefficients of 1-methoxybutadiene and 2-methoxybutadiene (*70, 73*), we can see that both substituents increase the same coefficient, that at C4 of **613**. The effect on the regiochemistry of a reaction with electron-attracting substituted olefins is shown in Figure 5.6.

These simple considerations rationalize the regiochemistry of the reaction of **613** with **625** and also allow the conclusion that a carbonyl group constitutes a more potent dienophile activator than does a sulfoxide (*see* **614**) or a sulfone (*see* **622**).

We must emphasize that a number of other analogs of **613** were also prepared (*98*), namely the methylated derivatives **631**, **632**, and **633**; the 4-phenylseleno derivative **634**; and the 1-methoxy derivative **635**. This last analog offered rapid access to different functionalized ring systems (Scheme 5.38).

The diene **632** was utilized in the synthesis of chamaecynone (**638**), a norsesquiterpene isolated from an essential oil of *Chamaecyparis formosensis* Matsum. Its reaction with 5-ethynyl-2-methylcyclohex-2-en-1-one (**636**) gave the key intermediate **637**, which has the same four chiral centers as **638** through an attack from the less hindered face of the dienophile (*99*, *100*) (Scheme 5.39).

Several substances were synthesized by using the diene **635**: lasiodiplodin (**639**) (*101*), a plant growth inhibitor; epigriseofulvin (**641**) (*102*); and griseofulvin (**643**) (*103*). In the last two cases control over the relative configuration of C6′ was secured by the use of different dienophiles, **640** and **642**, respectively (Scheme 5.40).

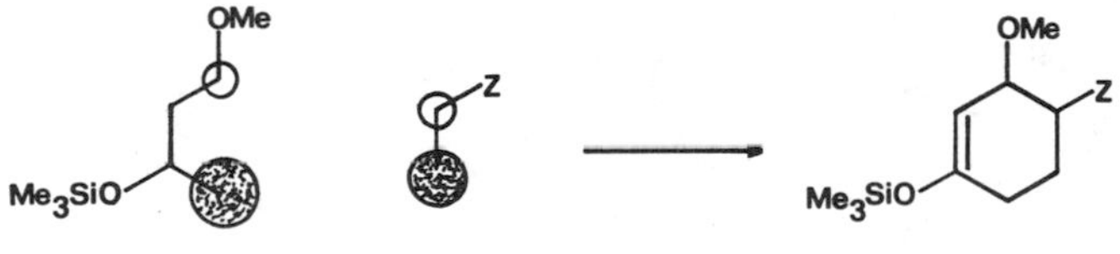

OMe Me$_3$SiO Z OMe Z Me$_3$SiO

Figure 5.6. HOMO coefficients of **613** *and regiochemistry with electrophilic olefins.*

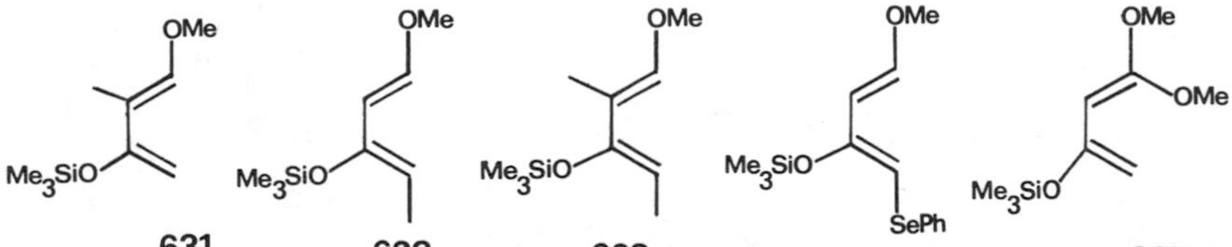

OMe Me$_3$SiO **631** OMe Me$_3$SiO **632** OMe Me$_3$SiO **633** OMe Me$_3$SiO SePh **634** OMe OMe Me$_3$SiO **635**

Scheme 5.38

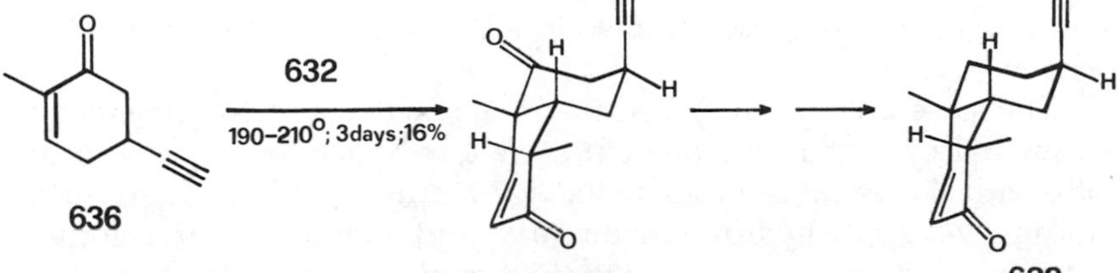

O **636** **632** 190–210°; 3days;16% O H H H O **637** H H H O **638**

Scheme 5.39

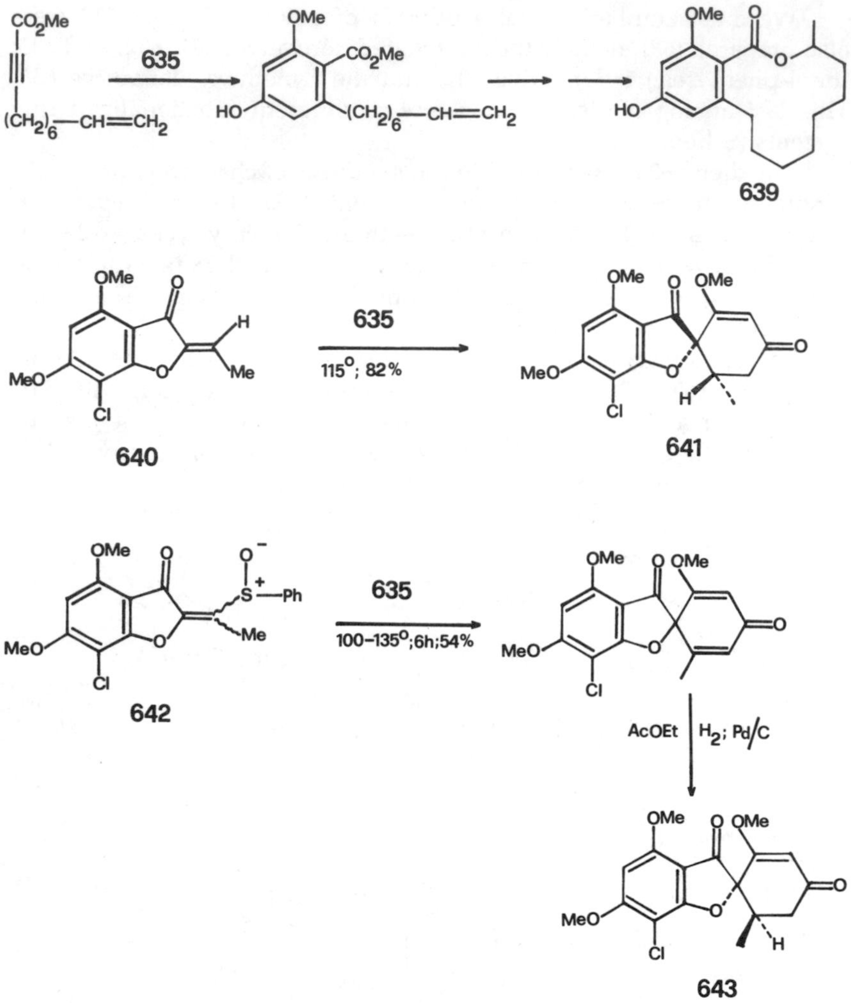

CO_2Me ≡ $(CH_2)_6$—CH=CH$_2$ → (635) → OMe, CO_2Me, HO, $(CH_2)_6$—CH=CH$_2$ → → OMe O, O, HO

639

OMe O, H, MeO, O, Me, Cl

640

635

115°; 82%

OMe O OMe, O, MeO, O, H, Cl

641

OMe O, O$^-$, S$^+$—Ph, MeO, O, Me, Cl

642

635

100–135°;6h;54%

OMe O OMe, O, MeO, O, Cl

AcOEt | H_2; Pd/C

OMe O OMe, O, MeO, O, Cl, H

643

Scheme 5.40

A 4-butyl derivative of **635** reacted with 2,6-dichloronaphthazarin (**644**) to give **645**, which was easily transformed into penta-*O*-methylrubrocomatulin (**646**), the tetramethyl ether of a crinoid pigment (*104*) (Scheme 5.41).

The same authors (*104*) described the reaction of 1,1,4-trimethoxy-3-trimethylsilyloxybuta-1,3-diene (**647**) (a 4-methoxy derivative of **635**), both with 6-acetyl-3-chloro-5-methoxy-7-methylnaphthoquinone (**648**) affording 2-acetyl-5-hydroxyemodin (**649**) and with 2-bromo-5-chloro-8-methoxy-6-methylnaphthoquinone (**650**) leading to the trimethyl ether of xanthorin (**651**) (Scheme 5.42).

The choice of the suitable diene opened the route to three coccid anthraquinones: deoxyerythrolaccin (**654**), laccaic acid D (**657**), and ker-

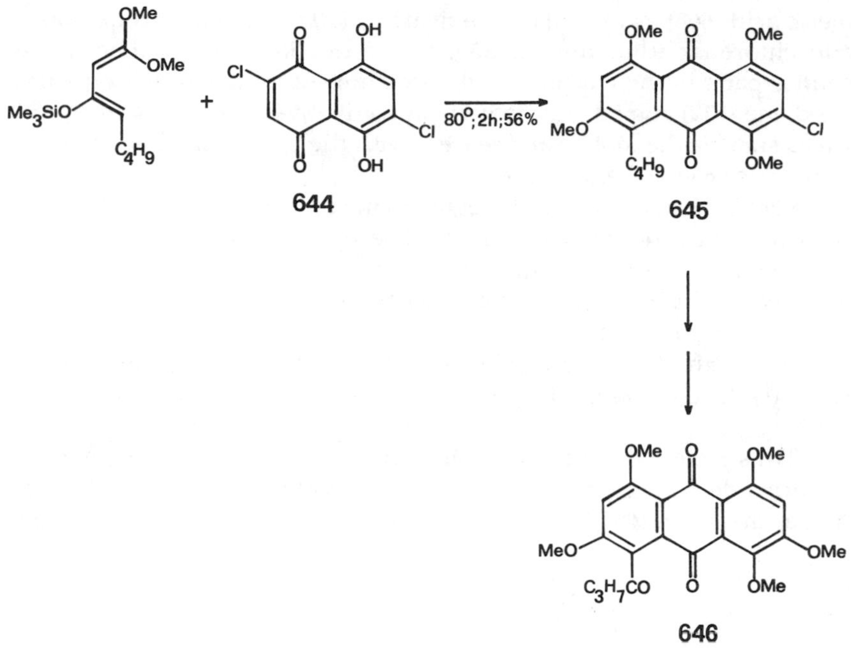

Scheme 5.41

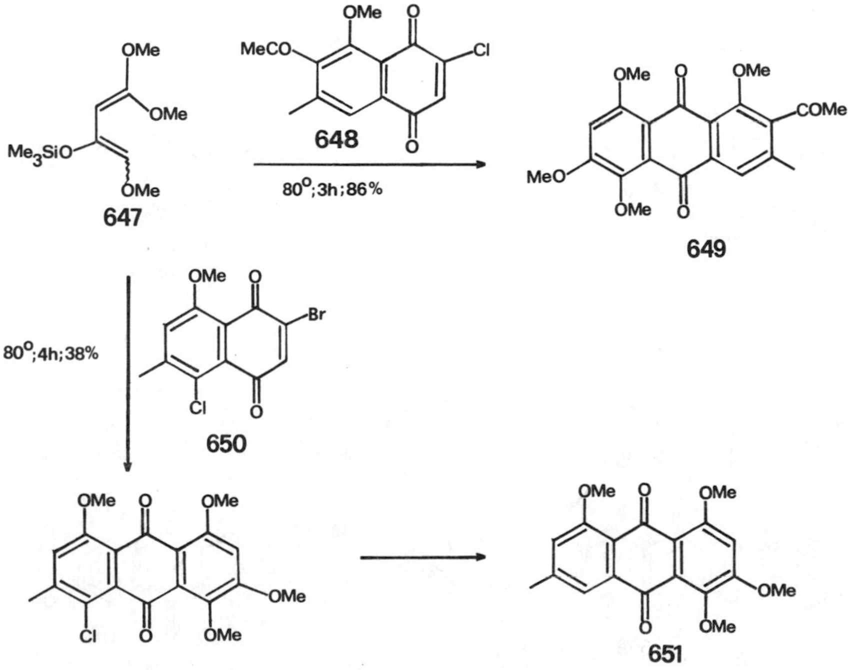

Scheme 5.42

mesic acid (**660**) as methylated derivatives (*105*). Chlorobenzoquinones and chloronaphthoquinones **653**, **656**, and **658** were the dienophile counterparts in the reactions with 4-methoxy-2-trimethylsilyloxypenta-1,3-diene (**652**) and its 3-carbomethoxy derivative (**655**). The adduct **659** was a suitable dienophile and reacted with the previously described **647** to produce **660** (Scheme 5.43).

For all these dienes a fully regiospecific cycloaddition was observed with the electron-poor dienophiles. The extensive use of halo-substituted benzo- and naphthoquinones as dienophiles **644**, **648**, **650**, **653**, **656**, **658**, and **659** might be explained by the regiochemical control induced by this kind of substituent.

Although the following dienes lack a methoxy group, they behave similarly to the Danishefsky diene because of their trimethylsilyloxy groups.

Thus 1-methyl-3-trimethylsilyloxycyclohexa-1,3-diene (**661**) and its 2-methyl derivative (**662**) were used in the synthesis of 7-desmethyl-11-norseychellanon (**663**) (*106*) and, more interestingly, in two syntheses

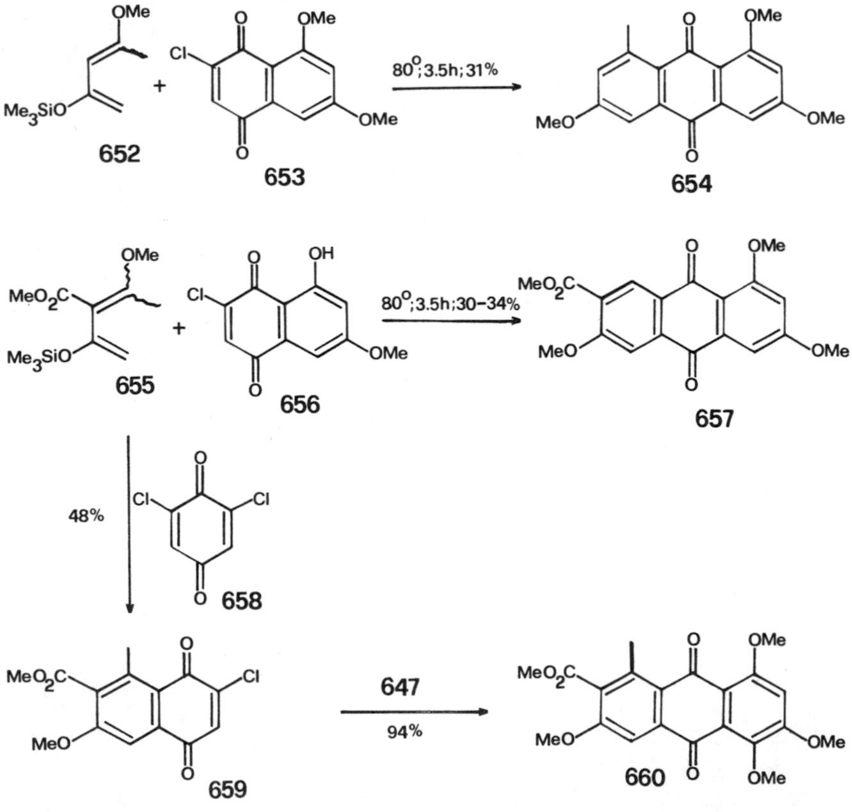

OMe
Me$_3$SiO
652
+
Cl
O
OMe
O
OMe
653
80°;3.5h;31%
O
OMe
MeO
OMe
O
654

OMe
MeO$_2$C
Me$_3$SiO
655
+
Cl
O
OH
OMe
O
656
80°;3.5h;30–34%
O
OMe
MeO$_2$C
MeO
OMe
O
657

48%
O
Cl
Cl
O
658

MeO$_2$C
MeO
O
Cl
O
659
647
94%
MeO$_2$C
MeO
O
OMe
OMe
OMe
O
660

Scheme 5.43

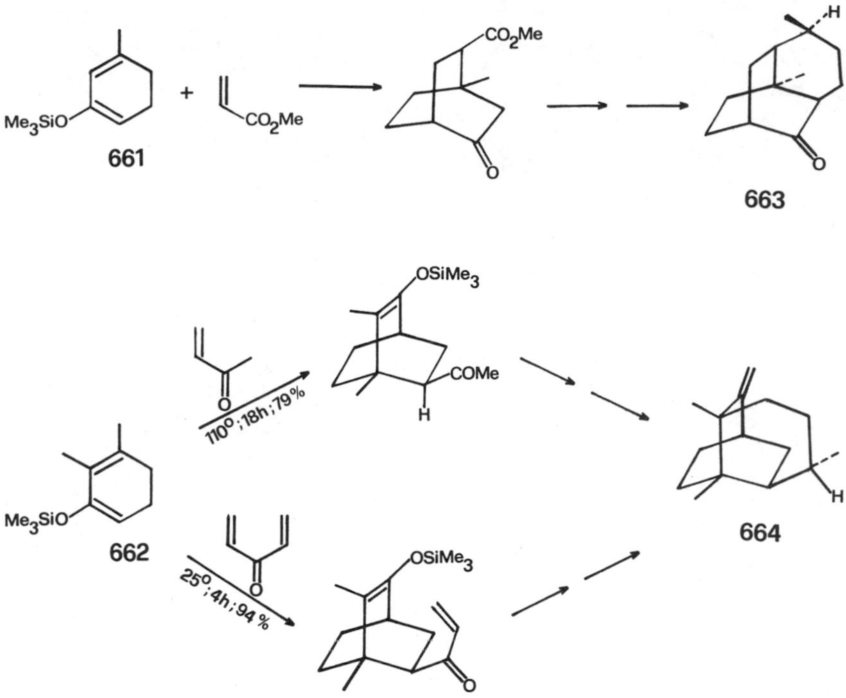

Scheme 5.44

of (±)-seychellene (**664**) (*107, 108*), the second of which is much easier despite containing the same number of steps (Scheme 5.44).

A fully stereospecific route to (±)-coriolin (**669**) and its congeners coriolin B (**670**) and diketocoriolin B (**671**) was realized by reaction of 3-trimethylsilyloxy-1,3-pentadiene (**665**) with enedione **666.** A single adduct **667** was obtained and subsequently transformed into the tricyclic derivative **668,** from which the desired compounds were prepared (*109, 110*) (Scheme 5.45). The enedione **666** has two possible orientations in the Diels–Alder reaction, the reacting double bond having two electron-attracting carbonyls at the ends. The observed regiochemistry was interpreted in terms of driving force for removal of unsaturation from the junction (*110*).

An acetylcyclopentene served as precursor of the diene **672,** which reacted with the cyclobutene **673** to give the adduct **674,** transformed in seven steps into (±)-illudol (**147**) (*112*), a natural sesquiterpene alcohol already mentioned in the section entitled "Photochemical [2 + 2] Cycloadditions" in Chapter 3 (Scheme 5.46). Interestingly, the dienophile **673** was obtained by a thermal [2 + 2] cycloaddition from ethyl propiolate and 1,1-diethoxyethylene.

Three dienes bearing the trimethylsiloxy group on C1 and a second alkoxy (methoxy or trimethylsiloxy) group on C3 behave like **613** from

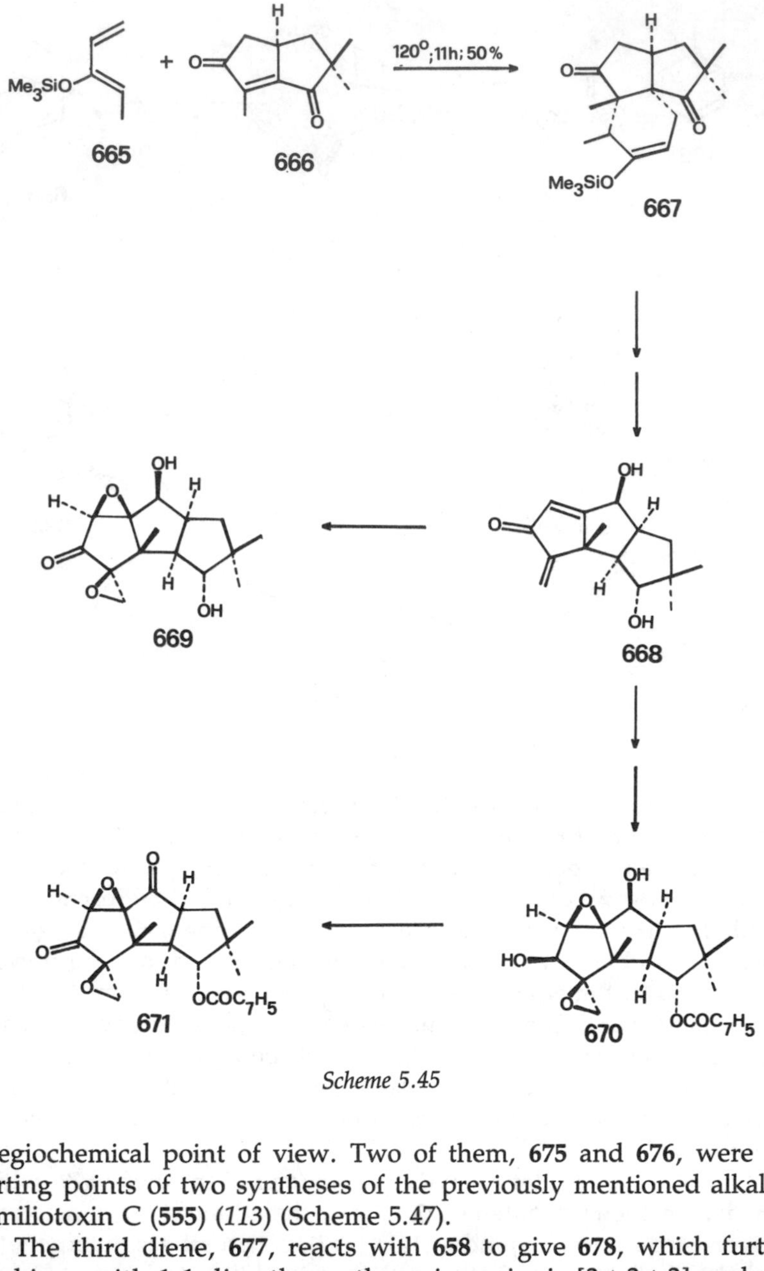

Me$_3$SiO 665 + 666 120°; 11h; 50% 667

669 668

671 670

Scheme 5.45

a regiochemical point of view. Two of them, **675** and **676**, were the starting points of two syntheses of the previously mentioned alkaloid pumiliotoxin C (**555**) (*113*) (Scheme 5.47).

The third diene, **677**, reacts with **658** to give **678**, which further combines with 1,1-dimethoxyethene in an ionic [2+2+2] cycloaddition (discussed in Chapter 6) to produce the insect dyestuff kermesic acid (**679**) (*114*) (Scheme 5.48).

Two other 1-silyloxy-substituted dienes, **680** and **682**, have been successfully utilized for the construction of different compounds. The first, **680**, was the starting material in an elegant route to the terpenoid antibiotic trichodermol (**681**) (*115*) (Scheme 5.49). The synthesis of six naturally occurring quinones [7-methyljuglone (**683**), phomarin (**684**), physcion (**685**), emodin (**686**), chrysophanol (**687**), and helminthosporin (**688**)] began with the reaction of **682** with different dienophiles. The overall sequence (*116*) also relies on the employment of other old and

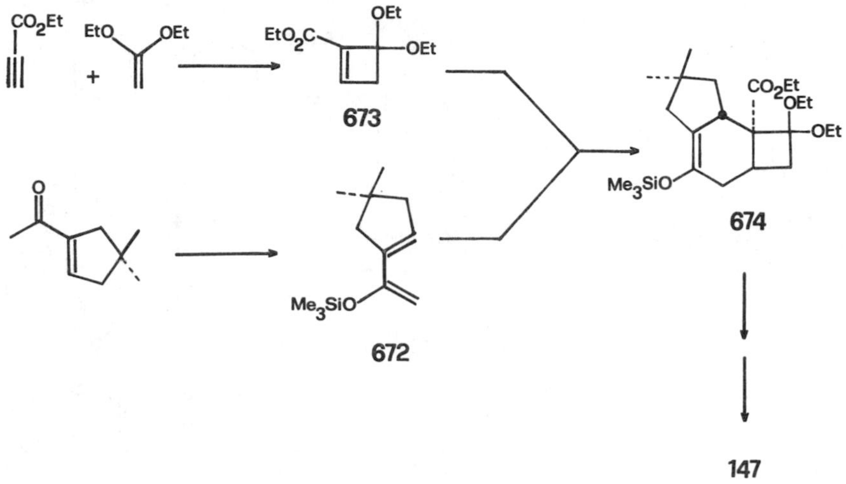

Scheme 5.46

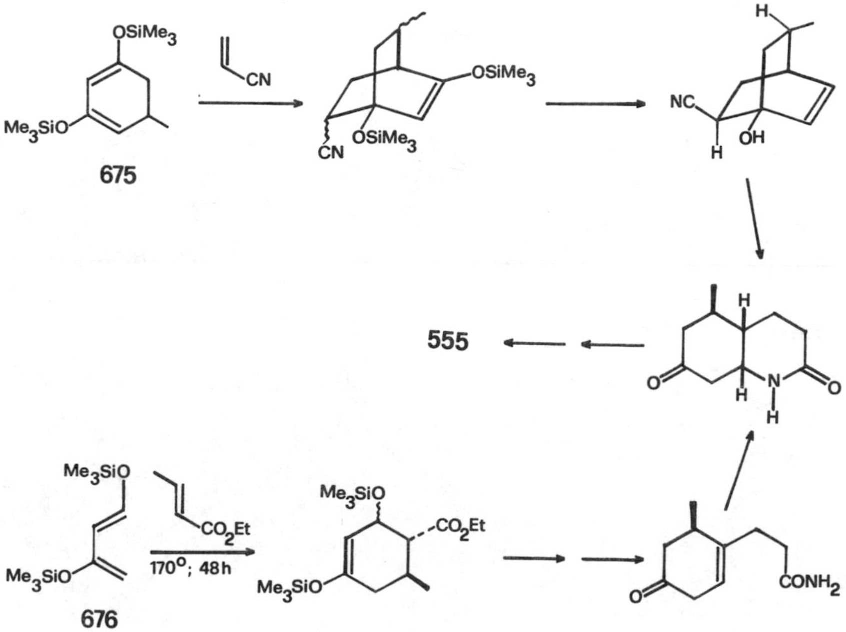

Scheme 5.47

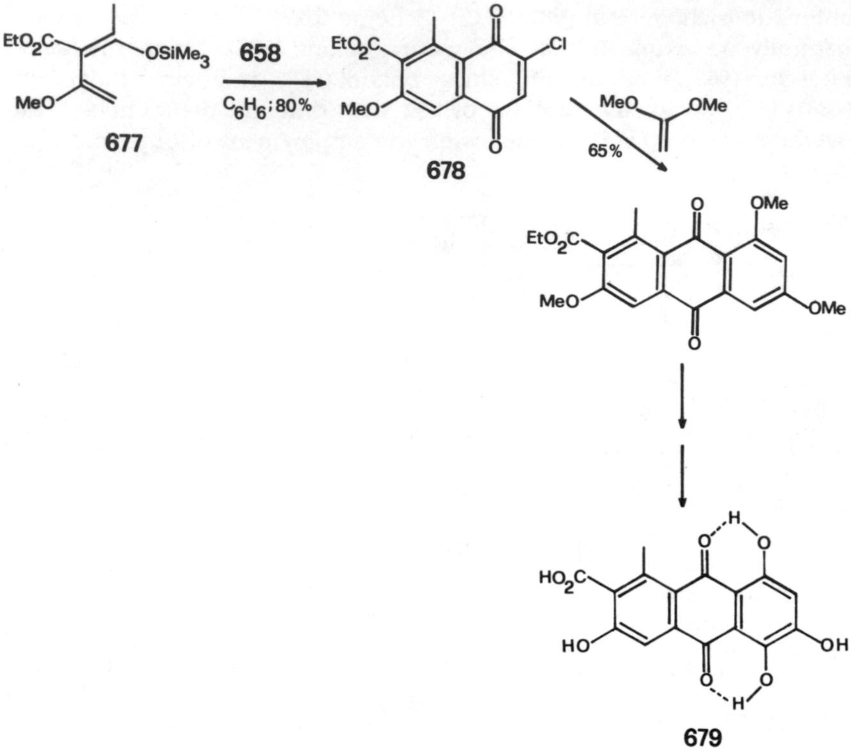

EtO_2C $OSiMe_3$ MeO **677** → **658** (C_6H_6; 80%) → EtO_2C Cl MeO O O **678** → MeO OMe (65%) → EtO_2C O OMe MeO O OMe → → HO_2C O H O HO O O H OH **679**

Scheme 5.48

OSi **680** + O O → (25°; 5days, 90%) → O O OSi → → O PhCOO → → H O O OH H **681**

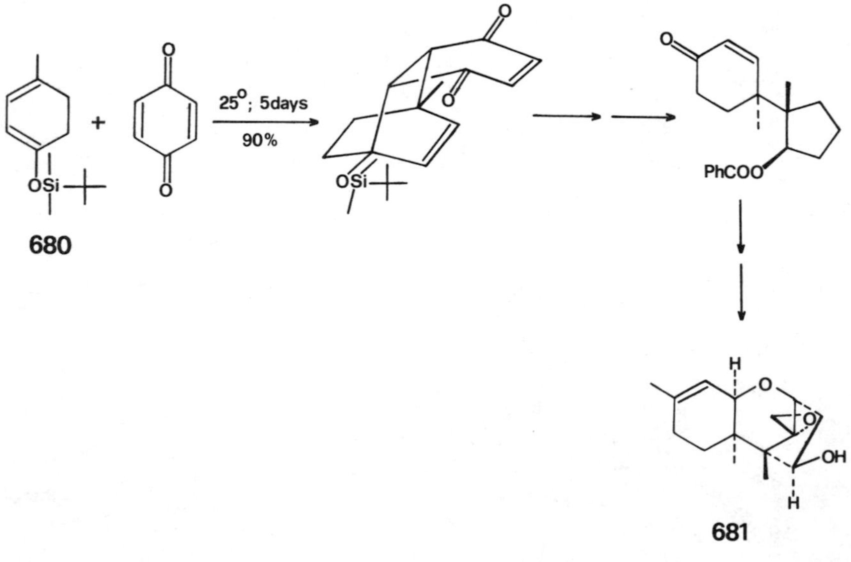

Scheme 5.49

new dienes, as outlined in Schemes 5.50 and 5.51, although these dienes are less fascinating than those previously mentioned.

The search for new dienes that secure regiochemical control of the Diels–Alder reaction is not limited to the field of silyloxy-substituted dienes. Thus, Trost et al. (*117*) proposed several 2-alkoxy-3-alkylthiobuta-1,3-dienes (**690**) obtained by electrocyclic ring opening (*see* the next chapter) of suitably substituted cyclobutenes **689**. The field seems very promising because MINDO/3 calculations of HOMO coefficients (*118*) have shown strongly polarized ends with the coefficient at C4 greater than that at C1 (about 0.5 vs. 0.3), as depicted in **691** (Scheme 5.52).

In Situ Generation of Dienes for Intermolecular Diels–Alder Reactions. A diene system for a Diels–Alder reaction can be easily generated in situ through several ways, including pericyclic reactions. This ap-

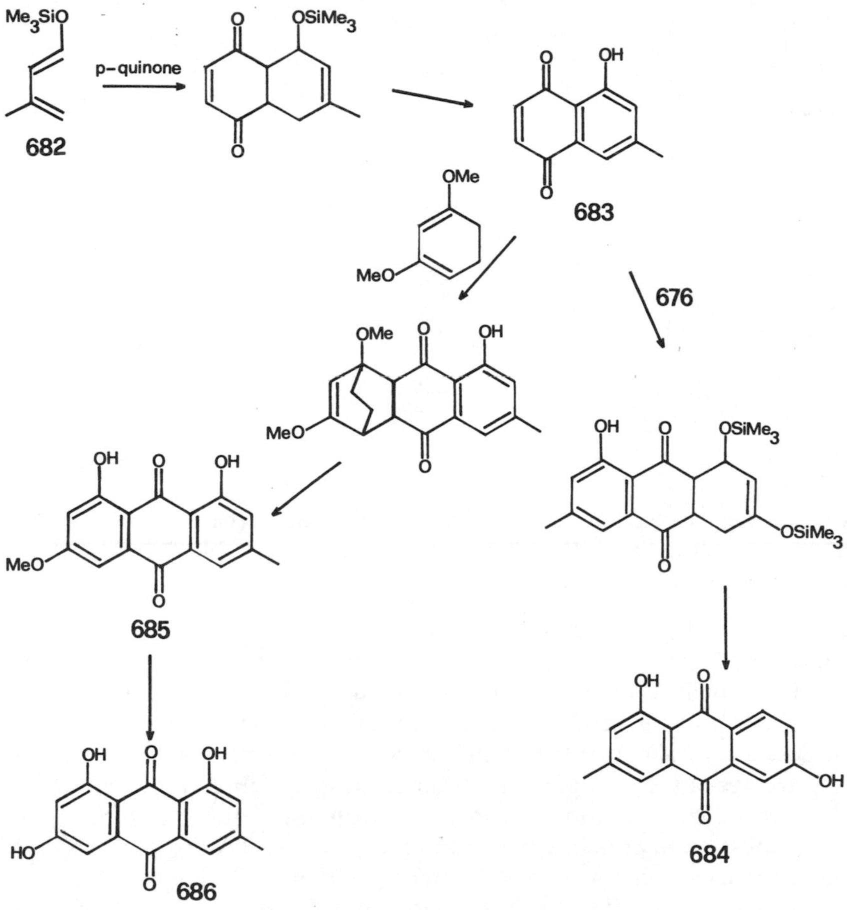

Scheme 5.50

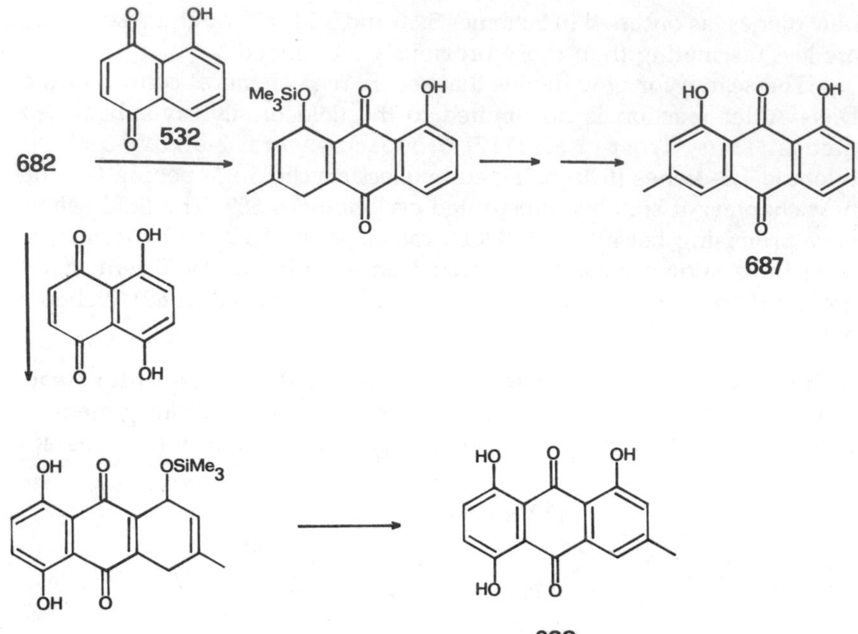

682 → 532 → 687

688

Scheme 5.51

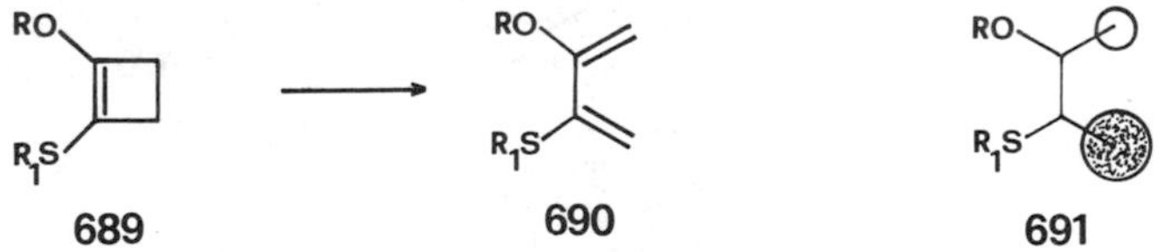

689 → 690 691

Scheme 5.52

proach has become more important in the last decade. Scheme 5.53 summarizes the various methods of generating unstable dienes such as *ortho*-xylylenes (*119*).

ELECTROCYCLIC CYCLOBUTENE RING OPENING. A widely used way of producing dienes is the thermolysis or photolysis of cyclobutenes (A) via a thermally allowed conrotatory or photochemically allowed disrotatory electrocyclic process. The theoretical approach will be considered in detail in Chapter 9. Nevertheless, we wish to point out the crucial importance of the configuration of the starting cyclobutene for obtaining the desired configuration of the product dienes. Thus exposure of *cis*-2,3-disubstituted cyclobutenes **692** to thermal conditions produces (*E*,*Z*) dienes **693**; the *trans* isomers **694**, which could give (*E*,*E*) or (*Z*,*Z*) dienes, open outward to afford only the less sterically hindered (*E*,*E*) dienes

695. Ring opening under photochemical conditions gives the opposite results. All these processes are reversible (Scheme 5.54).

An example of this approach in the syntheses of natural products was the stereocontrolled route to a model of aureolic acid aglycone (*120*) (**700**). Thus cyanobenzocyclobutene (**696**) reacted regioselectively (9:1) with the suitable sugar derivative **697** and stereoselectively (5:1) in the correct regioisomer **698a,b**, finally affording **699**, which possesses the critical stereochemistry of **700** (Scheme 5.55).

Even more interesting was the stereoselective route to pentacyclic triterpenes alnusenone (**704**) and friedelin (**705**): the starting step was the electrocyclic ring opening of alkoxy-substituted **696** to free the diene, which was subsequently trapped with isoprene (*121, 122*). The tetralin derivative **701** thus obtained was transformed into **702**. A second elec-

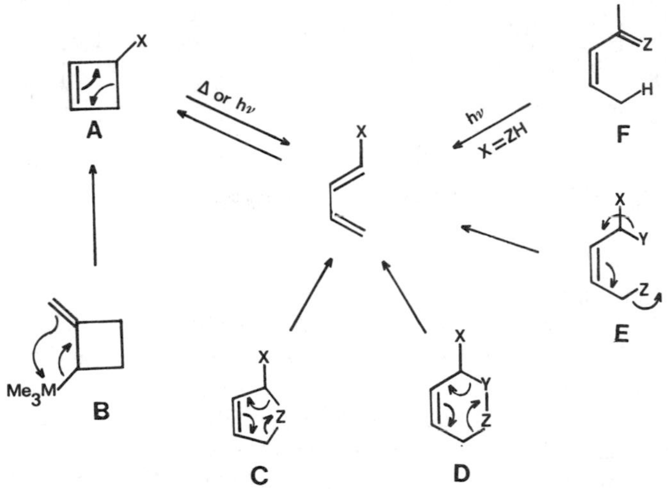

Scheme 5.53

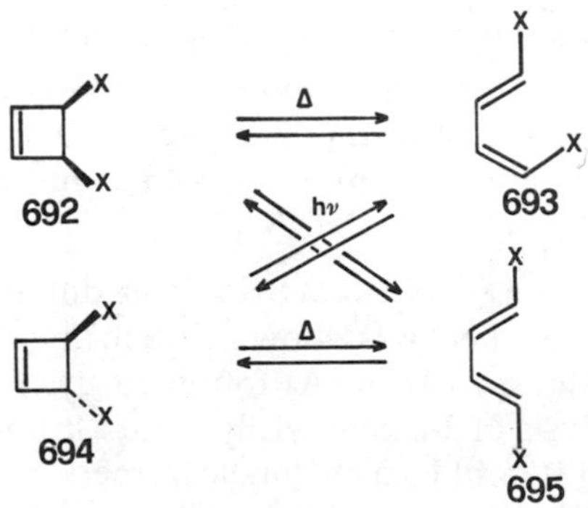

Scheme 5.54

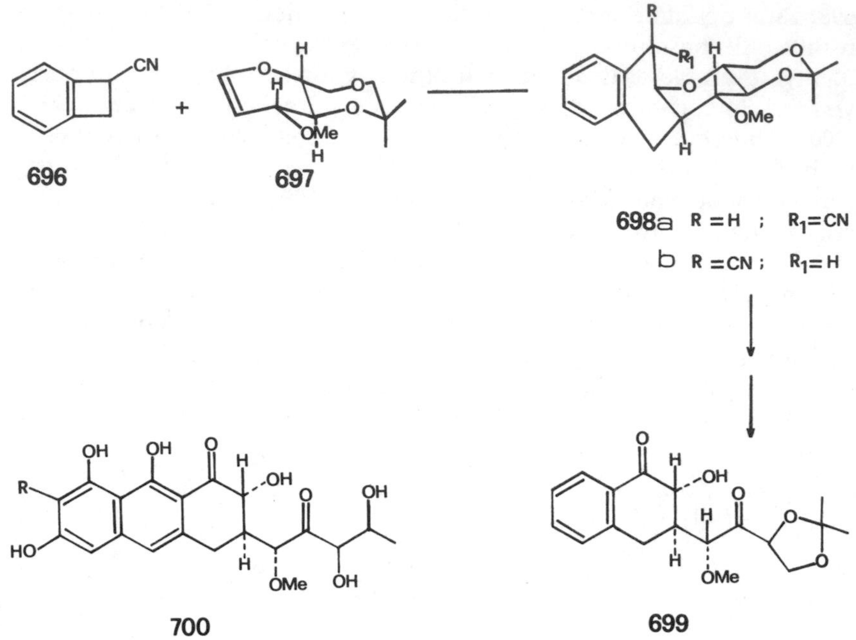

696 + 697 → 698a R = H ; R_1 = CN
b R = CN ; R_1 = H
→ 699 700

Scheme 5.55

trocyclic ring opening followed by intramolecular cycloaddition gives **703**, the key intermediate to **704** and **705** (Scheme 5.56). This brilliant intramolecular approach to polycyclic natural structures is a characteristic of Kametani's work and will be discussed in detail in the section entitled "Electrocyclic Cyclobutene Ring Opening" later in this chapter.

Photochemical ring opening of a benzocyclobutene derivative was a useful approach to the synthesis of islandicin (**710**), which is an anthraquinone of *Penicillum islandicum*, and digitopurpone (**711**) (*123*). Interestingly, the starting material was the Diels–Alder adduct **707** from anthracene and oxidized **706**, which, with vapor-phase pyrolysis, underwent a retro-Diels–Alder reaction to produce 3-(methoxymethyloxy)-benzocyclobuten-1,2-dione (**709**) upon thermal conrotatory ring closure of **708**. The photochemical disrotatory ring opening to **708** in the presence of 2-methylbenzoquinone results in a mixture of **710** and **711** (Scheme 5.57).

ALLYLIC ISOMERIZATION OF CYCLOBUTENES. A different approach to the in situ generation of dienes is shown in Scheme 5.53, which depicts allylic isomerization of (B) to (A) followed by its electrocyclic ring opening. Thermolysis of 2-trimethylsilyl- and 2-trimethylstannylmethylenecyclobutanes (**712a,b**) leads to allylic isomerization to **713a,b**, which open to **714a,b**. Their capture by methyl acrylate gave a mixture of regioisomers **715a,b** and **716a,b** in a ratio dependent upon the nature of metal (M). An easy work-up of the major isomer gave δ-terpineol (**717**) (*124*) (Scheme 5.58).

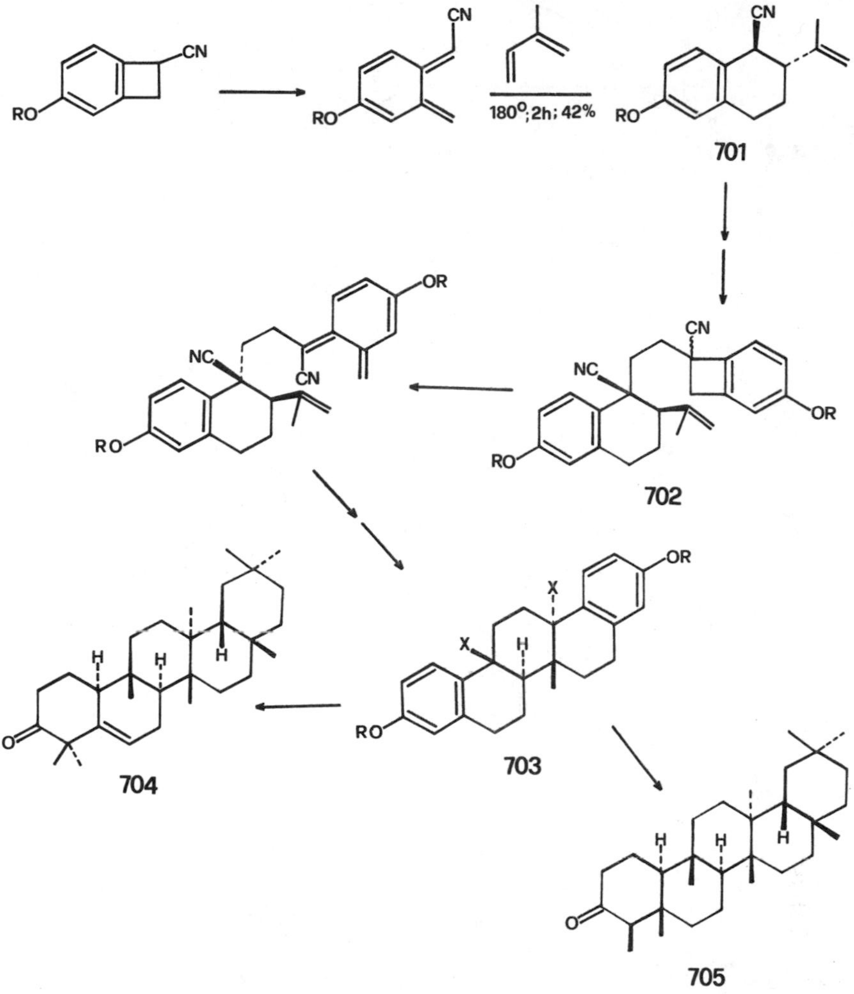

Scheme 5.56

The regioisomer distribution is rationalized in terms of the ability of silicon or tin to stabilize a positive charge on the β-carbon of **718a**. An alternative to the presence of zwitterionic intermediates can be proposed in terms of the electron-donating character of the trimethylstannyl versus trimethylsilyl substituents. The greater the electropositive character of M (Sn > Si), the greater the HOMO coefficient at C1 as depicted in **718b**, thus increasing the regioselectivity of the *para* adduct **540** in Scheme 5.17.

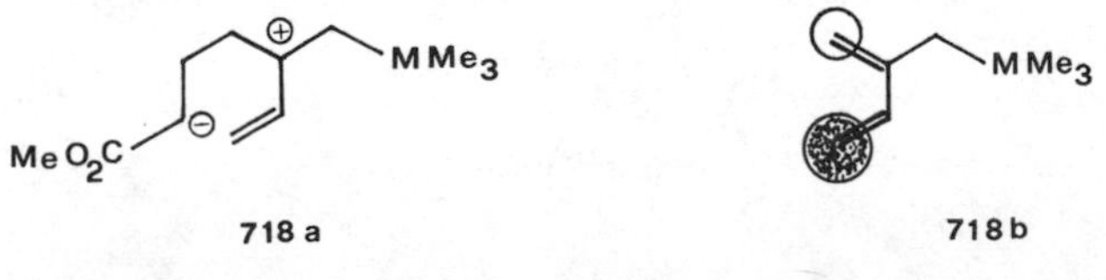

718 a　　718 b

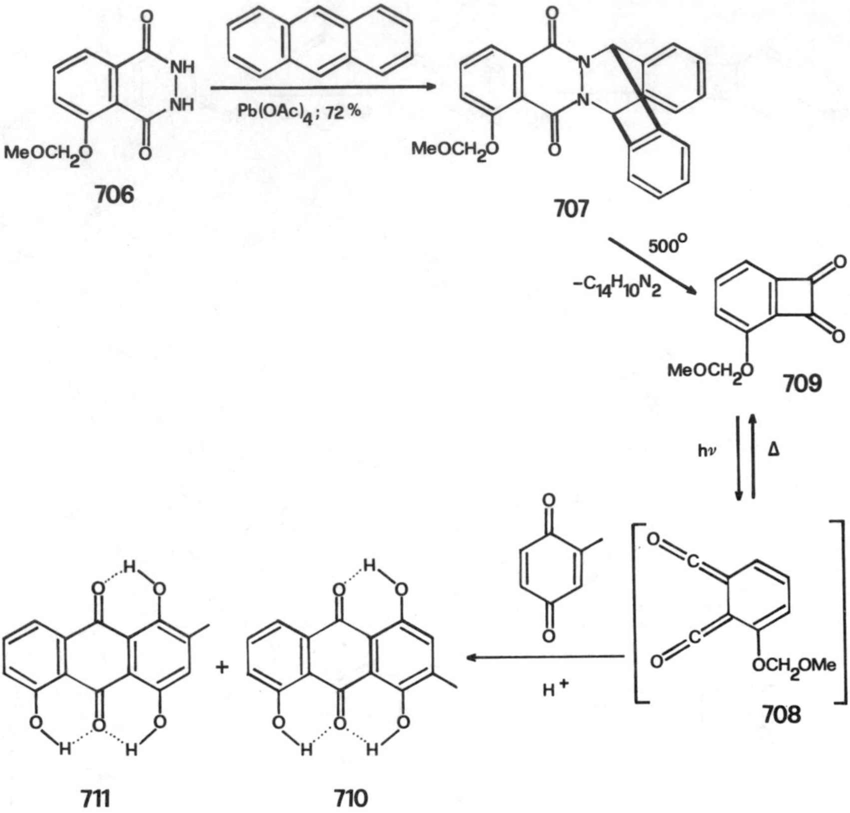

Scheme 5.57

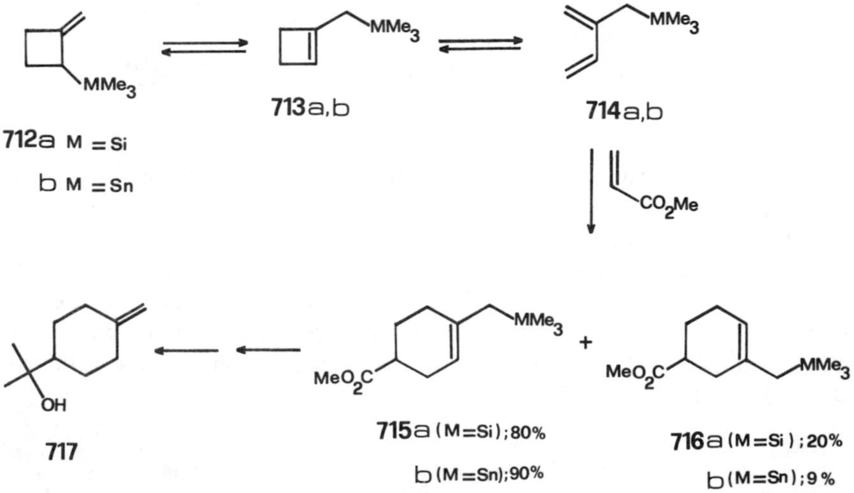

Scheme 5.58

CHELETROPIC REACTION. Several new methods of generating dienes utilize retrocycloadditions (from C and D in Scheme 5.53). In particular, the linear cheletropic extrusion of a molecule Z from a Δ^3-unsaturated five-membered ring is very useful, where Z can be sulfur dioxide from sulfones **91**, (discussed in Chapter 2 in the section entitled "Extrusion Reactions") or any other suitable group. As an example of this approach we report the generation of *o*-quinodimethane (**720**) and its reaction with 1,4-naphthoquinone to give **721**, a model for a straightforward route to tetracycline derivatives (*124*). Thus **720** results both from a linear cheletropic reaction of **719** (obtained by oxidation of *N*-aminodihydroisoindole with mercuric oxide) with loss of nitrogen (Scheme 5.59) and from *o*-xylyl dibromide (**722**) by elimination of bromine, as discussed in the next section.

ELIMINATION. The elimination of halogens (from E in Scheme 5.53), usually bromine, from 1,2-disubstituted *o*-xylenes such as **722** in Scheme 5.59, provides a useful route to *o*-xylylenes **720**, which are very reactive dienes. The choice of 2-bromomethyl-3-dibromomethylanisole (**723**) as a source of the bromomethoxy-substituted **720** and its reaction with 6-bromotoluquinone or 5-bromotoluquinone (**726**) allowed a regioselective synthesis of **725** and **727**. These two compounds were easily converted into islandicin (**710**) and digitopurpone (**711**) (*126*) (Scheme 5.60).

Under dry conditions both reactions are highly regioselective; a mixture of **725** and **727** in the ratio 92:8 or 8:92 was obtained from **724**

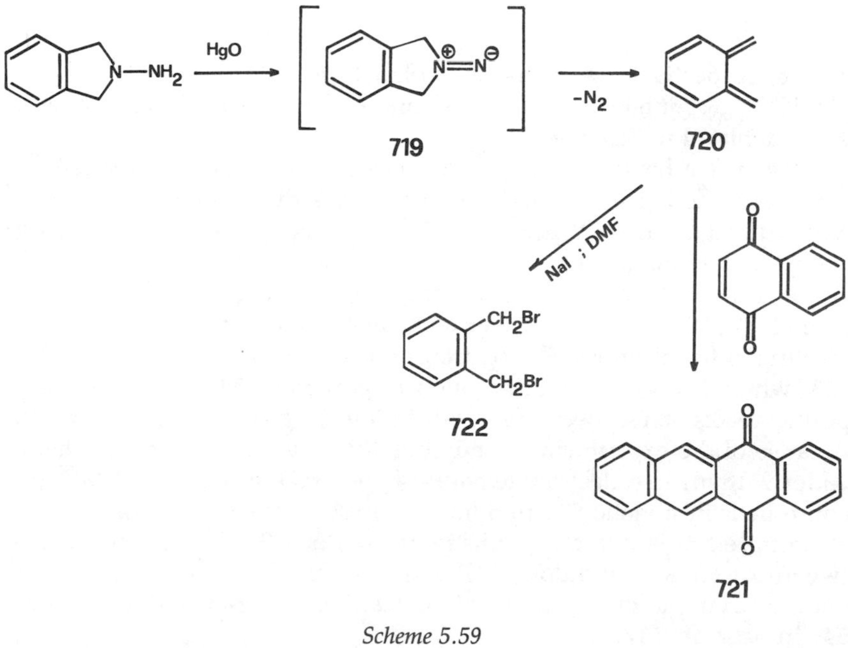

HgO

N—NH_2

N=N

$-N_2$

719

720

NaI ; DMF

CH_2Br

CH_2Br

722

721

Scheme 5.59

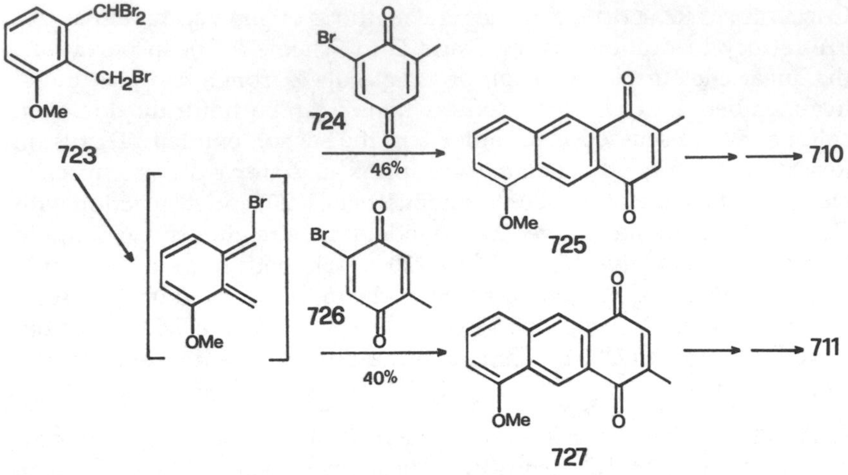

Scheme 5.60

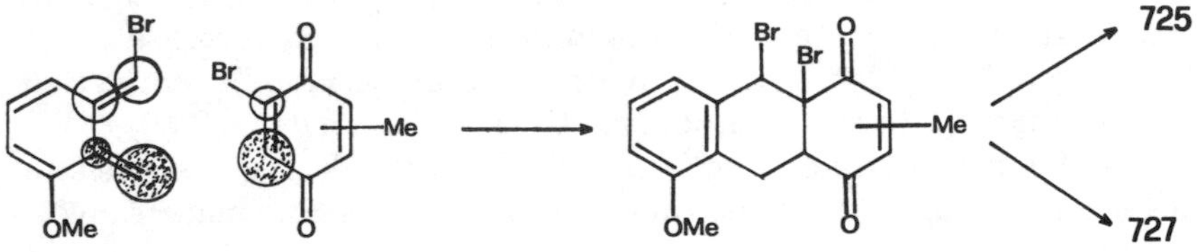

Scheme 5.61

or **726**, respectively. A plausible explanation in terms of $HOMO_{diene}/LUMO_{dienophile}$ interaction can be advanced if we assume their shape to be as depicted in Scheme 5.61.

Again we have an example of a Diels–Alder reaction between an electron-attracting substituted (Z)-diene and a dienophile leading to an *ortho*-adduct (such as **542** in Scheme 5.17) through primary [or secondary (*127*)] orbital interactions.

Another good example is the description of the synthesis of olivacine (*128*) (**731**), an alkaloid with antitumor and antileukemic activity. Heating of the alcohol **728** in hydrobromic acid produces the dibromide **729**, which is used without isolation to generate **730**. This latter compound reacts stereospecifically with indole to give **731** (Scheme 5.62). Because of the experimental conditions this is a nice example of Diels–Alder with inverse electron demand—hence $HOMO_{dienophile}/LUMO_{diene}$ controlled. Protonation to pyridinium ion **729** makes **730** a diene with a strong electron-attracting substituent (Z) at C2. The interaction between the HOMO of indole (*129*) and the LUMO of **730** (Figure 5.7) is another example of MO interactions leading to a *para*-adduct such as **545** in Scheme 5.17.

Elimination may also involve groups other than halogens. Thus methanol loss by refluxing hydroxyacetals **732a–d** in benzene with a trace of *p*-toluenesulfonic acid generates the isobenzofurans **733a–d**. These latter compounds condense with dimethyl acetylenedicarboxylate (DMAD) to yield (via the tricyclic adducts **734**) the naphthols **735**, which are easily converted into seven naturally occurring 1-arylnaphthalide lignins (*130*): dehydropodophyllotoxin (**736a**), taiwanin E (**736b**) and its methyl ether (**736c**), diphyllin (**736d**), justicidin A (**736e**), and chinensinaphthol (**736g**) and its methyl ether (**736g**) (Scheme 5.63).

PHOTOENOLIZATION. Under photochemical conditions, *ortho*-alkyl substituted aromatic aldehydes **737a,b** (or ketones) undergo intramolecular hydrogen abstraction (*131*) (*see* F in Scheme 5.53). Promoted by the triplet $n\pi^*$ state, this reaction gives a relatively long-lived dienol intermediate **738a,b** through a process that can be summarized as

$$^{0}\text{Aldehyde} \rightarrow {}^{1}\text{Aldehyde} \rightarrow {}^{3}\text{Aldehyde} \rightarrow {}^{3}\text{Enol} \rightarrow {}^{0}\text{Enol}$$

The intermediate can be efficiently trapped with suitable dienophiles, giving Diels–Alder adducts such as **739a** and **b**. This process was applied to the synthesis of four natural lignins: dehydropodophyllotoxin

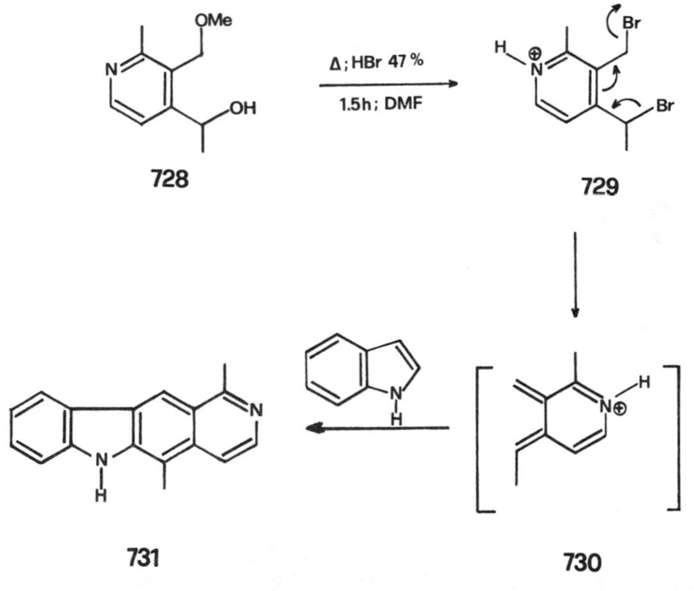

Scheme 5.62

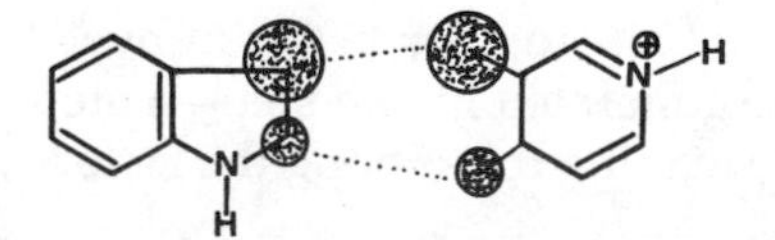

*Figure 5.7. FMO interactions between indole and **730**.*

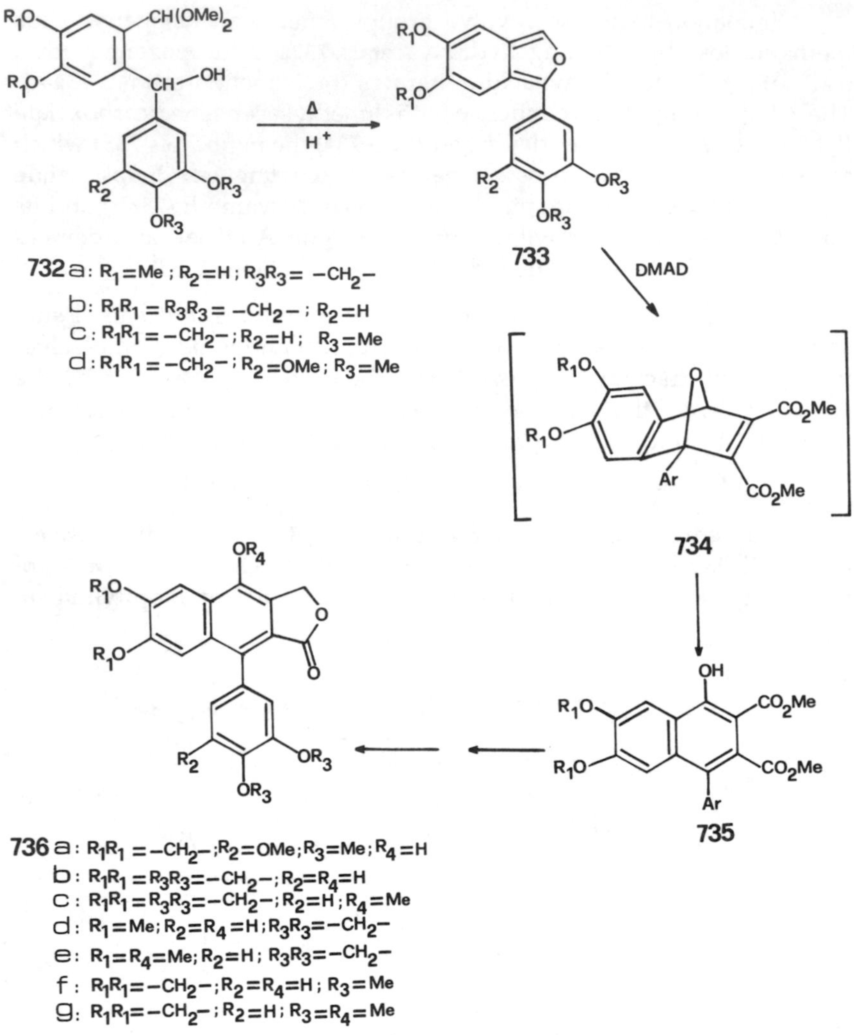

R_1O $CH(OMe)_2$ R_1O CH—OH R_2 OR_3 OR_3 Δ H^+ R_1O R_1O O R_2 OR_3 OR_3

732 a: $R_1=Me$; $R_2=H$; $R_3R_3= -CH_2-$
b: $R_1R_1 = R_3R_3 = -CH_2-$; $R_2=H$
c: $R_1R_1 = -CH_2-$; $R_2=H$; $R_3=Me$
d: $R_1R_1 = -CH_2-$; $R_2=OMe$; $R_3=Me$

733 DMAD

R_1O O CO_2Me R_1O Ar CO_2Me

734

OH R_1O CO_2Me R_1O CO_2Me Ar

735

OR_4 R_1O O R_1O O R_2 OR_3 OR_3

736 a: $R_1R_1 = -CH_2-$; $R_2=OMe$; $R_3=Me$; $R_4=H$
b: $R_1R_1 = R_3R_3 = -CH_2-$; $R_2=R_4=H$
c: $R_1R_1 = R_3R_3 = -CH_2-$; $R_2=H$; $R_4=Me$
d: $R_1=Me$; $R_2=R_4=H$; $R_3R_3 = -CH_2-$
e: $R_1=R_4=Me$; $R_2=H$; $R_3R_3 = -CH_2-$
f: $R_1R_1 = -CH_2-$; $R_2=R_4=H$; $R_3=Me$
g: $R_1R_1 = -CH_2-$; $R_2=H$; $R_3=R_4=Me$

Scheme 5.63

(**736a**), taiwanin E (**736b**), justicidin E (**740**), and taiwanin C (**511b**) (*132*) (Scheme 5.64).

Intramolecular Diels–Alder Reactions

A review of intramolecular Diels-Alder reactions (IDAs) is both easy and intriguing. In 1974 (*133*), 1977 (*134*), and 1980 (*135*) the field was reviewed, hence a large part of the material already has been arranged. However, at least three chiral centers are generated in one step—a step in which MO interactions do not play the leading role among the factors involved. This characteristic makes IDAs one of the most intriguing topics covered in this book.

As was clearly pointed out by Brieger and Bennet (*135*), the main feature of IDAs is reflected by the thermodynamic parameters of pentadienylacrylamide (**741**) cycloaddition, which yields equal amounts of *trans* and *cis* adducts **743a,b**, the former through an *exo* **742a**, the latter through an *endo* **742b** transition state (*136*) (Scheme 5.65).

Although the free energy of activation ($\Delta G^{\neq}_{298}$ = 25.3 kcal/mol) falls within the range of a typical Diels–Alder reaction, the entropy value ($\Delta S^{\neq} = -14.4$ e.u.) is less than half of the value usually found for an intermolecular cycloaddition. This finding indicates that some effort is required to adopt a highly ordered transition state.

The simultaneous generation of at least three chiral centers is the result of a preferred *exo* or *endo* transition state. Before discussing this point, however, we will consider the regioselectivity of IDAs through examples from the natural products field.

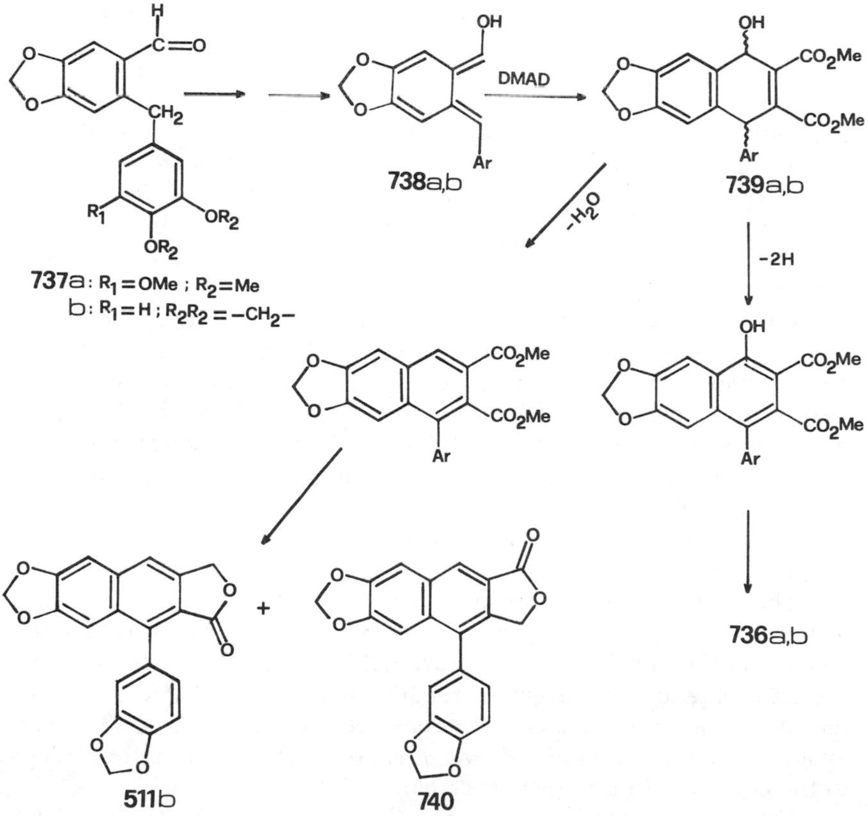

Scheme 5.64

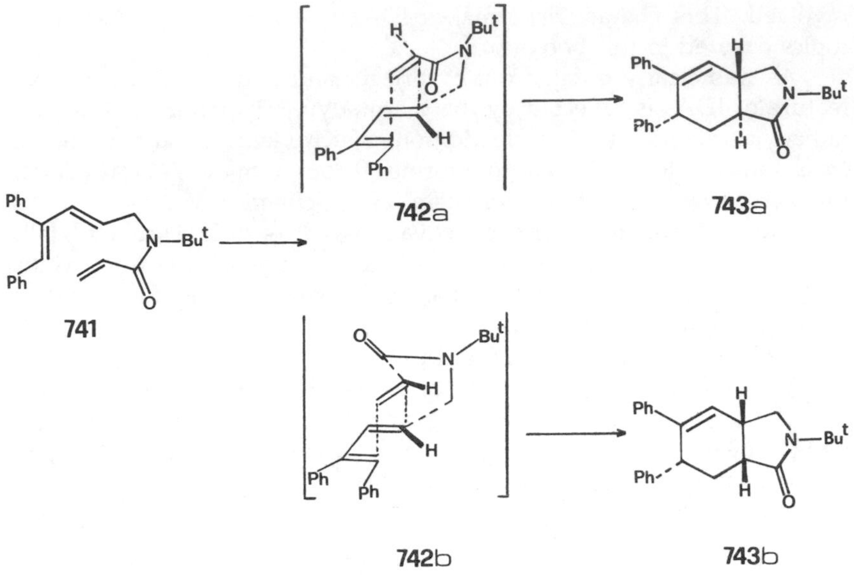

Scheme 5.65

Regioselectivity. The crucial point in determining regioselectivity lies in the configuration of the diene at the substituted end to which the ansa carrying the dienophile fragment is connected. We define "ansa" as the part of the molecule that connects diene and dienophile in IDAs.

For an (*E*)-diene **744**, the degree of freedom to produce **745** or **746** is much lower than that for a (Z)-diene **747** (Scheme 5.66).

For an (E)-diene the ansa must include about 12 atoms to afford both regioisomers, with the strained one, **745**, predominating. In the approach to the macrolide cytochalasin, whose isoindolone nucleus has also been synthesized by IDA (*137*) (this point will be considered in detail later), the diene-anhydride **748** was refluxed in toluene under high dilution conditions to minimize intermolecular reactions. This gave rise to a mixture of two products, **749** and **750**, in the ratio 27:5 (*138*). The primary product was the less strained one with the framework of cytochalasin B (Scheme 5.67).

This result seems to be to date the only example involving the formation of two regioisomers from an (*E*)-diene. Although (Z)-dienes give regioisomeric mixtures more easily, this result does not occur along the IDA route to cytochalasin C from (Z)-diene (*139*) **751**. Here the preferred conformation **751a** causes a regiospecific cyclization to **752**, thus avoiding the formation of **753**, which is not suitable as an intermediate to the cytochalasin ring (Scheme 5.68).

Two regioisomers are formed in the synthesis of (±)-seychellene (**640**) [an (*E*)-diene], where the IDA cyclization of **754** affords a 1:3 mixture of the regioisomers **755** and **756**. The latter substance can be transformed into **664** (*140*) (Scheme 5.69).

Three cyclohexadiene derivatives (**757**, **758**, and **759**) are the starting materials for a different approach to the same (±)-seychellene (**664**) (*141*) and for the synthesis of two components of patchouli oil, an important raw material for the perfume industry. The two components are patchouli alcohol (**760**) (*142*) and norpatchoulenol (**761**) (*143*). Although structures **757**–**759** are similar to that of **754**, only one regioisomer is formed. To further complicate the matter, the regioisomer

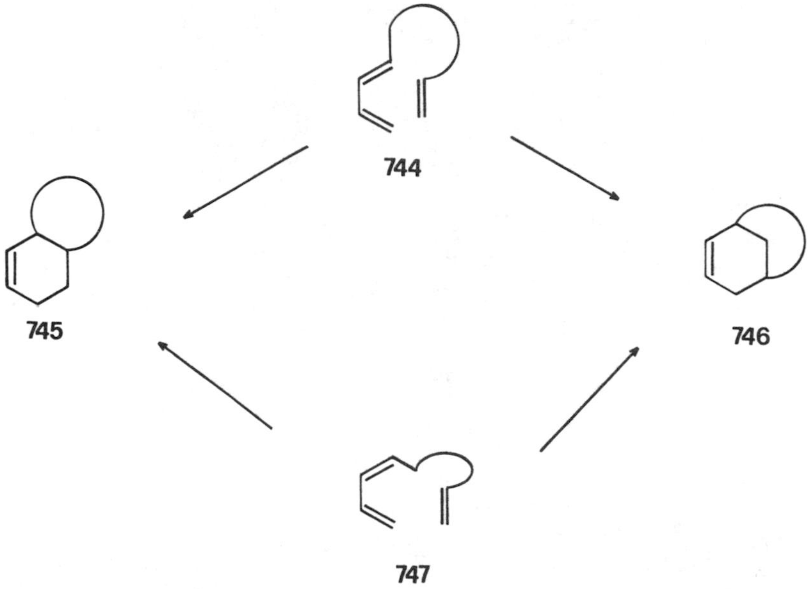

744

745

746

747

Scheme 5.66

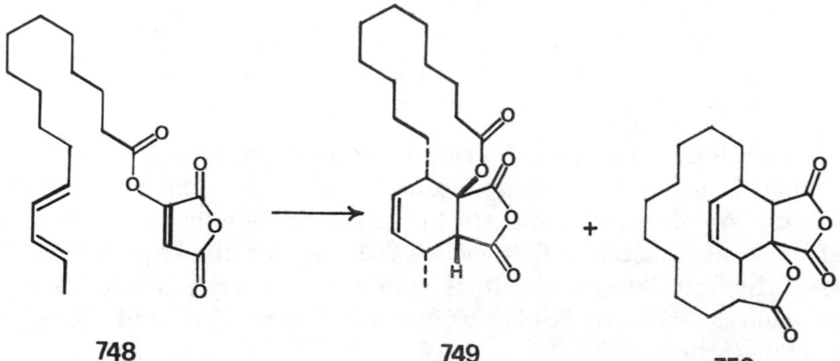

748

749

750

Scheme 5.67

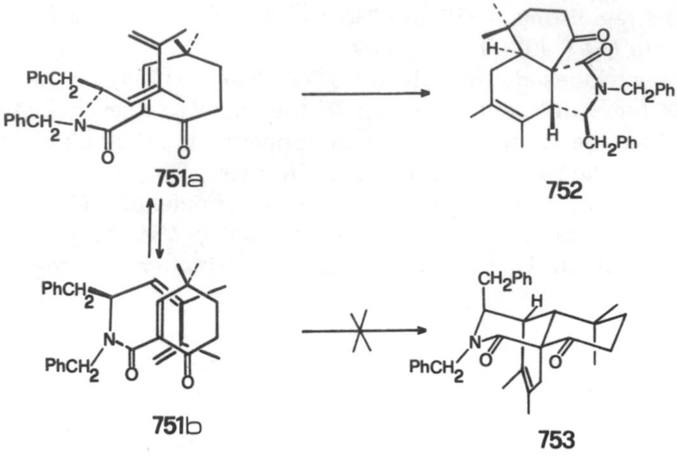

Scheme 5.68

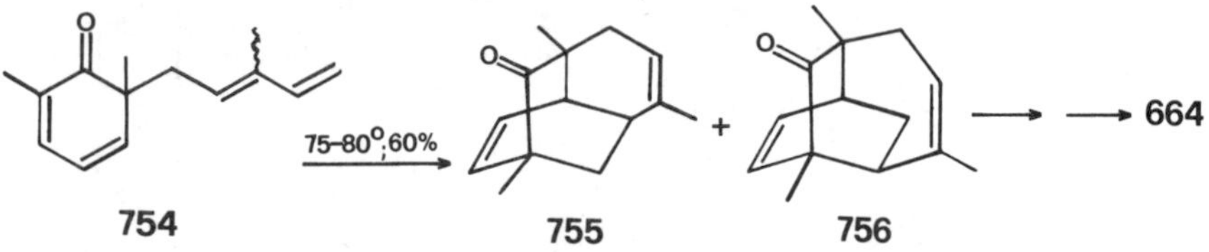

Scheme 5.69

formed is the one whose structure corresponds to that of the minor regioisomer **755** obtained from **754** (Scheme 5.70).

The main difference between **754** and **757–759** is the presence of a second double bond in the chain in the transition state leading to **756**. This bond, which lies above the cyclohexadiene carbonyl, could stabilize the nonbonding interaction between the HOMO of the C1 = C2 − C3 = C4 − C5 = O system and the LUMO of the C1′ = C2′ − C3′ = C4′ system. C5 and C4′ are the specific atoms involved. See Figure 5.8 for this transition state; a similar interaction can be found if the opposite LUMO/HOMO interaction is taken into account.

Stereochemistry. In general the *endo* versus *exo* preference in the intermolecular Diels–Alder reaction is due to the nonbonding interactions of the dienophile substituents. In the IDA such is not the case. Roush et al. (*144*) demonstrated that trienes **762** and **763** afford preferentially *trans*-perhydroindene cycloadducts, independent of the dienophile stereochemistry; the ratios **764**:**765** and **766**:**767** were 65:35 and 60:40, respectively (Scheme 5.71).

Therefore we will refer to *exo* or *endo* transition state based exclusively on the configuration of the ansa **742a** or **742b**, respectively.

Several transition states are too strained, for example, *endo* transition states for (*Z*)-dienes. This is the reason why **768** and **769** cyclize through *exo* transition states to give **770** and **771** in the key steps of the syntheses of (±)-9-isocyanopupukeanane (**772**) (*145*) and (±)-9-pupukeanone (**773**) (*146*). The former compound is a constituent of a defensive secretion of a marine invertebrate (Scheme 5.72). For the more flexible (*E*)-dienes **744**, both transition states are conceivable a priori, with a crucial point concerning the number of atoms forming the ansa.

If the ansa comprises three atoms, both transition states cán be developed without significant angle strain (*147*). Thus the choice between *exo* **774** and *endo* **775** transition states depends on conformational energy, even if the formation of the new bond in the *endo* transition state occurs with an unfavored partially eclipsed arrangement (*135*). Apart from other factors, the *exo* transition state should be favored (Scheme 5.73).

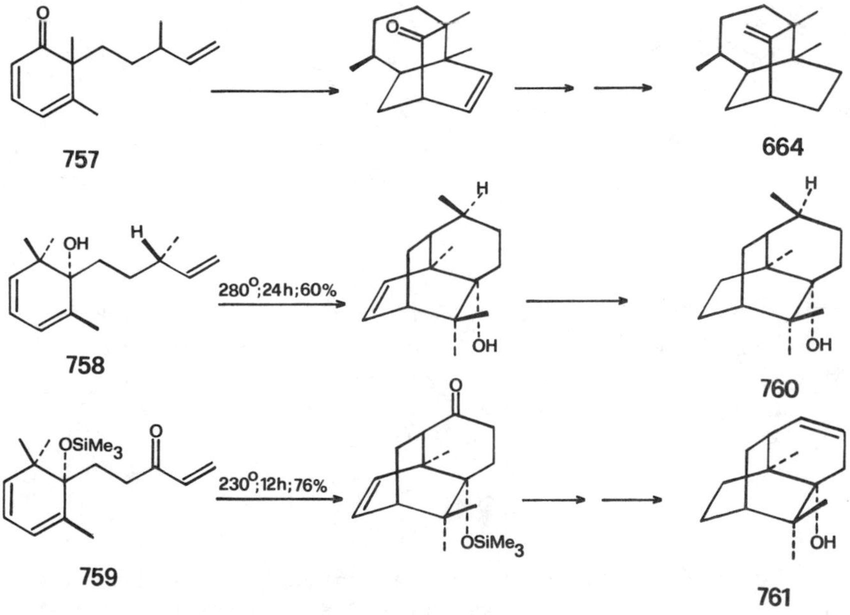

757 664

758 280°;24h;60% OH OH 760

759 230°;12h;76% OSiMe$_3$ OSiMe$_3$ OH 761

Scheme 5.70

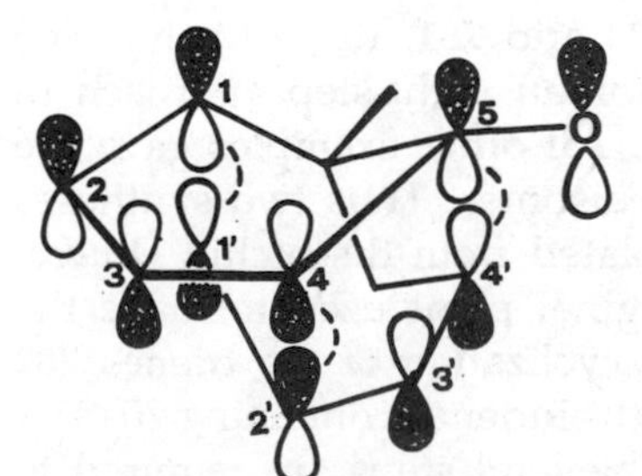

Figure 5.8. Transition state leading to **756** *with* endo *stabilizing, nonbonding interactions.*

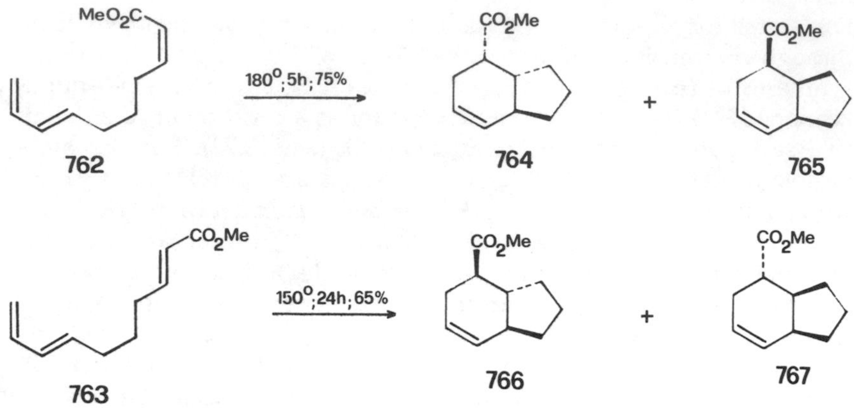

Scheme 5.71

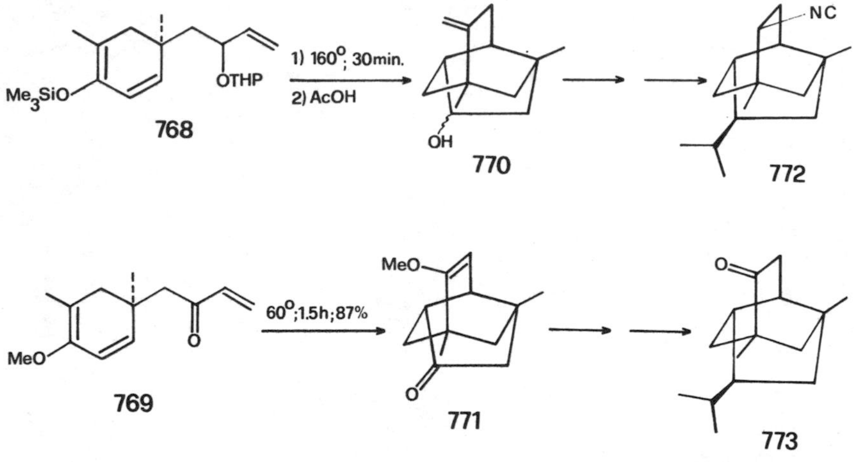

Scheme 5.72

The synthesis of (±)-pumiliotoxin (**555**) (*147*) via the *cis*-fused indanol **777**, which is the major isomer (**777**:**778** = 2:1) obtained from **776**, is the only example of predominating *endo* cycloaddition (Scheme 5.74).

The cyclization of **779** by IDA produces an approximately 1:1 ratio of *endo* and *exo* (isomerized) adducts (**780** and **781**, respectively). The former adduct is the starting material for an eight-step approach to marasmic acid (**525**) (*148*) (Scheme 5.75). All other examples gave the *exo* adduct as the major or the sole stereoisomer. Thus two syntheses of (±)-dendrobine (**786**), the alkaloid isolated from the orchid *Dendrobium nobile* Lindl (thought to be the original plant used to make the chinese drug Chin-Shih-Hu), utilize the cyclization of the trienes **782** and **783**. A mixture of *trans* and *cis* perhydroindenes containing 70–83% of the *exo* isomers **784** and **785** results. Several steps are required to transform the single or both stereoisomers into **786** (*149, 150*) (Scheme 5.76).

Analogously, the synthesis of the norsesquiterpene khusimone (**790**) proceeds by an IDA of the trienone ketal **787**, affording the two epimeric ketones **788** and **789** in the ratio 3:1. The low yield isomer (the *endo* one) gives **790** (*151*) (Scheme 5.77).

In the synthesis of cytochalasan, the octahydroisoindolone (**793**), which possesses adaptable functional groups for the attachment of the macrocyclic ring system, was obtained by IDA of **791** via **792** (*152*). The reaction occurs stereospecifically through an *exo* transition state; such is

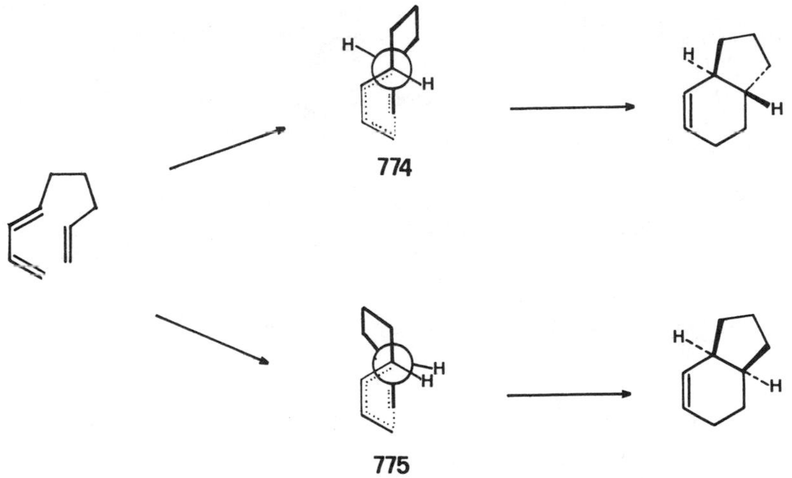

774

775

Scheme 5.73

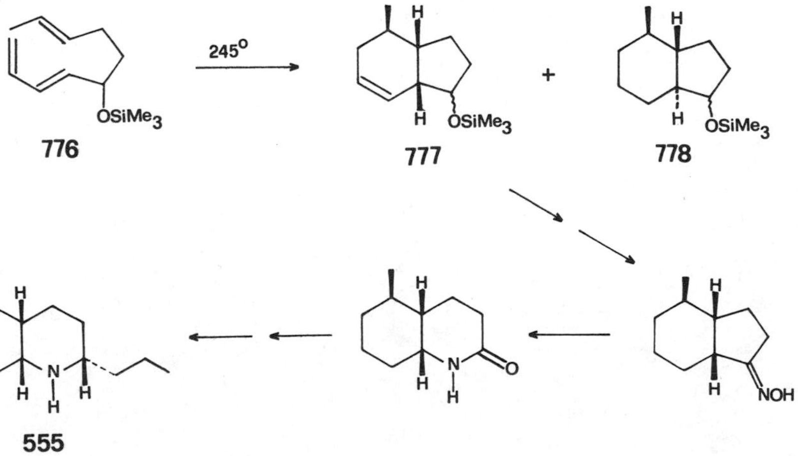

245°

OSiMe3

776

777

778

555

Scheme 5.74

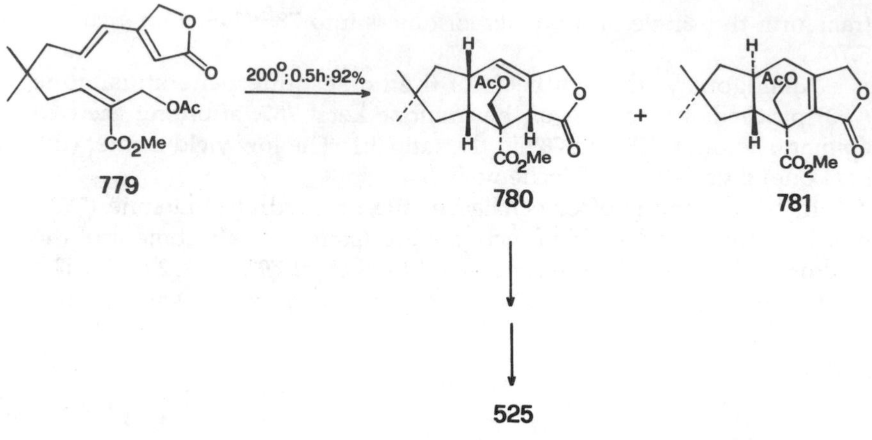

779 → (200°; 0.5h; 92%) → 780 + 781

780 → → 525

Scheme 5.75

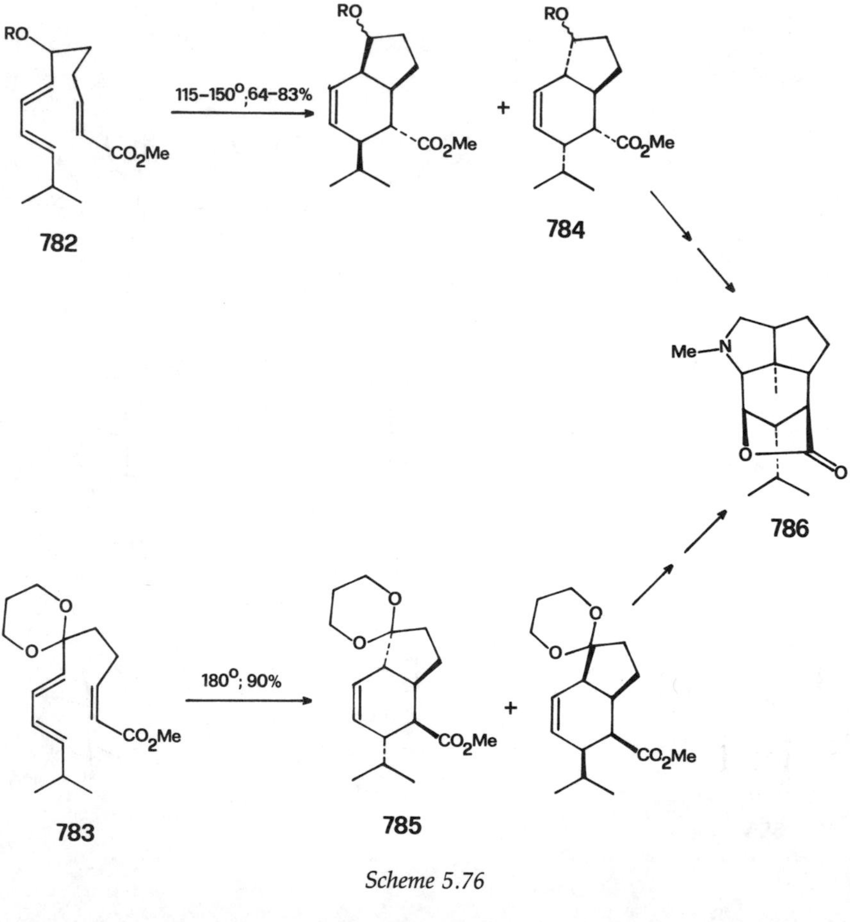

782 → (115–150°; 64–83%) → + 784 → → 786

783 → (180°; 90%) → 785 + → → 786

Scheme 5.76

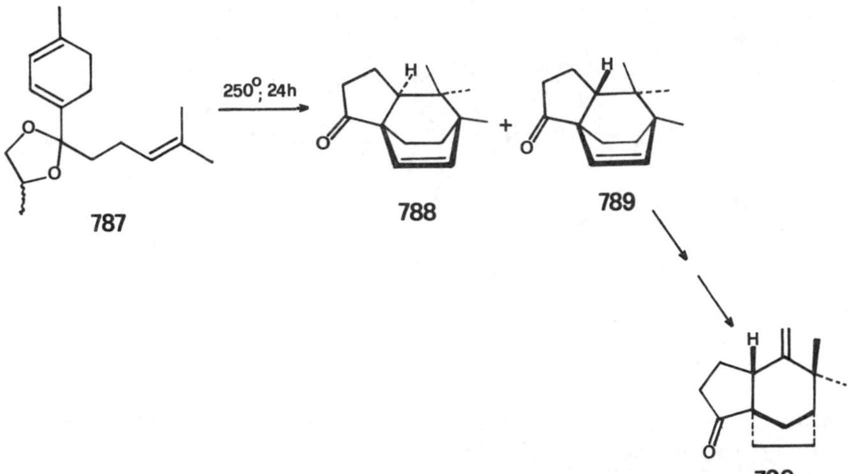

787 250°; 24h 788 + 789 790

Scheme 5.77

also the case in the total synthesis of (±)-cedrol (**796**) from the alkenyl cyclopentadiene **794**. This latter compound cyclizes to give **795** (*153, 154*). Dehydration afforded (±)-α-cedrene (**797**) and (±)-β-cedrene (**264**) in quantitative yield (ratio 80:20) (Scheme 5.78).

When the ansa has four atoms (with the diene still in the *E* configuration), analysis of the overall results leads to some general conclusions.

Aromatic systems do not permit easy predictions because similar substrates behave differently, illustrating the delicate nature of the interactions controlling the stereochemistry of IDAs. Two approaches to the galanthan ring system **803**, which forms the basic skeleton of alkaloid lycorine, were performed by Stork and coworkers (*155, 156*). One synthesis started from **798** and gave a mixture of **799** and **800**; the *endo* product **799** prevailed slightly (1:0.84). The other synthesis started from **801** and gave exclusively the *exo* product **802**. This latter product was easily converted into 7-oxo-α-lycorane (**803**), identical with the material prepared from (±)-galanthan (α-lycorane) (**552**) (Scheme 5.79).

If the originally formed adduct aromatizes spontaneously, as in the synthesis of lachnanthocarpone (**805**), the question of whether **804** is *exo* or *endo* becomes irrelevant from the synthetic point of view (*157, 158*) (Scheme 5.80).

In general the *exo* transition state is preferred if the triene is not suitably substituted (discussed later). The same reasons for favoring **774** over **775** may be applicable here for two versus three *gauche* interactions. In Scheme 5.81, several examples are reported that outline the syntheses of (±)-selina-3,7(11)-diene (**810**) (*159*), the sesquiterpene isolated from the steam-volatile oil of hops; (±)-epizonarene (**811**) (*160*); (±)-fichtelite (**812**) (*161*), a hydrocarbon isolated from Bavarian pines; and the key

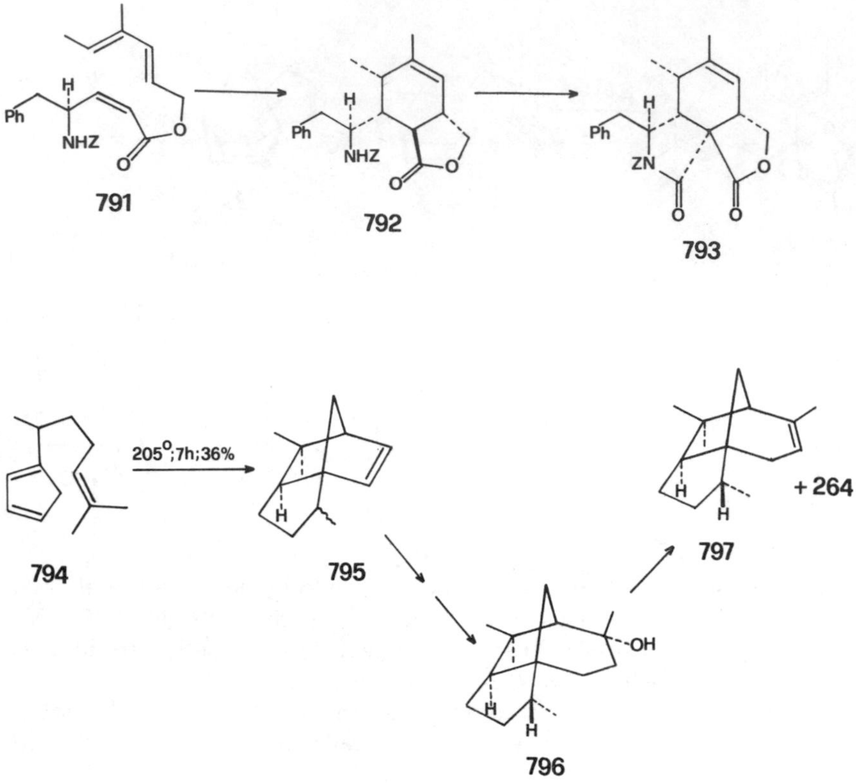

791 792 793

205°;7h;36%

794 795 796 797 + 264

Scheme 5.78

intermediate (**813**) used for a general stereoselective entry to eremophilane (**814**) and valencane (**815**) sesquiterpenes (*162*). Cyclization products **806–809** are *trans*-fused compounds, thus arising from *exo* transition state (Scheme 5.81).

Two specific points strongly favor the *endo* transition state. The first involves an amide group conjugated with the diene, with the nitrogen atom incorporated in the ansa. The *endo* transition state **817** allows a better overlap of the amide π-orbitals with the diene π-system. This situation is shown in the synthesis of (±)-pumiliotoxin C (**555**) starting from **816**, which cyclized to only **818** (*163, 164*). This approach allowed the enantioselective synthesis of (−)-pumiliotoxin C as well (*165*) (Scheme 5.82).

However, the *endo* transition state is strongly favored when the ansa carries a carbonyl group conjugated with the dienophile. An explanation that invokes secondary nonbonding interactions between the HOMO of the diene fragment and the LUMO of the α,β-unsaturated carbonyl seems appropriate (*166*) (Figure 5.9).

The cyclization of **819** to **820** and **821** in the ratio 1:9, a very favorable event in the synthesis of the sesquiterpene alcohol (±)-torreyol (**822**)

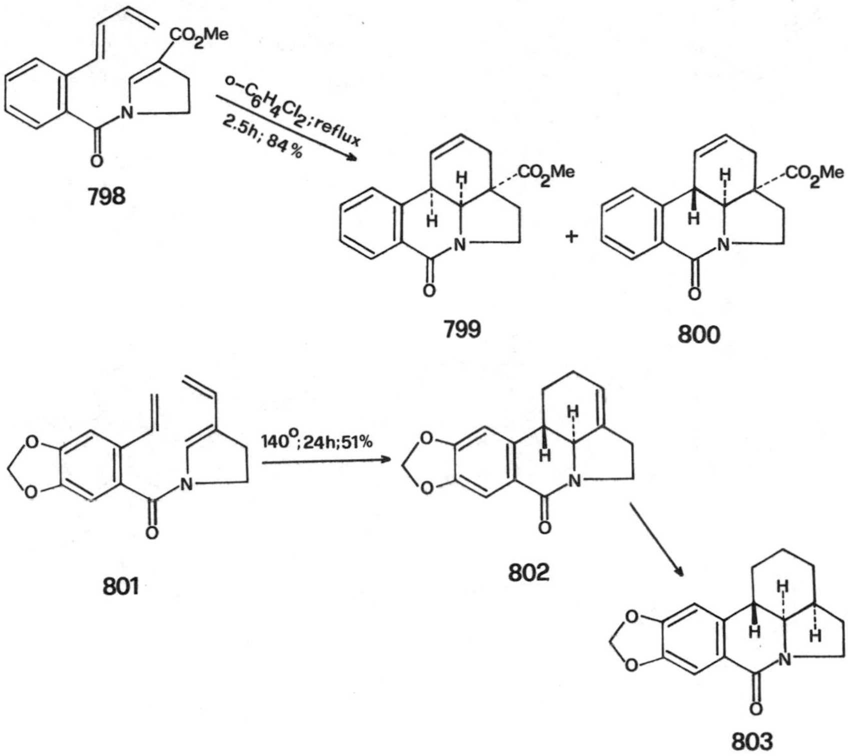

CO_2Me

o-$C_6H_4Cl_2$; reflux
2.5h; 84 %

798

799 + 800

140°; 24h; 51%

801

802

803

Scheme 5.79

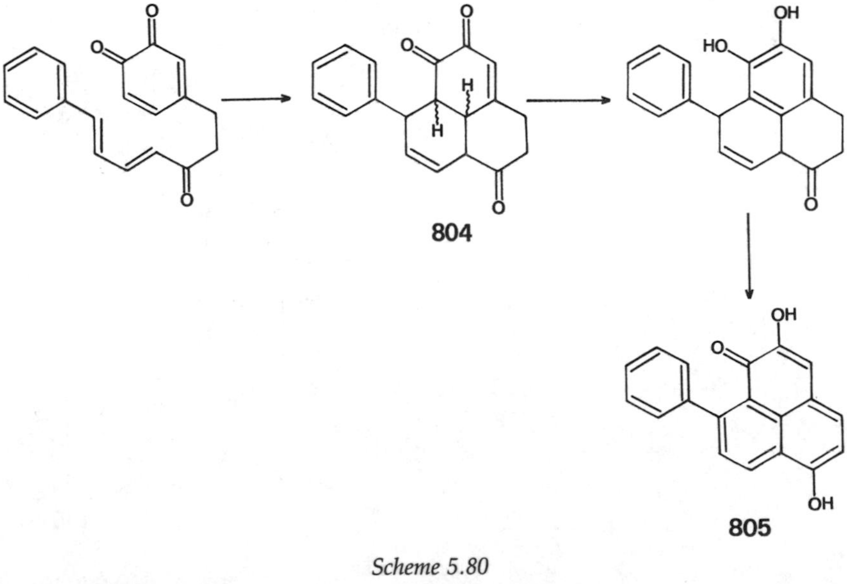

804

805

Scheme 5.80

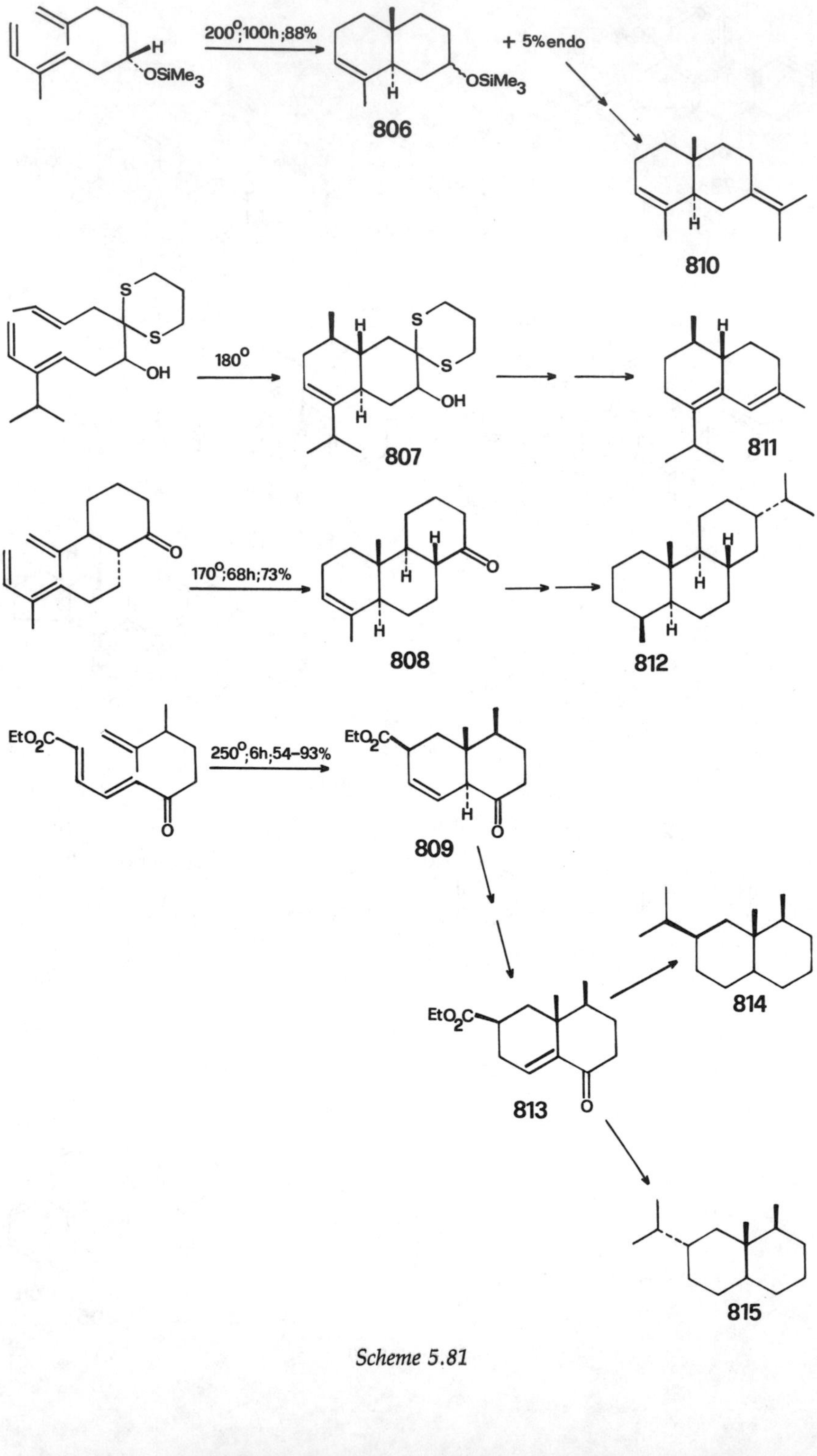

200°;100h;88%

H

OSiMe$_3$

H

OSiMe$_3$

806

+ 5% endo

H

810

S

S

OH

180°

H

S

S

OH

H

807

H

811

O

170°;68h;73%

H

H

O

H

808

H

H

H

812

EtO$_2$C

O

250°;6h;54–93%

EtO$_2$C

H

O

809

EtO$_2$C

O

813

814

815

Scheme 5.81

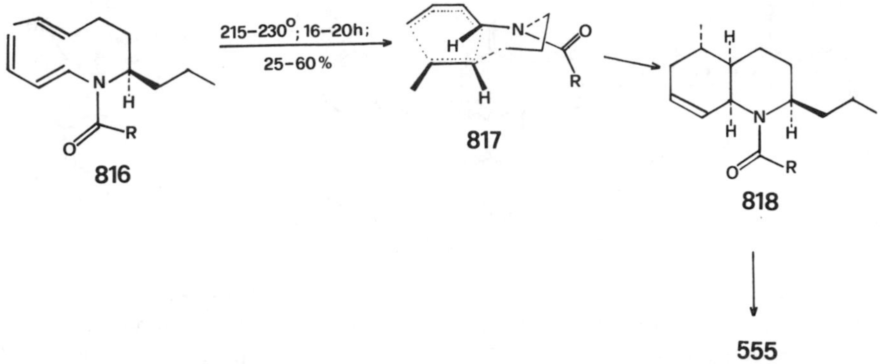

215–230°; 16–20h; 25–60%

816 817 818

555

Scheme 5.82

Figure 5.9. The endo *transition state that explains the preferred formation of* ***821*** *from* ***819.***

(*167*), and the *endo* cyclization of **586** to **587** in the previously mentioned synthesis of gibberellic acid (**589**) (*75, 76, 168*), are examples of such reactions (Scheme 5.83).

Biomimetic Synthesis of Alkaloids. Wenkert (*169*) first proposed and Scott (*170–173*) and Battersby et al. (*174*) subsequently proposed that the facile interconversion observed experimentally of the *Corynanthe, Strychnos, Aspidosperma,* and *Iboga* alkaloids takes place by way of a 2-vinylindolyl dihydropyridine or tetrahydropyridine (**826** and **831**, respectively), which can undergo IDA.

The presence of **831** along the metabolic pathway was supported by the isolation of **823**, an alkaloid with secodine skeleton, from *Rhazya orientalis* (*175*). Additional support was the formation of the dimeric products presecamine (**824**) or its regioisomer ($R=CH_2\text{-}CH_2\text{-}NC_7H_{12}$) from *R. stricta* by a Diels–Alder reaction (*176*).

The relationship between the *Aspidosperma* alkaloid tabersonine (**827**) and the *Iboga* alkaloid catharanthine (**828**) was indicated by in vitro transformations of stemmadenine (**825**) into **827** and **828** through the hypothetical dihydrosecodine intermediate **826** (*177–180*).

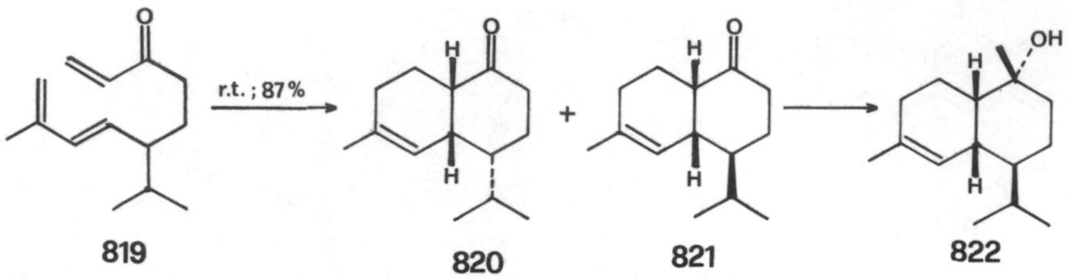

r.t. ; 87%

819 820 821 822

Scheme 5.83

Iboga- or *Aspidosperma*-like structures may alternatively arise, depending on whether the dihydropyridine fragment behaves as a diene or as a dienophile. In both cases, the reaction occurs through an *exo* transition state (Scheme 5.84).

When the open intermediate has a secodine-like structure, the pyridine fragment can act only as dienophile, blocking the *Iboga* route.

Thus **831**, obtained from **829** (*181, 182*) or **830** (*183*), undergoes IDA through an *exo* transition state, giving rise to the *Aspidosperma* alkaloids vincadifformine (**832a**), minovine (**832b**), and ervinceine (**832c**) (Scheme 5.85).

Two other alkaloids have been prepared through routes that, in spite of the harder conditions, could be considered biomimetic (*158*). This classification is particularly significant in the synthesis of andranginine (**835**), an indole alkaloid isolated from *Craspidospermum verticillatum* Boj. var. *petiolare,* which was obtained by thermolysis of the precondylocarpine acetate (**833**) through **834** (*184*) (Scheme 5.86).

Analogous relationships exist between two Piperaceae alkaloids: piperstachine (**836a**) and cyclopiperstachine (**837a**) (*185*), the latter substance the IDA product of the former. This cyclization parallels that of the ester **836b**, which gives a mixture of *endo* and *exo* products, **837b** and **838**, respectively. The former is easily transformed into the alkaloid cyclostachine A (**837c**) belonging to the same Piperaceae family (*188*) (Scheme 5.87).

This fascinating topic is not limited to IDAs. For example, consider the relationship of auroglaucin and neoechinulin B or C with criptoechinulins B or D (*187*), where the latter compounds can be considered the adducts of a Diels–Alder reaction between the former compounds. This area is very promising for further research.

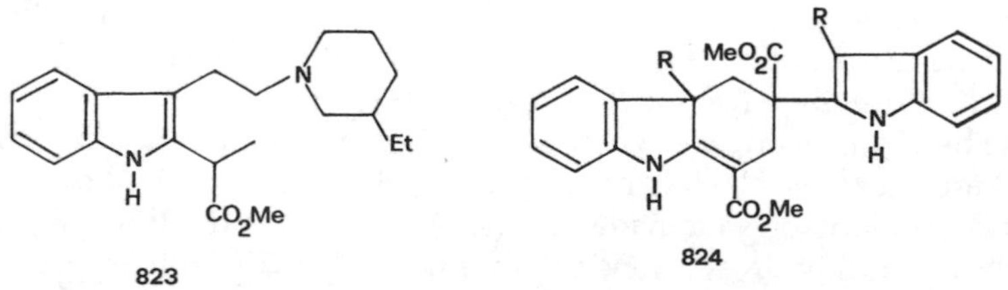

823 824

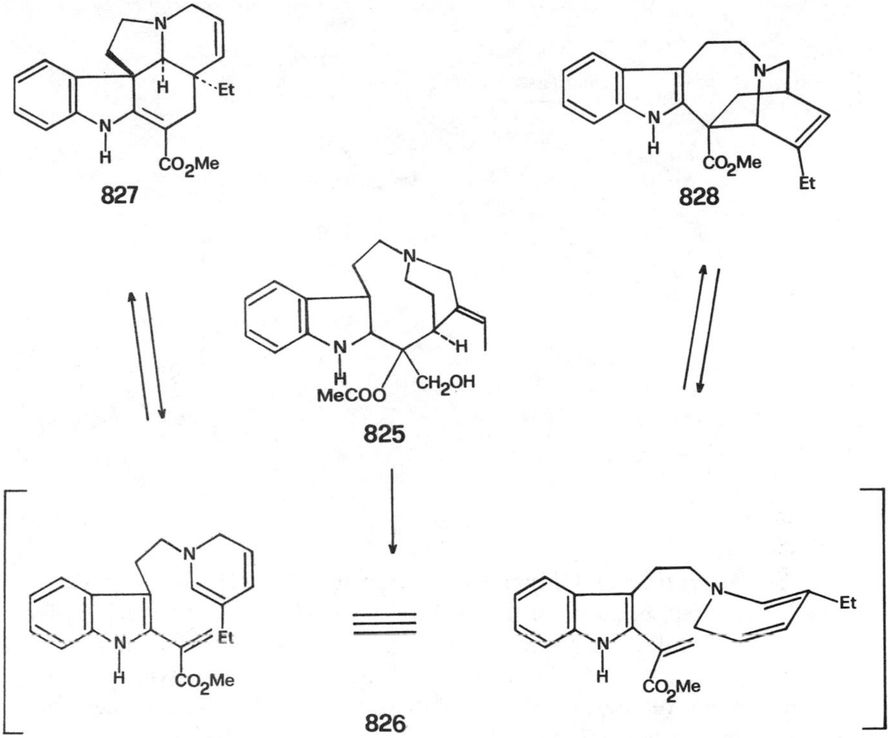

827 828

825

826

Scheme 5.84

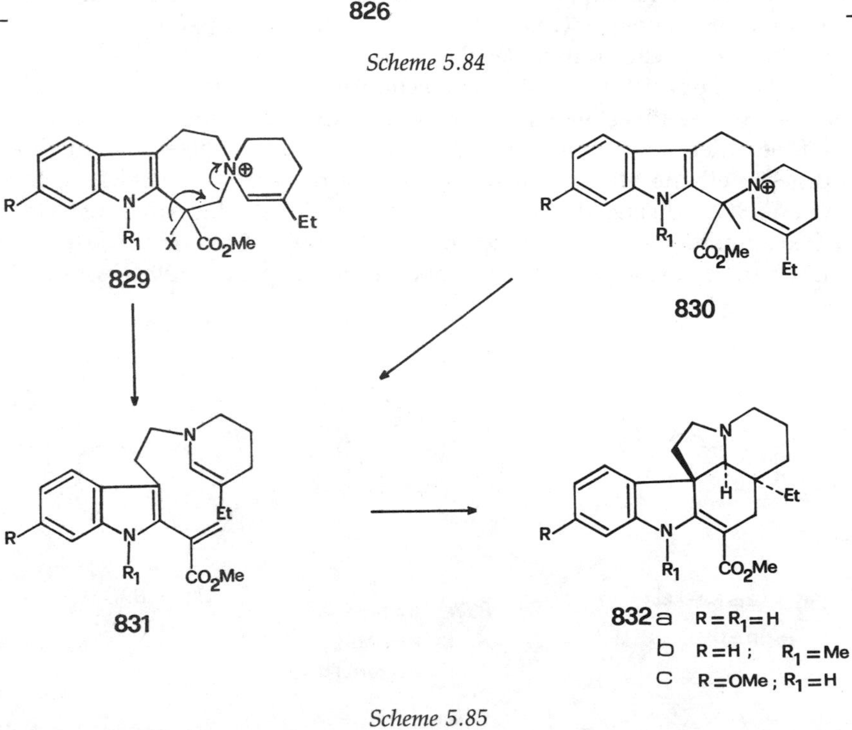

Scheme 5.85

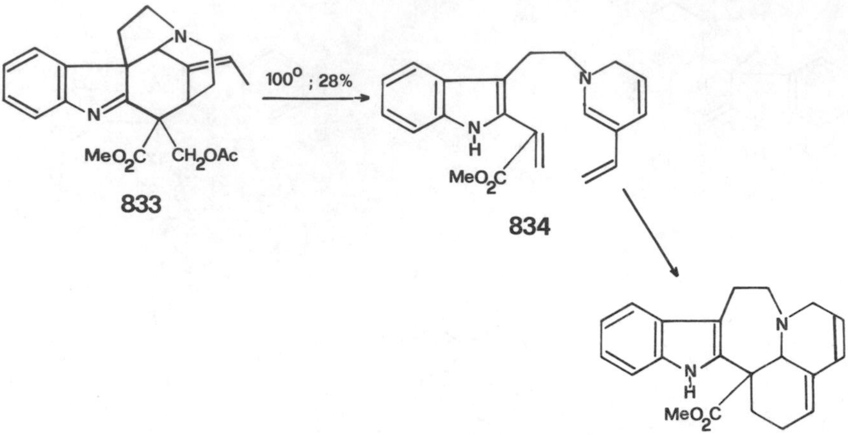

100°; 28%

MeO$_2$C CH$_2$OAc

833

834

835

Scheme 5.86

In Situ Generation of Dienes for Intramolecular Diels–Alder Reactions. This topic, like that of the section entitled "Nitrones" in Chapter 4 for the intermolecular processes, was mainly developed by the groups of Oppolzer, Kametani, Funk, and Vollhardt. This area was partially covered by two reviews (*119, 188*). The first papers in 1971 revealed that *o*-quinodimethanes, obtained by electrocyclic ring opening of benzocyclobutenes, react intramolecularly when a double bond is suitably placed in the substrate (*189*). The mechanism of the reaction of optically active **839** was investigated, and its rate of disappearance was found to be first order. Assuming that the sterically favored *trans*-substituted *o*-quinodimethane **840** is formed, the ratio of the *endo* product **841** to the *exo* one **842** indicates that the transition state for the formation of **841** is 1.0 kcal/mol lower than that for the formation of **842** and 0.2 kcal/mol higher than the barrier of the electrocyclic ring opening **839** → **840** (*190*) (Scheme 5.88).

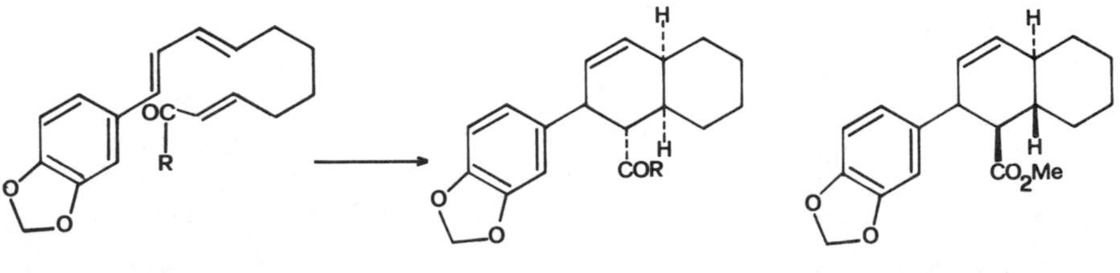

836a R= NH-i-But
b R=OMe

837a R=NH-i-But
b R=OMe
c R=pyrrolidinyl

838

Scheme 5.87

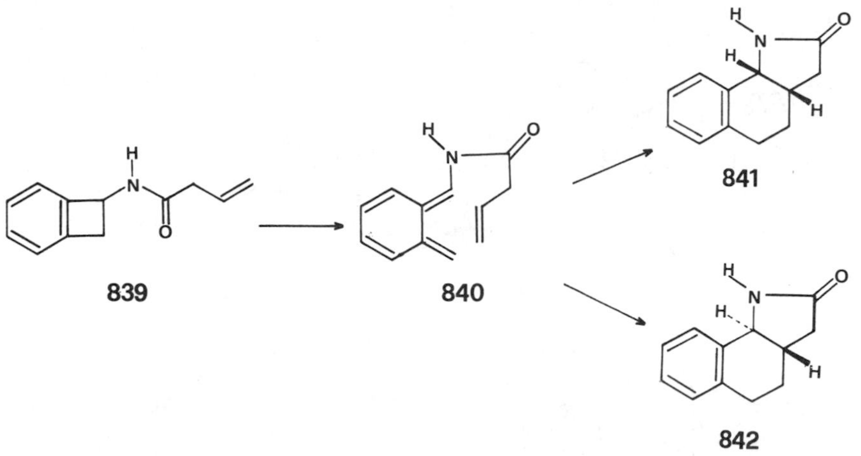

Scheme 5.88

ELECTROCYLIC CYCLOBUTENE RING OPENING. The sequence reported above is the most popular route for the in situ generation of dienes and has been extensively used by the Oppolzer and Kametani groups.

The first application in the natural product field involved the synthesis of (±)-chelidonine (**846**), the main alkaloid of *Chelidonium majus*, performed by Oppolzer (*191*) in 1971 by heating **843** at 120 °C for 1 h. Formation of **845** in 73% yield, via **844**, determined the success of the approach (Scheme 5.89).

In 1977 Kametani et al. reported the syntheses of alnusenone (**704**) and friedelin (**705**) (*121, 122*), which we have already illustrated in Scheme 5.56. These two groups also proposed several elegant syntheses of natural products in the last three years. Kametani achieved the stereoselective total synthesis of the tetracyclic diterpenes hibaol (**206b**) and hibane (**206d**). (A photochemical approach was described in Chapter 3 in the section entitled "Intermolecular Cycloadditions.") Kametani's synthesis proceeds by thermolysis of **847** through a transition state represented by **848**, which gives **849** exclusively (*192*) (Scheme 5.90).

Two other tetracyclic compounds, **851** and **853**, were obtained by similar routes. Starting from **850**, which contains the basic skeleton of aphidicolan-type terpenes, **851** was synthesized (*193*). The latter compound, **853**, was formed from **852**, a potential intermediate for the quassinoid klaineanone (**854**), the bitter principle of Simarubaceae species (*194*) (Scheme 5.91).

The only contribution by other workers in this area was the elegant synthesis of (±)-coronafacic acid (**514**). The key step is the electrocyclic cyclobutane ring opening with concomitant retro-Diels–Alder elimination of dimethylfulvene from **855** to produce **856**. Obtained in the *exo* conformation for steric reasons, **856** cyclizes to **857**, which is easily transformed into **514** (*195*) (Scheme 5.92).

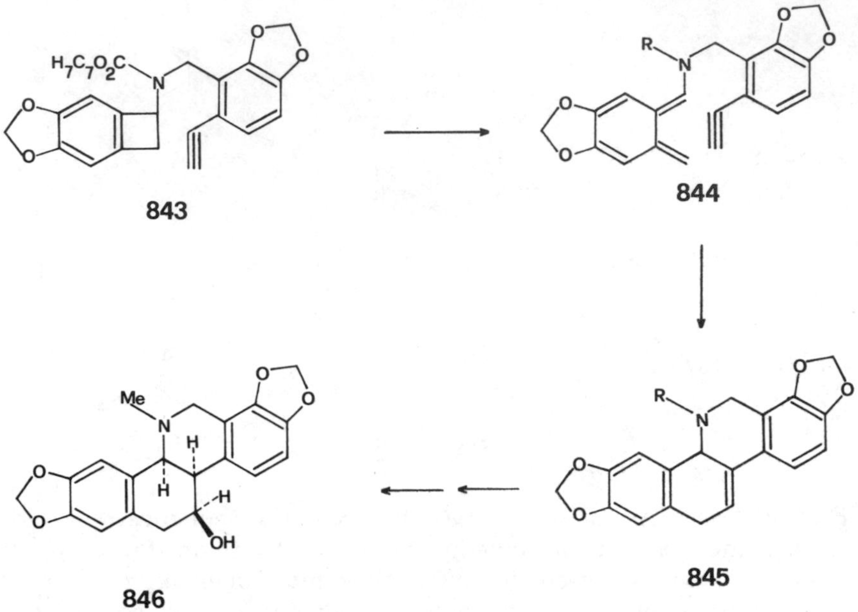

843 844

846 845

Scheme 5.89

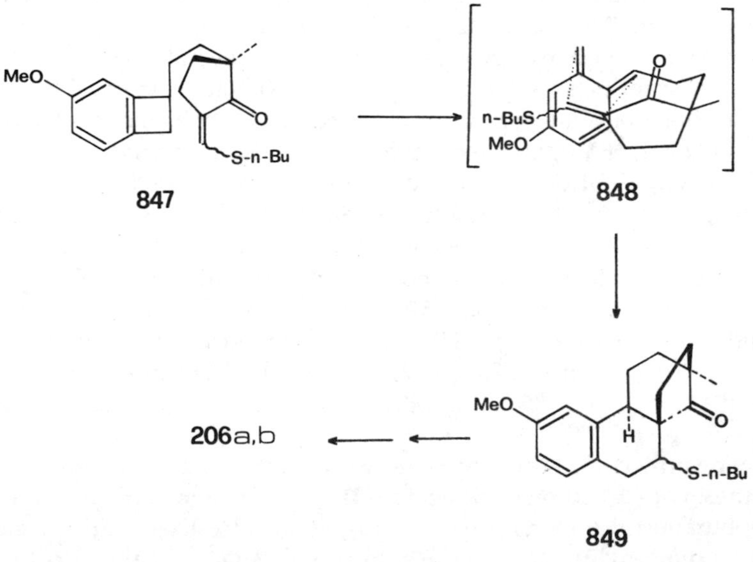

847 848

206a,b 849

Scheme 5.90

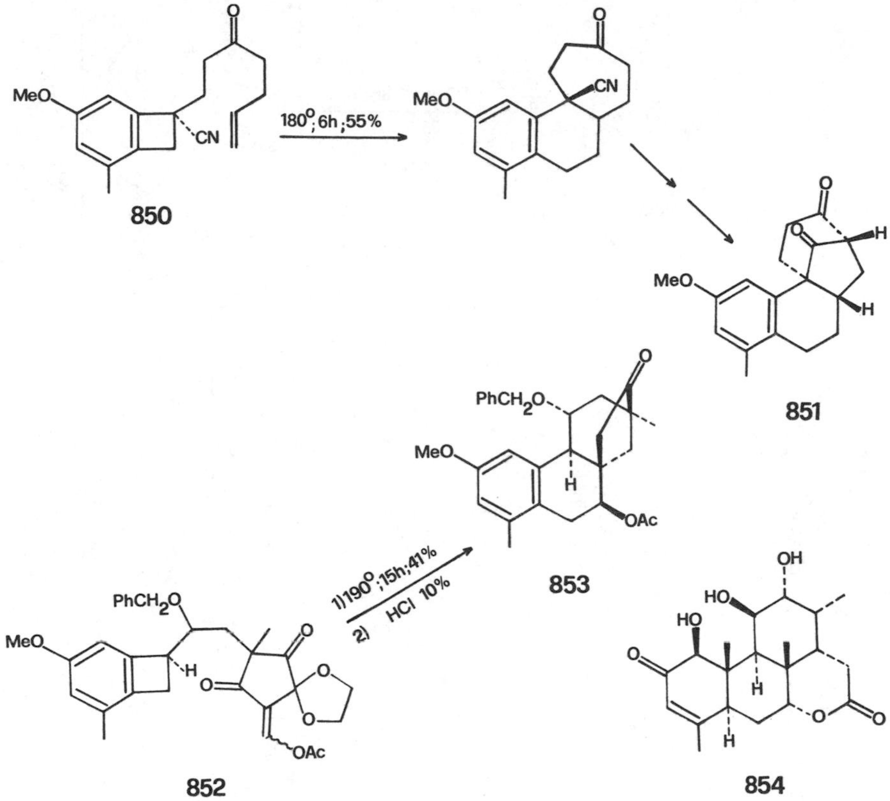

Scheme 5.91

These brilliant results were overshadowed by those obtained in the steroid field (*196*).

In 1977 Oppolzer and his group described the synthesis of two aromatic steroids where the substituents present in the ansa play a crucial role in the choice of the transition state. Either *endo* (**859**) or *exo* (**860**) compounds were obtained, depending on the nature of X in the substrate **858** (*197, 198*) (Scheme 5.93).

The field was fully developed by the Kametani group and the total syntheses of estrone and estradiol were two of the more outstanding results.

The stereoselective total synthesis of estrone (**864a**) was realized via *O*-methyl-D-homoestrone (**863**), which was obtained by ring opening of **861** to **862** (*199*). The cycloaddition proceeds stereospecifically through the less hindered *exo* transition state. Attempts to avoid the tedious route via homoestrone were, at that time, unsuccessful. This goal was eventually achieved by the Grieco group in 1980 (*200*), who successfully synthesized **865a** and converted it to 3-methoxyestra-1,3,5(10)-trien-17β-ol (**866a**) (Scheme 5.94).

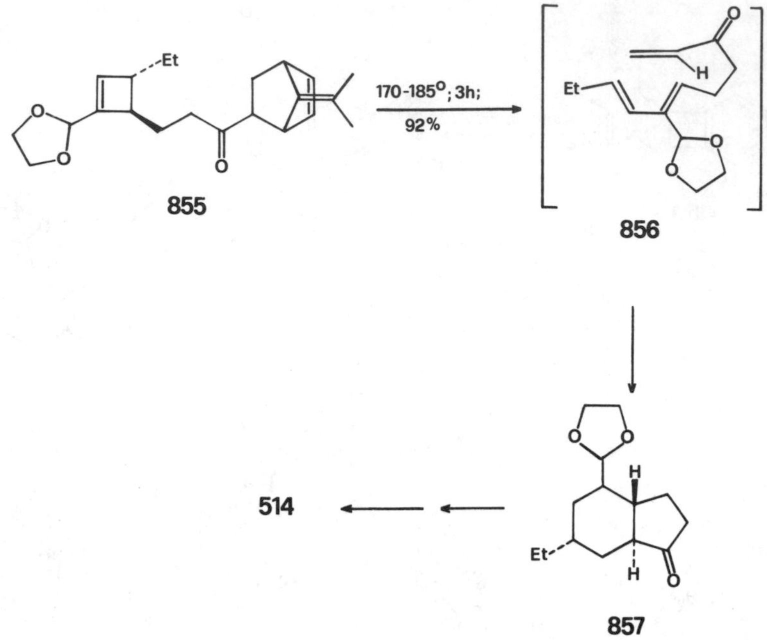

Et
O
O
O
855
170-185°; 3h;
92%
O
H
Et
O
O
856
O
O
H
Et
H
O
857
514

Scheme 5.92

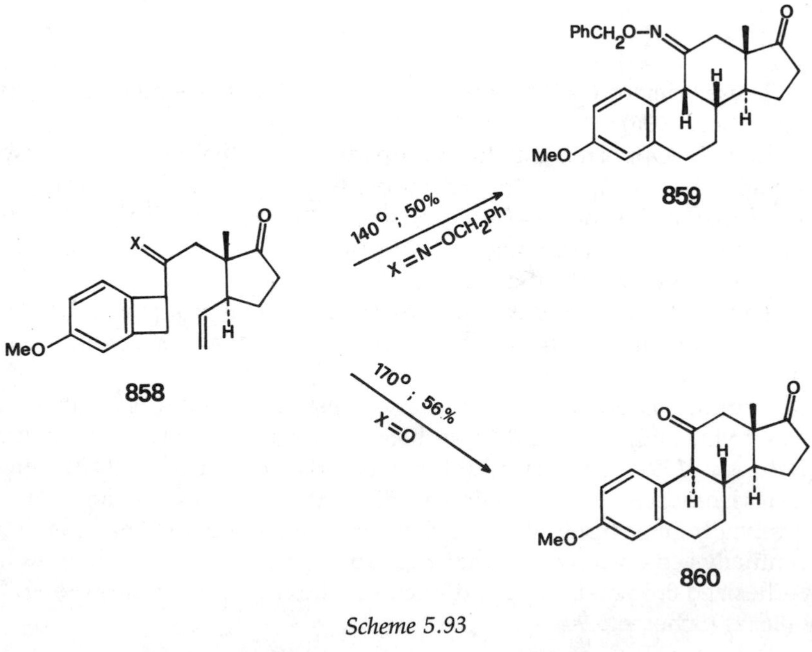

PhCH$_2$O–N
O
H
H
H
MeO
859
140° ; 50%
X =N–OCH$_2$Ph
X
O
MeO
H
858
170° ; 56%
X=O
O
O
H
H
H
MeO
860

Scheme 5.93

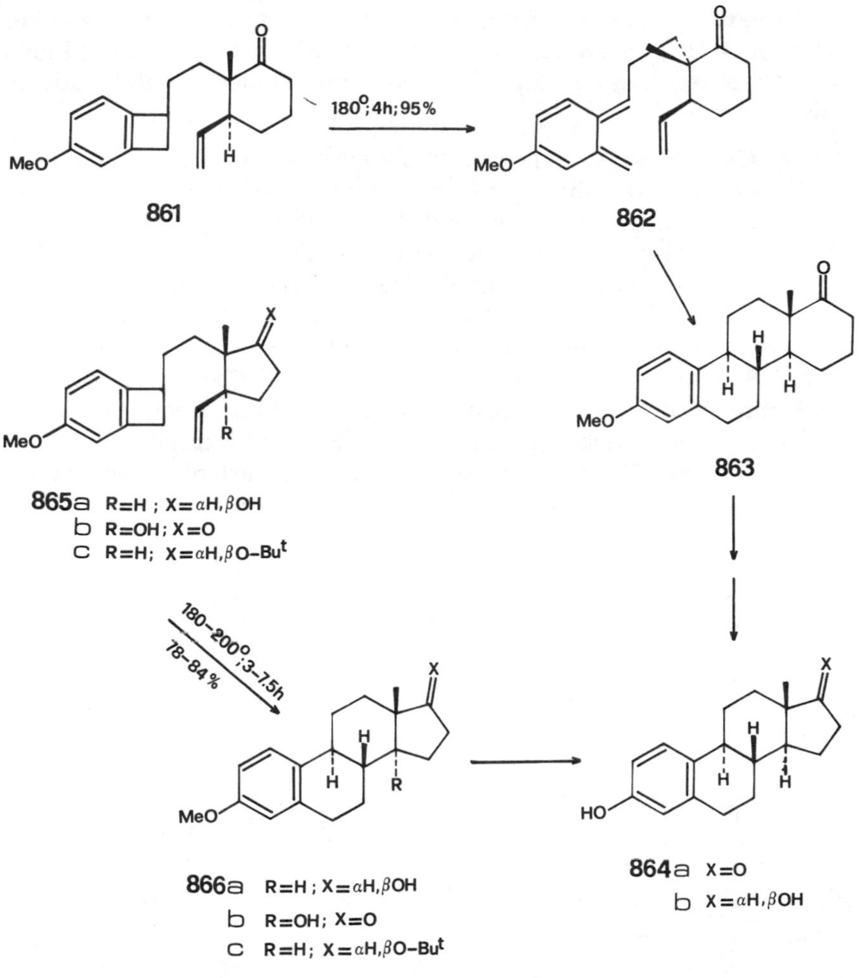

180°; 4h; 95%

861

862

863

865a R=H ; X=αH,βOH
b R=OH; X=O
c R=H; X=αH,βO-Bu^t

180–200°; 3–7.5h
78–84%

866a R=H ; X=αH,βOH
b R=OH; X=O
c R=H; X=αH,βO-Bu^t

864a X=O
b X=αH,βOH

Scheme 5.94

Kametani obtained 14α-hydroxyestrone-3-methyl ether (**866b**) via **865b** (*201*). 14α-Hydroxyestrone is an estrogen metabolite. The total synthesis of estradiol (**864b**) was realized along the same route (*202, 203*). The formation of **866c** from **865c**, in the absence of substituents in the ansa, supports previous evidence that steric interactions alone lead the reaction through an *exo* transition state. The topic was completed with the synthesis of some unnatural D-ring aromatic steroids. The product dramatically depends on the nature of the substituent on the cyclobutane ring. When R = H, (+)-3β-hydroxy-17-methoxy-D-homo-18-nor-5α-androsta-13,15,17-triene (**869**) is obtained (*204*) through an *exo* transition state. When R = CN, a *cis* configuration of the ring junction in **871** (*205, 206*) can account for a less probable *endo* transition state; a different configuration of the *o*-quinodimethane intermediate would lead to an *exo* transition state. An outward electrocyclic ring opening when R = H gives **868**; an inward process when R = CN would give **870**, thus resulting in the observed configuration of the adducts (Scheme 5.95).

METAL-CATALYZED ACETYLENE COOLIGOMERIZATION. The difficult synthetic approach to suitably substituted benzocyclobutenes limits the synthetic usefulness of the method described in the previous section. A promising alternative seems to be found with the discovery that hexa-1,5-diynes **872** cocyclize with sterically hindered alkynes such as bis(trimethylsilyl)acetylene (BTMSA) in the presence of a cobalt(I) catalyst (η^5-cyclopentadienyldicarbonylcobalt) (*207*). By this route Funk and Vollhardt (*119*) obtained 4,5-bis(trimethylsilyl)benzocyclopentene (**873**) through the plausible mechanism illustrated in Scheme 5.96.

The use of suitably substituted diynes **872** with R carrying the dienophile (such as **874**) allows facile synthesis of a diastereomeric mixture

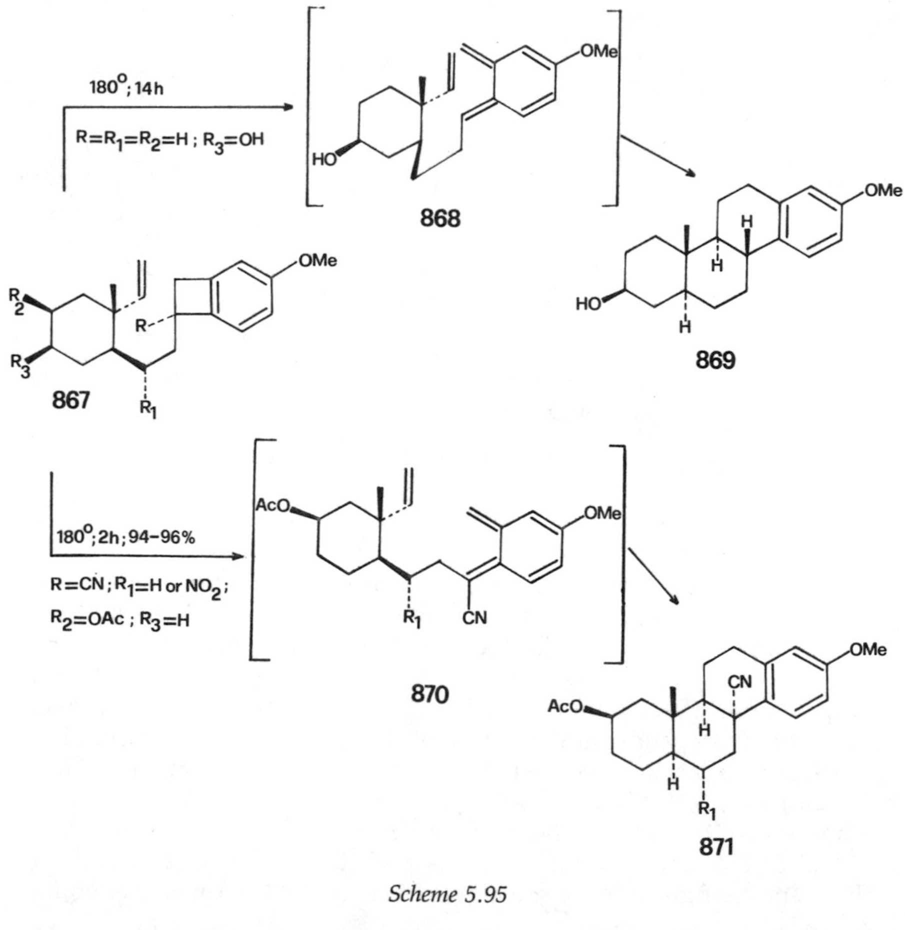

180°; 14h
$R=R_1=R_2=H$; $R_3=OH$

868 **869** **867**

180°; 2h; 94–96%
$R=CN$; $R_1=H$ or NO_2;
$R_2=OAc$; $R_3=H$

870 **871**

Scheme 5.95

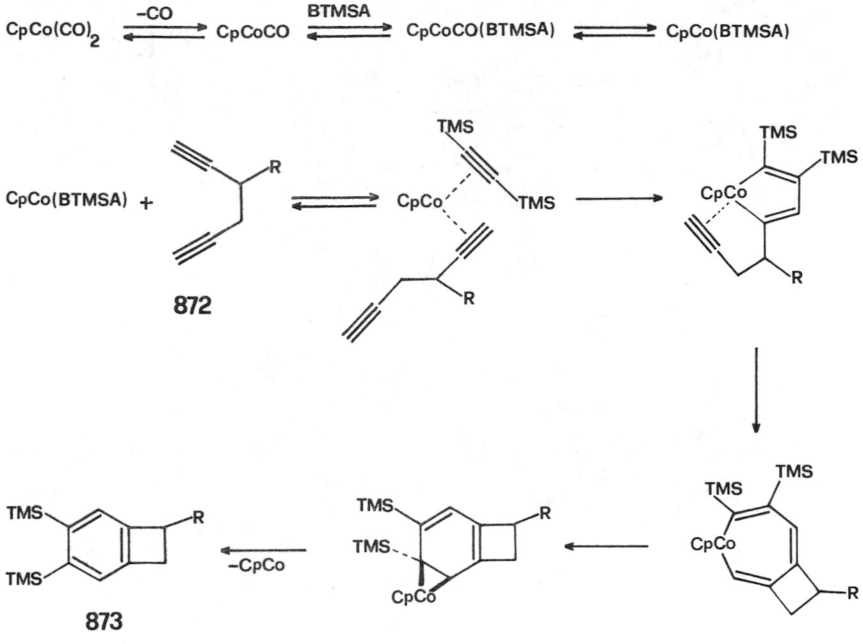

Scheme 5.96

of **875**. By conrotatory-outward opening of the four-membered rings, both isomers give the same *o*-xylylene intermediate **876**, whose favorable MO interactions (drawn in **876** as $HOMO_{xylylene}/LUMO_{ene}$) (*208*) allow IDA to occur through the less hindered *exo* transition state, producing **877**. By this route an important synthesis of *dl*-estrone (**864a**) was realized (*209, 210*) (Scheme 5.97).

A further promising development of the method is the one-step construction of chiral polycycles. This process involves an intramolecular cooligomerization of diynenes as exemplified by the transformation of **878** into **879** (*211*). When $n = 3$ and R = $SiMe_3$, only an *exo* transition state is involved; for larger rings or R = H, both *exo* and *endo* transition states are probable (Scheme 5.98).

OTHER ROUTES. The in situ generation of dienes for IDAs can be also performed following several other routes.

The regioselective electronic excitation and photoenolization of **880**, easily obtained from inexpensive chemicals, gives the phototransient *o*-xylylene derivative **881**. This latter compound cyclizes via an *exo* transition state to **882**, which is easily converted to (±)-estrone (**864a**) (*212*) (Scheme 5.99). The synthesis can be directed asymmetrically to the (+)-natural enantiomer (*213*).

An original method for generating *o*-xylylene intermediates under mild, neutral conditions illustrates another useful aspect of organosilicon chemistry. Treatment of the epoxide **883** with cesium fluoride at

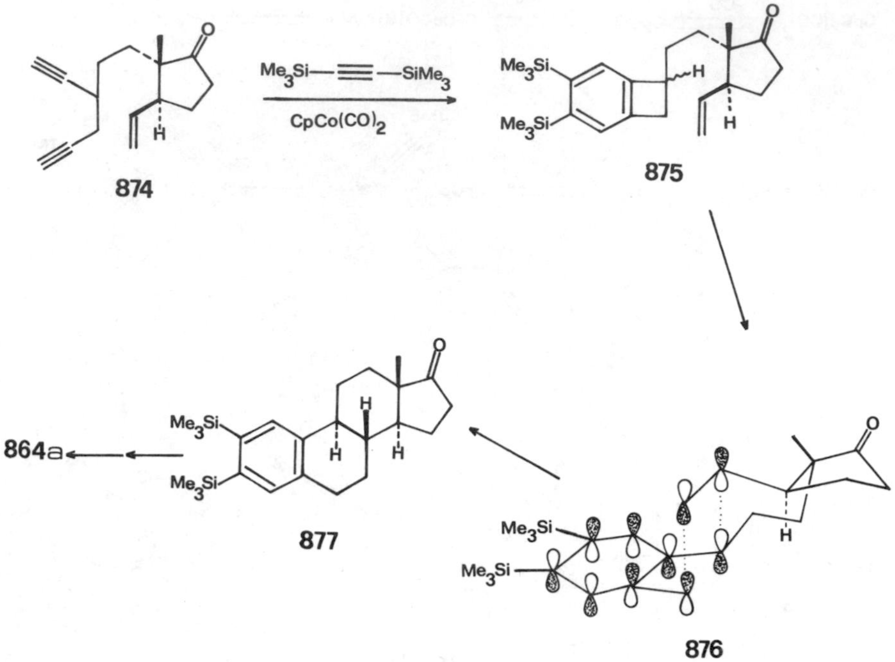

Scheme 5.97

room temperature gives 11α-hydroxyestrone methyl ether (**885**) in high yield through **884** in the *exo* transition state (*214*) (Scheme 5.100).

The cheletropic approach, via a sulfur dioxide extrusion from alkenoyl sulfones **886a,b** at 210–240 °C, gave *o*-quinodimethanes **887a,b** of obvious synthetic utility (*215*). Estra-1,3,5(10)-trien-17-one (**888**) (*216, 217*) and (+)-estradiol (**864b**) (*218*) were obtained by this route, the former from **886a**, the latter from **886b**, both via *exo* transition states (Scheme 5.101).

To complete this section we report the synthesis (*219*) of *Aspidosperma* alkaloid aspidospermine (**892**). Thermolysis at 600 °C of the enamide **889** causes sulfur dioxide extrusion, thus giving **890**, which cyclizes to the tricyclic synthon hydrolulolidine (**891**). Hydrolulolidine can

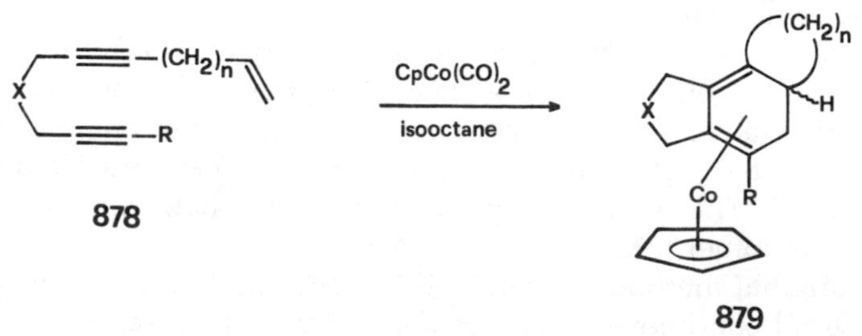

Scheme 5.98

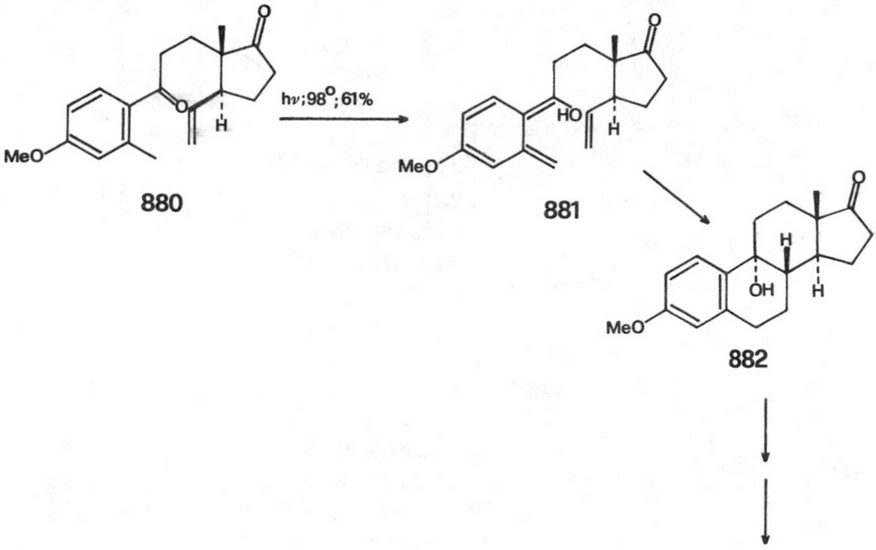

880 → (hν; 98°; 61%) 881 → 882 → → 864a

Scheme 5.99

be transformed into **892**, thereby completing a new formal total synthesis of aspidospermine (Scheme 5.102). We wish to point out that **891** has an all-*cis* configuration, clearly suggesting an *endo* transition state for the IDA. An explanation in terms of less energy strain in the *cis* ring junction than in the *trans* one is supported by an **893a**:**893b** ratio of 2:1 obtained in the six-membered ring homologous system leading to the hydrojulolidine system. Once again the *endo* transition state is operative, whereas this section has clearly shown that the *exo* transition state is always preferred in the absence of favorable MO interactions or significant development of conjugation.

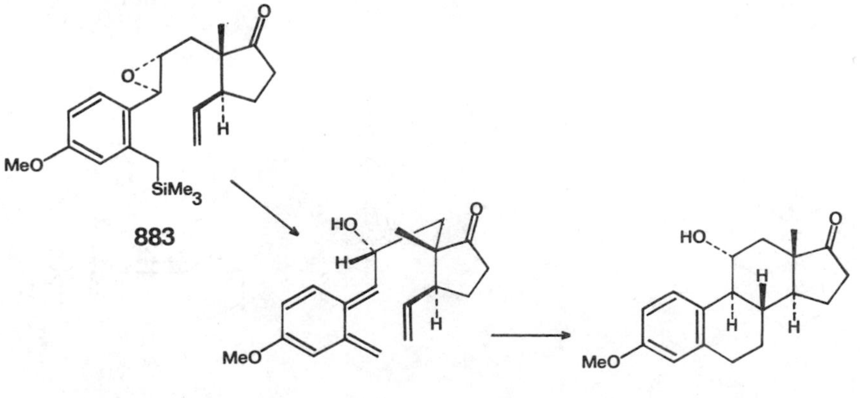

883 → 884 → 885

Scheme 5.100

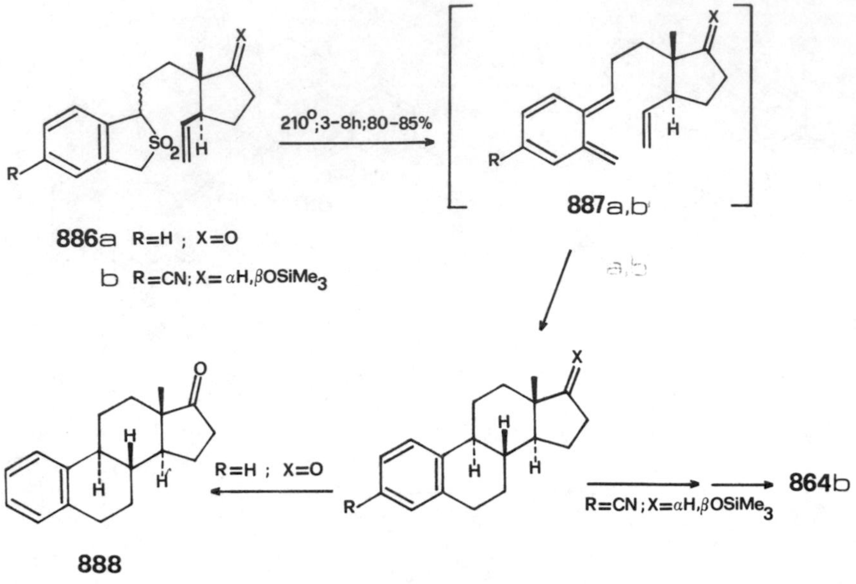

210°;3–8h;80–85%

886a R=H ; X=O

b R=CN; X= αH,βOSiMe$_3$

887a,b

R=H ; X=O

888

R=CN ; X=αH,βOSiMe$_3$

864b

Scheme 5.101

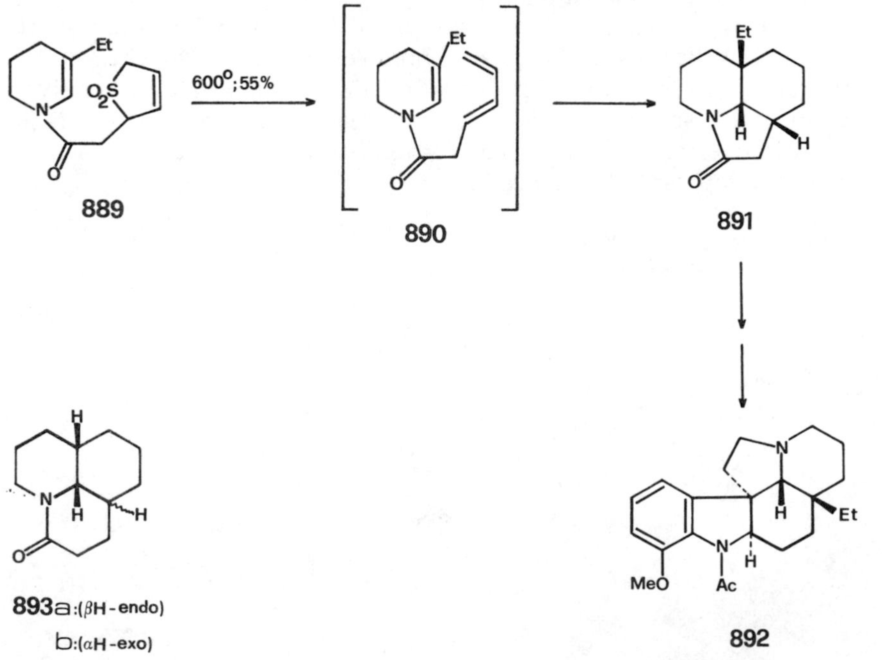

600°; 55%

889

890

891

892

893a :(βH - endo)

b:(αH -exo)

Scheme 5.102

An alternative explanation in terms of MOs can be proposed in which this reaction is considered as a Diels–Alder with inverse electron demand. This proposal is supported by the observation that the first ionization potential of butadiene (9.07 eV) is lowered in acylamino dienes (7.90–8.66 eV) (*220*), indicating that the amido-substituents significantly raise the energy of Ψ_2.

If the dominant interaction is $HOMO_{dienophile}/LUMO_{diene}$ and, to infer the coefficients' symmetry, the enamide is compared to a pentadienyl anion (*see* Ref. 3 in Chapter 1), the MO interactions leading to **891** are represented in Figure 5.10. An attractive nonbonding interaction between C2 in butadiene and the nitrogen stabilizes the *endo* transition state. Due to the asymmetry of the wave function in the $O{=}C{-}\overline{N}{-}C{=}C$ system, the nodal point does not coincide with the nuclear position 2′, which therefore becomes bonding from nonbonding.

Lewis Acid Catalyzed Diels–Alder Reactions

The Diels–Alder reaction is affected by the presence of a Lewis acid. This effect was interpreted in terms of a two-step mechanism for ring formation through a zwitterionic intermediate, which could have a long enough lifetime to be trapped (*221*).

Lewis acids affect the rate of the reaction, the regiochemistry, and the stereochemistry. Typical examples of these effects are listed here.

1. The rate constant of the reaction between butadiene and methyl acrylate (Scheme 5.103) is enhanced by a factor as large as 10^5 entirely due to lowering of the activation energy (*222*).
2. The synthesis of 1,4-dimethylcyclohex-3-enyl methyl ketone (**894**), a very simple natural product isolated from the fruit of *Juniperus communis* L., was regiospecifically realized from isoprene and isopropenyl methyl ketone. In the absence of catalyst (140 °C, 48 h, yield unknown) a mixture of *para* and *meta* adducts (**894** and **895**, respectively) was obtained. The former adduct is probably the main

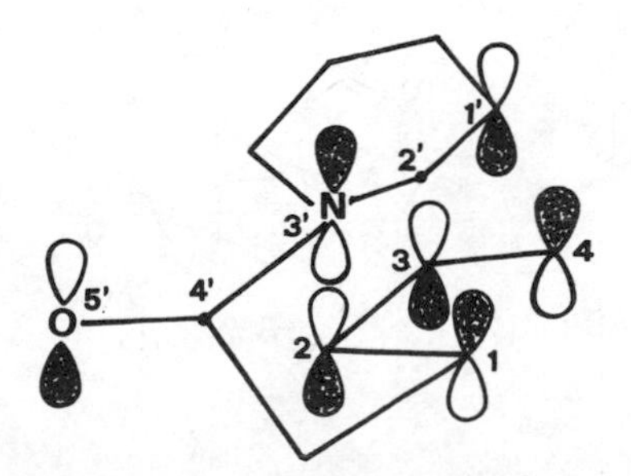

*Figure 5.10. Transition state leading to **891**.*

CO$_2$Me

CO$_2$Me

Scheme 5.103

product, and became the only product if the reaction was performed in the presence of stannic chloride (benzene, 0 °C, 1.3–4 h, 70% yield, **895** less than 4%) (*223*) (Scheme 5.104). The *para* adduct was sometimes reported to be the minor isomer. The reaction of isoprene with 2-methyl-2-cyclohexenone was described to give a 9:1 mixture of **896** and **897**, the former becoming the only reaction product in the presence of aluminum trichloride (*224*) (Scheme 5.105). This "unusual" behavior was due to a wrong structure assigned to the reaction products. A more careful investigation revealed that **897** is the major isomer, and it became the only product with aluminum chloride catalysis (*225*).

3. The reaction between cyclopentadiene and methyl acrylate (CH_2Cl_2, 0 °C, 22–51% yield) gave a mixture of *endo* **898** and *exo* **899** adducts in the ratio 82:18. If the reaction was performed (CH_2Cl_2, 0 °C) in the presence of 10% stannic chloride, not only was the overall yield increased to 67–79%, but the ratio **898**:**899** became 95:5. The reaction was more stereoselective (*226*) (Scheme 5.106).

Care must be taken in considering several reactions between butadienes and cycloalkenones and cycloalkadienones in the presence of aluminum trichloride. Under the reaction conditions, the primary *cis* adduct **900** may undergo partial or total isomerization to the *trans* isomer **901** (*227, 228*) (Scheme 5.107).

The perturbation approach was useful in understanding a large part of the above-mentioned effects. The principal function of a Lewis acid is the complexation of the carbonyl group eventually present in the reacting species. Hence it is important to distinguish between carbonyl- (or CN-) substituted dienophiles and carbonyl-substituted dienes. The first case is the most popular.

In a $HOMO_{diene}/LUMO_{dienophile}$-controlled Diels–Alder reaction

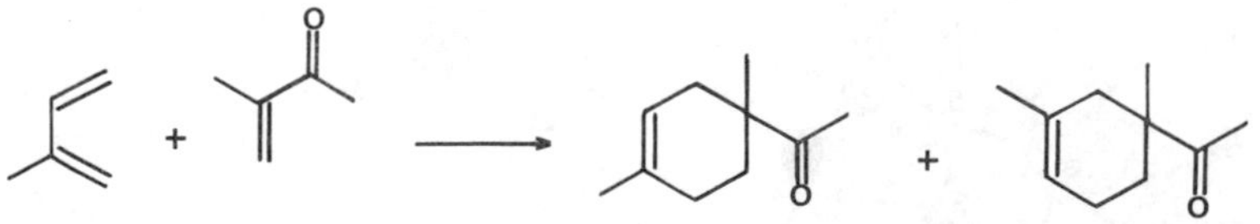

894 **895**

Scheme 5.104

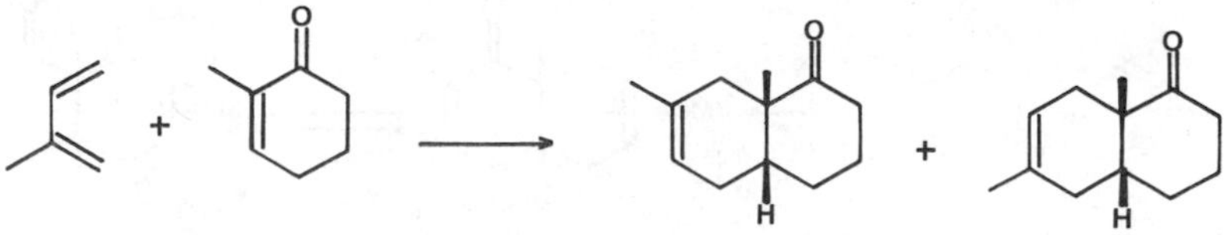

896 897

Scheme 5.105

(with normal electron demand), complexation of the dienophile drastically lowers both its HOMO and its LUMO (*229*). A lower energy separation between the LUMO of the dienophile and the HOMO of the diene results, thus increasing ΔE (the energy gained when the orbitals of one reagent overlap those of the other) in Eq. 2.

A direct proportionality between ΔE and the free energy is usually assumed through Eq. 4. However, by using the perturbation approach, the slope of the early part of the path along the reaction coordinate leading to the transition state can be estimated. With this information, the free energy can be calculated by using Eq. 5. Thus Eq. 6 can be assumed from the previously mentioned experimental observation for butadiene and methyl acrylate (*222*) (Scheme 5.103). As an example we report the variation of the energy levels and coefficients for the reaction between cyclopentadiene (*230*) and acrolein (*231*) (used as a model carbonyl-substituted dienophile) and protonated acrolein (*231*) (used as a model Lewis acid complexed molecule) (Figure 5.11).

$$\Delta E = a \cdot \log k = b \cdot \Delta G \tag{4}$$

$$\Delta G = \Delta H - T\,\Delta S \tag{5}$$

$$\Delta E = c \cdot \Delta E^{\neq} + d \tag{6}$$

Figure 5.11 illustrates the effect of the catalyst on the reaction rate. The lower $HOMO_{diene}/LUMO_{dienophile}$ energy separation diminishes the denominator of Eq. 2 and therefore increases the energy gained in the cycloaddition (ΔE).

The different LUMO coefficients of acrolein (A) and protonated acrolein (B) (Figure 5.12) explain the effect of the catalyst on the regiochemistry of the reactions with isoprene (Schemes 5.104 and 5.105).

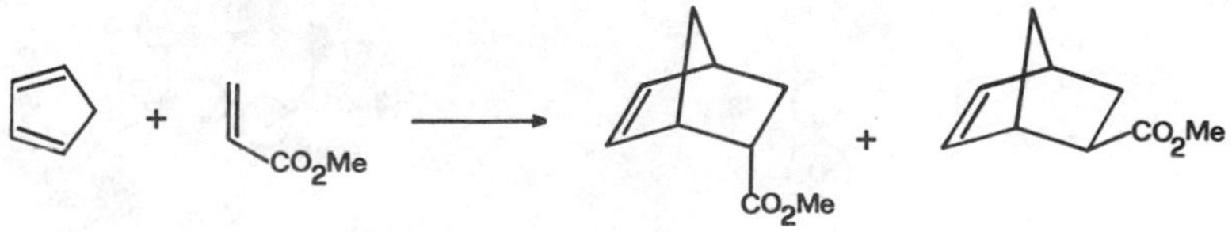

CO_2Me CO_2Me CO_2Me

898 899

Scheme 5.106

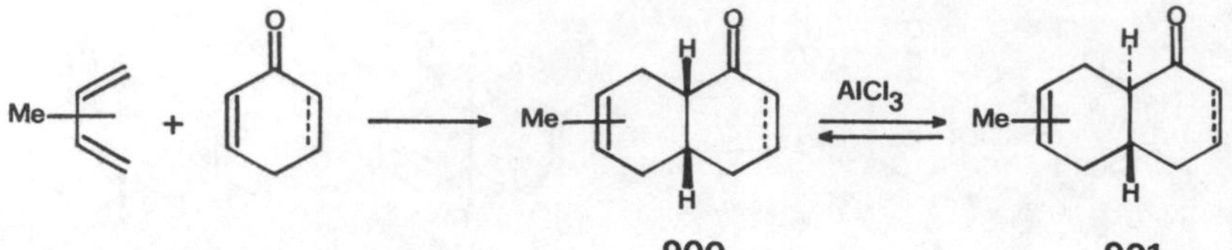

Scheme 5.107

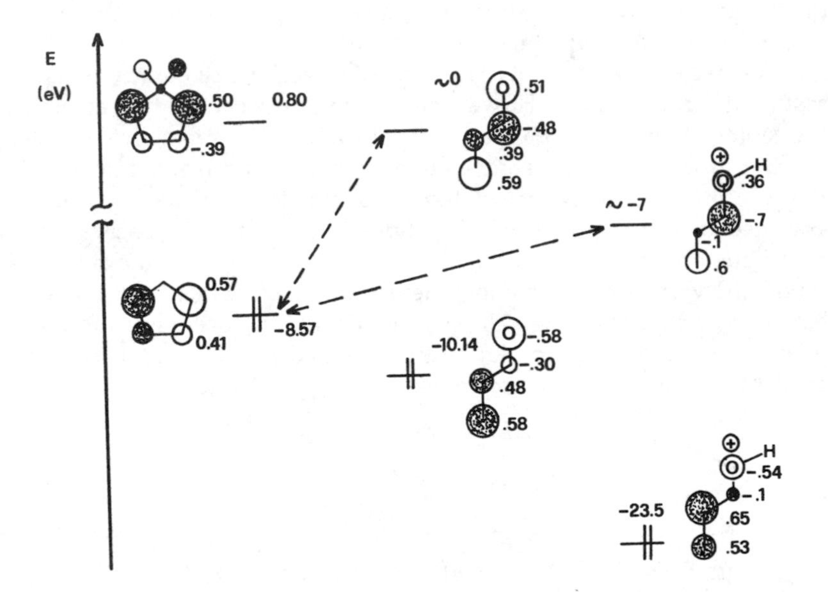

Figure 5.11. FMOs of cyclopentadiene with acrolein and protonated acrolein.

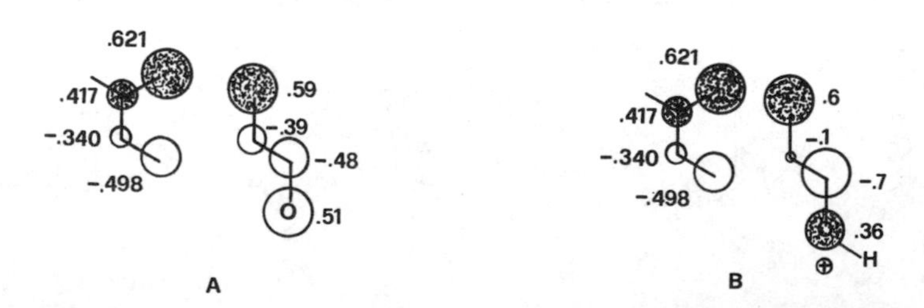

Figure 5.12. FMOs of isoprene with acrolein (A) and protonated acrolein (B).

The greater polarization of the coefficients in B increases the regioselectivity of the reaction in favor of the *para* isomer.

The increased LUMO coefficient on the carbonyl carbon atom increases the secondary nonbonding interactions in the acid-catalyzed reactions, thus favoring the *endo* transition state (Figure 5.13). This relationship accounts for the predominance of the *endo* isomer **898** in the reaction with cyclopentadiene (Scheme 5.106).

The previously mentioned approach concerns reactions where the Lewis acid forms a salt with the dienophile; if the diene has a substituent carbonyl group, the catalyst acts on it. Thus 4-methylhexa-3,5-dien-2-one (**902**) reacts with isobutylene to give a 20% yield of **903** (and several other products obtained by competitive carbonium ion reactions), which, by aldol condensation with acetaldehyde, gives α-damascenone (**904**) (*232, 233*) (Scheme 5.108).

Despite the presence of a zwitterionic intermediate, the Lewis acid coordination of an acyl butadiene can be considered to increase the electron-attracting character of the carbonyl group by increased polarization of the 1,4 coefficients and to decrease the energy of both HOMO and LUMO of acylbutadiene. Taking into account that the ionization potential and activation energy of isopentene are 9.23 and −2.19 eV (*234*), respectively, and first ionization potential$_{(c=c)}$ of pentadienal (used as a model of **902**) is 9.09 eV (*235*), the FMO representation of the uncatalyzed and catalyzed reaction can be derived (Figure 5.14). The catalyzed reaction definitely can be considered a Diels–Alder with inverse electron demand. Under conditions of maximum overlap, the $HOMO_{isopentene}/LUMO_{acylbutene}$ interactions favor the formation of the regioisomer **903**.

The last general point to consider concerns the nature and action of the Lewis acid catalyst. Each Lewis acid has its own complexation energy, which causes variation of the energies and coefficients of the LUMO and HOMO. As an example, Figure 5.15 depicts the LUMO energies and coefficients for acrolein complexes with Li^+ and Na^+ (*236*).

Furthermore, the Lewis acids used as catalysts can be divided into two classes:

1. Boron derivatives [boron trifluoride and the so-called boron triacetate, which actually is tetraacetyl diborate

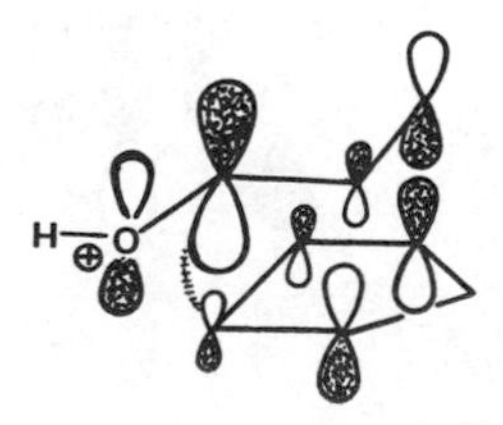

Figure 5.13. The endo *transition state of cyclopentadiene and protonated acrolein with secondary nonbonding interactions.*

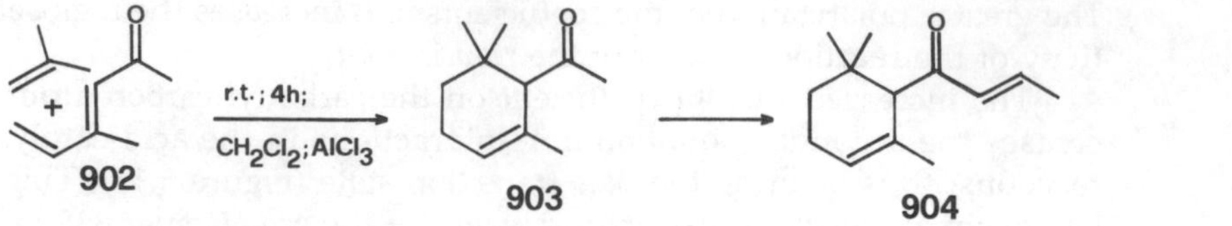

r.t.; 4h; CH_2Cl_2; $AlCl_3$

902 903 904

Scheme 5.108

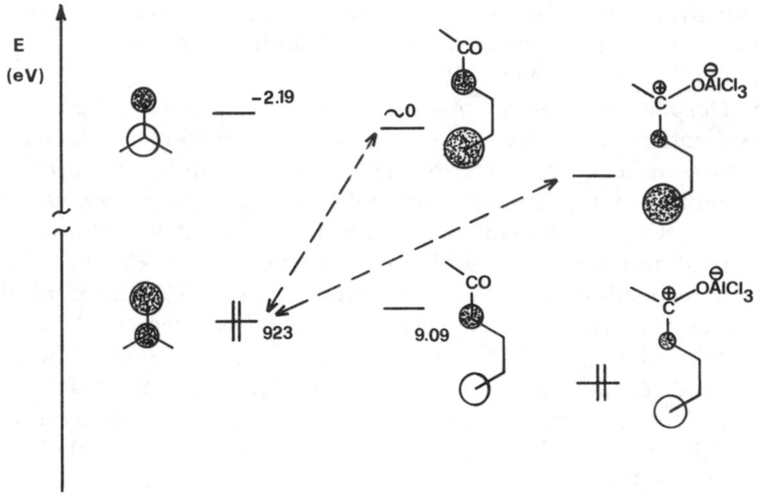

Figure 5.14. FMO interactions in diene-catalyzed Diels–Alder reaction.

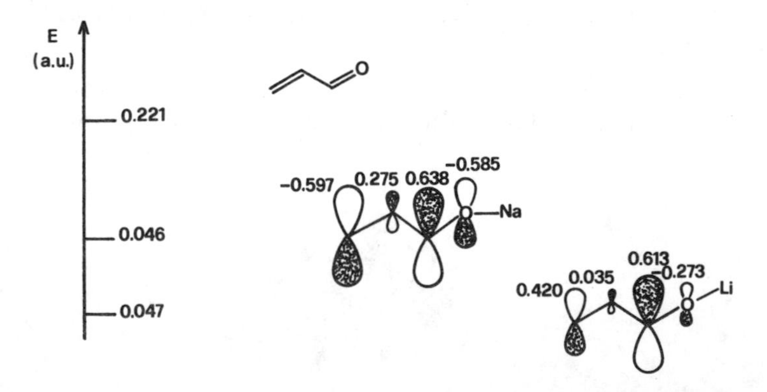

Figure 5.15. LUMO energies and coefficients (STO-3G) of acrolein complexes with lithium and sodium; LUMO energy for acrolein is 0.537 a.u.

(*237*)] having no *d* orbital available (protic acids are excluded);

2. Aluminum compounds (halides, alkyl and alkoxy derivatives), stannic chloride, titanium tetrachloride, etc., all with *d* orbitals available.

Members of both classes give normal acid–base interaction with substrates having a single coordination site. If the substrate is bidentate [e.g., juglone (**532**) and its derivatives], boron compounds (BR_3) can promote either coordination on both carbonyls to give **905** and **906** (if X ≠ H) or nucleophilic substitution at the boron atom (if X = H) to give a coordinate metal salt **907**. All other Lewis acids (AR_x) give coordinate complexes **908** involving the *d* orbital, even if X ≠ H (*238*) (Scheme 5.109).

Application of these concepts explains the different behavior of 2-methoxy-5-methylbenzoquinone with isoprene in the presence of stannic chloride or boron trifluoride (*239*).

Under thermal conditions a 1:1 mixture of **909** and **910** is obtained. With a boron trifluoride as a catalyst, the ratio is 1:24; with stannic chloride as a catalyst, the ratio is 20:1.

If the boron trifluoride interaction occurs at the less hindered site on the more basic carbonyl group, as in **911**, then **910** is favored because the greater LUMO coefficient is at C6. If coordination with stannic chloride gives the bidentate complex **912**, with the greater LUMO coefficient at C5, **909** prevails (Scheme 5.110).

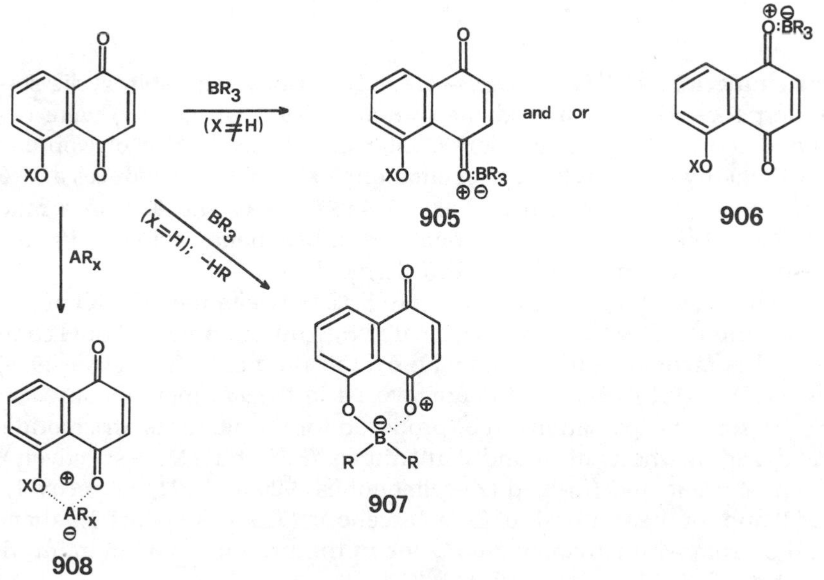

Scheme 5.109

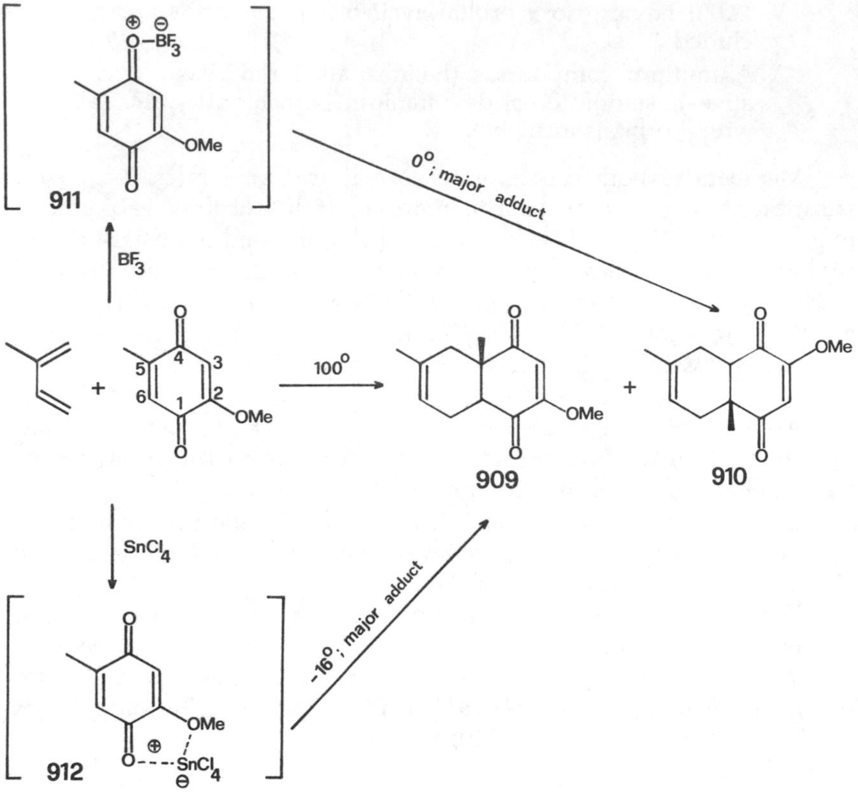

Scheme 5.110

Intermolecular Catalyzed Diels–Alder Reactions. The spirocyclic sesquiterpenes γ- and δ-acoradiene (**918** and **919**, respectively) were obtained through a stannic chloride catalyzed cycloaddition of isoprene and 3-methyl-2-methylenecyclopentanone (**913**), which produces a 39% yield of a mixture of four adducts, **914**, **915**, **916**, and **917**, in a ratio 69:27:3:1. The two major components are subsequently converted to the desired sesquiterpenes (*240, 241*) (Scheme 5.111).

The interaction reported in Figure 5.12 between the HOMO of isoprene and the LUMO of the coordinated α,β-unsaturated carbonyl compound is favored in this case by an overall ratio of 94:4 [(**914**) + (**915**) vs. (**916**) + (**917**)]; **914** and **916** are favored in the *anti*-methyl approach.

A similar explanation can be proposed for the aluminum trichloride-catalyzed synthesis of α- and β-atlantones (**505** and **922**, respectively), from isoprene and (*E*)- and (*Z*)-ocimenones (**920** and **921**, respectively) (*242*) and for the synthesis of β-damascenone (**923**) from penta-1,3-diene and 3-bromo-4-methylpent-3-en-2-one in the presence of aluminum trifluoride (*243, 244*) (Scheme 5.112).

For steric reasons, boron trifluoride should give a complex at the

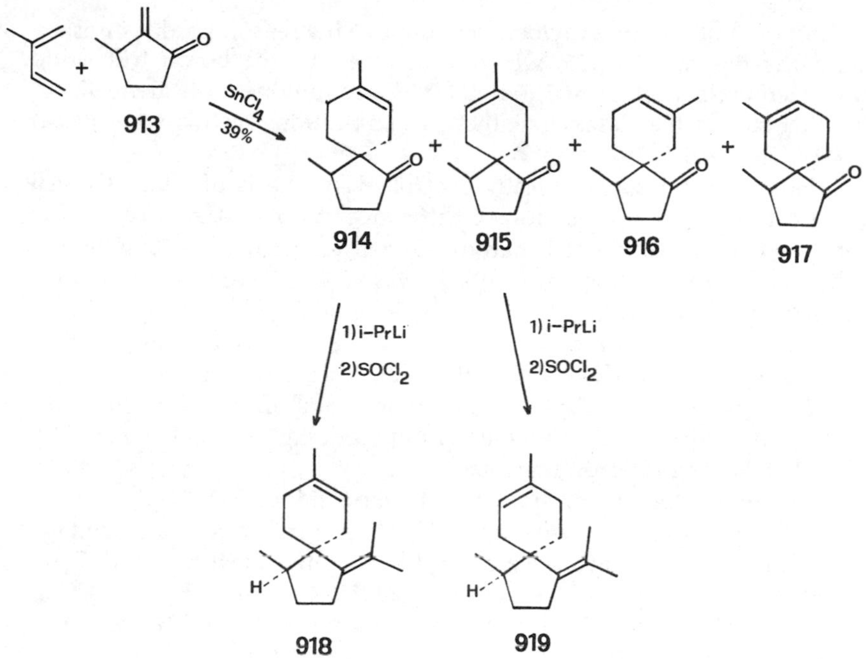

SnCl4 39%

913 914 915 916 917

1) i-PrLi 2) SOCl2

918 919

Scheme 5.111

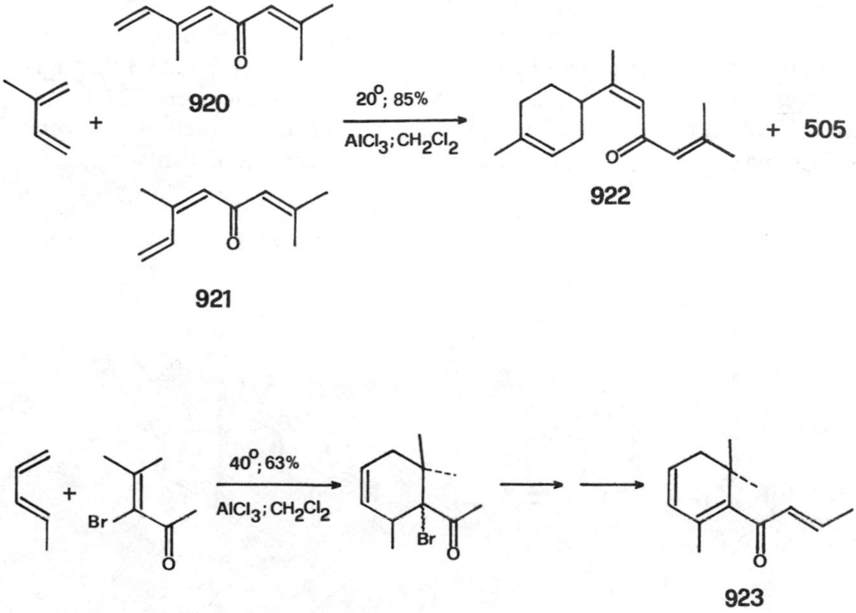

920 921

20°; 85% AlCl3; CH2Cl2

922 + 505

40°; 63% AlCl3; CH2Cl2

Br

923

Scheme 5.112

C4 carbonyl of 2,6-dimethylbenzoquinone. This reasoning does not explain the regioisomer **925**, which is obtained by the boron trifluoride catalyzed cycloaddition of 2,6-dimethylbenzoquinone with diene **924** in the first step of the classical synthesis of (±)-estrone methyl ether (**866a**) (*245*) (Scheme 5.113).

However, a regiochemically controlled synthesis of altersolanol B (**929**) offers a nice application of the concepts. The triacetoxyborane catalyzed cycloaddition of 1-methoxy-3-methylbutadiene (**926**) with 5,7-dihydroxy-1,4-naphthoquinone (**927**) gives 80% yield of **928** only, which in four steps is converted to **929** (*246*) (Scheme 5.114).

Obviously the complex of **927** involved, depicted in Figure 5.16 (with the greater LUMO coefficient at C2), interacts with the HOMO of **926** [with the coefficient at C4 greater than that at C1, because methoxy and methyl substituents act similarly in polarizing the coefficients (*70*)]. Thus **928** is the preferred regioisomer.

In the alkaloid field, syntheses of ibogamine (**933**) (*247*) and catharanthine (**828**) (*248*) were obtained by boron trifluoride catalyzed cycloaddition of acrolein to 1-acetoxy-1,3-hexadiene (**930a**: R = Me, R_1 = Et) or to 1,4-dipivaloxy-1,3-butadiene (**930b**: R = *t*-Bu, R_1 = OCO *t*-Bu). The cyclohexenes (**931**) react with tryptamine to give **932a,b**, which cyclize in several steps to the desired products (Scheme 5.115).

The total synthesis of (±)-quassin (**939**), a bitter principle of quassia wood, illustrates the best result of this strategy. The first approach of Valenta and coworkers (*249*) through a boron trifluoride-catalyzed addition of **934** to 2,6-dimethylbenzoquinone (**935** is the regioisomer predicted by coordination at C4 carbonyl) gave a fragment with four of the seven chiral centers of the target (Scheme 5.116).

The puzzle was solved by Grieco and coworkers (*250, 251*) through an aluminum trichloride-catalyzed cycloaddition of the diene **936** to the enone **937**. The regiochemistry predicted by FMO interaction, the preferred *endo* transition state, and the addition from the convex face of

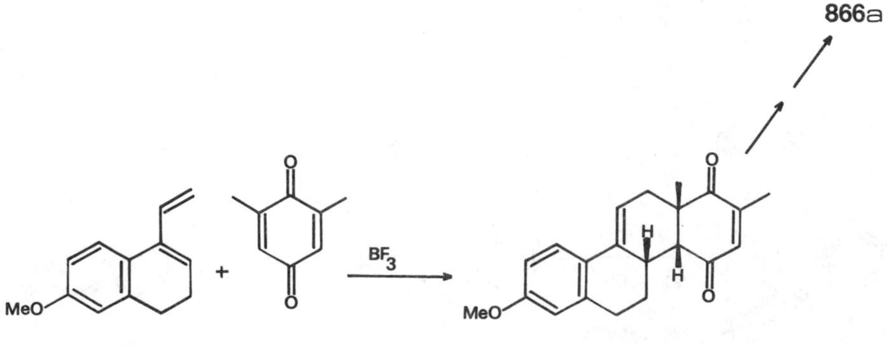

866a

MeO 924 + BF$_3$ → MeO H H 925

Scheme 5.113

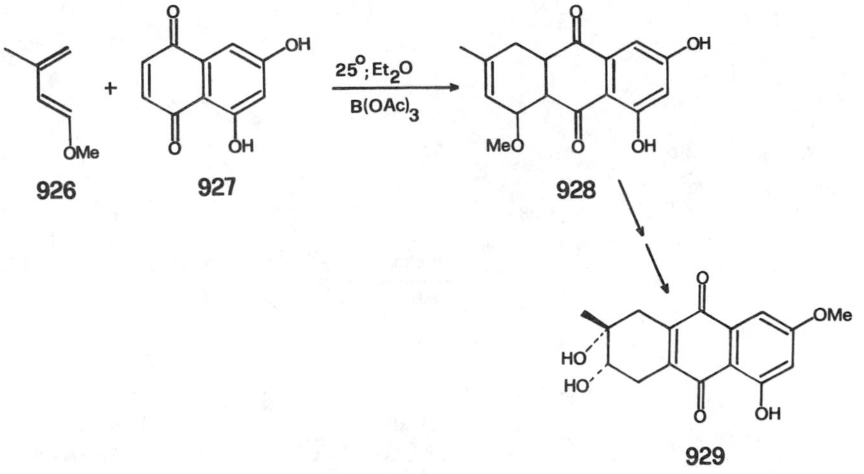

Scheme 5.114

the molecule add three chiral centers to those of **937**. Thus six of the seven chiral centers of quassin (**939**) are properly located in **938** (Scheme 5.117).

Intramolecular Catalyzed Diels–Alder Reactions. In IDAs, the rate-accelerating properties of the catalyst generally affect all conformational factors mentioned in the preceding section. Therefore, it is not surprising that trienes **762** and **763** (Scheme 5.71) undergo cyclization under milder conditions (23 °C instead of 150–180 °C) when the reaction is Lewis-acid catalyzed (*252*). Nevertheless the stereochemical influence of the catalyst is not constant. The all-*trans* triene **763** affords exclusively the *trans*-fused adduct **766**, but the triene isomer **762** gives a reaction mixture containing **764** and **765**. Both reactions occur under both thermal and catalyzed conditions. Moreover the low regioselectivity (ratio 65:35) observed without catalyst is lost in its presence (ratio 52:48). An aluminum trichloride-catalyzed IDA was the key step in the synthesis of α- and β-himachalenes (**942** and **190**, respectively). Trienone **940** cyclizes through an *endo* transition state to **941**, which was easily transformed to the required sesquiterpenes (*253*) (Scheme 5.118).

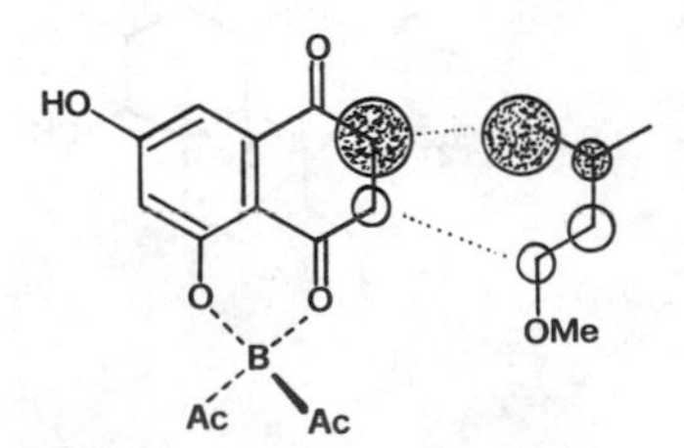

*Figure 5.16. FMO interactions of **926** with the boron acetate complex of **927**.*

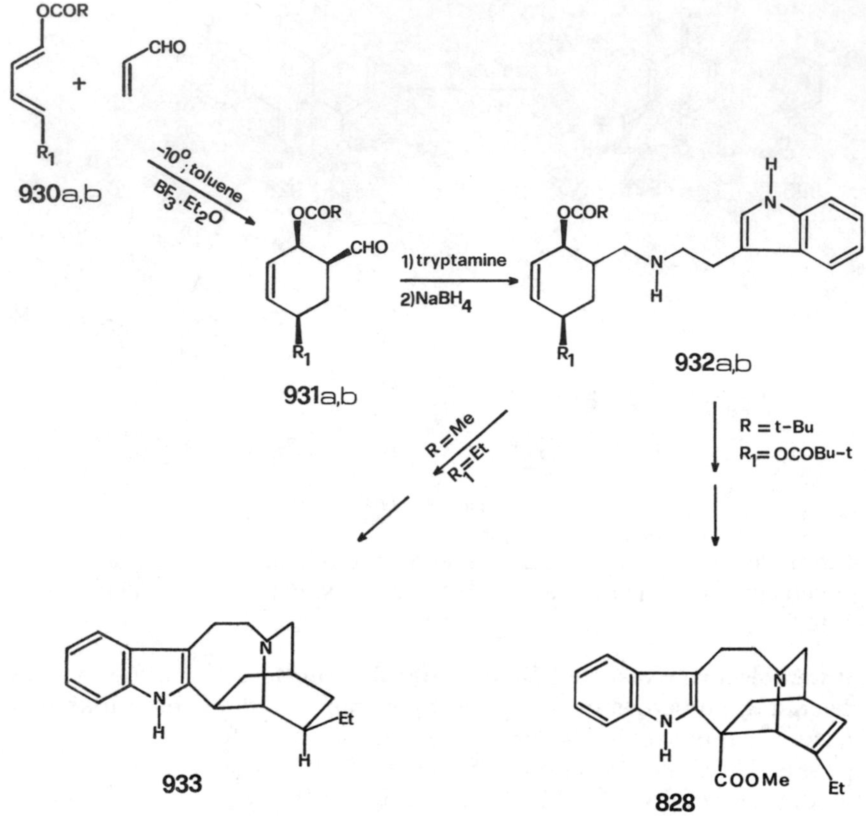

OCOR
CHO
930a,b
-10°; toluene
$BF_3 \cdot Et_2O$
931a,b
1) tryptamine
2) $NaBH_4$
932a,b
R = Me
R_1 = Et
R = t-Bu
R_1 = OCOBu-t
933
828
COOMe

Scheme 5.115

934
935
OAc
AcO
EtO_2C
OH

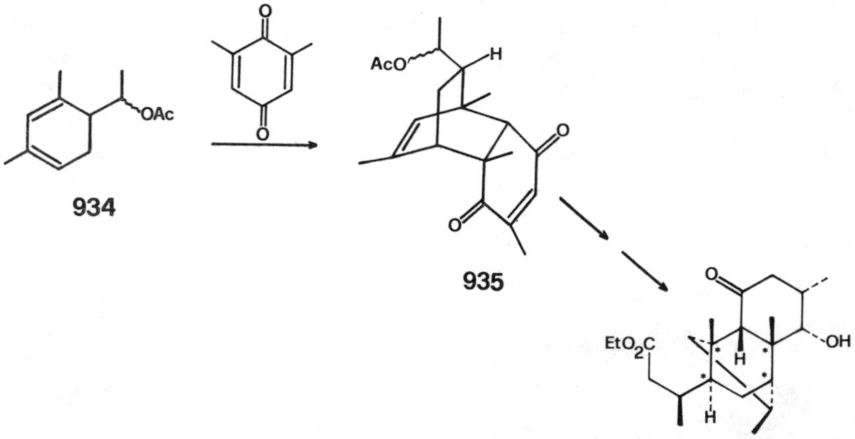

Scheme 5.116

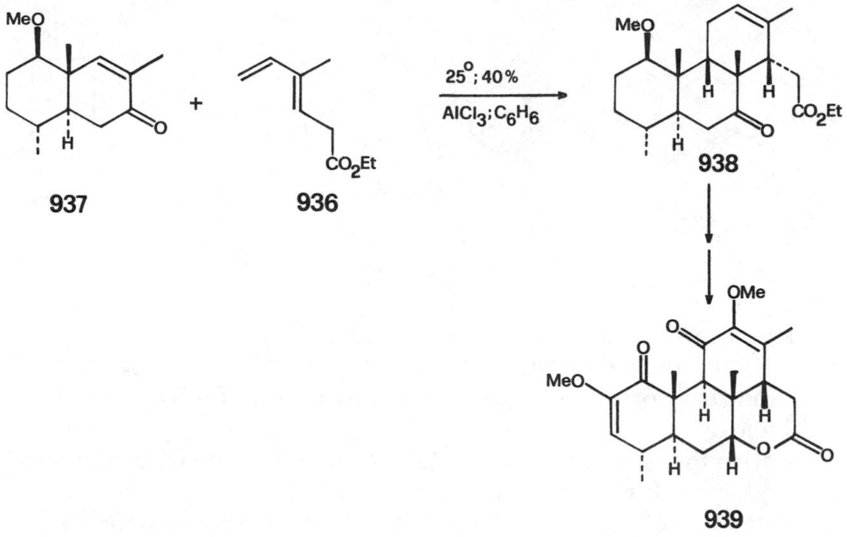

MeO H 937 + 936 CO_2Et 25°; 40% $AlCl_3$; C_6H_6 MeO H H H CO_2Et 938 OMe MeO 939

Scheme 5.117

The strain of the small ansa is responsible for the *exo* transition state in the IDA of *N*-furfuryl-*N*-acrylamide (**943**) catalyzed by internal coordination of metal salts (*254*). The adduct **944** was then transformed into the monoterpenoid karahana ether (**945**) (*255*), isolated from Japanese hop "Shinshu Wase" (Scheme 5.119).

Retro-Diels–Alder Reactions

The retro-Diels–Alder reaction is the best known example of [4+2] cycloreversion giving a diene and a dienophile from cyclohexene derivatives. The applications of the reaction in organic synthesis were reviewed in Reference 256.

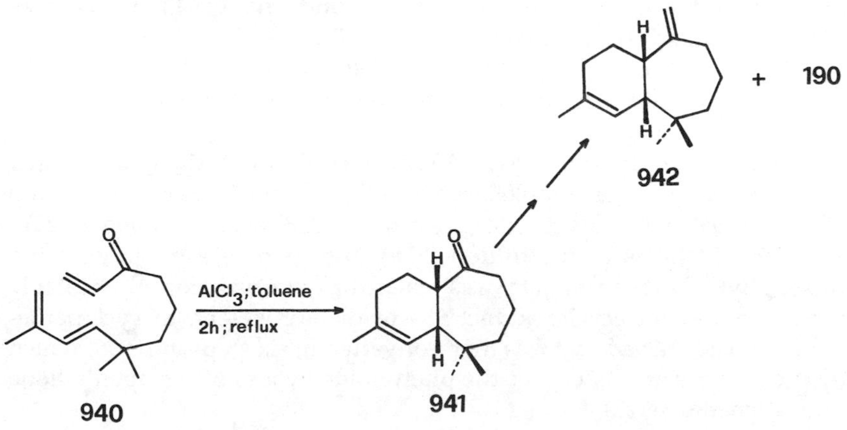

940 $AlCl_3$; toluene 2h; reflux 941 942 + 190

Scheme 5.118

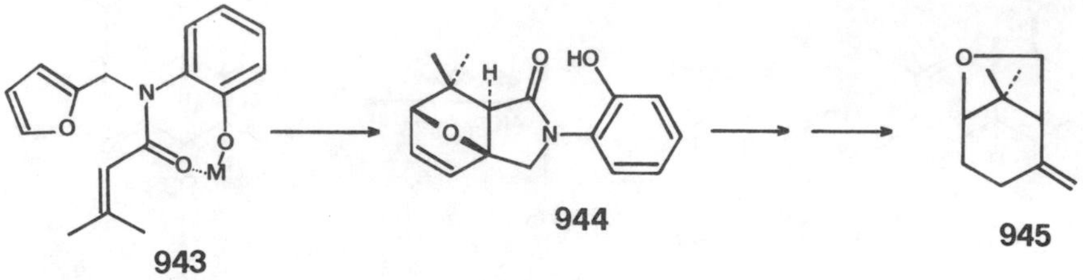

Scheme 5.119

The theoretical approach to this reaction was studied by Dewar et al. (*257*), and their calculations (MINDO/3) suggest an unsymmetrical transition state corresponding to weakening of one of the two breaking C–C bonds.

During the cycloreversion, the C3–C4 and C5–C6 bonds are broken. Although two chiral centers are lost, the relative configuration of the substituents at C3 and C6 and at C4 and C5 is maintained (microreversion of the *cis*-principle of the Diels–Alder reaction). This process is certainly not suitable for building up new chiral centers.

This process is generally utilized to mask a diene fragment (regenerated after elimination of the dienophile) or to protect a double bond (reformed after elimination of the diene). The material will be reviewed according to these classifications.

Elimination of the Diene. The masking of a double bond can be achieved by reaction with cyclopentadiene or its derivatives, and can be subsequently eliminated after elimination of the dienophile. Such elimination is the topic of this section.

Cyclopentadiene was used in the synthesis of (±)-methyl jasmonate (**949**) where *endo*-tricyclo[5,2,1,$0^{2,6}$]deca-3,8-dien-5-one (**946**) was transformed into **947**. Cycloreversion with loss of cyclopentadiene regenerates the cyclopentenone double bond giving **948**, which was easily elaborated to **949** (*258*) (Scheme 5.120).

Similarly, the last step of the synthesis of (±)-turmerone (**951**) involved the elimination of cyclopentadiene from **950** (*259*) (Scheme 5.121).

The cyclopentadiene adduct **952** of itaconic anhydride was utilized to protect its precious double bond and then eliminated in the last step of the synthesis of pheromone (±)-ipsenol (**953**) (*260*) (Scheme 5.122).

The synthesis of multifiden (**956**), the gamete attractant of the phaeophyte *Cutleria multifida*, was realized by starting from the suitably functionalized tetracyclic adduct **954** of dichloroketene to cyclopentadiene dimer. This adduct was then converted in six steps into **955**, which by thermolysis at 500 °C gave the pheromone by loss of cyclopentadiene (*261*) (Scheme 5.123).

The protection of a double bond can also be achieved by using dimethylfulvene. In this way, one of the double bonds of benzoquinone

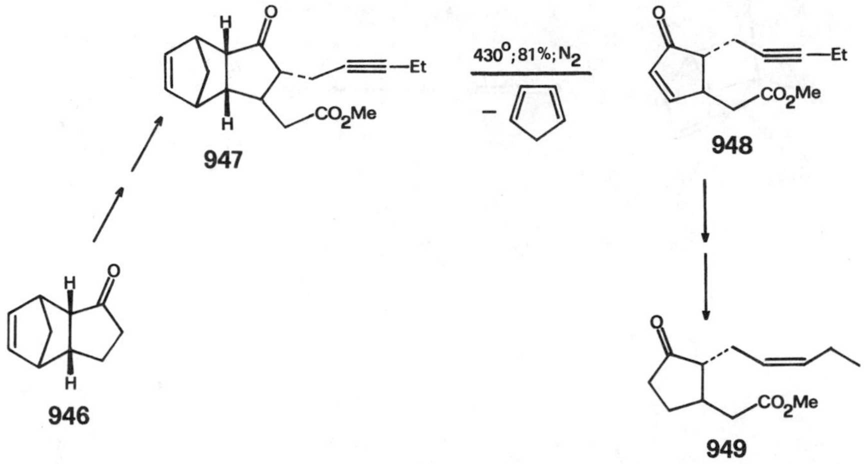

Scheme 5.120

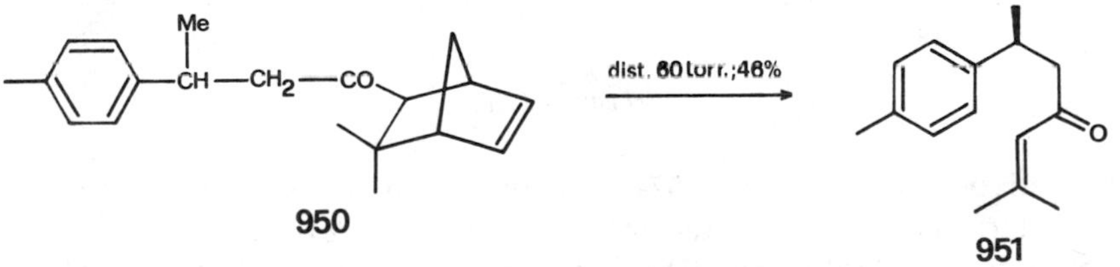

Scheme 5.121

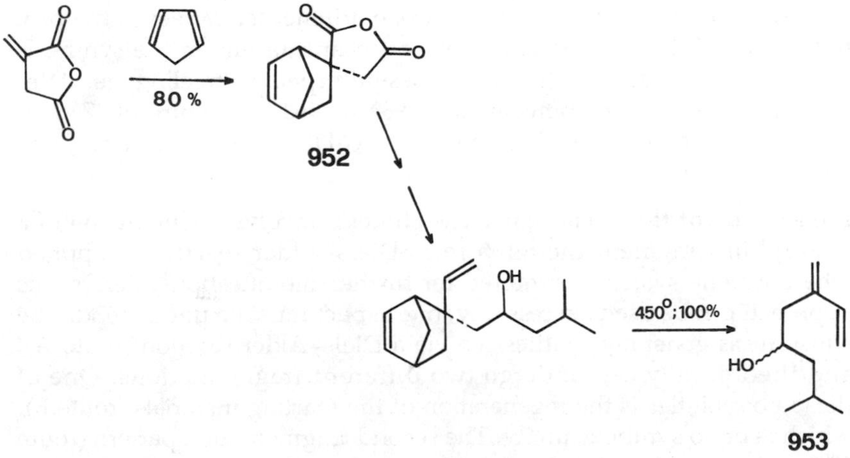

Scheme 5.122

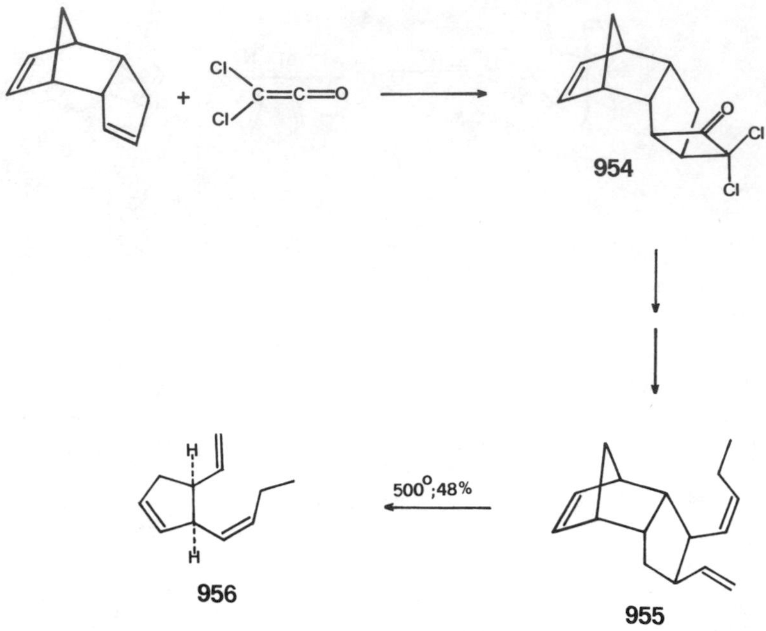

Cl
C=C=O
Cl
954
Cl
Cl
O
500°;48%
H
H
956
955

Scheme 5.123

derivatives was protected (**957a,b**), allowing the elaboration of the unprotected one. Thus several epoxycyclohexene derivatives were synthesized: (±)-phyllostine (**958**) (*262*); (±)-epoxydon (*262*) (**959**); crotepoxide (*263, 264*) (**960**), isolated from the fruits of *Croton macrostachys;* and senepoxyde (**961**) (*264, 265*), isolated from *Uvaria catcarpa,* a French plant used in folk medicine. The importance of these epoxides is due to their antitumoral activity (Scheme 5.124).

An exception to the use of cyclopentadiene derivatives is the protection of 2,5-dimethylbenzoquinone with anthracene in the synthesis of a key intermediate to the furano-sesquiterpenoid freelingyne. After the photochemical rearrangement of **962** to **963**, a mixture of (*Z*) and (*E*)-butenolide **964** was obtained by retro-Diels–Alder reaction (*266*) (Scheme 5.125).

Elimination of the Dienophile. The strategy involving elimination of a dienophilic fragment though a retro-Diels–Alder reaction is applied when a diene system is required for further modifications. Before this approach can be used successfully, one aspect must be understood. The substrate is generally synthesized via a Diels–Alder reaction (route A), and theoretically can undergo two different fragmentations. One of these possibilities is the regeneration of the starting materials (route B), which is of no synthetic utility. The second fragmentation pattern (route C) is followed only when its activation energy is lower than that of route B. This situation generally occurs if A=B is a very stable molecule (eth-

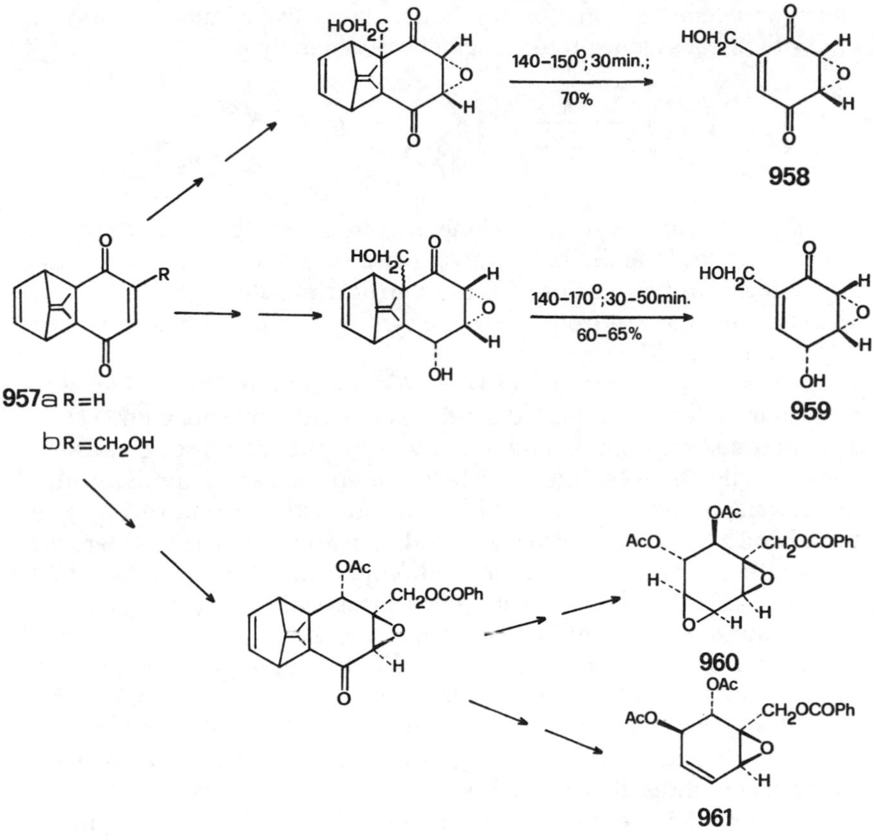
HOH2C
140–150°; 30min.;
70%
958
957a R=H
b R=CH2OH
140–170°; 30–50min.
60–65%
OH
959
OAc
CH2OCOPh
AcO
960
961

Scheme 5.124

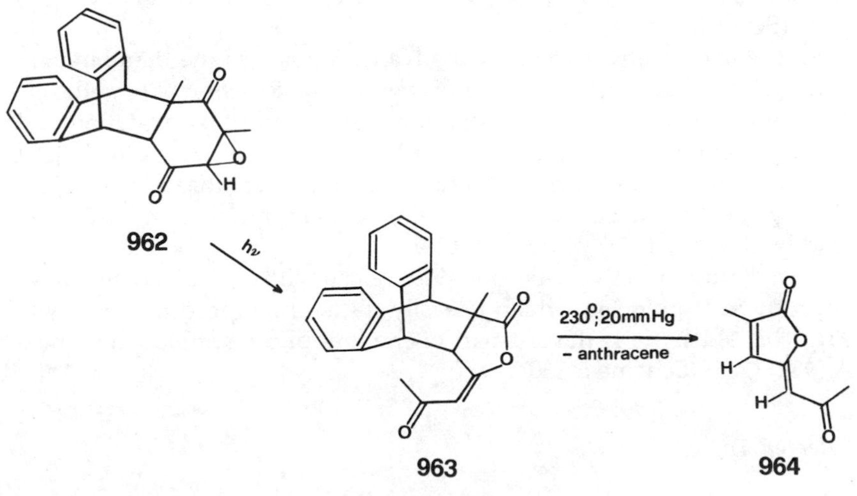
962
hν
230°; 20mmHg
– anthracene
963
964

Scheme 5.125

ylene, acetylene, carbon dioxide). In the alternative elimination of A = B or C = D, the less substituted fragment is generally lost.

Thus acetylene is usefully eliminated from **965** (the adduct between *N*-benzoylpyrrole and DMAD) giving rise to 3,4-disubstituted pyrrole **966**, which can be transformed into verrucarin E (**967**) (*267*). Ethylene was the fragment lost from **968** in the synthesis of emodinanthron (**969**) (*268*) (Scheme 5.126).

An interesting sequence of Diels–Alder and retro-Diels–Alder reactions is included in a reported synthesis of daunomycinone (**972**) (*269*). The intermediate anthraquinone **970** was synthesized through a Diels–Alder reaction between the Danishefsky diene **592** and naphthazarin. It underwent a second Diels–Alder reaction with 1-methoxy-1,3-cyclohexadiene to give **971**, which upon oxidation and retro-Diels–Alder reaction (with loss of ethylene) gave daunomycinone (**972**) (Scheme 5.127). The Diels–Alder reaction of **970** is not regiospecific, and the regioisomer of **971**, through the same reaction sequence, gave isodaunomycinone.

A similar sequence, in the field of anthraquinone pigments, was previously described in Scheme 5.50 in the route to emodin (**686**) (*116*).

3-Carbomethoxy-2-pyrone is an interesting diene because the initial adducts with dienophiles undergo decarboxylation (the retro-Diels–Alder stage) under the conditions of the Diels–Alder reaction. It reacts with 4-methyl-3-cyclohexenone (the enol form of the pyrone in the cyclo-coupling) to give the *cis*-adduct **973**, a valuable intermediate. This adduct can be converted into (±)-occidentalol (**974**) (*270*), α- and β-copaene (**975**) and (**256**) (*271*), and α- and β-ylangenes (**976**) and (**977**) (*271*) (Scheme 5.128).

Pyrone derivatives were used advantageously as the diene moiety in the Diels–Alder approach to natural products. Reaction of 6-methoxy-4-methylpyran-2-one with naphthoquinone, followed by oxidation and demethylation, gave pachybasin (**978**) in 64% overall yield. When naphthoquinone was substituted for juglone or naphthazarin, the previously mentioned chrysophanol (**687**) or helminthosporin (**688**) was readily obtained (*272*) (Scheme 5.129).

The optically active α-pyrone (**979**) reacts with *N*,*N*-dibenzyl-1-aminopropyne to give **980**, which is easily transformed into the aldehyde **981**. This aldehyde is the left part of the ionophore antibiotic lasalocid A (**982**) (*273*) (Scheme 5.130).

Selected Topics

The following section is devoted to homogeneous arguments, all involving Diels–Alder strategic approaches.

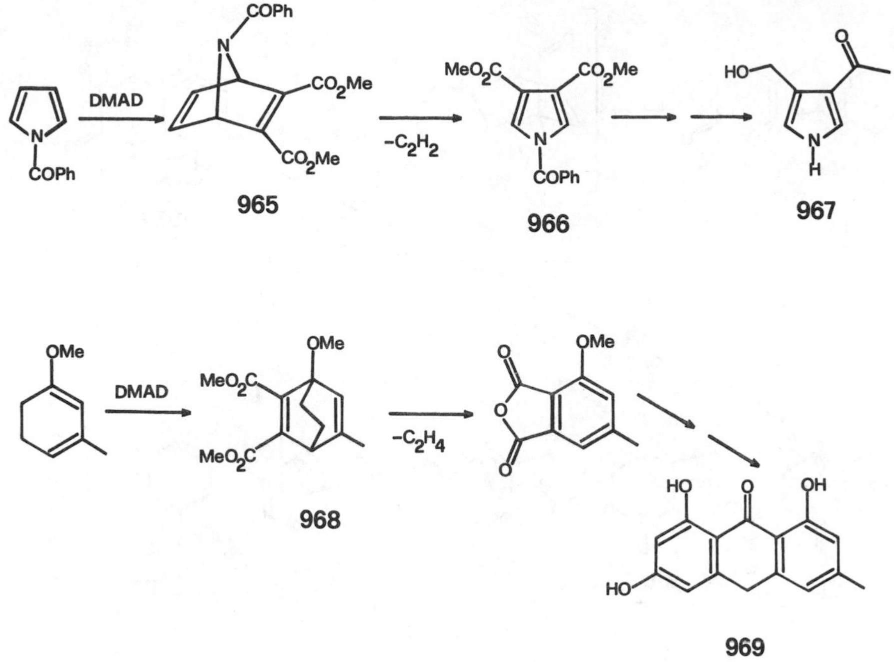

DMAD
$-C_2H_2$
965
966
967

DMAD
$-C_2H_4$
968
969

Scheme 5.126

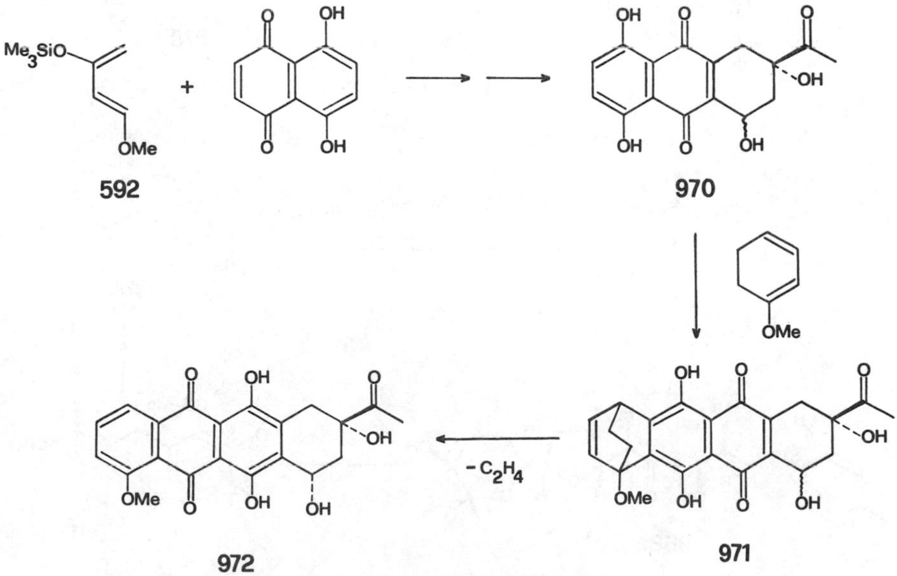

592
970
971
972
$-C_2H_4$

Scheme 5.127

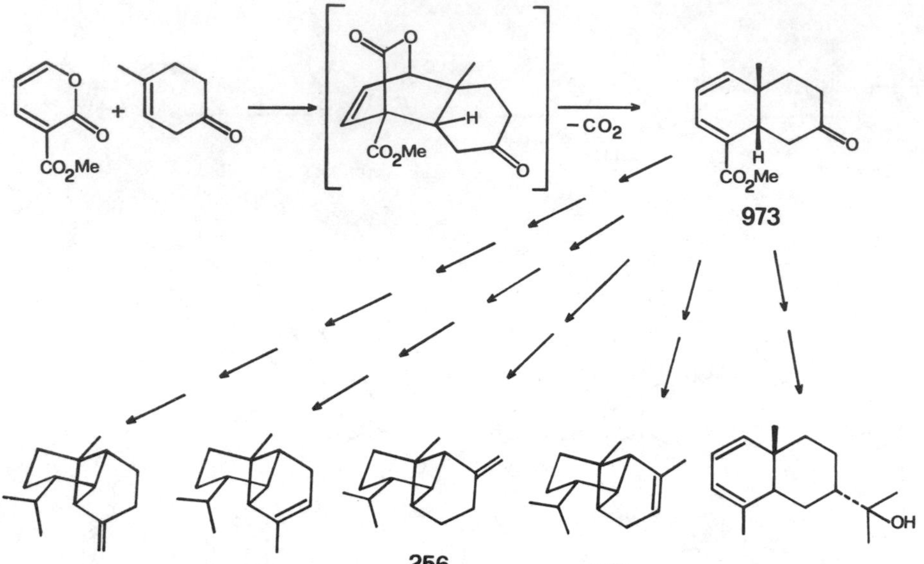

CO_2Me + → [CO_2Me H] $-CO_2$ → H CO_2Me

973

977 976 256 975 974

OH

Scheme 5.128

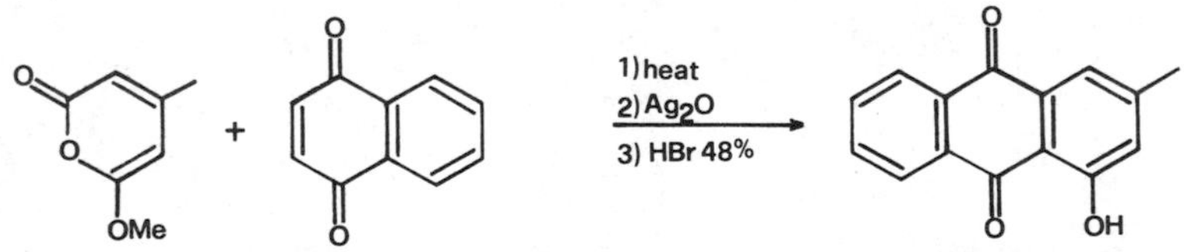

1) heat
2) Ag_2O
3) HBr 48%

OMe

OH

978

Scheme 5.129

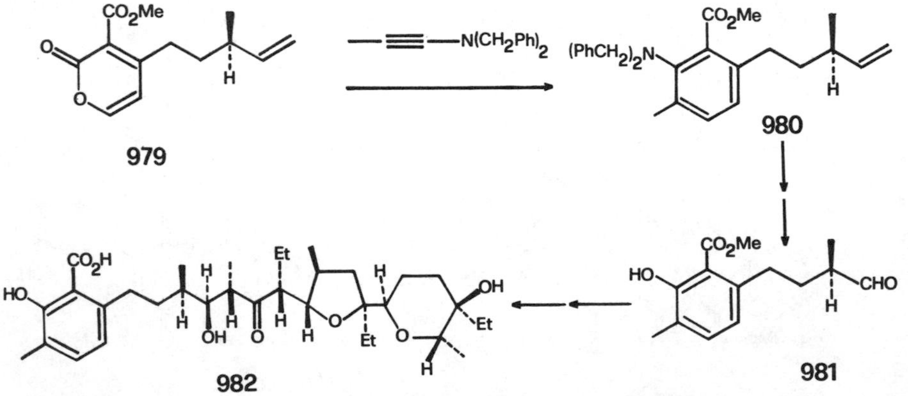

CO_2Me H

979

—≡—$N(CH_2Ph)_2$

$(PhCH_2)_2N$ CO_2Me H

980

HO CO_2Me H CHO

981

HO CO_2H H H OH H O Et H H O Et OH Et H

982

Scheme 5.130

Anthracycline Antibiotics. The discovery of the potent antitumor properties of anthracycline antibiotics such as adriamycin (**983a**), daunomycin (**983b**), carminomycin (**983c**), and to a lesser extent aklacynomycines sparked tremendous efforts to synthesize aglycones of these molecules and their analogs.

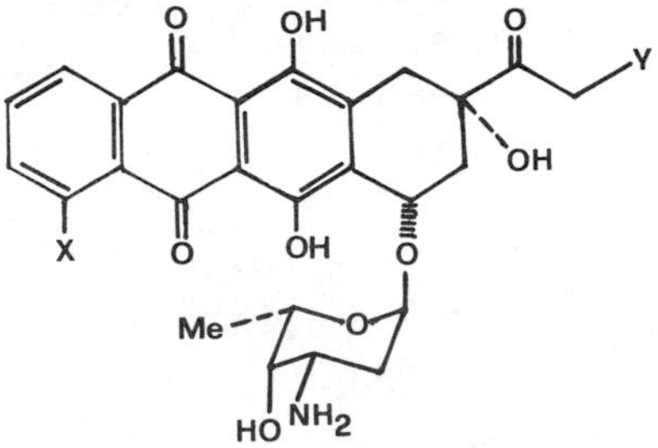

983 a ; X = OMe ; Y = OH
b ; X = OMe ; Y = H
c ; X = OH ; Y = H

Although practical syntheses are available for the sugar portion of the molecule as well as for its coupling with the appropriate aglycone, difficulties have been encountered in the development of strategies for an efficient construction of the tetracyclic-substituted skeleton of the aglycones.

The entire field was reviewed by Ross Kelly (*274*), an outstanding researcher in this area. Here, we briefly update only those synthetic efforts based on the ubiquitous Diels–Alder reaction that played an important role by offering solutions to almost all retrosynthetically conceivable disconnections.

CYCLOADDITIONS WITH QUINIZARINEQUINONES AND NAPHTHAZARINEQUINONES AS DIENOPHILES. The Diels–Alder reaction between 5-methoxy-1,4,9,10-anthradiquinone (quinizarinequinone) (**984a**) and 2-acetoxybuta-1,3-diene (**985a**) was used as the key step in one of the first syntheses (*275*) of the aglycone **972** of daunomycin (**983b**). Later (*276–278*) the same strategy was applied to the preparation of (±)-4-demethoxydaunomycinone (**987**), an analog with 8–10 times the biological activity of the natural compound. The starting materials for this latter synthesis were unsubstituted quinizarinequinone (**984b**) and the dienes **985b** and **985c**. The tetracyclic adducts **986a–c** were further elaborated to the final compounds (Scheme 5.131).

This approach has some important drawbacks. First, the reaction of many dienes with quinizarinequinone, a bifunctional dienophile, can involve unwanted addition to the internal double bond. Several solutions to ensure sitoselective addition have been suggested (*279–282*). No

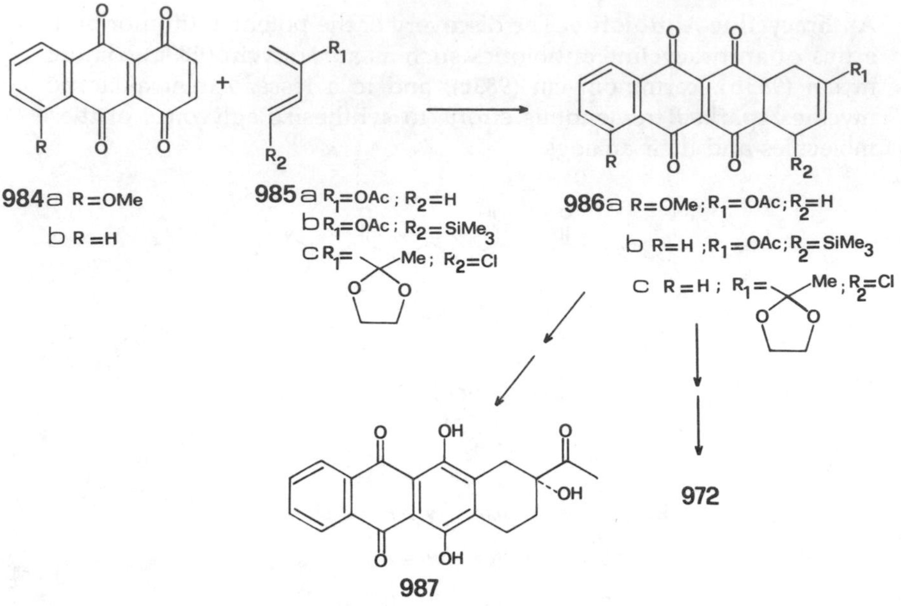

984a R=OMe
b R=H

985a R_1=OAc; R_2=H
b R_1=OAc; R_2=$SiMe_3$
c R_1= (2-methyl-1,3-dioxolan-2-yl); R_2=Cl

986a R=OMe; R_1=OAc; R_2=H
b R=H; R_1=OAc; R_2=$SiMe_3$
c R=H; R_1= (2-methyl-1,3-dioxolan-2-yl); R_2=Cl

987

972

Scheme 5.131

obvious correlations exist between the structure of the diene and the site of its cycloaddition.

A second problem in achieving regioselectivity in anthracyclinone syntheses based on Diels–Alder reaction was solved by Krohn and Tolkiehn (*269*) (*see* Scheme 5.127) and Ross Kelly et al. (*283*), by drawing on the extensive exploratory work in the field (*89, 116, 284–287*). A 10-step synthesis (*283*), which proceeds in 36% overall yield and involves two Diels–Alder reactions and one retro-Diels–Alder reaction, provides an easy access to (±)-**972**. Complete regiochemical control of the relative orientation of the A and D ring substituents is obtained (Scheme 5.132).

The Diels–Alder reaction between *p*-nitrocarbobenzoxy (*p*-NCBz) naphthazarin (**988**) and 1-methoxycyclohexa-1,3-diene, followed by oxidation, gave **989**. This latter compound reacted as its more dienophile tautomer **990** through intramolecular transfer of the directing group, with the diene **676** to produce regiospecifically **991**, further elaborated to **972**.

Krohn and Tolkiehn (*269*) obtained both (±)-**972** and its 4-demethoxy analog **987** starting from naphthazarin (shown in Scheme 5.127).

CYCLOADDITIONS WITH DIENES OF *o*-QUINONEDIMETHANE TYPE. An alternative approach to the tetracyclic linear structure of aglycones entails generation of dienes from different suitable substrates (*see also* the section in this chapter entitled "In Situ Generation of Dienes for Intermolecular Diels–Alder Reactions").

Thus the 4-demethoxy aglycone **987** was obtained (*288*) from *o*-quinonedimethane (**993**), which was generated from **992** and trapped with methyl vinyl ketone. The tetracyclic skeleton **994** was formed and was further elaborated to the final target (Scheme 5.133).

The same compound (**987**) was also obtained (*289*) through Diels–Alder reaction between *o*-quinonedimethane generated from tetra-bromo-*o*-xylene and the quinone **995**. The result was an adduct that spontaneously loses hydrobromic acid producing **996**, which was transformed into (±)-**987** (Scheme 5.134).

This sequence was also extended to the methoxy series corresponding to adryamicin (**983a**) and carminomycin (**983c**) starting from suitable *o*-quinonedimethane. However, regiochemical problems were encountered.

The quinone **995** as its acetyl **998** was used as the dienophile counterpart to trap the diene generated from **997** in the so called isobenzofuran route (*290*). This sequence includes several Diels–Alder reactions (three normal and two retro), and allows an original entry to (±)-4-demethoxy derivative **987**, as well as to (±)-**972**, but with usual regiochemical problems (Scheme 5.135).

The possibility offered by this strategy was perceived by Kametani et al. (*291, 292*), who demonstrated the feasibility of approaching linear tetracyclic structures without extension to total synthesis. Related ap-

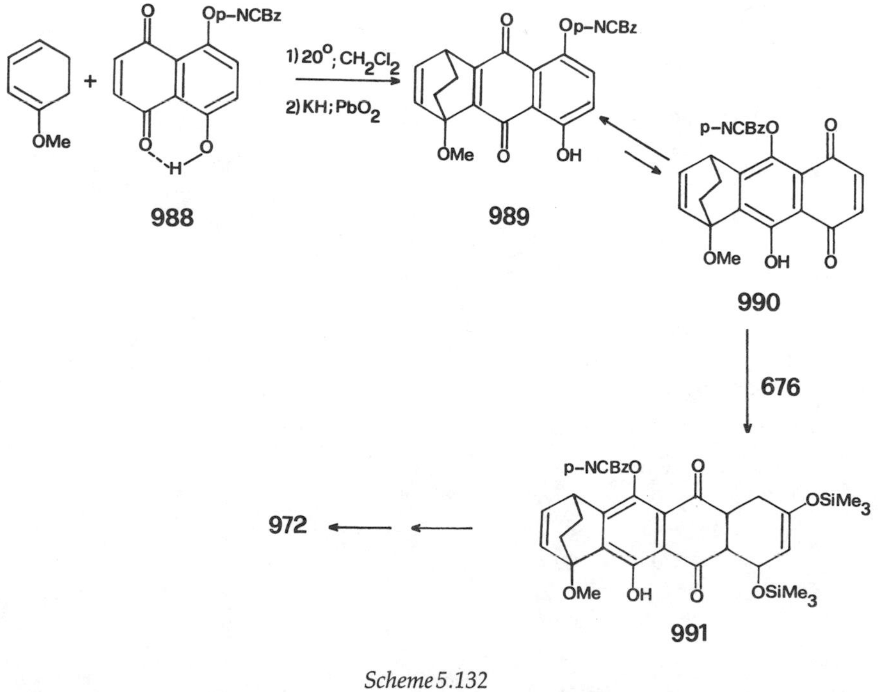

OMe + 988 → 1) 20°; CH_2Cl_2 2) KH; PbO_2 → 989 ⇌ 990 → 676 → 991 → → 972

Op-NCBz, OMe, OH, p-NCBzO, $OSiMe_3$

Scheme 5.132

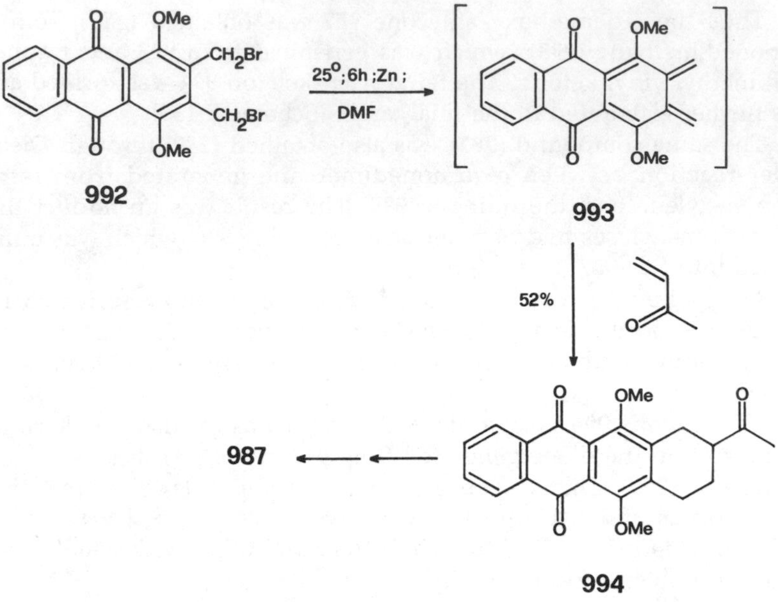

OMe
CH_2Br
CH_2Br
OMe
992

25°; 6h ; Zn ;
DMF

OMe
OMe
993

52%

OMe
OMe
994

987

Scheme 5.133

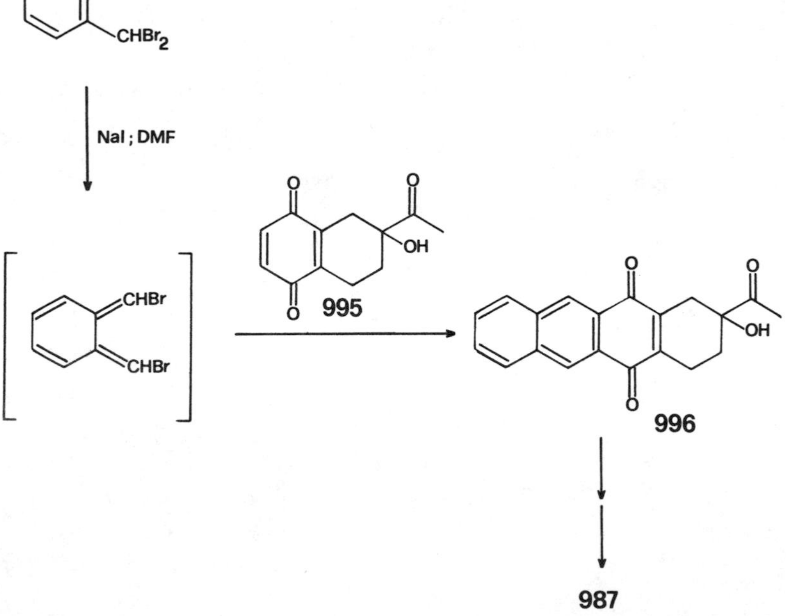

$CHBr_2$
$CHBr_2$

NaI ; DMF

CHBr
CHBr

OH
995

OH
996

987

Scheme 5.134

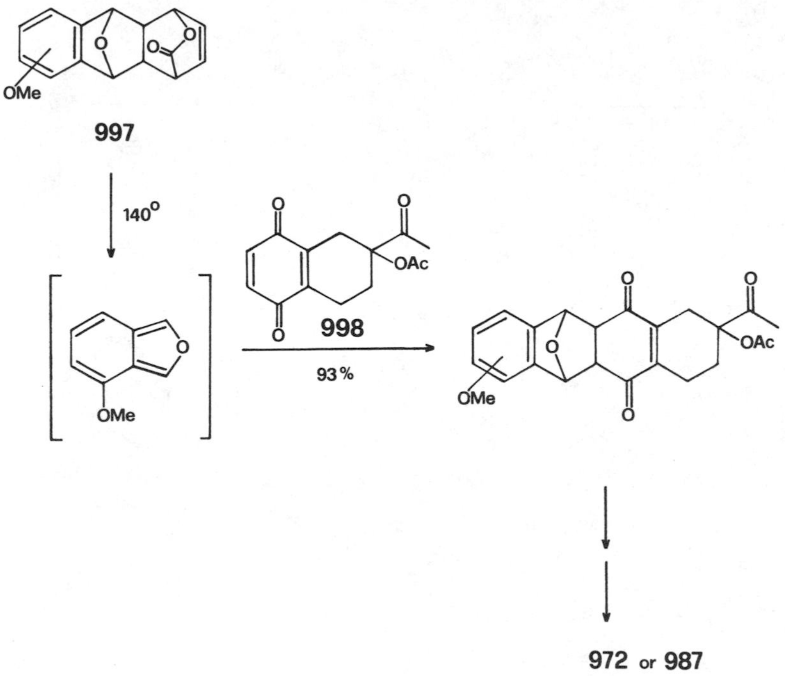

997

140°

998

93%

972 or 987

Scheme 5.135

proaches as well as several syntheses of intermediates have also been reported (*293–296*).

A double Diels–Alder addition to 2,3,5,6-tetramethylidene-7-oxanorbornane (**999**), which can add sequentially two dienophiles, has been used (*297, 298*) to generate a wide variety of anthracyclinone precursors and the usual 4-demethoxy analog **987**. An alternative approach (*299*) culminating in the synthesis of (±)-**987** utilizes benzocyclobutenedione monoketal (**1000**) as a 1,4-dipole equivalent. This latter compound couples with **1001** in the presence of butyllithium to give **1002**, readily converted to the desired target in high yield (Scheme 5.136).

The ring closure process can be considered an intramolecular Diels–Alder reaction proceeding through a negatively charged *o*-quinonedimethane intermediate.

CYCLOADDITIONS WITH DIENES OF VINYLKETENE ACETAL TYPE. The final group of papers we discuss (*300, 301*) deals with a Diels–Alder type regiospecific cycloaddition of vinylketene acetals to halogenated naphthoquinones as a new way of assembling the tetracyclic skeleton. This approach offers some advantages over the previous ones; the reaction is highly regiospecific and displays the flexibility of incorporating a variety of peripheral substitution patterns.

Thus cycloaddition of both vinylketene enol-ether **1003** (*300*) and vinylketene acetal **1004** (*301*) to 3-bromo-5-methoxy-1,4-naphthoqui-

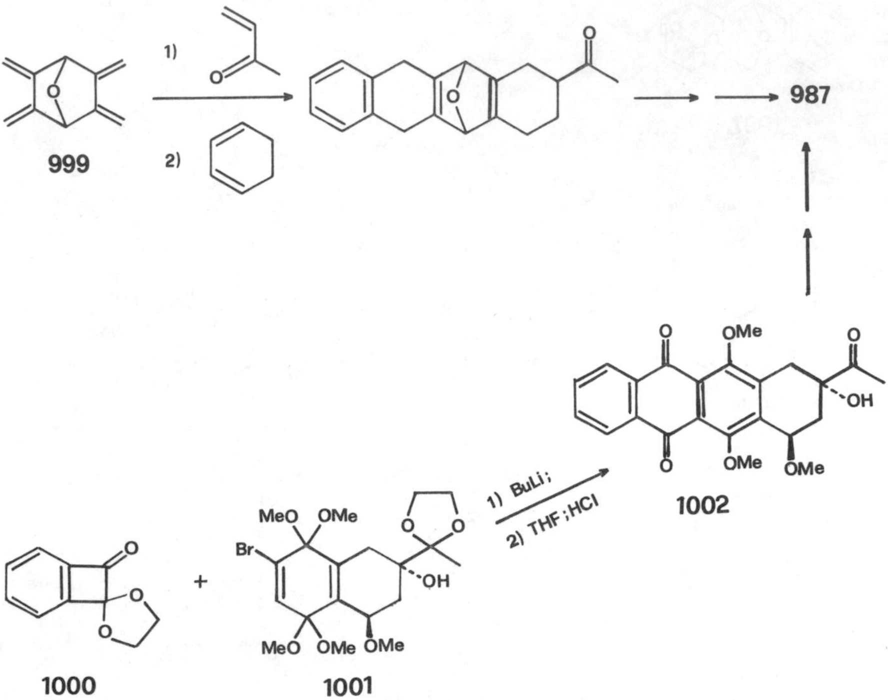

999 1) 2) 987 1000 + 1001 1) BuLi; 2) THF; HCl 1002

Scheme 5.136

none produced the tetracyclic cycloadduct **1005**, which can be further elaborated to 11-deoxy anthracyclines. Both dienes **1003** and **1004** are readily accessible from commercially available material (Scheme 5.137).

***Iboga* Alkaloids.** Through common usage, the term "*Iboga* alkaloids" has come to include all of the bases from diverse Apocyanaceae species having the ibogamine (**933**), epiibogamine (**1006**), or ibogaine (**1007**) skeleton, or simple variations of it such as catharanthine (**828**) and other natural bases. The syntheses of these representative members were achieved by several researchers, who all considered a suitable functionalized isoquinuclidine system as an obvious intermediate. Buchi et al. (*302, 303*) completed the first total syntheses of **933**, **1006**, and **1007** by basing their synthetic strategy on the assumption that the isoquinuclidine ring could be prepared by Diels–Alder cycloaddition of the dihydropyridine **1008** with methyl vinyl ketone as dienophile. The resulting cycloadduct **1009** was then elaborated and finally condensed with the indole counterpart (Scheme 5.138).

A similar approach was proposed by Wakamatsu and coworkers (*304, 305*), who obtained the substituted isoquinuclidine system **1011** by cycloaddition of the cyanodihydropyridine **1010** and acrylonitrile. Several steps completed the synthesis of ibogamine (**933**) and of the corresponding C4 epimer **1006** (Scheme 5.138).

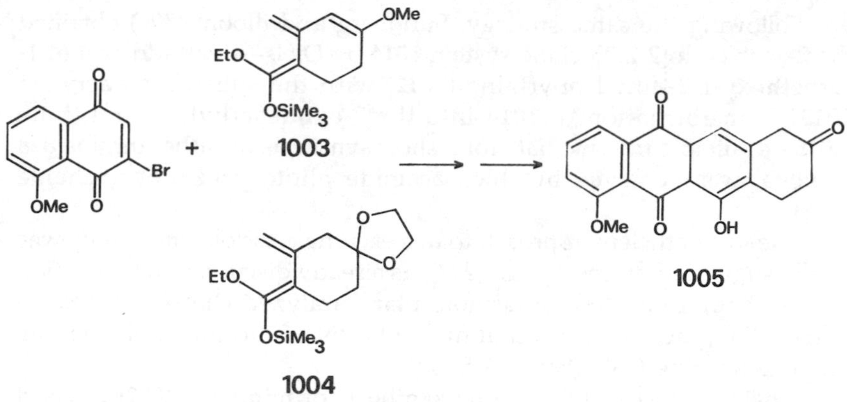

EtO OMe OSiMe3 O O Br OMe + 1003 → → O O OMe O OH 1005

EtO OSiMe3 O O 1004

Scheme 5.137

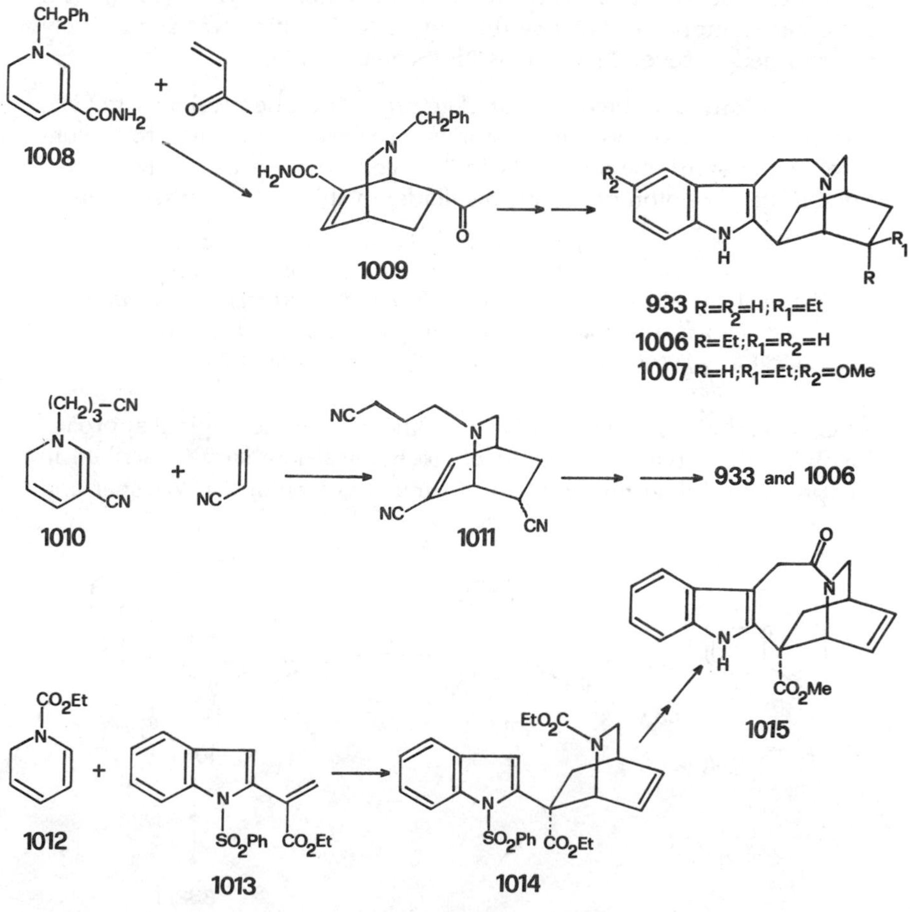

CH_2Ph N $CONH_2$ + O 1008 → H_2NOC N CH_2Ph O 1009 → → R_2 N H N R_1 R

933 $R=R_2=H; R_1=Et$

1006 $R=Et; R_1=R_2=H$

1007 $R=H; R_1=Et; R_2=OMe$

$(CH_2)_3$–CN N CN 1010 + NC → NC N NC CN 1011 → → 933 and 1006

CO_2Et N 1012 + N SO_2Ph CO_2Et 1013 → EtO_2C N N SO_2Ph CO_2Et 1014 → O N N H CO_2Me 1015

Scheme 5.138

Following the same strategy, Sundberg and Bloom (*306*) obtained the 2-azabicyclo[2.2.2]octane system **1014** by Diels–Alder addition of 1-carbethoxy-1,2-dihydropyridine (**1012**) with the substituted acrylate **1013**. Transformation of **1014** into the 7-oxodesethylcatharanthine (**1015**), a suitable intermediate for a short synthesis of catharanthine **828** analogs, was achieved by chloroacemide photocyclization (Scheme 5.138).

The most efficient approach to this exciting class of compounds was finally suggested by Trost et al. (*247*), as already discussed in the section in this chapter entitled "Intermolecular Catalyzed Diels–Alder Reactions." They described a beautiful synthesis of racemic and optically active ibogamine (**933**) (Scheme 5.115).

Finally, we describe the total synthesis of minovine (**832b**), based upon the biogenetic considerations of Wenkert (*169*) concerning the origin of *Aspidosperma* and *Iboga* alkaloids. Ziegler and Spitzner (*307, 308*) performed the critical condensation of the indolylacrylate (**1016**) with the cyclic enamine **1017** to give the adduct **1018**, which was successfully transformed in several steps to **832b** (Scheme 5.139).

Aconite Alkaloids. The Centenary Lecture of the Chemical Society (*309*) offered a review of systematic studies of Wiesner and his group concerning the synthetic strategies for the construction of the hexacyclic polybridged skeleton of aconite alkaloids. In this lecture, Wiesner said:

> *It was my belief that this exercise, i.e. a systematic search for the simplest possible method to construct a complicated compound, would significantly contribute to the art of synthesis and advance the day when compounds of the complexity of delphinine (if sufficiently important and expensive) might be produced on the industrial scale.*

After several years of progressive simplification of the original approach involving the already described photochemical key step (described in Chapter 3 in the section entitled "Cumulated Systems"), Wiesner and

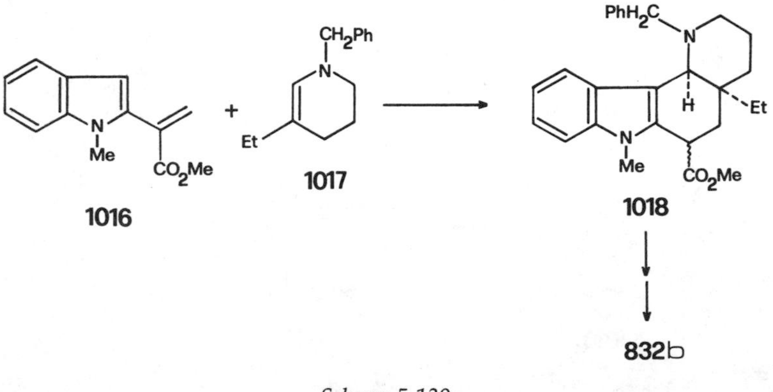

1016 + 1017 → 1018 → 832b

Scheme 5.139

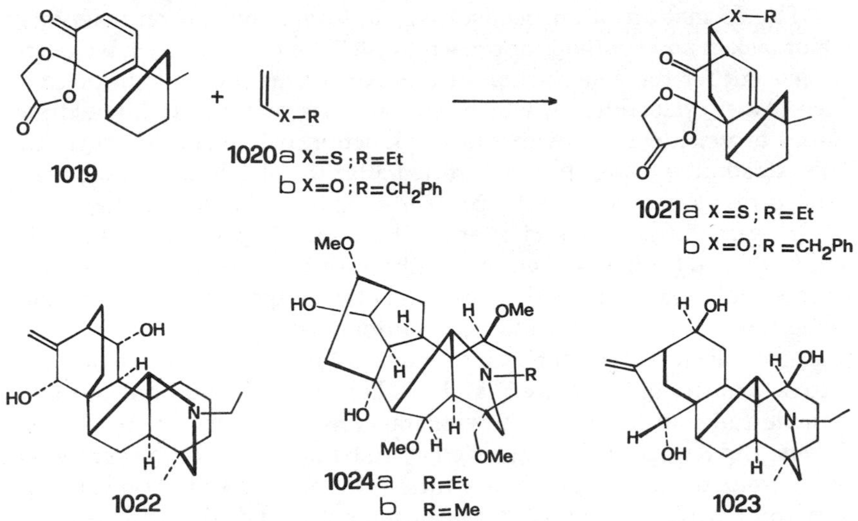

Scheme 5.140

his associates proposed the "fourth generation" approach, which entailed a crucial Diels–Alder stage.

After preliminary studies on model compounds (*310*), the reaction of the masked *o*-quinone **1019** with ethyl vinyl sulfide (**1020a**) (*311*) or benzyl vinyl ether (**1020b**) (*312*) (producing the functionalized systems **1021a,b**, respectively) disclosed a shorter synthesis of the skeleton of denudatine (**1022**) (*311*) and napelline (**1023**) (*312*). This new route was also applicable to the total stereospecific synthesis (*313, 314*) of chasmanine (**1024a**) and 13-deoxydelphonine (**1024b**) (Scheme 5.140).

Cytochalasans. The complex structures of a large family of mold metabolites known as the cytochalasans are exemplified by cytochalasin B (**1025**), a naturally occurring cytostatic substance.

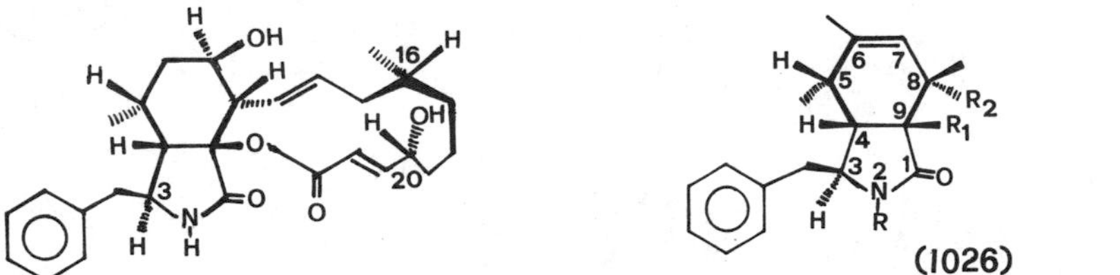

(1025)

(1026)

a R = $COCH_2Cl$	R_1 = H	R_2 = $CH_2OCOC_2H_5$
b R = CH_2CH_3	R_1 = OH	R_2 = CH_2OCH_2Ph
c R = $COCH_3$	R_1 = $OCOCH_3$	R_2 =

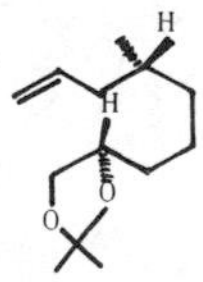

The formation of an octaisohydroindolone system represents a major task in any synthetic approach to **1025**. Several studies have been carried out for the construction of this part of the molecule through a Diels–Alder cycloaddition, which requires a *cis* ring-fusion. Some difficulties arose, however, with the introduction of the functionality R_1. This substituent either may be inserted after the elaboration of the condensed ring system or may be present already in the dienophile.

The first approach was attempted by several researchers. Vedejs and Gadwood (*315*) described the synthesis of **1026a**; this synthesis used an intermolecular Diels–Alder reaction between the diene **1027** and the dienophile **1028**. The cycloaddition controls the regiochemistry and defines four asymmetric centers in the adduct having the cytochalasin B stereochemistry at C3, C4, C5, and C8. However, the introduction of a suitable functionality R_1 at C9 with the correct stereochemistry failed. Owens and Raphael (*316*) suggested a similar approach for the synthesis of the isoindolone moiety. Their synthesis involved the reaction of (*E*,*E*)-4-methylhexa-2,4-dienol (**1029**) with maleic anhydride, with further elaboration of the adduct **1030** to produce **1031** with the wrong stereochemistry at C4 and C9 (Scheme 5.141).

Kim and Weinreb (*317*) effected the cycloaddition of benzyl ether **1032** with *N*-ethylmaleimide to furnish the adduct **1033**, which was elaborated to **1026b** (Scheme 5.141).

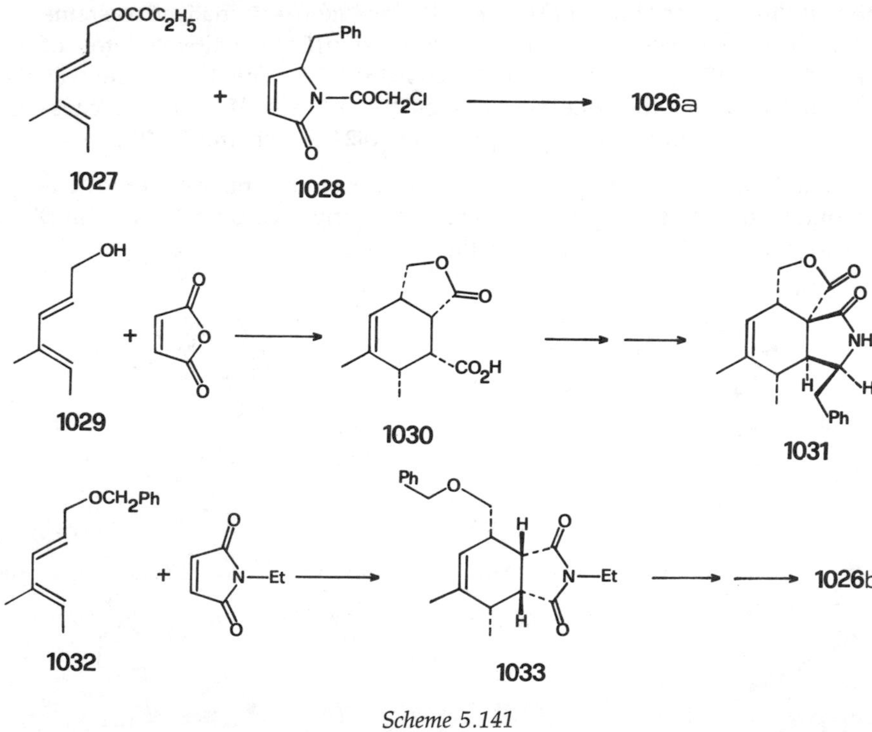

$OCOC_2H_5$ Ph N—$COCH_2Cl$
1027 + 1028 → 1026a

OH CO$_2$H NH H H Ph
1029 + maleic anhydride → 1030 → → 1031

OCH_2Ph N—Et Ph O H H
1032 + N—Et → 1033 → → 1026b

Scheme 5.141

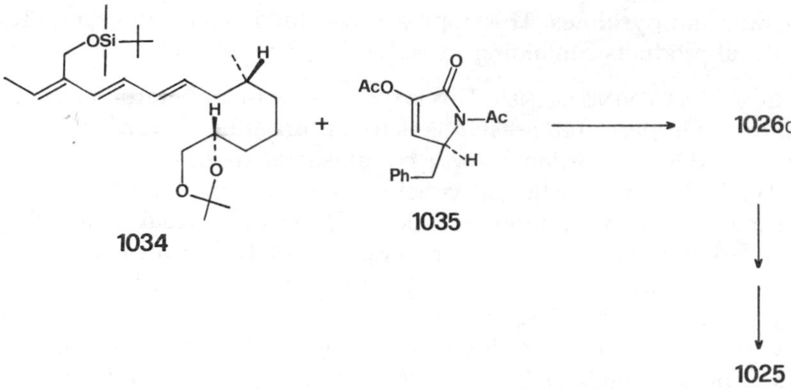

Scheme 5.142

However, the stereochemical assignment at C4 and C9 of isoindolone **1026b** is based only upon mechanistic considerations.

A second concept was well applied by Stork et al. (*318*) in the total synthesis of cytochalasin B (**1025**). This synthesis is an example of absolute asymmetric synthesis, where the isolated stereocenters at C3, C16, and C20 were derived from three distinct optically active starting materials. The synthetic project was designed by assuming the usual preference for an *endo* transition state. Therefore the triene **1034** would be expected to approach the α,β-unsaturated lactam dienophile **1035** from the side opposite to the C3 benzyl substituent. Furthermore the synthesis was based upon the assumption that the triene **1034**, for steric as well as electronic reasons, should undergo regioselective [4+2] cycloaddition (Scheme 5.142).

Vedejs and Gadwood (*315*) also observed that formation of tautomers from pyrrolinone dienophile could be avoided by introducing on nitrogen an electron-withdrawing substituent that destabilizes the aromatic hydroxypyrrole intermediate. This prototropic phenomenon was later studied by Schmidlin and Tamm (*319*), who approached (*152*) the problem as described earlier in this chapter. With all these considerations in mind, the cycloaddition of **1034** with **1035** in xylene (170 °C, 4 days) gave rise to the isoindolone **1026c** (40% conversion), which was elaborated to **1025**.

Heterodiene Reactions

The name heterodiene reactions can be generally assigned to each [4+2] thermal cycloaddition that has at least one heteroatom either on the four-atom fragment (heterodiene) or on the two-atom fragment (heterodienophile). We will review the papers for homogeneous classes of heteroatoms.

Heterodienes in Intermolecular Reactions. A Diels–Alder reaction of dienes bearing oxygen or nitrogen atoms provides an easy preparation of pyrans and pyridines. This approach was followed for the syntheses of natural products containing these heterocyclic systems.

OXYGEN-CONTAINING DIENES. This is the most popular heterodiene reaction, owing in part to the easy access to α,β-unsaturated carbonyl compounds and to α-dicarbonyls, which are used as dienes.

The MO characteristics of acrolein and glyoxal (*320*), and their benzoderivatives *o*-benzoquinone methide (*321*) and *o*-benzoquinone (*322*), are compared with those of ethylene and of two typical electron-donating and -attracting substituted ethanes in Figure 5.17.

As shown in the diagram, the π-HOMOs of acrolein and glyoxal are very stable, hence the dominating interaction occurs between the HOMO of dienophile and the LUMO of diene. (First ionization potentials for acrolein and glyoxal are due to nonbonding electrons, hence these orbitals are not fruitful for a cycloaddition.) These heterodiene reactions require an inverse electron demand; the more useful dienophiles are the nucleophilic olefins and acetylenes (e.g., enamines and ynamines) in addition to the vinyl ethers, an example of which is reported in Figure 5.17.

The conjugation in *o*-benzoquinone derivatives effects a lowering of the LUMO and a significant increasing of the HOMO. The π-ionization potential of acrolein is enhanced to 8.80 eV in *o*-benzoquinone methide and becomes the first ionization potential. Similarly the π-ionization potential of glyoxal is increased to 9.98 eV in *o*-benzoquinone (the π-ionization potentials of both systems are not significantly changed upon conjugation). As a result of these variations, the reactivity of *o*-benzoquinone methide and *o*-benzoquinone is HOMO- or LUMO-controlled depending on the nature of the dienophiles. Good reactivity with electrophilic olefins can be predicted, particularly for *o*-benzoquinone methide.

This aspect of α,β-unsaturated carbonyl compounds was reviewed in 1975 (*323*), and the syntheses of several natural products successfully approached in this way were reported: frontalin (**1036**), brevicomin (**1037**), valerianine (**1038**), and adaline (**1039**).

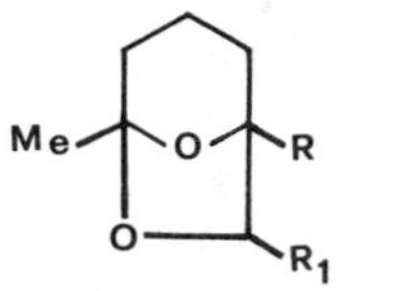

(**1036**); R = Me; R_1 = H

(**1037**); R = H; R_1 = Et

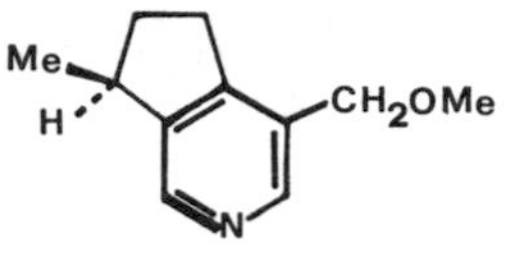

(**1038**)

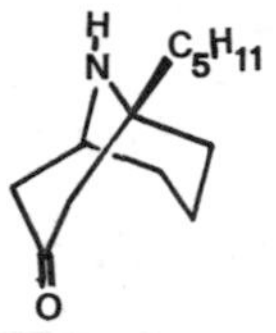

(**1039**)

Some further papers concerning the syntheses of insect pheromones (*324*), frontalin (*325*) (**1036**), and brevicomin (*325, 326*) (**1037**) were detailed. The synthesis of α-multistriatin (**1040**) (*325, 327*), one of three essential components of an aggregation pheromone for the Eu-

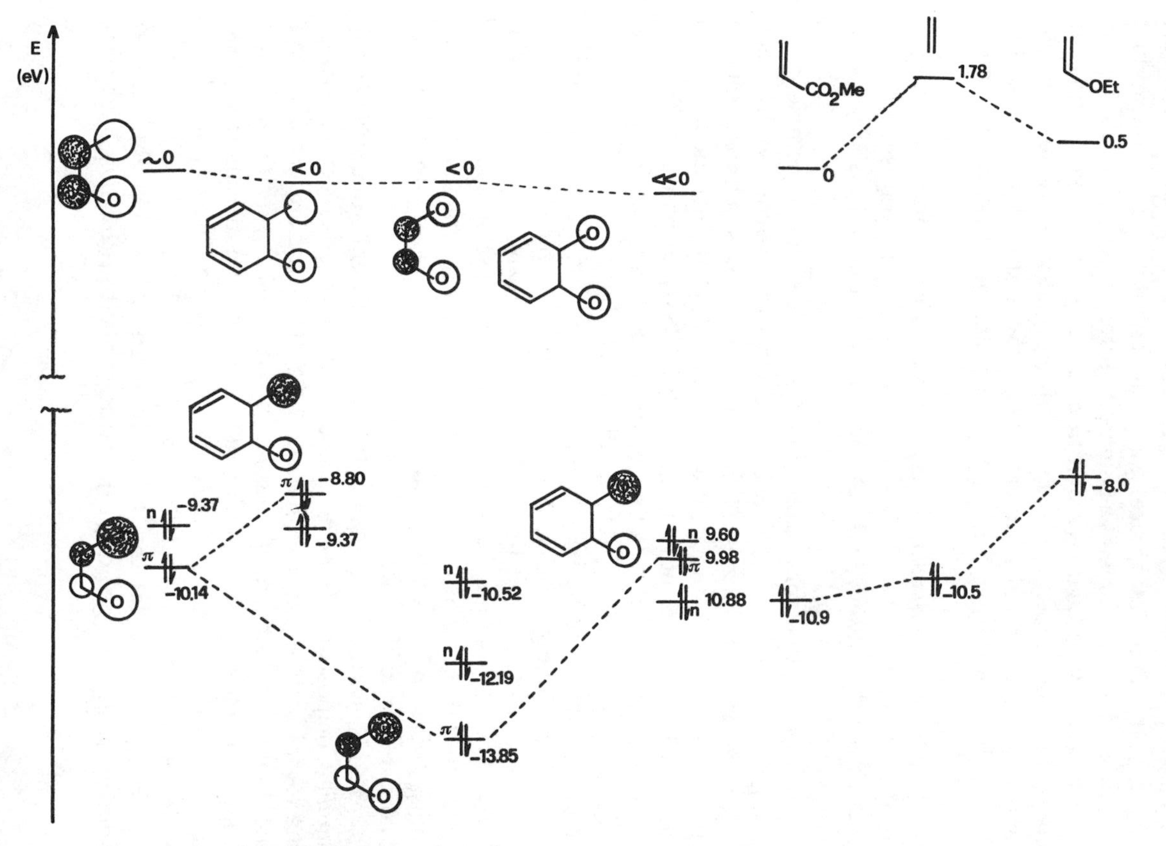

Figure 5.17. FMO interactions of oxygen-containing dienes with ethenes.

ropean elm bark beetle *Scolytus multistriatus* (Marsham) (Scheme 5.143), was investigated.

In the same field, the principal aggregation pheromone, (±)-chalcogran (**1042**), of the beetle "Kupferstecher" [*Pityogenes chalcographus* (L.)], a pest of Norway spruce, was obtained starting from the adduct **1041** of acrolein and 2-methylenetetrahydrofuran (*328*) (Scheme 5.144).

Civet, the glandular secretion from the civet cat (*Viverra civetta*), is one of the most expensive animal perfume materials. A new component, (*cis*-6-methyltetrahydropyran-2-yl)acetic acid (**1043**), was isolated and synthesized starting from methyl vinyl ketone and ethyl vinyl ether (*329*) (Scheme 5.145).

Two 3,6-dideoxy-*arabino*-hexopyranoses (**1045**) (D-form, tyvelose; L-form, ascarylose) and a *ribo*-isomer (**1046**) (D-form, paratose) were obtained from the adduct **1044** of methyl vinyl ketone and isobutyl vinyl ether (*330*) (Scheme 5.146).

A new versatile annelation procedure utilized the α-methylene ketone **1047** as a heterodienophile in the Diels–Alder reaction with several methyl α-substituted acrylates. The product, **1048**, was transformed in several steps to the tetracyclic compound **1049**, an intermediate with the structural features of fusidic acid (**1050**) (*331*) (Scheme 5.147).

Heterodiene syntheses between α,β-unsaturated carbonyl derivatives and *o*-quinones provide potential biosynthetic pathways to several natural products. Dimerization of a single component or cycloaddition between two natural cycloaddends could be involved.

The dimerization of 1-oxo-1,2,3,4-tetrahydroanthraquinones (**1051**) proceeds via a [4+2] cycloaddition through an *exo* transition state, and yields **1052** (Scheme 5.148). Analogously, a heterodiene reaction can be suggested in the biosynthetic route to (−)flavoskyrin (**1053**), a yellow colorant of *Penicillum islandicum* (*332*).

The biosynthesis of silybin (**1056**), a constituent from *Sylibum marianum* (Compositae) can be described as a heterodiene reaction between the α-dicarbonyl system of quinone **1054**, derived from taxifolin, with the double bond of coniferyl alcohol (**1055**) (*170–173*). Incidentally, silydianin (**1057**), another natural product isolated from the same plant, can result if the diene system of the quinone is involved in a standard Diels–Alder reaction (Scheme 5.149).

Fuerstione (**1058a**) and its 3β-acetoxy derivative (**1058b**) were isolated from the leaf glands of the African plant *Plectranthus nilgherricus*

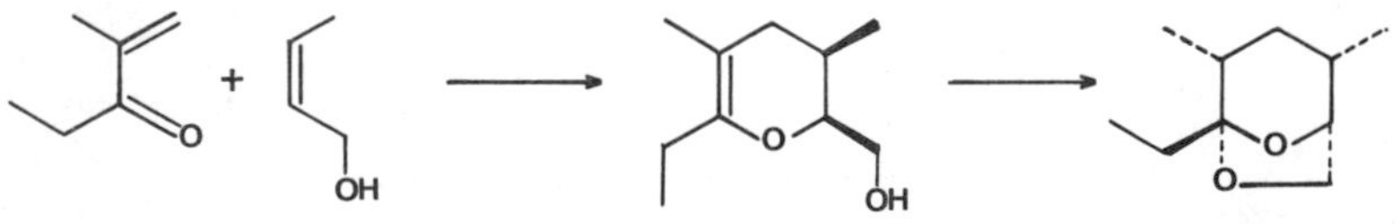

O OH OH O O **1040**

Scheme 5.143

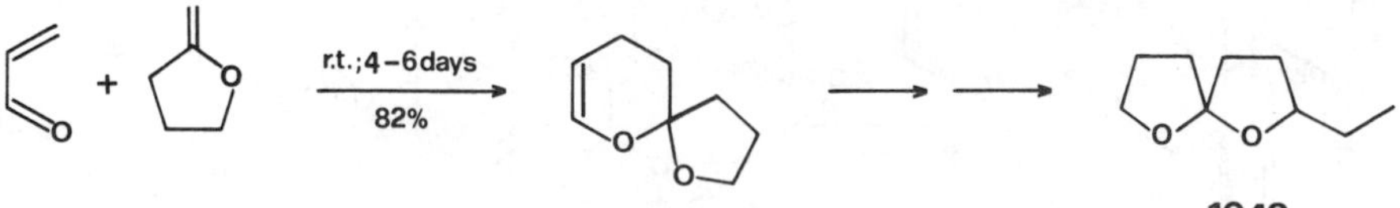

r.t.;4-6days
82%

1041

1042

Scheme 5.144

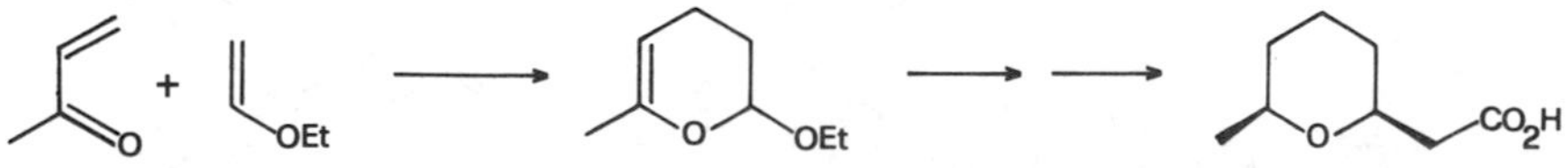

1043

Scheme 5.145

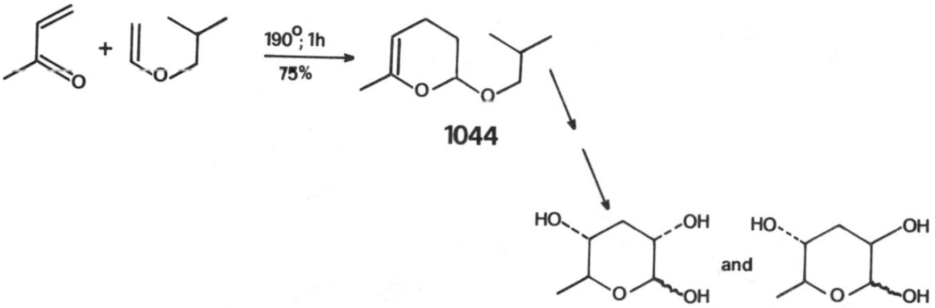

190°; 1h
75%

1044

1045 and 1046

Scheme 5.146

Benth. (Labiatae) together with two dimeric diterpenoid quinones, nilgherron A (**1060a**) and B (**1060b**).

Retro-heterodiene reaction of the last two compounds under mild thermolysis conditions gave **1058a** and **1058b** in addition to the *o*-benzoquinone **1059**.

Model experiments showed the easy formation of dihydrodioxin derivatives from *o*-benzoquinones and *p*-quinomethanes, suggesting a biosynthetic scheme for **1060a** and **1060b** that involves a heterodiene cycloaddition (*333*).

NITROGEN-CONTAINING DIENES. The introduction of nitrogen in a diene moiety increases the π-ionization potential because of its electronegativity. Therefore, inverse electron demand in regard to the Diels—Alder reaction is enhanced, as shown by the comparison of the photoelectron spectra of methanal azine (*334*) ($CH_2{=}N{-}N{=}CH_2$) and butadiene.

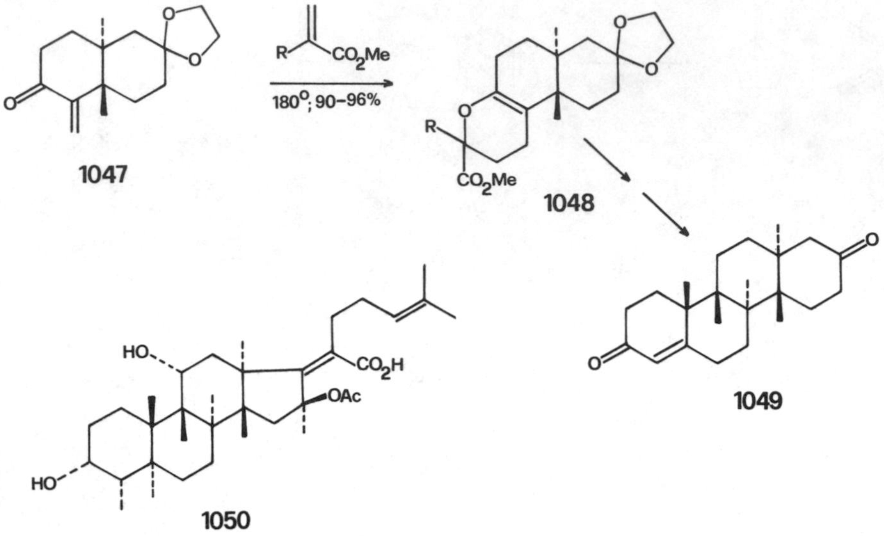

R

CO_2Me

180°; 90–96%

1047

R

CO_2Me

1048

1049

HO

CO_2H

OAc

HO

1050

Scheme 5.147

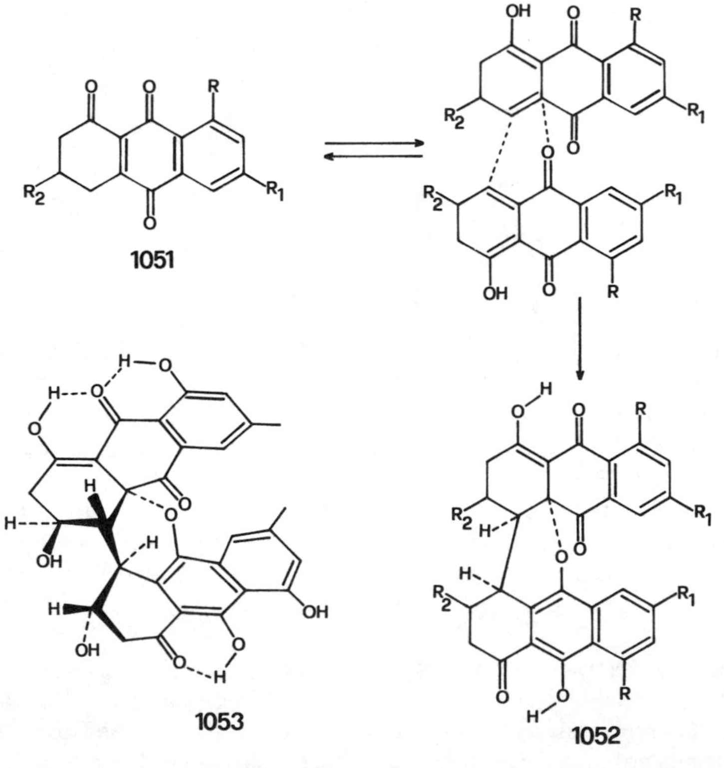

OH O R

R_2 R_1

O R

R_2 R_1

O

1051

R_2 R_1

OH O R

H—O

H---O

O

H

H

O

H

OH

H

OH

OH

O

H

1053

O H

O R

R_2 H R_1

O

H

R_2 R_1

O

O R

H

1052

Scheme 5.148

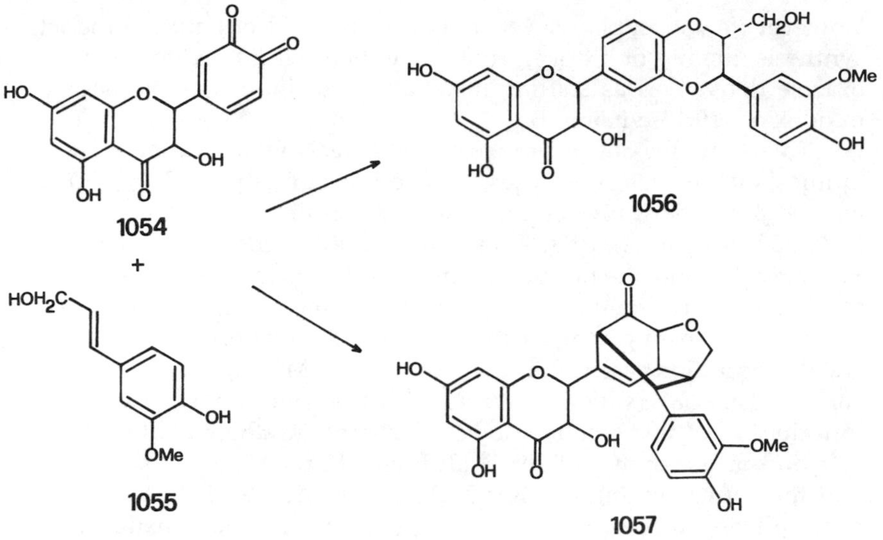

HO O OH OH O 1054 + HOH$_2$C OH OMe 1055 → 1056 (HO, O, CH$_2$OH, OMe, OH, OH, HO, O) ; → 1057 (O, O, HO, OH, OMe, OH, O)

Scheme 5.149

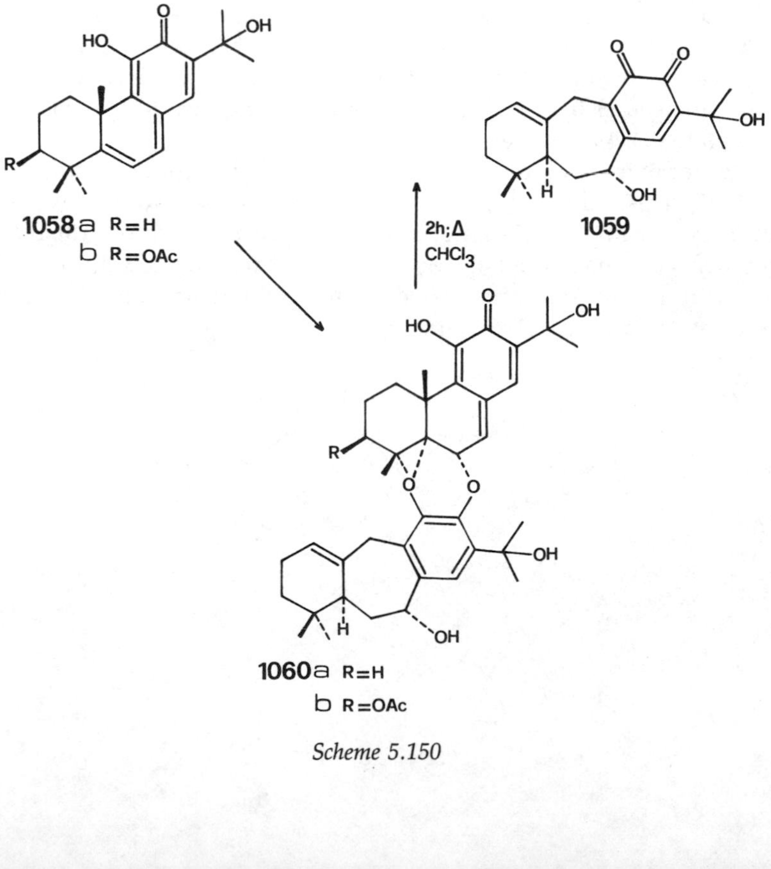

1058 a R=H
b R=OAc

→ 1060 a R=H
b R=OAc

2h; Δ CHCl$_3$ → 1059

Scheme 5.150

However, few examples can be found in the field of natural products synthesis; the use of nitrogen-containing dienes is limited to the use of oxazole derivatives as starting materials of an industrial synthesis of pyridoxine (**1062**), vitamin B_6.

Thus 4-methyl-5-alkoxyoxazoles (**1061**) react with a variety of dienophiles such as esters, nitriles, maleic acid anhydride (*335, 336*), 2-buten-1,4-diol derivatives (*335, 336*), and 2,5-dihydrofuran derivatives (*336, 337*) to give adducts. These adducts are sometimes isolated (in Scheme 5.151 only the maleic anhydride adduct is reported), then easily converted to the required product **1062** (Scheme 5.151).

It is interesting that electron-attracting-substituted dienophiles make the reaction easier, thus suggesting a $HOMO_{oxazole}/LUMO_{dienophile}$ control. This observation contrasts with the predictions made in the introduction of this section. However, comparing the HOMOs of oxazole and maleic anhydride [9.83 (*338*) and 12.0 (*339*) eV, respectively] and the corresponding LUMOs [-2.25 and 1.28 eV (*340*), that of the oxazole being higher, as generally occurs for electron-rich five-membered heterocycles], the experimental observation becomes fully understood (Figure 5.18).

Heterodienes in Intramolecular Reactions. To introduce this section, we use the Chapman synthesis of carpanone (**1065**) by palladium-catalyzed coupling of 2-(*trans*-1-propenyl)-4,5-methylenedioxyphenol (**1063**). The intramolecular cycloaddition of **1064** to **1065** still represents the most significant example of the generation of six chiral centers by an intramolecular heterodiene reaction occurring through an *endo* transition state (*341*) (Scheme 5.152).

Recent works in the field are less spectacular, although compounds like **1067**, a useful intermediate in the synthesis of iridoids, were obtained from 1-allylic-2,2-dimethylethylenetricarboxylates (**1066**) (*342*) (together with ene adducts). Compound **1069** was formed through **1068** (*343*) with a 100% stereocontrolled configuration corresponding to that of tetrahydrocannabinol (Scheme 5.153).

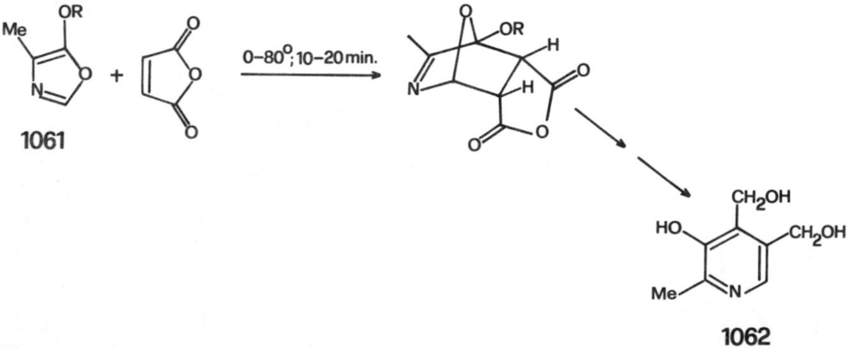

Me OR O N + O O O 0–80°; 10–20 min. O OR H O N H O O O CH2OH HO CH2OH Me N

1061

1062

Scheme 5.151

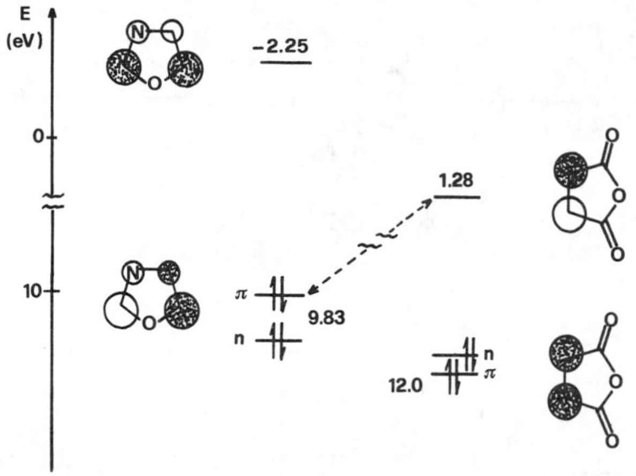

Figure 5.18. FMO interactions of oxazole and maleic anhydride.

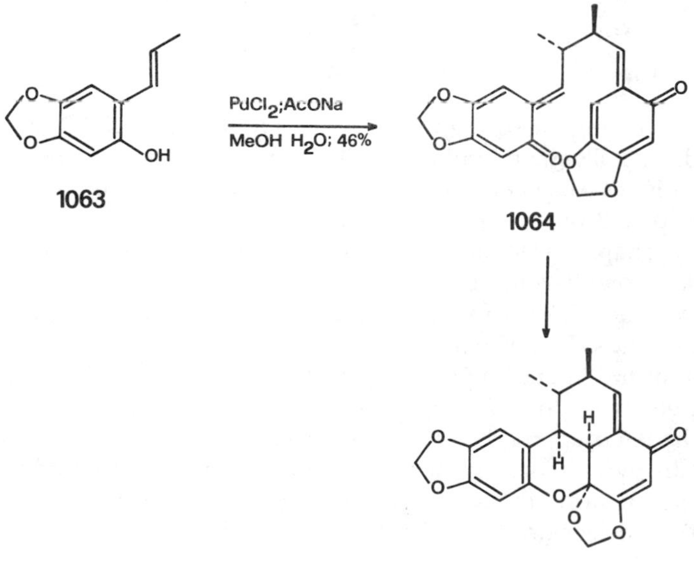

Scheme 5.152

Heterodienophiles in Intermolecular and Intramolecular Reactions. This relatively new field seems promising inasmuch as various heterocyclic rings can be approached by using C=O, C=N−, −N=N−, and −N=O dienophiles.

The presence of electronegative atoms leads to a lowering of π-ionization potential, as shown by acetaldehyde (*344*), acetaldehyde-methylimine (*345*), azomethane (*346*), and nitrosomethane (*347, 348*) compared to that of *trans*-butene (*349*). In general, all the heterodiene

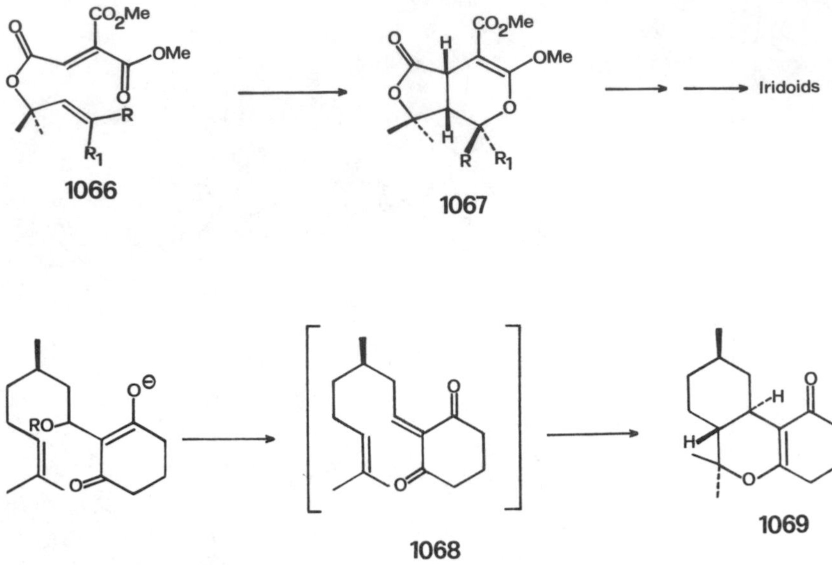

1066 1067 Iridoids

1068 1069

Scheme 5.153

reactions involving these dienophiles are $HOMO_{diene}/LUMO_{dienophile}$-controlled, as is illustrated in Figure 5.19, where butadiene is used as a model for dienes (the energies of the LUMOs are taken from Reference 146 of Chapter 3 or are estimated from substituent effects), with the exception perhaps of electron-rich imines [hydrazones $C=N-NR_2$ have π-ionization potentials up to 9.44 eV (*350*)].

Because the LUMOs for $C=O$ and $N=O$ dienophiles are lower, these compounds should be the most reactive. Alternatively, the energy separation of the interacting orbitals can be lowered by conjugation or by using electron-poor dienophiles as well as strong electron-rich dienes. The latter approach is the more effective.

Dihydropyranes, easily derived from carbonyl compounds and electron-rich dienes, are useful synthons for the synthesis of carbohydrates **1070**, nucleoside antibiotics **1071**, or tromboxanes (*351*) **1072** (*see* Scheme 5.154).

Periodate oxidation of benzhydroxamic acid generates nitrosocarbonylbenzene **1073**, which adds at room temperature to cyclohexadiene (a dienophile with low LUMO) to give **1074**. This compound can be transformed in two steps into **1075** (*352*), which has all functional groups in the correct stereochemical dispositions at the four asymmetric centers of the narcissus alkaloid lycoricidine (**1076**) (Scheme 5.155).

The key step of the synthesis of streptonigrin (**1082**), a metabolite with antitumor activity produced by a few species of *Streptomyces* and *Actinomyces*, provides a beautiful example of addition to a $C=N$ dienophile. We must emphasize that severe experimental conditions are required, despite the fact that the reaction occurs between an electron-

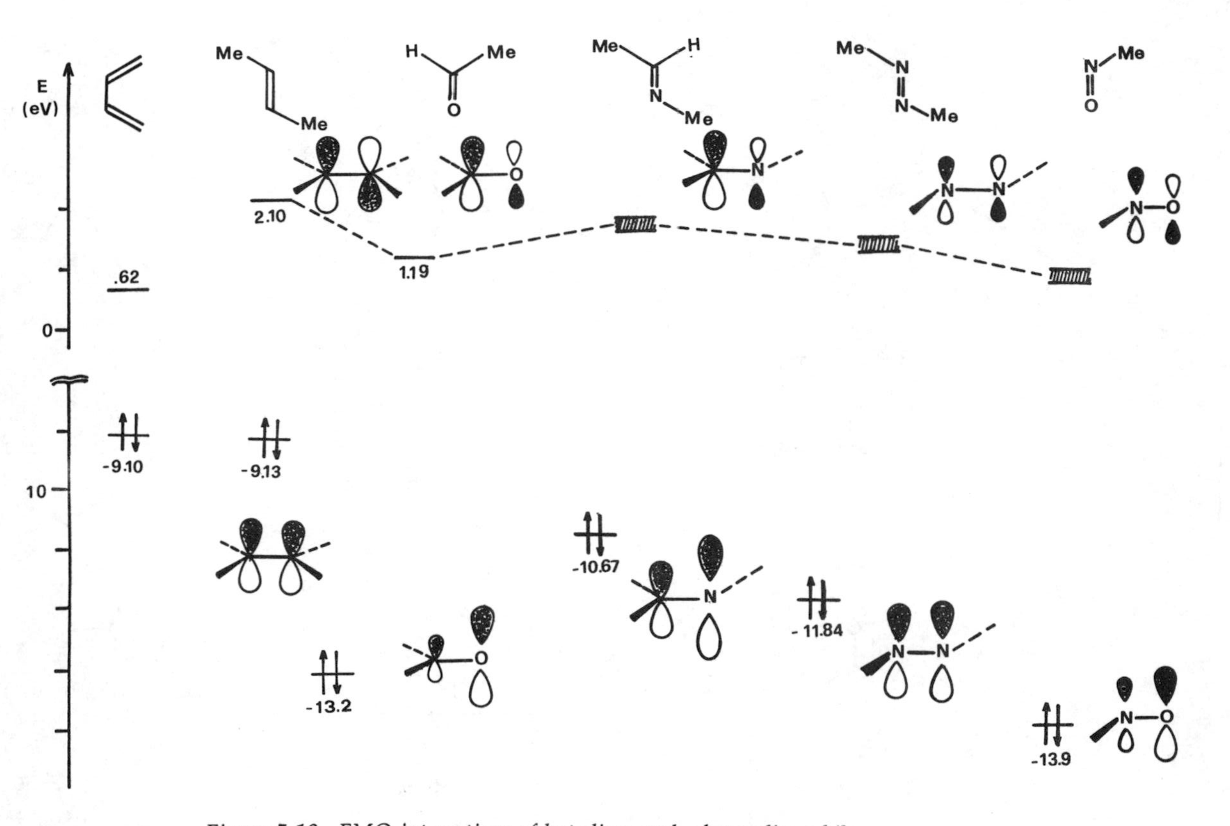

Figure 5.19. FMO interactions of butadiene and a heterodienophile.

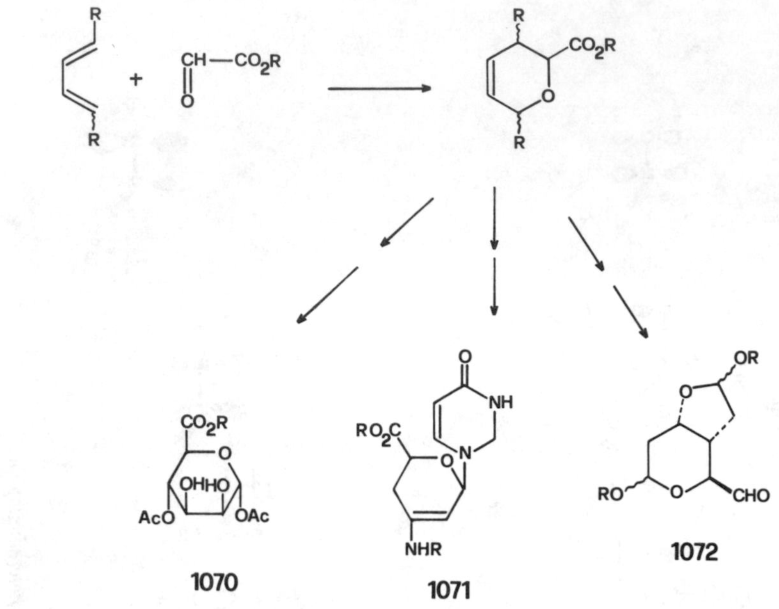
CH—CO_2R
CO_2R
NH
RO₂C
NHR
OR
CHO
AcO
OAc
1070
1071
1072

Scheme 5.154

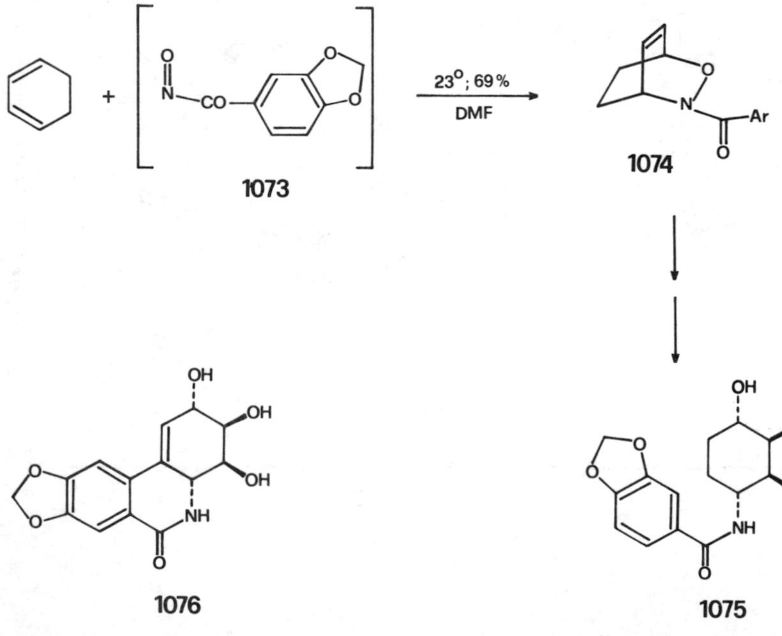
23°; 69%
DMF
Ar
1073
1074
1075
1076

Scheme 5.155

rich diene **1077** and 2,5-imidazolinedione **1078** bearing electron-attracting groups that lower its LUMO. The hetero-Diels–Alder adduct **1079** is suitably functionalized to yield the required tetrahydropyridine intermediate **1080** (*353, 354*), which is further converted in several steps through **1081** to the final target **1082** (*355*) (Scheme 5.156).

Unusual azadienophiles were used by Kametani and coworkers for regioselective syntheses of xylopinine (**1086b**) (*356, 357*) and hexadehydro-17-methoxyyohimbane (**1084b**). Thermolysis of 1,2-dihydro-5-methoxybenzocyclobutene-1-carbonitrile in the presence of equimolar amounts of 3,4-dihydro-β-carboline (**1083**) afforded the cycloaddition product **1084a** of the intermediate *o*-quinodimethane on the C=N bond. Reductive decyanation produced the yohimbane derivative **1084b** (*356*). Similarly, 1,2-dihydro-4,5-dimethoxycyclobutene, upon ring opening, underwent hetero-Diels–Alder cycloaddition with 3,4-dihydro-6,7-dimethoxyisoquinoline (**1085**) through an *endo* transition state to give

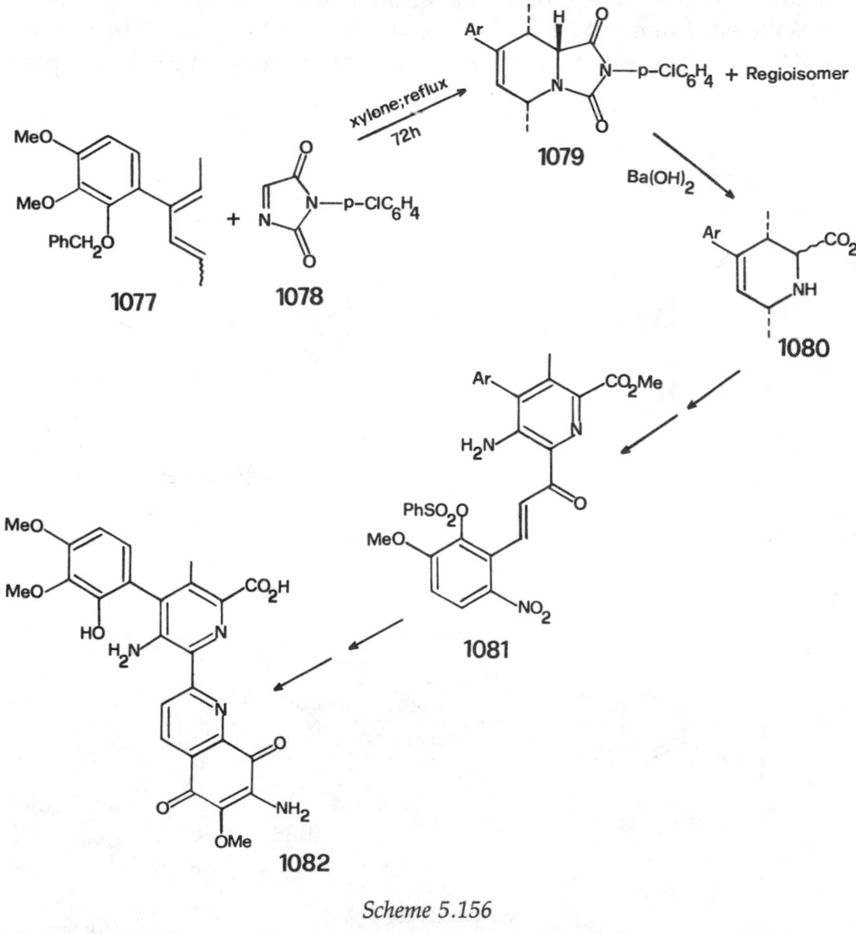

xylene;reflux 72h

Ar H O N—p-ClC$_6$H$_4$ + Regioisomer

MeO MeO PhCH$_2$O + N N—p-ClC$_6$H$_4$ O O

1077 1078 1079

Ba(OH)$_2$

Ar CO$_2$H NH

1080

Ar CO$_2$Me H$_2$N N O PhSO$_2$O MeO NO$_2$

1081

MeO MeO HO H$_2$N CO$_2$H N N O O NH$_2$ OMe

1082

Scheme 5.156

1086a (*357*). Decyanation gave the previously mentioned protoberberine alkaloid xylopinine (**1086b**) (Scheme 5.157).

The high stereoselectivity shown by intramolecular hetero-Diels–Alder reaction (not very usual indeed) with a C=N group as dienophile is difficult to rationalize. The dienophile configuration is obviously influenced by the possible (*E*) ⇄ (*Z*) inversion of the imino group (*358*).

Because the overall diene–azadienophile system is unstable, giving rise to immediate cyclization, all examples involve an intermediate generation of one of the two partners through a pericyclic process.

The cheletropic extrusion of sulfur dioxide from the sulfone **1087** afforded the diene group suitably placed to undergo intramolecular heterocycloaddition with the imine dienophile. Thus **1088** was isolated in 68% yield and easily converted to elaeokanine A (**420**) (*359*) (Scheme 5.158).

The previous process also involves the generation of the C=N dienophile upon thermal cleavage of a methylol acetate group. This fragmentation can also be regarded as a retro-ene reaction; the leaving acetic acid is the ene fragment (*see* the section in Chapter 8 entitled "The Retro-Ene Reaction"). The overall process is illustrated in detail by the simple

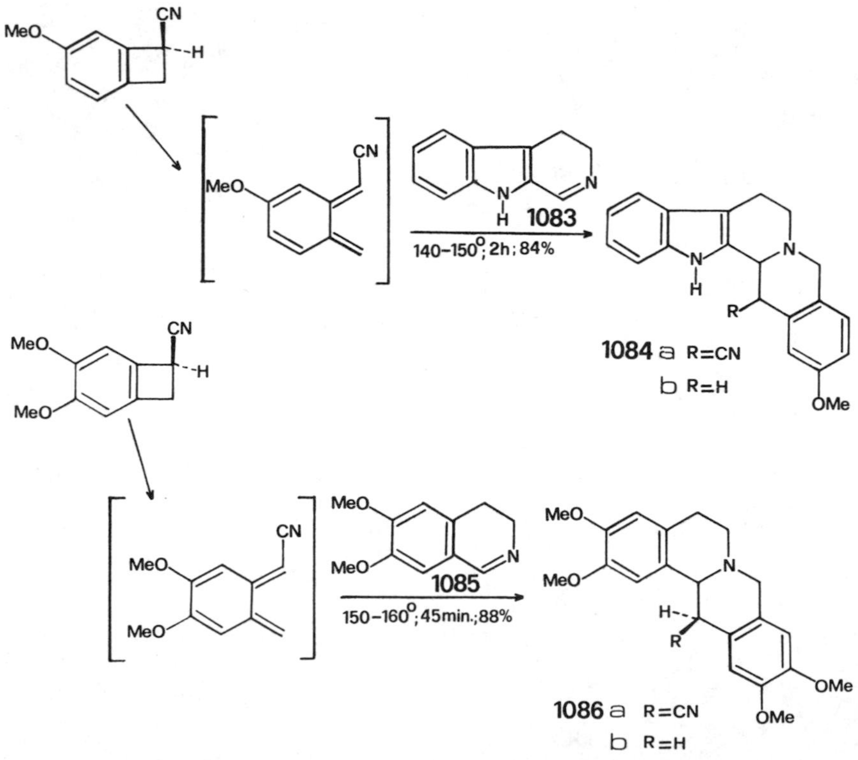

Scheme 5.157

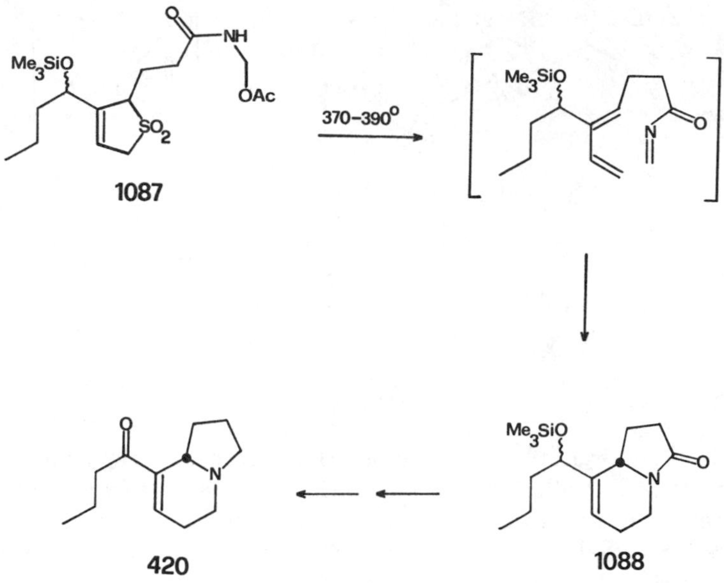

Scheme 5.158

synthesis of the indolizine alkaloid δ-coniceine (**1090**); its formation requires the thermal cleavage of the diene derivative **1089** (*360*) (Scheme 5.159).

Both diene and dienophile groups are generated, but at different stages, in the synthesis of another indolizine alkaloid, tylophorine (**1093**). Whereas the C=N dienophile was generated through the previously mentioned methylol acetate pyrolysis, the diene was formed from **1091** through a Claisen [3,3] sigmatropic rearrangement following the Johnson orthoacetate variant (*see* Chapter 7, "The Johnson-Orthoester Variant"). Thus, three pericyclic processes (Claisen, retro-ene, and hetero-Diels–Alder) involve a single synthetic pathway (*360*) (Scheme 5.160).

Retro-Hetero-Diels–Alder Reactions. As discussed earlier in this chapter for retro-Diels–Alder reactions, the retro-hetero-Diels–Alder (RHDA) process is mainly concerned with generating or regenerating the diene or the dienophile groups. The dienophile groups are used as a means of protection during some stages of a synthetic scheme. Nice examples of generation of dienes by RHDA are the syntheses of two alkaloids: evodiamine (**1096a**), (R=H), and rutecarpine (**1096b**), (R=Me). The starting sulfinamide anhydrides (**1094a,b**) lose sulfur dioxide to give **1095a,b**, which act as azadienophiles in the presence of 3,4-dihydro-β-carboline (**1083**). The overall process occurs at room temperature in very good yields (65% and 80%, respectively) (*361*) (Scheme 5.161). However, a stepwise mechanism involving the nucleophilic attack of **1083** directly on **1094a,b** cannot be ruled out.

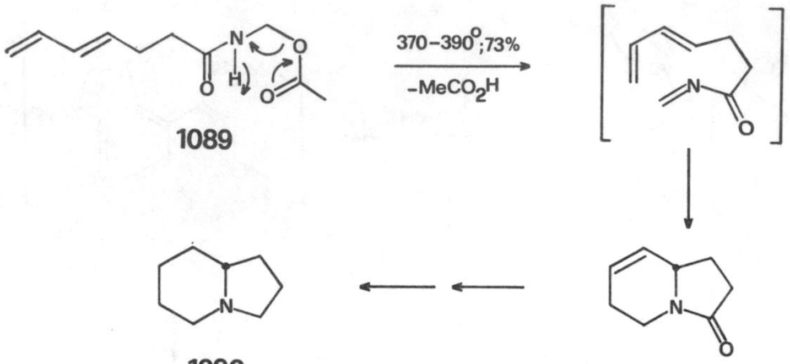

370–390°; 73%
$-MeCO_2H$

1089

1090

Scheme 5.159

We found examples of intramolecular processes in the syntheses of two pyrrolizidine alkaloids, (±)-heliotridine (**1098**) and (±)-retronecine (**406**). The syntheses included cycloaddition of a diene onto an N=O group, generated by RHDA from **1097** with loss of 9,10-dimethylanthracene (*362, 363*) (Scheme 5.162).

A nice sequence of a hetero-Diels–Alder reaction followed by a retro process distinguishes the synthesis of a benzofuran derivative **1101**, which seems a promising synthon for highly oxygenated sesquiterpenes. The acetylenic oxazole **1099** undergoes intramolecular cycloaddition giving **1100**, which loses acetonitrile under the experimental conditions through an RHDA process, giving **1101** (*364*) (Scheme 5.163).

A useful example of diene protection as a Diels–Alder adduct and its regeneration by the retro process was reported by Oppolzer et al. in the synthesis of (±)-lysergic acid (**1105**) (*365*). A true RHDA process is not involved, but a retro-Diels–Alder stage generates the diene undergoing a hetero-Diels–Alder reaction. The overall process occurs when **1102** is warmed in 1,2,4-trichlorobenzene under conditions of infinite dilution. The reaction proceeds through loss of cyclopentadiene and formation of **1103**, which cyclizes to **1104a**. Attempts to use more concentrated solutions afforded dimeric materials only, due to the spontaneous dimerization of the 2-carbomethoxy-1,3-diene derivatives. This side reaction prevents the utilization of the preformed dienic system and fully justifies the devised device (Scheme 5.164). The process can also be regarded as a synthesis of paspalic acid (**1104b**). This natural product, isolated from saprophytic cultivation of appropriate ergot strains in fermenters, is a nonisolated intermediate because it isomerizes readily to the thermodynamically more stable **1105**.

Photochemical [4+2] Cycloadditions

There is no obvious reason to force a diene and a dienophile, which react easily under thermal conditions in a suprafacial way, to adopt a more strained antarafacial transition state to react photochemically (*366*).

MeC(OEt)$_3$

130–135°; 2h; 84%

1091

1092

220°; 5h

1093

Scheme 5.160

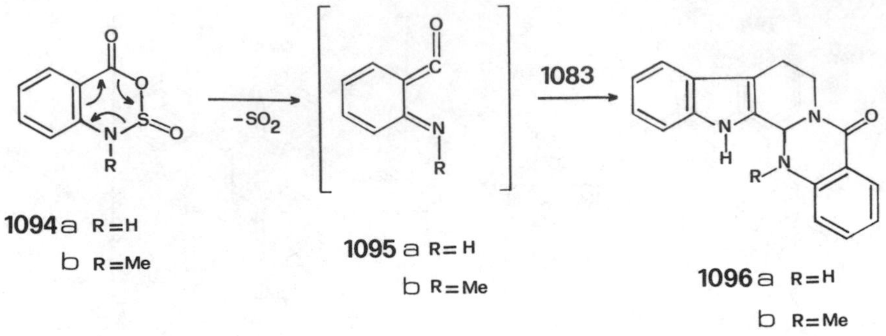

O O N S O R

1094 a R=H
b R=Me

$-SO_2$

O C N R

1095 a R=H
b R=Me

1083

N H N O R N

1096 a R=H
b R=Me

Scheme 5.161

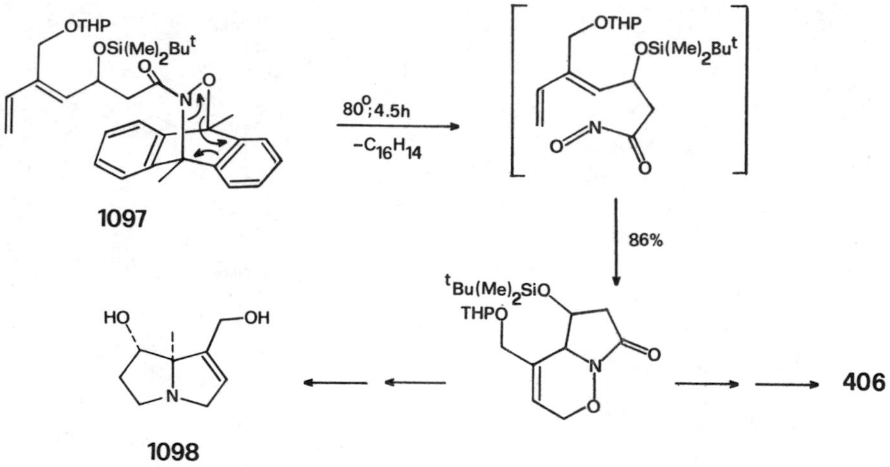

OTHP
$OSi(Me)_2Bu^t$
O N O

1097

80°; 4.5h
$-C_{16}H_{14}$

OTHP
$OSi(Me)_2Bu^t$
O N O

86%

$^tBu(Me)_2SiO$
THPO
N O O

HO OH N

1098

406

Scheme 5.162

N O
EtO
$MeOCH_2$
OMe

1099

O OMe N OEt OMe

1100

-MeCN

OMe OMe OEt O

1101

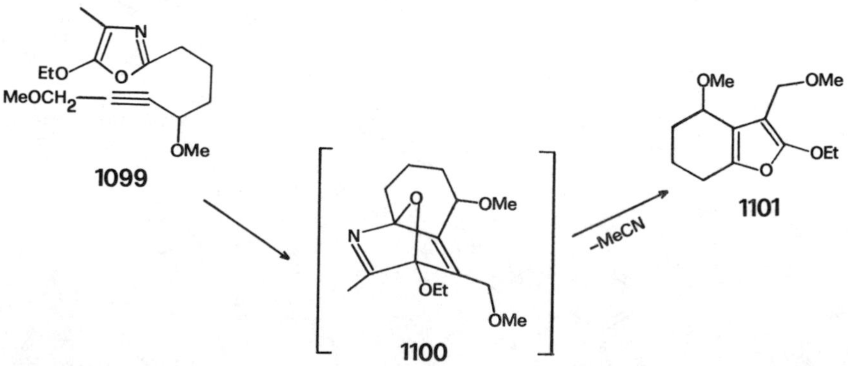

Scheme 5.163

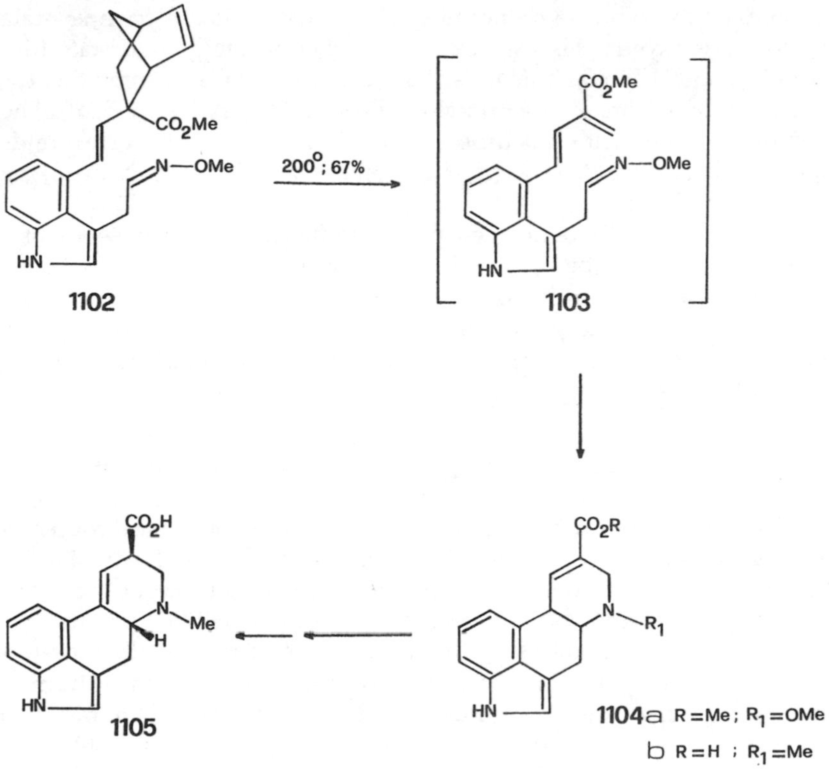

1102 → 200°; 67% → [1103] → 1104 a R=Me; R_1=OMe; b R=H; R_1=Me → 1105

Scheme 5.164

Hence these reaction conditions are used for only particular reagents. Examples of such reactions include those where molecular oxygen is used as a dienophile and conjugated trienes as substrates; the cycloaddition occurs intramolecularly between 4π and 2π electron fragments.

Diels–Alder Reactions with Singlet Oxygen. Photooxygenation was discovered accidentally during the synthesis of chlorophyll and consists of the reaction of singlet oxygen with an olefin or a diene. In this section we consider only the latter because it occurs through a [4+2] cycloaddition.

Two reviews on the use of singlet oxygen (*367*) (the reactant) and of endoperoxides (*368*) (the products) in organic syntheses covered nearly all the literature. Therefore, we mention only the papers concerning syntheses of natural products and in particular only those where the endoperoxide was actually isolated.

We must point out that this reaction is in this section for traditional reasons only. It is not a photochemical cycloaddition, but when performed under photochemical conditions, the process consists of photochemical generation of a dienophile, which then reacts with a diene under thermal conditions.

Singlet oxygen (the dienophile) is the first excited electronic state of molecular oxygen. The two electrons, which in the ground state triplet have parallel spin and lie in two degenerate orbitals, assume the less stable configuration on one orbital with opposite spin (Figure 5.20). The energy associated with this process is 22.4 kcal/mol, which corresponds to a wavelength of 1278 nm (or a frequency of 7823 cm^{-1}), hence lying in the near IR region.

The excitation is performed photochemically in the presence of a sensitizer (e.g., rose bengal, eosin, hematoporphyrin, chlorophyll). Because two electrons shift from an antibonding to a bonding state, we can represent the change in a way that is theoretically not strictly correct, but which nevertheless has the advantage of being easily understandable.

$$\overset{\uparrow}{O}-\overset{\uparrow}{O} \xrightarrow{22.4\ \text{kcal/mol}} O=O$$

Singlet oxygen can be formed through nonphotochemical routes as well as by using light. Besides the thermolysis of ozonides or the decomposition of peroxides, a chemical route to singlet oxygen from hydrogen peroxide and sodium hypochlorite works well. This route was used in the total synthesis of (±)-15-deoxyprostaglandin E_1 (**1109**), a potential intermediate for microbiological transformation into the natural PGE_1. When the cyclopentadiene derivative **1106** was allowed to react at −10 °C with chemically generated singlet oxygen ($^1O_2$), **1109** was obtained (together with its regioisomer) via **1107**. Four further steps gave **1109** (*369*) (Scheme 5.165).

The first synthesis of a natural product via 1,4-addition of $^1O_2$ to a diene was published by Schenk and Ziegler in 1945 (*370*). By shaking an α-terpinene (**1110**) solution in the presence of chlorophyll in the light, they isolated ascaridole (**1111**), a naturally occurring terpenoid peroxide found in chenopodium oil. Ascaridole possesses antihelminthic activity (Scheme 5.166).

The approach of $^1O_2$ occurs from the less hindered face of the substrate as illustrated by the syntheses of α-agarofuran (**1113**), a furanoid sesquiterpene isolated from fungus-infected agarwood, and (±)-cybullol (**1115**), a metabolite of the gastromycetous *Cyathus bulleri*. In both hexahydronaphthalenes **1112** and **1114**, the addition is guided by the angular C10 methyl (*371, 372*) (Scheme 5.167).

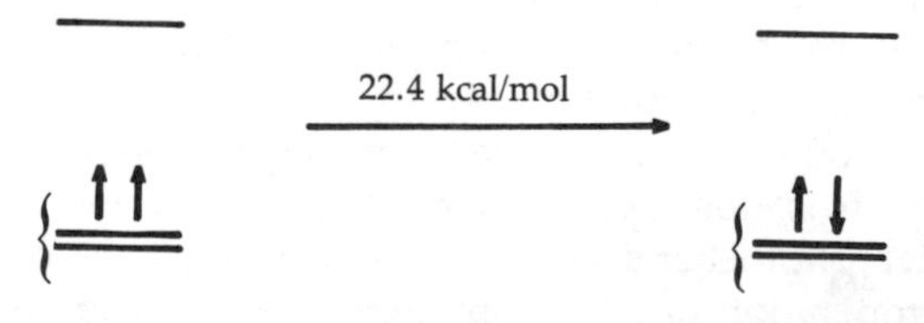

Figure 5.20 Singlet oxygen generation from the ground state triplet.

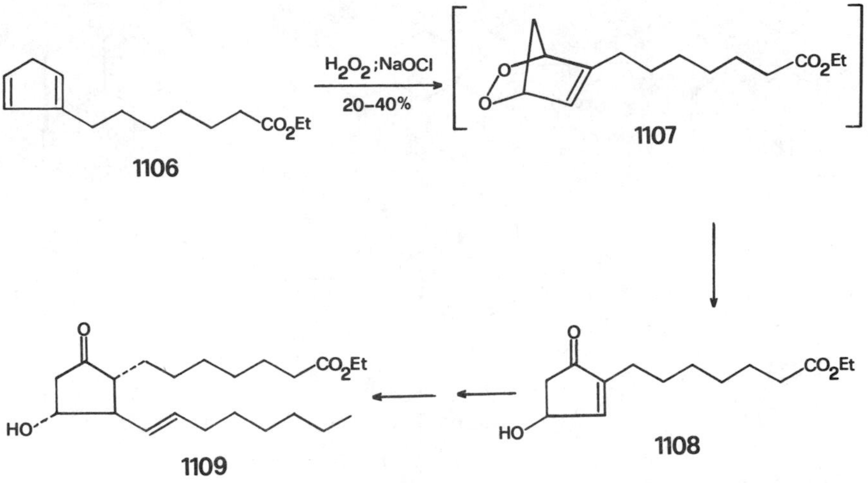

Scheme 5.165

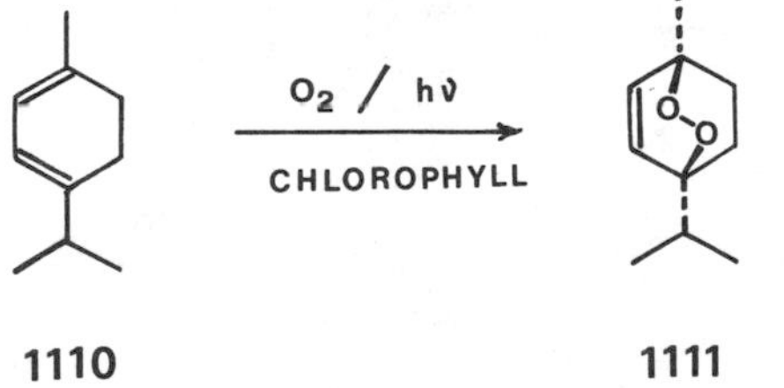

Scheme 5.166

Senepoxyde, **961**, and crotepoxide, **960**, two natural highly oxygenated cyclohexane epoxides displaying significant antitumoral activity, were synthesized by 1,4-cycloaddition of $^1O_2$. The former product was obtained by photooxygenation of the epoxydiene **1116**, producing a *trans*-endoperoxide **1117**. This intermediate was easily converted within three steps into **961** and further developed into seneol, **1118**, which is a metabolite of senepoxide in *Uvaria catocarpa* (*373*) (Scheme 5.168).

Crotepoxide was obtained by $^1O_2$ photoaddition either of the diol **1119** (*374*) to give the endoperoxide **1120** (along with its stereoisomer) or of the styrene **1121** (*375*) to give **1122** (along with its stereoisomer). Both intermediates are converted in a few steps to **960** (Scheme 5.168).

Treatment of the endoperoxides derived by photooxygenation of a diene with lithium *t*-butoxide followed by dehydration with sulfuric acid is an approach to the furan ring. Several naturally occurring furanoterpenes were thus obtained from **1123** (*376*) and **1124** (*377*). The furanoterpenes so formed were perillene (**1125**), dendrolasin (**1126a**), neotorreyol (**1126b**), torreyal (**1126c**), perillaketone (**1127**), α-clausenane (**1128**), ipomeamarone (**1129a**), and epiipomeamarone (**1129b**). All of these com-

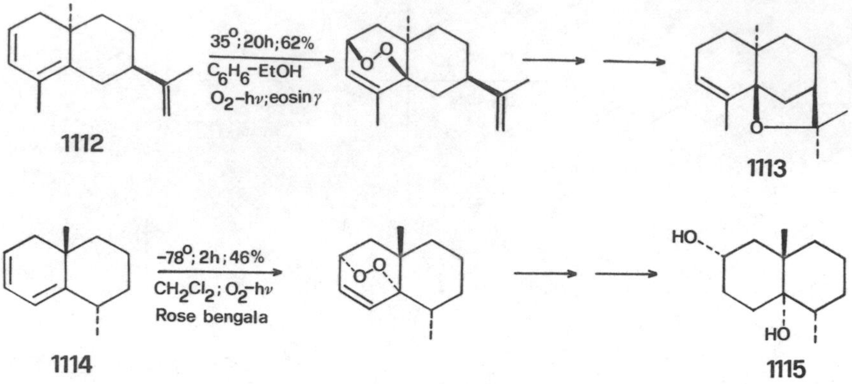

Scheme 5.167

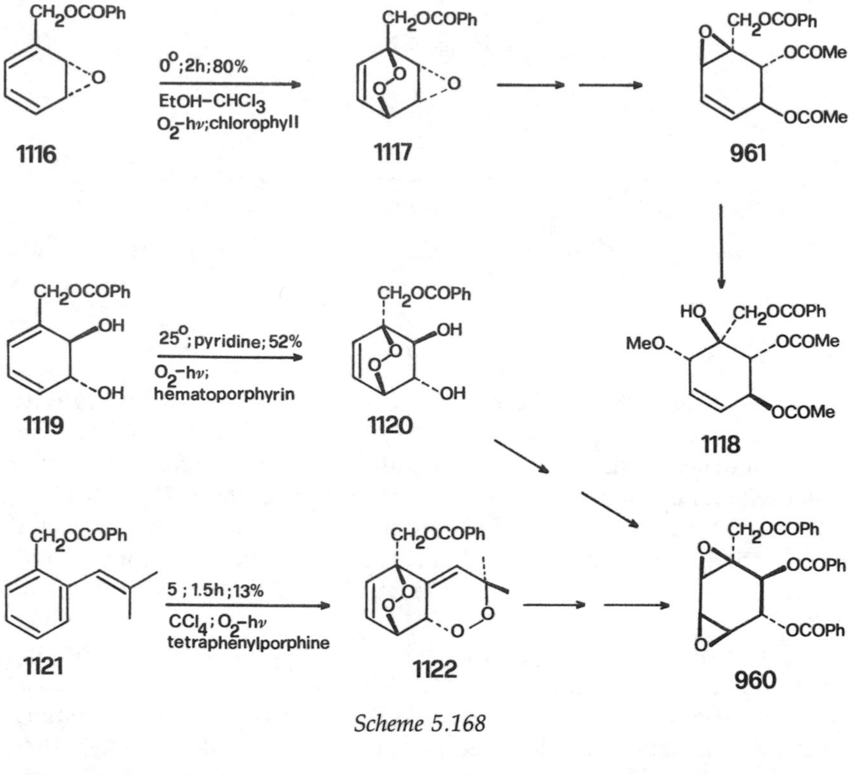

Scheme 5.168

pounds are found in species of the animal and vegetable world (Scheme 5.169).

The So-Called Photochemical Diels–Alder Reaction. Irradiation of conjugated trienes leads to bicyclo[3.1.0]hexenes. For example, allocimene (**1130**) (*378*) gives **1131** (along with other photoproducts) and α-phellandrene (**1132**) (*379*) produces **1133**, a precursor of **1134**. (Scheme 5.170).

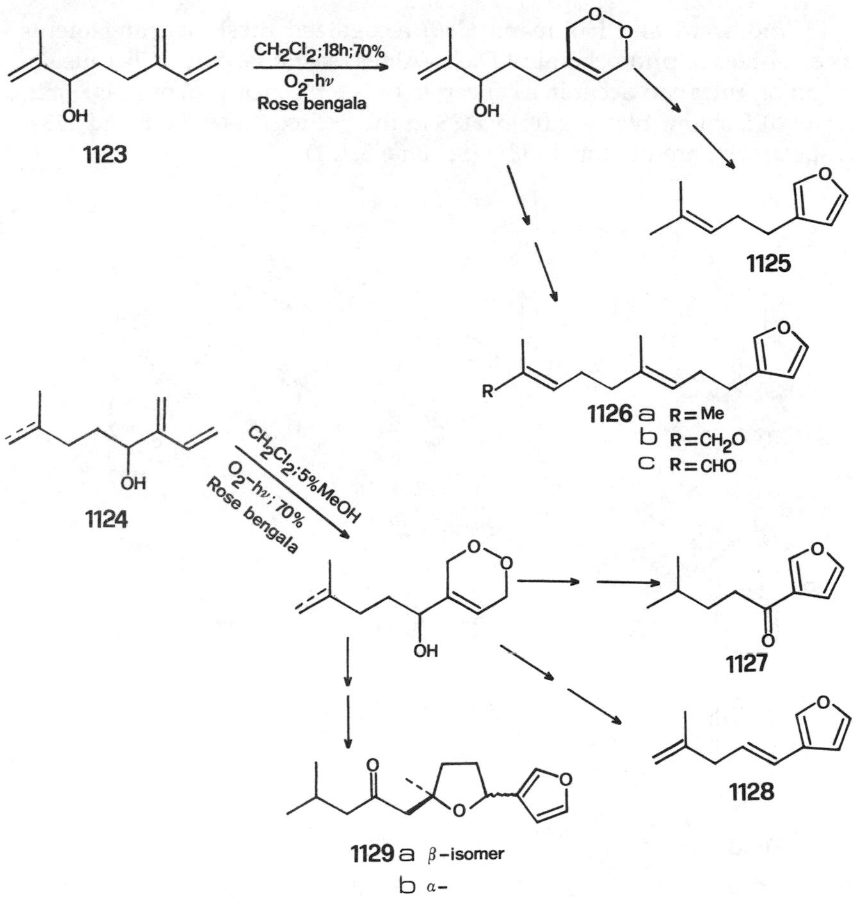

CH_2Cl_2;18h;70%
O_2-hν
Rose bengala

OH

1123

1125

R

1126 a R=Me
b R=CH_2O
c R=CHO

OH

1124

CH_2Cl_2;5%MeOH
O_2-hν; 70%
Rose bengala

OH

1127

1128

1129 a β-isomer
b α-

Scheme 5.169

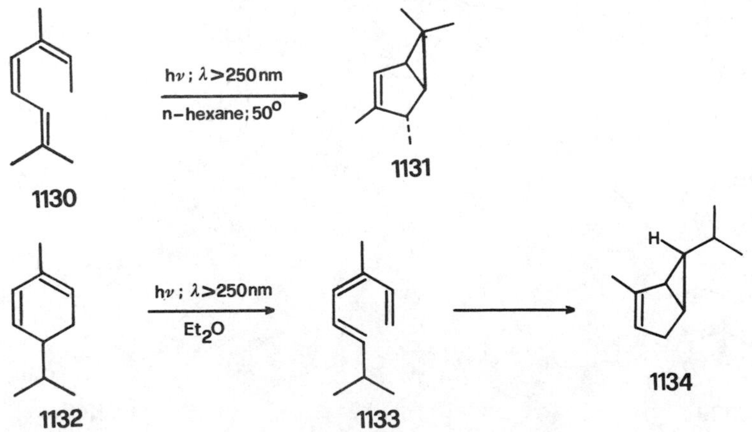

hν; λ>250 nm
n-hexane; 50°

1130

1131

hν; λ>250nm
Et_2O

1132

1133

H

1134

Scheme 5.170

Woodward and Hoffmann (*380*) recognized these rearrangements as examples of photochemical Diels–Alder reactions, which because of selection rules can occur in a concerted ($\pi 4s + \pi 2a$, route a) or ($\pi 4a + \pi 2s$, route b) fashion. Starting from **1135** in the excited state, **1136** and **1137**, respectively, are obtained (*381*) (Scheme 5.171).

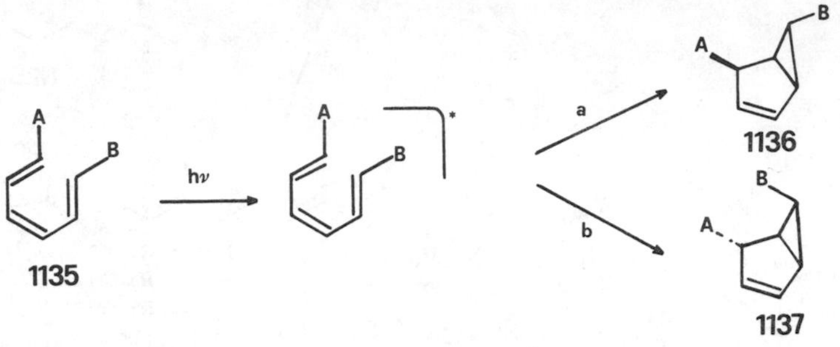

Scheme 5.171

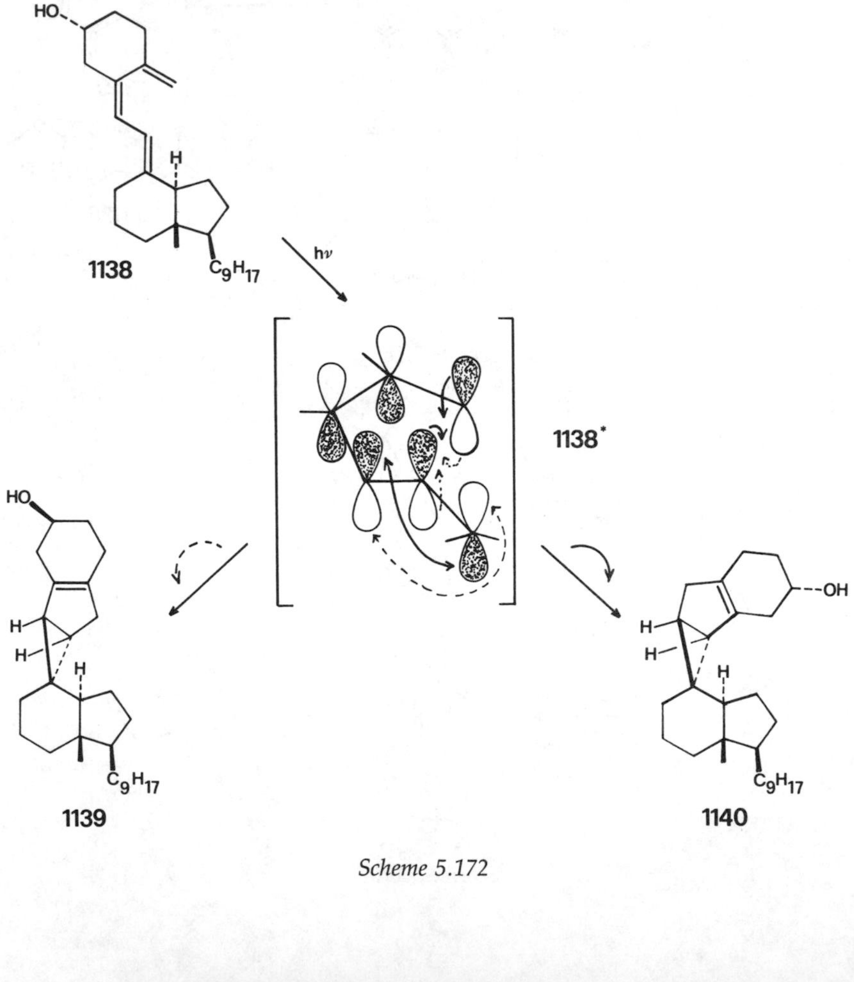

Scheme 5.172

The topic would be of marginal interest in this review except that this reactivity is analogous to the photochemical conversion of calciferol (vitamin D_2, **1138**) to suprasterols I and II (**1139** and **1140**, respectively) (*382–384*) (Scheme 5.172). Their formation can be nicely rationalized if the nodal structures of the LUMO of **1138** (*385*) (reported in **1138*** for the triene part only) are considered. The $[\pi 4s + \pi 2a]$ process, schematized by solid and dotted arrows, gives **1139** and **1140**.

Few compounds have had the same importance as the vitamin D series in the development of the Woodward–Hoffmann rules. In this chapter we introduced these compounds; further sections emphasize the role played by rearrangements of these compounds in understanding the importance of orbital symmetry in steric control of the reactions.

In this and later sections we consider only the main photoproducts of vitamin D. The field was authoritatively covered in a review by Jacobs and Havinga (*384*). Those interested in this intriguing and fascinating topic are referred to this review, which describes the novel photoproducts that have been isolated as well as their pathways of formation.

Literature Cited

1. Woodward, R. B. "Perspectives in Organic Chemistry"; Todd, A. R., Ed.; Interscience: New York, 1956; p. 155.
2. Woodward, R. B.; Sondheimer, F.; Taub, D.; Heusler, K.; McLamore, W. M. *J. Am. Chem. Soc.* **1952**, *74*, 4223.
3. Stork, G.; Stotter, P. L. *J. Am. Chem. Soc.* **1969**, *91*, 7780.
4. Gates, M.; Tschudi, G. *J. Am. Chem. Soc.* **1956**, *78*, 1380.
5. Woodward, R. B.; Bader, F. E.; Bickel, H.; Frey, A. J.; Kierstead, R. W. *Tetrahedron* **1958,** *2*, 1.
6. Pearlman, B. A. *J. Am. Chem. Soc.* **1979**, *101*, 6404.
7. Wender, P. A.; Schaus, J. M.; White, A. W. *J. Am. Chem. Soc.* **1980**, *102*, 6157.
8. Schreiber, J.; Leimgruber, W.; Pesaro, M.; Schudel, P.; Threlfall, T.; Eschenmoser, A. *Helv. Chim. Acta* **1961**, *44*, 540.
9. Dauben, W. G.; Kessel, C. R.; Takemura, K. H. *J. Am. Chem. Soc.* **1980**, *102*, 6893.
10. Brown, E. D.; Clarkson, R.; Leeney, T. J.; Robinson, G. E. *J. Chem. Soc., Perkin Trans. 1* **1978**, 1507.
11. Caton, M. P. L. *Tetrahedron* **1979**, *35*, 2705.
12. Sauer, J. *Angew. Chem., Int. Ed. Engl.* **1966**, *5*, 211.
13. Ibid., **1967**, *6*, 16.
14. Babler, J. M.; Olsen, D. O.; Arnold, W. H. *J. Org. Chem.* **1974**, *39*, 1656.
15. Monti, S. A.; Yang, Y. L. *J. Org. Chem.* **1978**, *43*, 4062.
16. Dastur, K. P. *J. Am. Chem. Soc.* **1973**, *95*, 6509.
17. Dauben, W. G.; Kozikowski, A. P.; Zimmerman, W. T. *Tetrahedron Lett.* **1975**, 515.
18. Büchi, G.; Carlson, J. A. *J. Am. Chem. Soc.* **1968**, *90*, 5336.
19. Kato, T.; Suzuki, T.; Ototani, N.; Kitahara, Y. *Chem. Lett.* **1976**, 887.
20. Kato, T.; Suzuki, T.; Ototani, N.; Maeda, H.; Yamada, K.; Kitahara, Y. *J. Chem. Soc., Perkin Trans. 1* **1977**, 206.
21. Osmund de Silva, S.; St. Denis, C.; Rodrigo, R. *J. Chem. Soc., Chem. Commun.* **1980**, 995.
22. Tanis, S. P.; Nakanishi, K. *J. Am. Chem. Soc.* **1979**, *101*, 4398.
23. Rodrigo, R. *J. Org. Chem.* **1980**, *45*, 4538.
24. Ichihara, A.; Kimura, R.; Moriyasu, K.; Sakamura, S. *Tetrahedron Lett.* **1977**, 4331.

25. Kraus, G. A.; Frazier, K. *J. Org. Chem.* **1980**, *45*, 4820.
26. Kraus, G. A.; Roth, B. *J. Org. Chem.* **1980**, *45*, 4825.
27. Torii, S.; Oie, T.; Tanaka, H.; White, J. D.; Furuta, T. *Tetrahedron Lett.* **1973**, 2471.
28. Tsuda, Y.; Isobe, K.; Ukai, A. *J. Chem. Soc., Chem. Commun.* **1971**, 1555.
29. Tsuda, Y.; Isobe, K. *J. Chem. Soc., Chem. Commun.* **1971**, 1554.
30. Kreiser, W.; Janitschke, L. *Tetrahedron Lett.* **1978**, 601.
31. Kreiser, W.; Janitschke, L.; Voss, W.; Ernst, L.; Sheldrick, W. S. *Chem. Ber.* **1979**, *112*, 397.
32. Tokoroyama, T.; Matsuo, K.; Kubota, T. *Tetrahedron* **1978**, *34*, 1907.
33. Irie, H.; Nishitani, Y.; Sugita, M.; Uyeo, S. *J. Chem. Soc., Chem. Commun.* **1970**, 1313.
34. Hendrickson, J. B.; Bogard, T. L.; Fisch, M. E. *J. Am. Chem. Soc.* **1970**, *92*, 5538.
35. Kelly, R. C.; Schletter, I. *J. Am. Chem. Soc.* **1973**, *95*, 7156.
36. Ichimara, A.; Ubukata, M.; Sakamura, S. *Tetrahedron* **1980**, *36*, 1547.
37. Greenlee, W. J.; Woodward, R. B. *Tetrahedron* **1980**, *36*, 3661.
38. Ibid., 3367.
39. Oida, S.; Ohashi, Y.; Ohki, E. *Chem. Pharm. Bull.* **1973**, *21*, 528.
40. Goldsmith, D. J.; Srouji, G.; Kwong, C. *J. Org. Chem.* **1978**, *43*, 3182.
41. Kishi, Y.; Nakatsubo, F.; Aratani, M.; Goto, T.; Inoue, S.; Kakoi, H.; Sugiura, S. *Tetrahedron Lett.* **1970**, 5127.
42. Kishi, Y.; Nakatsubo, F.; Aratani, M.; Goto, T.; Inoue, S.; Kakoi, H. *Tetrahedron Lett.* **1970**, 5129.
43. Kishi, Y.; Aratani, M.; Fukuyama, T.; Nakatsubo, F.; Goto, T.; Inoue, S.; Tanino, S.; Sugiura, S.; Kakoi, H. *J. Am. Chem. Soc.* **1972**, *94*, 9218.
44. Kishi, Y.; Fukuyama, T.; Aratani, M.; Nakatsubo, F.; Goto, T.; Inoue, S.; Tanino, H.; Sugiura, S.; Kakoi, H. *J. Am. Chem. Soc.* **1972**, *94*, 9219.
45. Harayama, T.; Ohtani, M.; Oki, M.; Inubushi, Y. *Chem. Pharm. Bull.* **1973**, *21*, 1061.
46. Harayama, T.; Ohtani, M.; Oki, M.; Inubushi, Y. *J. Chem. Soc., Chem. Commun.* **1974**, 827.
47. Ferguson, R. N.; Whidby, J. F.; Sanders, E. B.; Levins, R. J.; Katz, T.; DeBardeleben, J. F.; Einolf, W. N. *Tetrahedron Lett.* **1978**, 2645.
48. Ross Kelly, T.; Magee, J. A.; Weibel, F. R. *J. Am. Chem. Soc.* **1980**, *102*, 798.
49. Herndon, W. C.; Hall, L. H. *Theor. Chim. Acta* **1967**, *7*, 4.
50. Sustmann, R. *Tetrahedron Lett.* **1971**, 2721.
51. Trong Anh, N.; Canadell, E.; Eisenstein, O. *Tetrahedron* **1978**, *34*, 2283.
52. Kwart, H.; Burchuk, I. *J. Am. Chem. Soc.* **1952**, *74*, 3094.
53. Eisenstein, O.; Lefour, J. M.; Trong Anh, N. *J. Chem. Soc., Chem. Commun.* **1971**, 969.
54. Eisenstein, O.; Lefour, J. M.; Trong Anh, N.; Hudson, R. F. *Tetrahedron* **1977**, *33*, 523.
55. Houk, K. N. *J. Am. Chem. Soc.* **1973**, *95*, 4092.
56. Fleming I.; Gianni, F. L.; Mah, T. *Tetrahedron Lett.* **1976**, 881.
57. Tegmo-Larsson, I. M.; Rozeboom, M. D.; Rondan, N. G.; Houk, K. N. *Tetrahedron Lett.* **1981**, 2047.
58. Hill, R. K.; Joule, J. A.; Loeffler, L. J. *J. Am. Chem. Soc.* **1962**, *84*, 4951.
59. Overman, L. E. *Acc. Chem. Res.* **1980**, *13*, 218.
60. Overman, L. E.; Jessup, P. J. *J. Am. Chem. Soc.* **1978**, *100*, 5179.
61. Overman, L. E.; Petty, C. B.; Doedens, R. J. *J. Org. Chem.* **1979**, *44*, 4183.
62. Overman, L. E.; Fukaya, C. *J. Am. Chem. Soc.* **1980**, *102*, 1454.
63. Corey, E. J.; Snider, B. B. *J. Am. Chem. Soc.* **1972**, *94*, 2549.
64. Ban, Y.; Honma, Y.; Oishi, T. *Tetrahedron Lett.* **1976**, 1111.
65. Birch, A. J.; MacDonald, P. L.; Powell, V. M. *Tetrahedron Lett.* **1969**, 351.
66. Ibid., *J. Chem. Soc. C* **1970**, 1469.
67. Deslongchamps, P. *Pure Appl. Chem.* **1977**, *49*, 1329.
68. Soffer, M. D.; Burk, L. A. *Tetrahedron Lett.* **1970**, 211.
69. Tanaka, A.; Uda, H.; Yoshikoshi, A. *J. Chem. Soc., Chem. Commun.* **1967**, 188.
70. Alston, P. V.; Ottenbrite, R. M. *J. Org. Chem.* **1975**, *40*, 1111.

71. Corey, E. J.; Anderson, N. H.; Carlson, R. M.; Paust, J.; Vedejs, E.; Ulattas, I.; Winter, R. E. K. *J. Am. Chem. Soc.* **1968**, *90*, 3245.
72. Titov, Y. A. *Russ. Chem. Rev.* (*Engl. Transl.*) **1962**, *31*, 267.
73. Alston, P. V.; Ottenbrite, R. M.; Shillady, D. D. *J. Org. Chem.* **1973**, *38*, 4075.
74. Fleming, I.; Michael, J. P.; Overman, L. E.; Taylor, G. F. *Tetrahedron Lett.* **1978**, 1313.
75. Corey, E. J.; Danheiser, R. L.; Chandrasekaran, S.; Siret, P.; Keck, G. E.; Gras, J. L. *J. Am. Chem. Soc.* **1978**, *100*, 8031.
76. Corey, E. J.; Danheiser, R. L.; Chandrasekaran, S.; Keck, G. E.; Godalan, B.; Larsen, S. D.; Siret, P.; Grass, J. L. *J. Am. Chem. Soc.* **1978**, *100*, 8034.
77. Mori, K.; Shiozaki, M.; Itaya, N.; Matusi, M.; Sumiki, Y. *Tetrahedron* **1969**, *25*, 1293.
78. Nagata, W.; Wakabayashi, T.; Narisada, M.; Mayase, Y.; Kamata, S. *J. Am. Chem. Soc.* **1971**, *93*, 5740.
79. When this section was already completed, an interesting paper by Rozeboom, M. D.; Tegmo-Larsson, I.; Houk, K. N. appeared in *J. Org. Chem.* **1981**, *46*, 2338, on FMO effects on cycloadditions to benzoquinones and naphthoquinones. LUMO coefficients of 2-methoxy-1,4-benzoquinone have C1 > C4, in full accordance with our naive considerations.
80. Stipanovic, R. D.; Bell, A. A.; O'Brien, D. H.; Lukefahr, M. J. *Tetrahedron Lett.* **1977**, 567.
81. Alston, P. V.; Ottenbrite, R. M.; Cohen, T. *J. Org. Chem.* **1978**, *43*, 1864.
82. Nazarov, I. N.; Kuznetsova, A. I.; Kuznetsov, N. V. *Zh. Obshch. Khim.* **1955**, *25*, 88; reported in Ref. 72.
83. Kraus, G. A.; Taschner, M. J. *J. Org. Chem.* **1980**, *45*, 1175.
84. Nouye, Y.; Karisawa, H. *Bull. Chem. Soc. Jpn.* **1969**, *42*, 3318.
85. Yamakawa, K.; Satoh, T. *Chem. Pharm. Bull.* **1977**, *25*, 2535.
86. Danishefsky, S.; Kitahara, T. *J. Am. Chem. Soc.* **1974**, *96*, 7807.
87. Danishefsky, S.; Kitahara, T. *J. Org. Chem.* **1975**, *40*, 538.
88. Danishefsky, S.; Kitahara, T.; Yan, C. F.; Morris, J. *J. Am. Chem. Soc.* **1979**, *101*, 6996.
89. Danishefsky, S.; Singh, R. K.; Harayama, T. *J. Am. Chem. Soc.* **1977**, *99*, 5810.
90. Danishefsky, S.; Harayama, T.; Singh, R. K. *J. Am. Chem. Soc.* **1979**, *101*, 7008.
91. Danishefsky, S.; Hirama, M. *J. Am. Chem. Soc.* **1977**, *99*, 7740.
92. Danishefsky, S.; Hirama, M.; Frisch, N.; Clardy, J. *J. Am. Chem. Soc.* **1979**, *101*, 7013.
93. Danishefsky, S.; Hirama, M.; Gombatz, K.; Harayama, T.; Berman, E.; Schuda, P. *J. Am. Chem. Soc.* **1978**, *100*, 6536.
94. Ibid., **1979**, *101*, 7020.
95. Danishefsky, S.; Morris, J.; Mullen, G.; Gammill, R. *J. Am. Chem. Soc.* **1980**, *102*, 2838.
96. Danishefsky, S.; Schuda, P. F.; Kitahara, T.; Etheredge, S. J. *J. Am. Chem. Soc.* **1977**, *99*, 6066.
97. Ritchie, C. D.; Sager, W. F. *Prog. Phys. Org. Chem.* **1964**, *2*, 334.
98. Danishefsky, S.; Yan, C. F.; Singh, R. K.; Gammill, R. B.; McCurry, P. M., Jr.; Fritsch, N.; Clardy, J. *J. Am. Chem. Soc.* **1979**, *101*, 7001.
99. Harayama, T.; Cho, H.; Inubushi, Y. *Tetrahedron Lett.* **1977**, 3273.
100. Ibid., *Chem. Pharm. Bull.* **1978**, *26*, 1201.
101. Danishefsky, S.; Etheredge, S. J. *J. Org. Chem.* **1979**, *44*, 4716.
102. Ibid., **1978**, *43*, 4604.
103. Danishefsky, S.; Walker, F. J. *J. Am. Chem. Soc.* **1979**, *101*, 7018.
104. Grandmaison, J. L.; Brassard, P. *J. Org. Chem.* **1978**, *43*, 1435.
105. Roberge, G.; Brassard, P. *J. Chem. Soc., Perkin Trans. 1*, **1978**, 1041.
106. Spitzner, D. *Tetrahedron Lett.* **1978**, 3349.
107. Jung, M. E.; McCombs, C. A. *J. Am. Chem. Soc.* **1978**, *100*, 5207.
108. Jung, M. E.; Pan, Y. G. *Tetrahedron Lett.* **1980**, 3127.
109. Danishefsky, S.; Zamboni, R.; Kahn, M.; Etheredge, S. J. *J. Am. Chem. Soc.* **1980**, *102*, 2097.

110. Danishefsky, S.; Zamboni, R. *Tetrahedron Lett.* **1980**, 3439.
111. Danishefsky, S.; Kahn, M. *Tetrahedron Lett.* **1981**, 489.
112. Semmelhack, M. F.; Tomoda, S.; Hurst, K. M. *J. Am. Chem. Soc.* **1980**, *102*, 7567.
113. Ibuka, T.; Mori, Y.; Inubushi, Y. *Chem. Pharm. Bull.* **1978**, *26*, 2442.
114. Cameron, D. W.; Feutrill, G. I.; Griffiths, P. G.; Hodder, D. J. *J. Chem. Soc., Chem. Commun.* **1978**, 688.
115. Clark Still, W.; Tsai, M. Y. *J. Am. Chem. Soc.* **1980**, *102*, 3653.
116. Krohn, K. *Tetrahedron Lett.* **1980**, 3557.
117. Trost, B. M.; Vladuchick, W. C.; Bridges, A. J. *J. Am. Chem. Soc.* **1980**, *102*, 3548.
118. Ibid., 3554.
119. Funk, R. L.; Vollhardt, K. C. *Chem. Soc. Rev.* **1980**, 41.
120. Franck, R. W.; John, T. V. *J. Org. Chem.* **1980**, *45*, 1170.
121. Kametani, T.; Hirai, Y.; Satoh, F.; Fukumoto, K. *J. Chem. Soc., Chem. Commun.* **1977**, 16.
122. Kametani, T.; Hirai, Y.; Shiratori, Y.; Fukumoto, K.; Satoh, F. *J. Am. Chem. Soc.* **1978**, *100*, 554.
123. Jung, M. E.; Lowe, J. A. *J. Org. Chem.* **1977**, *42*, 2371.
124. Wilson, S. R.; Phillips, L. R.; Natalie, K. J., Jr. *J. Am. Chem. Soc.* **1979**, *101*, 3340.
125. Kametani, T.; Takahashi, T.; Kajiwara, M.; Hirai, Y.; Ohtsuka, C.; Satoh, F.; Fukumoto, K. *Chem. Pharm. Bull.* **1974**, *22*, 2159.
126. Wiseman, J. R.; Pendery, J. P.; Otto, C. A.; Chiong, K. G. *J. Org. Chem.* **1980**, *45*, 516.
127. Alston, P. V.; Ottenbrite, R. M. *J. Org. Chem.* **1975**, *40*, 1111.
128. Kametani, T.; Suzuki, T.; Ichikawa, Y.; Fukumoto, K. *J. Chem. Soc., Perkin Trans. 1* **1975**, 2102.
129. Caramella, P.; Coda Corsico, A.; Corsaro, A.; Del Monte, D.; Marinone Albini, F. *Tetrahedron* **1982**, *38*, 173.
130. Plumann, H. P.; Smith, J. G.; Rodrigo, R. *J. Chem. Soc., Chem. Commun.* **1980**, 354.
131. Sammes, P. G. *Tetrahedron* **1976**, *32*, 405.
132. Arnold, B. J.; Mellows, S. M.; Sammes, P. G. *J. Chem. Soc., Perkin Trans. 1* **1973**, 1266.
133. Carlson, R. G. *Annu. Rep. Med. Chem.* **1974**, *9*, 270.
134. Oppolzer, W. *Angew. Chem., Int. Ed. Engl.* **1977**, *16*, 10.
135. Brieger, G.; Bennett, J. N. *Chem. Rev.* **1980**, *80*, 63.
136. Gschwend, H. W.; Lee, A. O.; Meier, H. P. *J. Org. Chem.* **1973**, *38*, 2169.
137. Auerbach, J.; Weinreb, S. M. *J. Org. Chem.* **1975**, *40*, 3311.
138. Bailey, S. J.; Thomas, E. J.; Turner, W. B.; Jarvis, J. A. J. *J. Chem. Soc., Chem. Commun.* **1978**, 474.
139. Pyne, S. G.; Hensel, M. J.; Byrn, S. R.; McKenzie, A. T.; Fuchs, P. L. *J. Am. Chem. Soc.* **1980**, *102*, 5960.
140. Frater, G. *Helv. Chim. Acta* **1974**, *57*, 172.
141. Fukamiya, N.; Kato, M.; Yoshikoshi, A. *J. Chem. Soc., Perkin Trans. 1* **1973**, 1843.
142. Näf, F.; Ohloff, G. *Helv. Chim. Acta* **1974**, *57*, 1868.
143. Oppolzer, W.; Snowden, R. L. *Tetrahedron Lett.* **1978**, 3505.
144. Roush, W. R.; Ko, A. I.; Gillis, H. R. *J. Org. Chem.* **1980**, *45*, 4264.
145. Yamamoto, H.; Shamm, H. L. *J. Am. Chem. Soc.* **1979**, *101*, 1609.
146. Schiehser, G. A.; White, J. D. *J. Org. Chem.* **1980**, *45*, 1864.
147. Oppolzer, W.; Fehr, C.; Warneke, J. *Helv. Chim. Acta* **1977**, *60*, 48.
148. Boeckman, R. K., Jr.; Sung Ko, S. *J. Am. Chem. Soc.* **1980**, *102*, 7146.
149. Roush, W. R. *J. Am. Chem. Soc.* **1980**, *102*, 1390.
150. Roush, W. R.; Gillis, H. R. *J. Org. Chem.* **1980**, *45*, 4283.
151. Büchi, G.; Hauser, A.; Limacher, J. *J. Org. Chem.* **1977**, *42*, 3323.
152. Schmislin, T.; Tamm, C. *Helv. Chim. Acta* **1978**, *61*, 2096.
153. Breitholle, E. G.; Fallis, A. G. *Can. J. Chem.* **1976**, *54*, 1991.

154. Ibid., *J. Org. Chem.* **1978**, *43*, 1965.
155. Morgans, D. J., Jr.; Stork, G. *Tetrahedron Lett.* **1979**, 1959.
156. Stork, G.; Morgans, D. J., Jr. *J. Am. Chem. Soc.* **1979**, *101*, 7110.
157. Bazan, A. C.; Edwards, J. M.; Weiss, U. *Tetrahedron Lett.* **1977**, 147.
158. Ibid., *Tetrahedron* **1978**, *34*, 3005.
159. Wilson, S. R.; Mao, D. T. *J. Am. Chem. Soc.* **1978**, *100*, 6289.
160. Wilson, S. R.; Misra, R. N. *J. Org. Chem.* **1980**, *45*, 5080.
161. Taber, D. F.; Saleh, S. A. *J. Am. Chem. Soc.* **1980**, *102*, 5085.
162. Näf, F.; Decorzant, R.; Thommen, W. *Helv. Chim. Acta* **1979**, *62*, 114.
163. Oppolzer, W.; Fröstl, W. *Helv. Chim. Acta* **1975**, *58*, 590.
164. Oppolzer, W.; Fröstl, W.; Weber, H. P. *Helv. Chim. Acta* **1975**, *58*, 593.
165. Oppolzer, W.; Flaskemp, E. *Helv. Chim. Acta* **1977**, *60*, 204.
166. Gras, J. L.; Bertrand, M. *Tetrahedron Lett.* **1979**, 4549.
167. Taber, D. F.; Gunn, B. P. *J. Am. Chem. Soc.* **1979**, *101*, 3992.
168. Corey, E. J.; Danheiser, R. L. *Tetrahedron Lett.* **1973**, 4477.
169. Wenkert, E. *J. Am. Chem. Soc.* **1962**, *84*, 98.
170. Qureshi, A. A.; Scott, A. I. *J. Chem. Soc., Chem. Commun.* **1968**, 945.
171. Ibid., 947.
172. Ibid., 948.
173. Scott, A. I. *Acc. Chem. Res.* **1970**, *3*, 151.
174. Battersby, A. R.; Byrne, J. C.; Kapil, R. S.; Martin, J. A.; Payne, T. G.; Arigoni, D.; Loew, P. *J. Chem. Soc., Chem. Commun.* **1968**, 951.
175. Brown, R. T.; Smith, G. F.; Stapleford, K. S. J.; Taylor, D. A. *J. Chem. Soc., Chem. Commun.* **1970**, 190.
176. Cordell, G. A.; Smith, G. F.; Smith, G. N. *J. Chem. Soc., Chem. Commun.* **1970**, 191.
177. Scott, A. I. *J. Am. Chem. Soc.* **1972**, *94*, 8262.
178. Scott, A. I.; Wei, C. C. *J. Am. Chem. Soc.* **1972**, *94*, 8266.
179. Ibid., 8263.
180. Ibid., 8264.
181. Kuehne, M. E.; Roland, D. M.; Hafter, R. *J. Org. Chem.* **1978**, *43*, 3705.
182. Kuehne, M. E.; Matsko, T. H.; Bohnert, J. C.; Kirkemo, C. L. *J. Org. Chem.* **1979**, *44*, 1063.
183. Kuehne, M. E.; Huebner, J. A.; Matsko, T. H. *J. Org. Chem.* **1979**, *44*, 2477.
184. Kan-Fan, C.; Massiot, G.; Ahond, A.; Das, B. C.; Husson, H. P.; Potier, P.; Scott, A. I.; Wei, C. C. *J. Chem. Soc., Chem. Commun.* **1974**, 164.
185. Joshi, B. A.; Viswanathan, N.; Gawad, D. H.; Von Philipsborn, W. *Helv. Chim. Acta* **1975**, *58*, 1551.
186. Joshi, B. S.; Viswanathan, N.; Gawad, D. H.; Balakrishnan, V.; Von Philipsborn, W.; Quick, A. *Experientia* **1975**, *31*, 880.
187. Inoue, S.; Takamatsu, N.; Hashizume, K.; Kishi, Y. *Yakugaku Zasschi* **1977**, *97*, 582.
188. Oppolzer, W. *Synthesis* **1978**, 793.
189. Oppolzer, W. *J. Am. Chem. Soc.* **1971**, *93*, 3833.
190. Ibid., 3834.
191. Oppolzer, W.; Keller, K. *J. Am. Chem. Soc.* **1971**, *93*, 3836.
192. Kametani, T.; Suzuki, K.; Nemoto, H.; Fukumoto, K. *J. Org. Chem.* **1979**, *44*, 1036.
193. Kametani, T.; Honda, T.; Shiratori, Y.; Fukumoto, K. *Tetrahedron Lett.* **1980**, 1389.
194. Kametani, T.; Chihiro, M.; Honda, T.; Fukumoto, K. *Chem. Pharm. Bull.* **1980**, *28*, 2468.
195. Ichihara, A.; Kimura, R.; Yamada, S.; Sakamura, S. *J. Am. Chem. Soc.* **1980**, *102*, 6353.
196. Kametani, T.; Nemoto, H. *Tetrahedron* **1981**, *32*, 3.
197. Oppolzer, W.; Petrzilka, M.; Bättig, K. *Helv. Chim. Acta* **1977**, *60*, 2964.
198. Oppolzer, W.; Bättig, K.; Petrzilka, M. *Helv. Chim. Acta* **1978**, *61*, 1945.
199. Kametani, T.; Nemoto, H.; Ishikawa, H.; Shiroyama, K.; Matsumoto, H.; Fukumoto, K. *J. Am. Chem. Soc.* **1977**, *99*, 3461.

200. Grieco, P. A.; Takigawa, T.; Schillinger, W. J. *J. Org. Chem.* **1980**, *45*, 2247.
201. Kametani, T.; Nemoto, H.; Tsubuki, M.; Purvaneckas, G. E.; Aizawa, M.; Nishiuchi, M. *J. Chem. Soc., Perkin Trans. 1* **1979**, 2830.
202. Kametani, T.; Matsumoto, H.; Nemoto, H.; Fukumoto, K. *Tetrahedron Lett.* **1978**, 2425.
203. Kametani, T.; Matsumoto, H.; Nemoto, H.; Fukumoto, K. *J. Am. Chem. Soc.* **1978**, *100*, 6218.
204. Kametani, T.; Suzuki, K.; Nemoto, H. *J. Chem. Soc., Chem. Commun.* **1979**, 1127.
205. Kametani, T.; Nemoto, H. *Tetrahedron Lett.* **1979**, 3309.
206. Kametani, T.; Tsubuki, M.; Nemoto, H. *J. Org. Chem.* **1980**, *45*, 4391.
207. Vollhardt, K. P. C. *Acc. Chem. Res.* **1977**, *10*, 1.
208. Funk, R. L.; Vollhardt, K. P. C. *J. Am. Chem. Soc.* **1977**, *99*, 5483.
209. Ibid., **1979**, *101*, 215.
210. Ibid., **1980**, *102*, 5253.
211. Sternberg, E. D.; Vollhardt, K. P. C. *J. Am. Chem. Soc.* **1980**, 4839.
212. Quinkert, G.; Weber, W. D.; Schwartz, U.; Dürner, G. *Angew. Chem., Int. Ed. Engl.* **1980**, *19*, 1027.
213. Quinkert, G.; Schwartz, U.; Stark, H.; Weber, W. D.; Baier, H.; Adam, F.; Dürner, G. *Angew. Chem., Int. Ed. Engl.* **1980**, *19*, 1029.
214. Djuric, S.; Sarkar, T.; Magnus, P. *J. Am. Chem. Soc.* **1980**, *102*, 6885.
215. Oppolzer, W.; Roberts, D. A.; Bird, T. G. C. *Helv. Chim. Acta* **1979**, *62*, 2017.
216. Nicolau, K. C.; Bernette, W. E. *J. Chem. Soc., Chem. Commun.* **1979**, 1119.
217. Nicolau, K. C.; Bernette, W. E.; Ma, P. *J. Org. Chem.* **1980**, *45*, 1463.
218. Oppolzer, W.; Roberts, D. A. *Helv. Chim. Acta* **1980**, *63*, 1703.
219. Martin, S. F.; Desai, S. R.; Phillips, G. W.; Miller, A. C. *J. Am. Chem. Soc.* **1980**, *102*, 3294.
220. Overman, L. E.; Taylor, G. F.; Houk, K. N.; Domelsmith, L. N.; *J. Am. Chem. Soc.* **1978**, *100*, 3182.
221. Thompson, H. W.; Melillo, D. G. *J. Am. Chem. Soc.* **1970**, *92*, 3218.
222. Inukai, T.; Kojima, T. *J. Org. Chem.* **1967**, *32*, 872.
223. Kreiser, W.; Haumesser, W.; Thomas, A. F. *Helv. Chim. Acta* **1974**, *57*, 164.
224. Nagakura, I.; Ogata, H.; Ueno, M.; Kitahara, Y. *Bull. Chem. Soc. Jpn.* **1975**, *48*, 2995.
225. Fringuelli, F.; Taticchi, A. Private communication.
226. Sauer, J.; Kredel, J. *Tetrahedron Lett.* **1966**, 731.
227. Fringuelli, F.; Pizzo, F.; Taticchi, A. *Synth. Commun.* **1979**, *9*, 391.
228. Fringuelli, F.; Pizzo, F.; Taticchi, A. Abstracts ESOC II, Stresa, June, 1981, p. 100A.
229. Trong Anh, N.; Seyden-Penne, J. *Tetrahedron* **1973**, *29*, 3259.
230. Caramella, P.; Cellerino, G.; Corsico Coda, A.; Gamba Invernizzi, A.; Grünanger, P.; Houk, K. N.; Marinone Albini, F. *J. Am. Chem. Soc.* **1976**, *41*, 3349.
231. Houk, K. N.; Strozier, R. W. *J. Am. Chem. Soc.* **1973**, *95*, 4094.
232. Cookson, R. C.; Tuddenham, R. M. *J. Chem. Soc., Chem. Commun.* **1973**, 742.
233. Cookson, R. C.; Tuddenham, R. M. *J. Chem. Soc., Perkin Trans. 1* **1978**, 678.
234. Giordan, J. C.; McMillan, M. C.; Moore, J. H.; Staley, S. W. *J. Am. Chem. Soc.* **1980**, *102*, 4870.
235. Klessinger, M.; Gunkel, E. *Tetrahedron* **1978**, *34*, 3591.
236. Lefour, J. M.; Loupy, A. *Tetrahedron* **1978**, *34*, 2597.
237. Dal Negro, A.; Ungaretti, L.; Perotti, A. *J. Chem. Soc., Dalton Trans.* **1972**, 1639.
238. Ross Kelly, T.; Montury, M. *Tetrahedron Lett.* **1978**, 4311.
239. Tou, J. S.; Reusch, W. *J. Org. Chem.* **1980**, *45*, 5012.
240. Marx, J. N.; Norman, L. R. *Tetrahedron Lett.* **1973**, 4375.
241. Marx, J. N.; Norman, L. R. *J. Org. Chem.* **1975**, *40*, 1602.
242. Adams, D. R.; Bhatnagar, S. P.; Cookson, R. C.; Tuddenham, R. M. *J. Chem. Soc., Perkin Trans. 1* **1975**, 1741.

243. Ayyar, K. S.; Cookson, R. C.; Kagi, D. A. *J. Chem. Soc., Chem. Commun.* **1973**, 161.
244. Ayyar, K. S.; Cookson, R. C.; Kagi, D. A. *J. Chem. Soc., Perkin Trans. 1* **1975**, 1727.
245. Dickinson, R. A.; Kubela, R.; MacAlpine, G. A.; Stojanac, Z.; Valenta, Z. *Can. J. Chem.* **1972**, *50*, 2377.
246. Ross Kelly, T.; Montury, M. *Tetrahedron Lett.* **1978**, 4309.
247. Trost, B. M.; Godleski, S. A.; Genet, J. P. *J. Am. Chem. Soc.* **1978**, *100*, 3930.
248. Trost, B. M.; Godleski, S. A.; Belletire, J. L. *J. Org. Chem.* **1979**, *44*, 2052.
249. Stojanac, N.; Sood, A.; Stojanac, Z.; Valenta, Z. *Can. J. Chem.* **1975**, *53*, 619.
250. Grieco, P. A.; Vidari, G.; Ferriño, S.; Haltiwanger, R. C. *Tetrahedron Lett.* **1980**, 1619.
251. Grieco, P. A.; Ferriño, S.; Vidari, G. *J. Am. Chem. Soc.* **1980**, *102*, 7586.
252. Roush, W. R.; Gillis, H. R. *J. Org. Chem.* **1980**, *45*, 4267.
253. Wenkert, E.; Naemura, K. *Synth. Commun.* **1973**, *3*, 45.
254. Mukaiyama, T.; Iwasawa, N. *Chem. Lett.* **1979**, 697.
255. Mukaiyama, T.; Iwasawa, N.; Tsuji, T.; Narasaka, K. *Chem. Lett.* **1979**, 1175.
256. Ripoll, J. L.; Rouessac, A.; Rouessac, F. *Tetrahedron* **1978**, *34*, 19.
257. Dewar, M. S.; Olivella, S.; Rzepa, H. S. *J. Am. Chem. Soc.* **1978**, *100*, 5650.
258. Ducos, P.; Rouessac, F. *Tetrahedron* **1973**, *29*, 3233.
259. Ho, T. L. *Synth. Comm.* **1974**, *4*, 189.
260. Haslouin, J.; Rouessac, F. *Bull. Soc. Chim. Fr.* **1977**, 1242.
261. Boland, W.; Jaenicke, L. *Chem. Ber.* **1978**, *111*, 3262.
262. Ichihara, A.; Kimura, R.; Oda, K.; Sakamura, S. *Tetrahedron Lett.* **1976**, 4741.
263. Oda, K.; Ichihara, A.; Sakamura, S. *Tetrahedron Lett.* **1975**, 3187.
264. Ichihara, A.; Oda, K.; Kobayashi, M.; Sakamura, S. *Tetrahedron* **1980**, *36*, 183.
265. Ichimura, A.; Oda, K.; Kobayashi, M.; Sakamura, S. *Tetrahedron Lett.* **1974**, 4235.
266. Kitamura, T.; Kawakami, Y.; Imagawa, T.; Kiwanisi, M. *Tetrahedron Lett.* **1978**, 4297.
267. Groves, J. K.; Cundasawmy, N. E.; Anderson, H. J. *Can. J. Chem.* **1973**, *51*, 1089.
268. Franck, B.; Bockhaus, H.; Rolf, M. *Tetrahedron Lett.* **1980**, 1185.
269. Krohn, K.; Tolkiehn, K. *Chem. Ber.* **1979**, *112*, 3453.
270. Watt, D.; Corey, E. J. *Tetrahedron Lett.* **1972**, 4651.
271. Corey, E. J.; Watt, D. S. *J. Am. Chem. Soc.* **1973**, *95*, 2304.
272. Jung, M. E.; Lowe, J. A. *J. Chem. Soc., Chem. Commun.* **1978**, 95.
273. Ireland, R. E.; McGarvey, G. J.; Anderson, R. C.; Badoud, R.; Fitzsimmons, B.; Thaisrivongs, S. *J. Am. Chem. Soc.* **1980**, *102*, 6178.
274. Ross Kelly, T. *Annu. Rep. Med. Chem.* **1979**, *14*, 288.
275. Kende, A. S.; Tsay, Y.; Mills, J. E. *J. Am. Chem. Soc.* **1976**, *98*, 1967.
276. Garland, R. B.; Palmer, J. R.; Schulz, J. A.; Sollman, P. B.; Pappo. R. *Tetrahedron Lett.* **1978**, 3669.
277. Ross Kelly, T.; Goerner, R. N., Jr.; Gillard, J. W.; Prazak, B. K. *Tetrahedron Lett.* **1976**, 3869.
278. Ross Kelly, T.; Tsang, W. G. *Tetrahedron Lett.* **1978**, 4457.
279. Inhoffen, H. H.; Muxfeldt, H.; Koppe, V.; Heimann-Trosien, J. *Chem. Ber.* **1957**, *90*, 1448.
280. Chandler, M.; Stoodley, R. J. *J. Chem. Soc., Chem. Commun.* **1978**, 997.
281. Chandler, M.; Stoodley, R. J. *J. Chem. Soc., Perkin Trans. 1* **1980**, 1007.
282. Birch, A. M.; Mercer, A. J. H.; Chippendale, A. M.; Greenhalgh, C. W. *J. Chem. Soc., Chem. Commun.* **1977**, 745.
283. Ross Kelly, T.; Vaya, J.; Ananthasubramanian, L. *J. Am. Chem. Soc.* **1980**, *102*, 5983.
284. Ross Kelly, T.; Gillard, J. W.; Goerner, R. N., Jr. *Tetrahedron Lett.* **1976**, 3873.

285. Ross Kelly, T.; Gillard, J. W.; Goerner, R. N., Jr.; Lyding, J. *J. Am. Chem. Soc.* **1977**, *99*, 5513.
286. Krohn, K.; Tolkiehn, K. *Tetrahedron Lett.* **1978**, 4023.
287. Boeckman, R. K., Jr.; Dolak, T. M.; Culos, K. O. *J. Am. Chem. Soc.* **1978**, *100*, 7098.
288. Kedersky, F. A. J.; Cava, M. P. *J. Am. Chem. Soc.* **1978**, *100*, 3635.
289. Wiseman, J. R.; French, N. I.; Hallmark, R. K.; Chiong, K. G. *Tetrahedron Lett.* **1978**, 3765.
290. Kende, A. S.; Cuzzan, D. P.; Tsay, Y.; Mills, Y. E. *Tetrahedron Lett.* **1977**, 3537.
291. Kametani, T.; Hirai, Y.; Satoh, F.; Fukumoto, K. *Chem. Pharm. Bull.* **1974**, *22*, 2159.
292. Kametani, T.; Chihiro, M.; Takeshita, M.; Takahashi, K.; Fukumoto, K.; Takano, S. *Chem. Pharm. Bull.* **1978**, *26*, 3820.
293. Boeckman, R. K., Jr.; Delton, M. H.; Nagasaka, T.; Watanabe, T. *J. Org. Chem.* **1977**, *42*, 2946.
294. Boeckman, R. K., Jr.; Delton, M. H.; Dolak, T. M.; Watanabe, T.; Glick, M. D. *J. Org. Chem.* **1979**, *44*, 4396.
295. Amaro, A.; Carreño, M. C.; Fariña, F. *Tetrahedron Lett.* **1979**, 3983.
296. Fariña, F.; Primo, J.; Torres, T. *Chem. Lett.* **1980**, 77.
297. Carrupt, P. A.; Vogel, P. *Tetrahedron Lett.* **1979**, 4533.
298. Bessière, Y.; Vogel, P. *Helv. Chim. Acta* **1980**, *63*, 232.
299. Jackson, D. K.; Narasimhan, L.; Swenton, J. S. *J. Am. Chem. Soc.* **1979**, *101*, 3989.
300. Gesson, J. P.; Jacquesy, J. C.; Mondon, M. *Tetrahedron Lett.* **1980**, 3351.
301. Bauman, J. G.; Barber, R. B.; Gless, R. D.; Rapoport, H. *Tetrahedron Lett.* **1980**, 4777.
302. Büchi, G.; Coffen, D. L.; Kocsis, K.; Sonnet, P. E.; Ziegler, F. E. *J. Am. Chem. Soc.* **1965**, *87*, 2073.
303. Ibid., **1966**, *88*, 3099.
304. Ban, Y.; Wakamatsu, T.; Fujimoto, Y.; Oishi, T. *Tetrahedron Lett.* **1968**, 3383.
305. Ikezaki, M.; Wakamatsu, T.; Ban, J. *J. Chem. Soc., Chem. Commun.* **1969**, 88.
306. Sundberg, R. J.; Bloom, J. D. *Tetrahedron Lett.* **1978**, 5157.
307. Ziegler, F. E.; Spitzner, E. B. *J. Am. Chem. Soc.* **1970**, *92*, 3492.
308. Ibid., **1973**, *95*, 7146.
309. Wiesner, K. *Chem. Soc. Rev.* **1977**, *6*, 413.
310. Wiesner, K.; Ho, P. T.; Oida, S. *Can. J. Chem.* **1974**, *52*, 1042.
311. Wiesner, K.; Tsai, T. Y. R.; Dmitrienko, G. I.; Nambiar, K. P. *Can. J. Chem.* **1976**, *54*, 3307.
312. Atwal, K. S.; Marini-Bettolo, R.; Sachez, I. H.; Tsai, T. Y. R.; Wiesner, K. *Can. J. Chem.* **1978**, *56*, 1102.
313. Wiesner, K.; Tsai, T. Y. R.; Nambiar, K. P. *Can. J. Chem.* **1978**, *56*, 1451.
314. Tsai, T. Y. R.; Nambiar, K. P.; Krikorian, D.; Botta, M.; Marini-Bettolo, R.; Wiesner, K. *Can. J. Chem.* **1979**, *57*, 2124.
315. Vedejs, E.; Gadwood, R. C. *J. Org. Chem.* **1978**, *43*, 376.
316. Owens, C.; Raphael, R. A. *J. Chem. Soc., Perkin Trans. 1* **1978**, 1504.
317. Kim, M. Y.; Weinreb, S. M. *Tetrahedron Lett.* **1979**, 579.
318. Stork, G.; Nakahara, Y.; Nakahara, Y.; Greenlee, W. J. *J. Am. Chem. Soc.* **1978**, *100*, 7775.
319. Schmidlin, T.; Tamm, C. *Helv. Chim. Acta* **1980**, *63*, 121.
320. Dougherty, D.; Brint, P.; McGlynn, S. P. *J. Am. Chem. Soc.* **1978**, *100*, 5597.
321. Eck, V.; Schweig, A.; Vermeer, H. *Tetrahedron Lett.* **1978**, 2433.
322. Schang, P.; Gleiter, R.; Rieker, A. *Ber. Bunsenges. Phys. Chem.* **1978**, *82*, 629.
323. Desimoni, G.; Tacconi, G. *Chem. Rev.* **1975**, *75*, 651.
324. Rossi, R. *Synthesis* **1978**, 413.
325. Lipkowitz, K. B.; Scarpone, S.; Mundy, B. P.; Bornmann, W. G. *J. Org. Chem.* **1979**, *44*, 486.
326. Chaquin, P.; Morizur, J. P.; Kossanyi, J. *J. Am. Chem. Soc.* **1977**, *99*, 903.
327. Gore, W. E.; Pearce, G. T.; Silverstein, R. M. *J. Org. Chem.* **1975**, *40*, 1705.
328. Ireland, R. E.; Häbich, V.D. *Tetrahedron Lett.* **1980**, 1389.

329. Maurer, B.; Grieder, A.; Thommen, W. *Helv. Chim. Acta* **1979**, *62*, 44.
330. Berti, G.; Catellani, G.; Magi, S.; Monti, L. *Gazz. Chim. Ital.* **1980**, *110*, 173.
331. Ireland, R. E.; Beslin, P.; Giger, R.; Hengartner, U.; Kirst, H. A.; Maag, H. *J. Org. Chem.* **1977**, *42*, 1267.
332. Seo, S.; Sankawa, U.; Ogihara, Y.; Iitaka, Y.; Shibata, S. *Tetrahedron* **1973**, *29*, 3721.
333. Miyase, T.; Rüedi, P.; Eugster, C. H. *Helv. Chim. Acta* **1977**, *60*, 2789.
334. Cole, K. C.; Ogilvie, J. F. *Can. J. Spectrosc.* **1975**, *20*, 162; *Chem. Abstr.* **1976**, *84*, 104849.
335. Harris, E. E.; Firestone, R. A.; Fister, K.; Boettcher, R. R.; Cross, F. J.; Currie, R. B.; Monaco, M.; Peterson, E. R.; Reuter, W. *J. Org. Chem.* **1962**, *27*, 2705.
336. Firestone, R. A.; Harris, E. E.; Reuter, W. *Tetrahedron* **1967**, *23*, 943.
337. Harris, E. E.; Rosenburg, D. W.; Chamberlin, E. C. U.S. Pat. 3 381 014; *Chem. Abstr.* **1968**, *69*, 52023d.
338. Palmer, M. H.; Findlay, R. H.; Egdell, R. G. *J. Mol. Struct.* **1977**, *40*, 191.
339. Almemark, M.; Bäckvall, J. E.; Moberg, C.; Akermark, B.; Asbrink, L.; Roos, B. *Tetrahedron* **1974**, *30*, 2503.
340. Younkin, J. M.; Smith, L. J.; Compton, R. N. *Theor. Chim. Acta* **1976**, *41*, 157.
341. Chapman, O. L.; Engel, M. R.; Springer, J. P.; Clardy, J. C. *J. Am. Chem. Soc.* **1971**, *93*, 6696.
342. Snider, B. B.; Roush, D. M.; Killinger, T. A. *J. Am. Chem. Soc.* **1979**, *101*, 6024.
343. Tietze, L. F.; Von Kiedrowski, G.; Harms, K.; Clegg, W.; Shedrick, G. *Angew. Chem., Int. Ed. Engl.* **1980**, *19*, 134.
344. Meeks, J. L.; Maria, H. J.; Brint, P.; McGlynn, S. P. *Chem. Rev.* **1975**, *75*, 603.
345. Haselbach, E.; Heilbronner, E. *Helv. Chim. Acta* **1970**, *53*, 684.
346. Houk, K. N.; Chang, Y. M.; Engel, P. S. *J. Am. Chem. Soc.* **1975**, *97*, 1824.
347. Bergmann, H.; Bock, H. *Z. Naturforsch. B: Anorg. Chem., Org. Chem.* **1975**, *30B*, 629.
348. Bergmann, H.; Elbel, S.; Demuth, R. *J. Chem. Soc., Dalton Trans.* **1977**, 401.
349. Levitt, L. S.; Levitt, B. W.; Parkanyi, C. *Tetrahedron* **1972**, *28*, 3369.
350. Rademacher, P.; Pfeffer, H. U.; Enders, D.; Eichenauer, H.; Weuster, P. *J. Chem. Res., Synop.* **1979**, 222.
351. Schmidt, R. R.; Angerbauer, R.; Abele, W. *Abstr. Int. Congr. Heterocycl. Chem., 8th* **1981**, 75.
352. Keck, G. E.; Fleming, S. A. *Tetrahedron Lett.* **1978**, 4763.
353. Kim, D.; Weinreb, S. M. *J. Org. Chem.* **1978**, *43*, 121.
354. Ibid., 126.
355. Basha, F. Z.; Hibino, S.; Kim, D.; Pye, W. E.; Wu, T. T.; Weinreb, S. M. *J. Am. Chem. Soc.* **1980**, *102*, 3962.
356. Kametani, T.; Kajiwara, M.; Takahashi, T.; Fukumoto, K. *J. Chem. Soc., Perkin Trans. 1* **1975**, 737.
357. Kametani, T.; Takahashi, T.; Honda, T.; Ogasawara, K.; Fukumoto, K. *J. Org. Chem.* **1974**, *39*, 447.
358. Nader, B.; Frank, R. W.; Weinreb, S. M. *J. Am. Chem. Soc.* **1980**, *102*, 1154.
359. Schmitthenner, H. F.; Weinreb, S. M. *J. Org. Chem.* **1980**, *45*, 3373.
360. Weinreb, S. M.; Khatri, N. A.; Shringarpure, J. *J. Am. Chem. Soc.* **1979**, *101*, 5073.
361. Kametani, T.; Higa, T.; Van Loc, C.; Ihara, M.; Koizumi, M.; Fukumoto, K. *J. Am. Chem. Soc.* **1976**, *98*, 6186.
362. Keck, G. E. *Tetrahedron Lett.* **1978**, 4767.
363. Keck, G. E.; Nickell, D. G. *J. Am. Chem. Soc.* **1980**, *102*, 3632.
364. Jacobi, P. A.; Craig, T. *J. Am. Chem. Soc.* **1978**, *100*, 7748.
365. Oppolzer, W.; Francotte, E.; Bättig, K. *Helv. Chim. Acta* **1981**, *64*, 478.
366. Epiotis, N. D.; Yates, R. L. *J. Org. Chem.* **1974**, *39*, 3150.
367. Wasserman, H. H.; Ives, J. L. *Tetrahedron* **1980**, *37*, 1825.
368. Balci, M. *Chem. Rev.* **1981**, *81*, 91.
369. Sih, C. J.; Salomon, R. G.; Price, P.; Peruzzoti, G.; Sood, R. *J. Chem. Soc., Chem. Commun.* **1972**, 240.

370. Schenck, G. P.; Ziegler, K. *Naturwiss.* **1945**, *32*, 157.
371. Barrett, H. C.; Büchi, G. *J. Am. Chem. Soc.* **1967**, *89*, 5665.
372. Ayer, W. A.; Rowne, L. M.; Fung, S. *Can. J. Chem.* **1976**, *54*, 3276.
373. Holbert, G. W.; Ganem, B. *J. Am. Chem. Soc.* **1978**, *100*, 352.
374. Demuth, M. R.; Garrett, P. E.; White, J. D. *J. Am. Chem. Soc.* **1976**, *98*, 634.
375. Matsumoto, M.; Dobashi, S.; Kuroda, K. *Tetrahedron Lett.* **1977**, 3361.
376. Kondo, K.; Matsumoto, M. *Tetrahedron Lett.* **1976**, 391.
377. Ibid., 4363.
378. Crowley, K. J. *Tetrahedron Lett.* **1965**, 2863.
379. Meinwald, J.; Eckell, A.; Erickson, K. L. *J. Am. Chem. Soc.* **1965**, *87*, 3532.
380. Woodward, R. B.; Hoffman, R. *Angew. Chem., Int. Ed. Engl.* **1969**, *8*, 781.
381. Seeley, D. A. *J. Am. Chem. Soc.* **1972**, *94*, 4378.
382. Dauben, W. G.; Bell, I.; Hutton, T. W.; Laws, G. F.; Rheiner, A.; Urscheler, H. *J. Am. Chem. Soc.* **1958**, *80*, 4116.
383. Dauben, W. G.; Baumann, P. *Tetrahedron Lett.* **1961**, 565.
384. Jacobs, H. J. C.; Havinga, E. *Adv. Photochem.* **1979**, *11*, 305.
385. Padwa, A.; Bodsky, L.; Clough, S. *J. Am. Chem. Soc.* **1972**, *94*, 6767.

6

$[m + n]$, Polar, and $[2 + 2 + 2]$ Cycloadditions

The field of cycloadditions is characterized by a wide variety of species involving both $[4n]$ or $[4n + 2]$ electrons in the transition state. A common example in this area is the cycloaddition of heptafulvalene to tetracyanoethylene, which under thermal conditions occurs in a $[\pi 14a + \pi 2s]$ fashion. In addition to these multiatom neutral cycloaddends, cations and anions are often involved in polar cycloadditions. In these cases, the number of atoms involved does not coincide with that of the electrons; the latter remains responsible for the concertedness of the reaction.

[m + n] Cycloadditions

This class of cycloaddition is represented in this review by a $[6 + 4]$ example producing chamazulene (**1144a**) and guaiazulene (**1144b**), the degradation products of hydroazulene sesquiterpenes. 6-Dimethylaminofulvene (**1141**), the six-electron partner, reacts with the suitable theophene dioxides **1142a,b**, the four-electron partners, to give **1143a,b** (along with their regioisomers). The introduction of the methyl group completes the syntheses (*1*) (Scheme 6.1).

The $[\pi 6s + \pi 4s]$ character of the cycloaddition is impossible to demonstrate because of the spontaneous loss of dimethylamine and sulfur dioxide from the original adduct.

Polar Cycloadditions

The most popular polar cycloadditions used in natural products syntheses are those involving the oxyallyl-Fe(II) cations **1146** as three-atom–two-electron cycloaddends, which were developed in the 1970s by Noyori. α,α′-Dibromoketones react with iron carbonyls [both mononuclear $Fe(CO)_5$ and dinuclear $Fe_2(CO)_9$ may be used] forming **1146** through the enolate **1145** (Scheme 6.2). The preparations are performed in situ, in the presence of a cycloaddend. An account of use of these reagents in organic syntheses was published by Noyori in 1979 (*2*).

0065-7719/83/0180-0255 $06.00/1

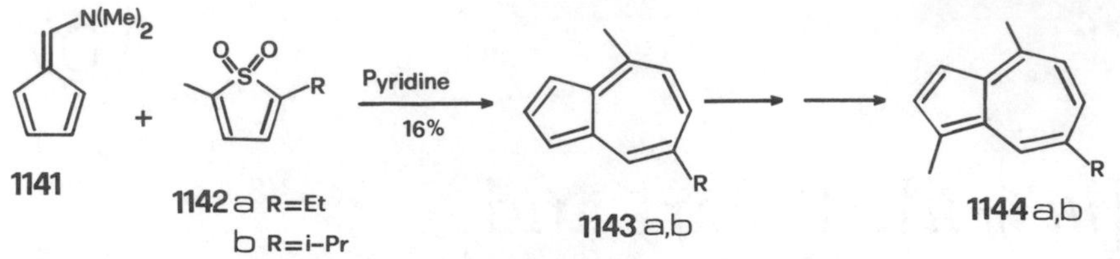

1141 1142 a R=Et b R=i-Pr Pyridine 16% 1143 a,b 1144 a,b

Scheme 6.1

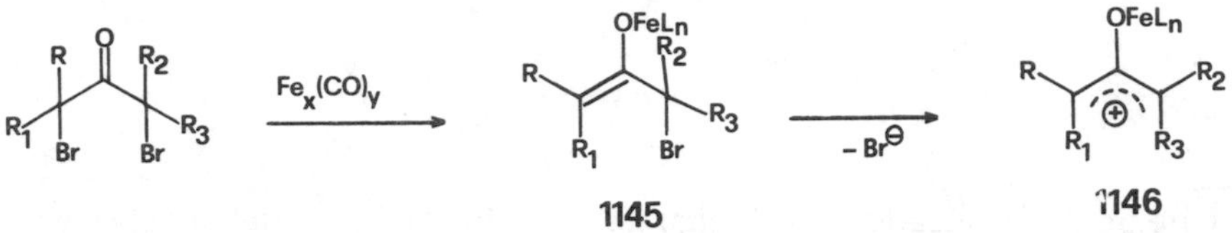

$Fe_x(CO)_y$ 1145 $-Br^{\ominus}$ 1146

Scheme 6.2

Cationic [3+2] Cycloadditions. The reaction of oxyallyl cations with an olefin is a [3+2] polar cycloaddition involving a four-electron transition state, and therefore is symmetry forbidden. The reaction proceeds via a two-step mechanism involving a discrete intermediate. A thorough investigation using *cis*-β-deuteriostyrene (**1147**) as olefin indicated that the [3+2] cycloaddition proceeds stereospecifically, regardless of the stepwise mechanism (*3*). The stereospecificity can be ascribed to a *cisoid* conformation of the zwitterion **1148**, which is fixed rigidly by charge-transfer or coulombic interactions (Scheme 6.3).

The FMOs do not account for the regiochemistry of the adduct; the relative stability of the intermediate controls this aspect. When α,α'-dibromobenzyl methyl ketone (**1149**) reacts with styrene in the presence of $Fe_2(CO)_9$, **1151** is the only reaction product (*4*) (Scheme 6.4). By considering the more stabilized intermediate **1150**, the formation of **1151** is explained. The interaction of the LUMO of the oxyallyl cation with the HOMO of styrene, under conditions of maximum overlap of the coefficients, would predict the opposite regioisomer (Figure 6.1).

Applications of this reaction to the terpene field are exemplified by a one-step synthesis of (±)-α-cuparenone (**4**) (*5*). Thus 1,3-dibromo-3-methyl-2-butanone reacts with 2-*p*-tolylpropene [55 °C in benzene, 17 h, in the presence of $Fe_2(CO)_9$] to give an 18% yield of **4** through the intermediate **1152** (Scheme 6.5).

Intramolecular cycloadditions occur analogously. Thus the dibromo-derivative **1153** yields a complex mixture; the major components are (±)-camphor and (±)-dihydrocarvone. However, **1154** affords (±)-campherenone (**1155**) along with minor amounts of its epimer (*6*) (Scheme 6.6).

Cationic [4+3] Cycloadditions. The Noyori cations react with dienes through [4+3] cationic cycloaddition involving a symmetry-allowed six-electron transition state. The FMOs can be used to explain the regioselectivity of the reaction. For example, the reaction of **1149** with methyl 3-methyl-2-furoate (**1156**) affords **1157** and **1158** in a 92:8 ratio (4) (Scheme 6.7). The interaction of the LUMO of the oxyallyl cation with the HOMO of the furan **1156**, under conditions of maximum overlap of the coefficients, favors the formation of the major isomer **1157**; the more stable intermediate would give the second isomer **1158** (Figure 6.2).

Other oxyallyl cations show less regioselectivity, which is also connected with a lower polarization of the coefficients. Several terpenes

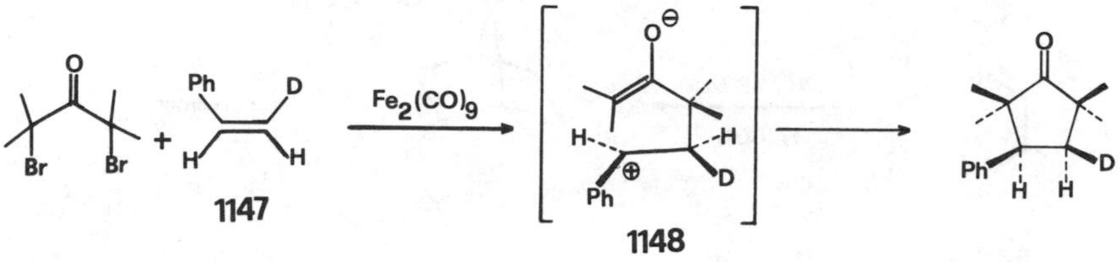

Scheme 6.3

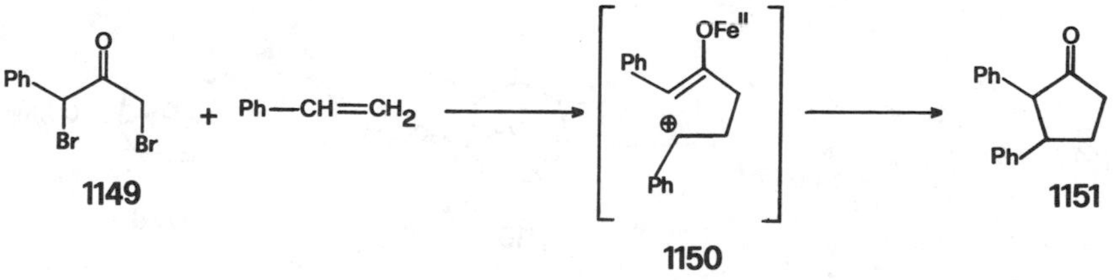

Scheme 6.4

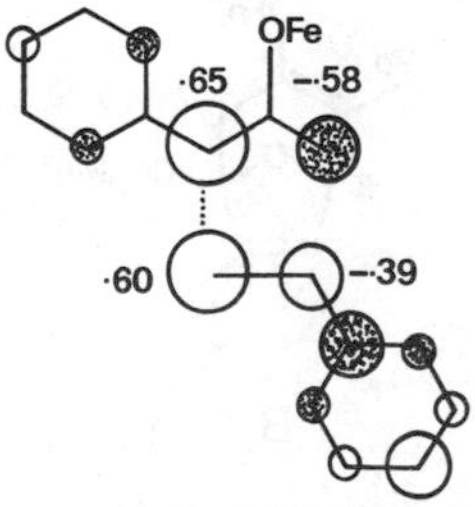

Figure 6.1. FMO interactions between $LUMO_{oxyallyl\ cation}$ and $HOMO_{styrene}$.

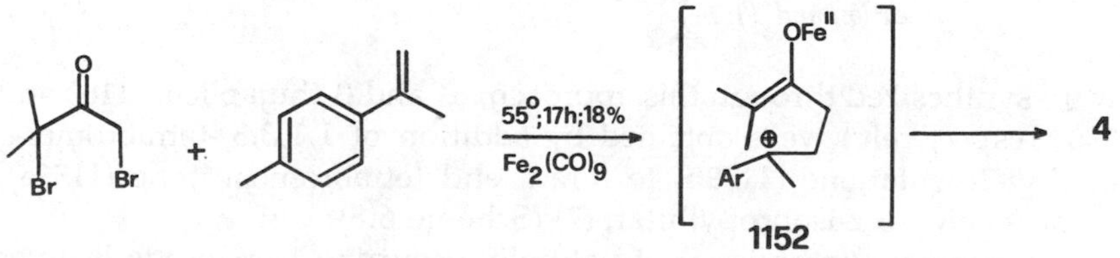

Scheme 6.5

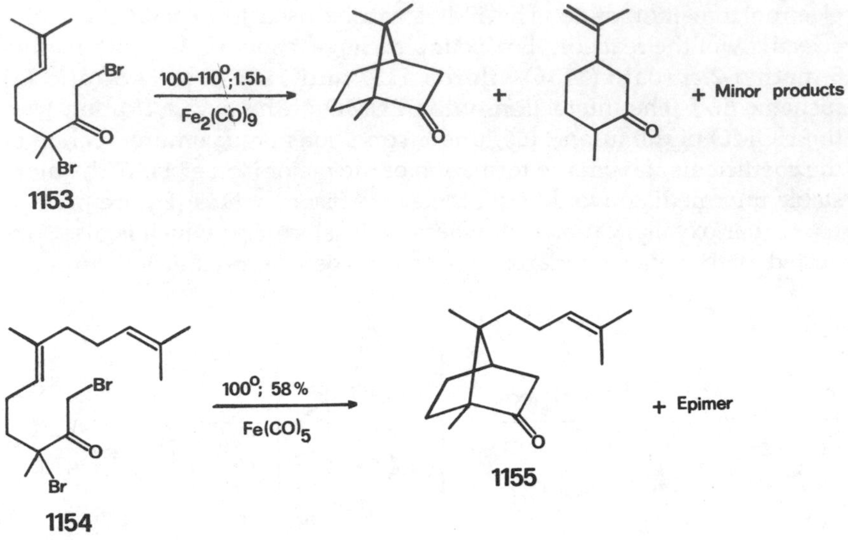
Br
100–110°;1.5h
Fe2(CO)9
O
Br
1153
+
+ Minor products
Br
100°; 58%
Fe(CO)5
+ Epimer
Br
O
1154
1155

Scheme 6.6

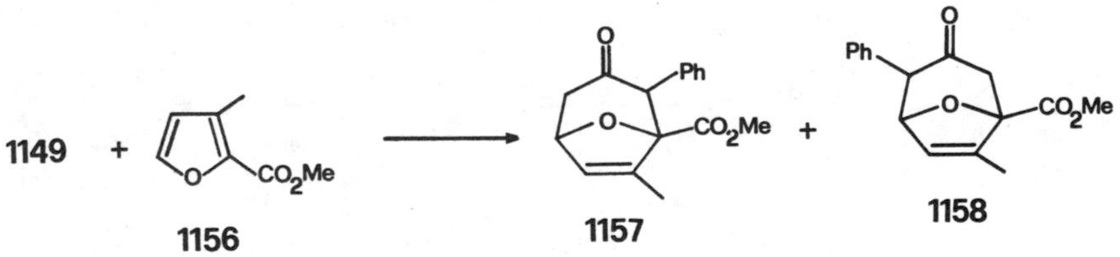
1149 +
CO2Me
1156
Ph
CO2Me
1157
+
Ph
CO2Me
1158

Scheme 6.7

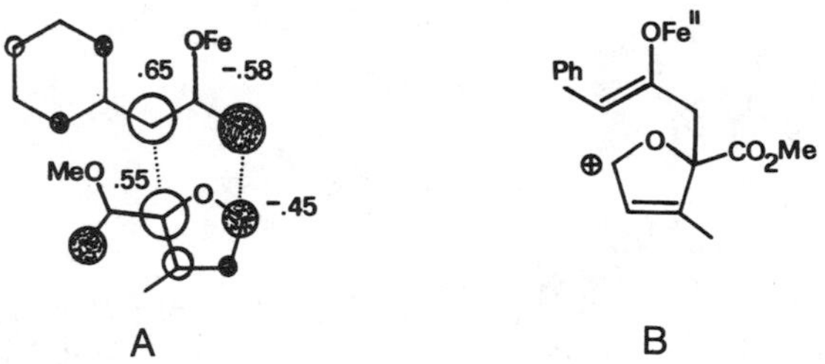
OFe
.65
-.58
MeO
.55
-.45
OFe''
Ph
CO2Me
A
B

*Figure 6.2. FMO interactions (A) and the more stable polar intermediate (B) in the reaction between oxyallyl cation and **1156**.*

were synthesized through this approach; α- and β-thujaplicin (**1160** and **390**, respectively) were obtained by addition of 1,1,3,3,-tetrabromo-4-methyl-2-pentanone (**1159b**) to furan, and tetrabromoacetone (**1159a**), respectively, to 2-isopropylfuran (*7*) (Scheme 6.8).

A general synthesis (*8*) of naturally occurring tropane derivatives begins with the cyclocoupling of α,α,α′,α′-tetrabromoacetone and *N*-

carbomethoxypyrrole. The intermediate is *N*-carbomethoxy-2,4-dibromo-8-azabicyclo[3.2.1]oct-6-en-3-one (**1161**), which can be further converted through reductive treatments to 6,7-dehydrotropine (**1162**) and then to tropine (**1163**), tropanediol (**1164**), scopine (**1165**), and teloidine (**1166**) (Scheme 6.9).

Three *C*-nucleosides were first prepared starting from noncarbohydrate materials and utilizing as the key step the iron carbonyl-promoted reaction of α,α,α′,α′-tetrabromoacetone and furan followed by zinc–copper couple reduction of the primary adduct **1167**. This adduct is easily transformed into pseudouridine (**1168**), pseudocytidine (**1169**), and showdomycin (**1170**) (*9*) (Scheme 6.10)

The formation of oxyallyl cations is not limited to the reaction of α,α′-dihaloketones with iron carbonyls. Zinc–copper couple reduction gives zinc–oxyallyl cations whose behavior seems to be identical with that of iron(II) derivatives. From 3-bromo-1-iodo-3-methyl-2-butanone (**1171**) and isoprene, karahanaenone (**1172**), a seven-membered monoterpene isolated from hop oil, was obtained in 54% yield (*10*) (Scheme 6.11).

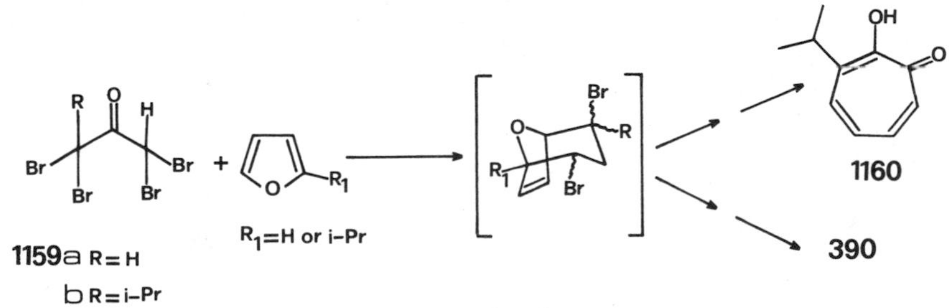

Scheme 6.8

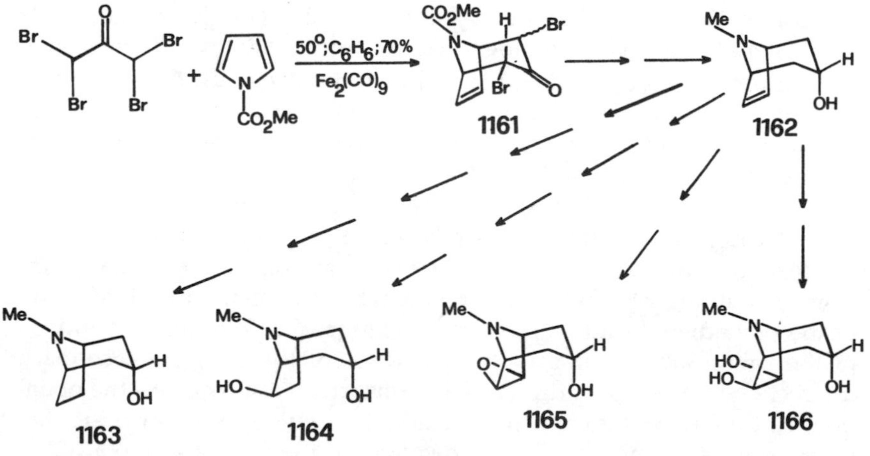

Scheme 6.9

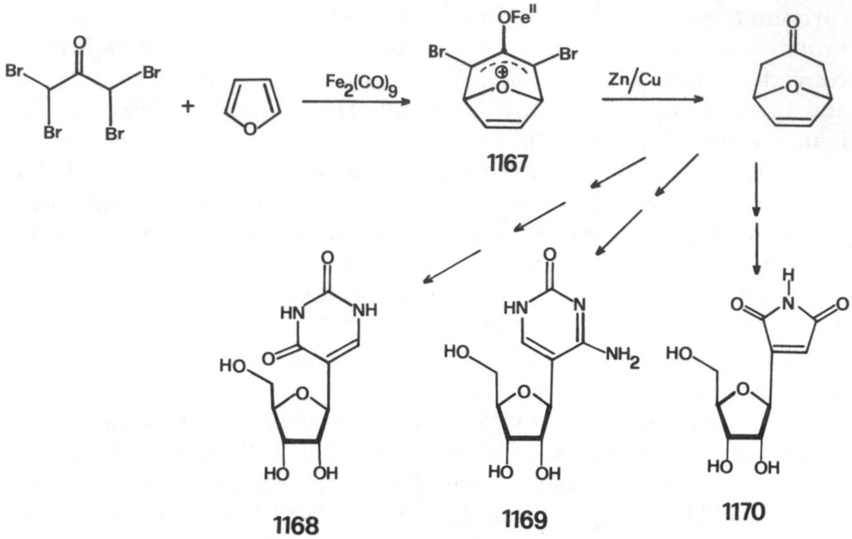

Scheme 6.10

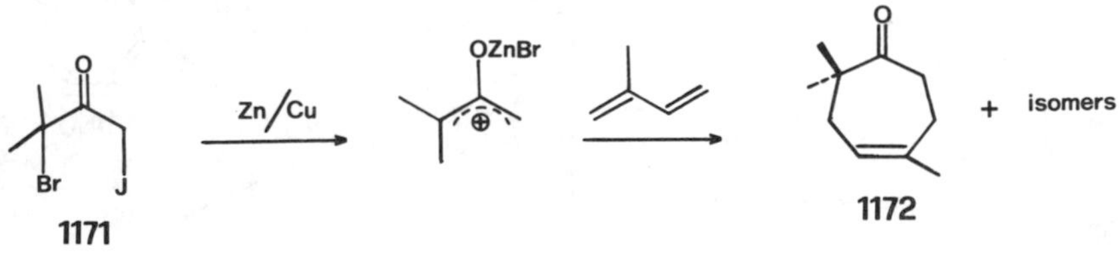

Scheme 6.11

Cationic [5 + 2] Cycloadditions. In the presence of Meerwein salts, silver fluoborate, or stannic chloride, *p*-quinone ketals **1173** give quinone cations **1174**. These five-atom–four-electron intermediates add to olefins through a symmetry-allowed [5 + 2] cationic cycloaddition. An *endo* transition state seems to be favored; a polar adduct (**1175**) is formed. The fate of the adduct depends on the experimental conditions. It can rearrange to the benzofurans **1176** (route a) or react with nucleophiles (e.g., water) to give bicycloenones **1177** (route b) (Scheme 6.12).

Although FMO interactions rationalize the concerted character of the reaction, they do not explain the eventual preferred *endo* transition state. The interaction between the HOMO of the olefin and the LUMO of the pentadienyl cation is reported in Figure 6.3. Whereas C1 and C5 of the cation are bonding with C1 and C2 of the olefin, a secondary interaction between **X** and C2 or C4 seems irrelevant. Because the nodal points of the wave function are coincident with these nuclear positions, these orbitals are non-bonding. The effect of the alkoxy group at C2 is not taken into account, and its eventual influence could become crucial.

This type of cycloaddition worked well in the field of terpenes and neolignans. Gymnonitrol (**1180**), a sesquiterpene isolated from *Gymnomitrion obtusum* (Lindb.) Pears, was synthesized by cycloaddition of the cation obtained from **1178** with stannic chloride to 1,2-dimethylcyclopentene. The major adduct **1179** obtained, together with minor amounts of the stereoisomer, was reduced and subsequently converted to **1180** (*11*) (Scheme 6.13).

A group of neolignans, secondary plant metabolites characterized by two arylpropanoid units, was synthesized by reaction of the cation **1182**, obtained from **1181**, with *E*-isosafrole (**1183**). The original adduct **1184** undergoes either reaction with water to give guianin (**1185**) or rearrangement to **1186**. The C,O ring closure of **1186** gives burchellin (**1187**); C,C ring closure gives futoenone (**1188**) (*12*) (Scheme 6.14).

If the substrate itself (e.g., 2-hydroxy-1,4-benzoquinone) can undergo tautomeric protonation to form the pentadienyl cation, a [5 + 2] polar cycloaddition takes place simply with heating. Such is the case of perezone (**1189**), which produces two isomeric sesquiterpenes: α- and β-pipitzol (**1190** and **1191**, respectively). These isomers were previously

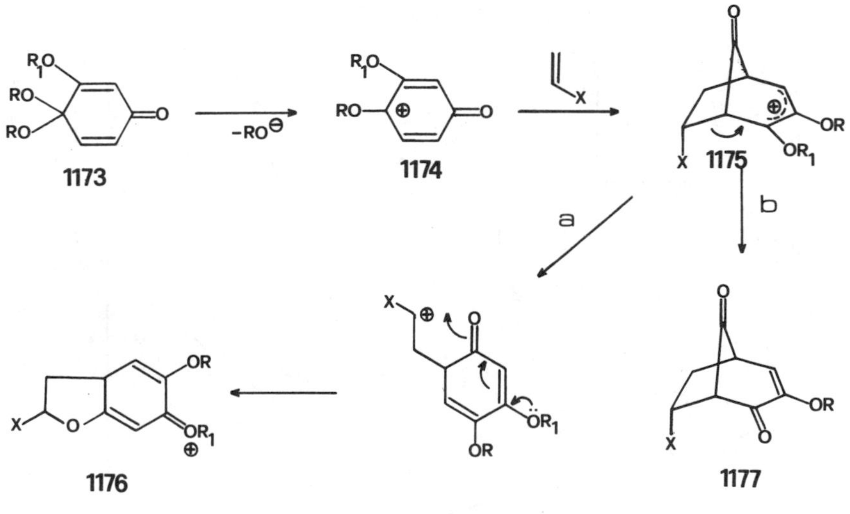

Scheme 6.12

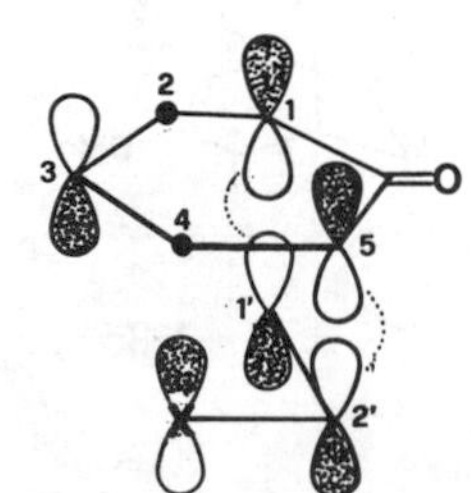

*Figure 6.3. FMO interactions between the LUMO of quinone cation **1174** and a nucleophilic olefin.*

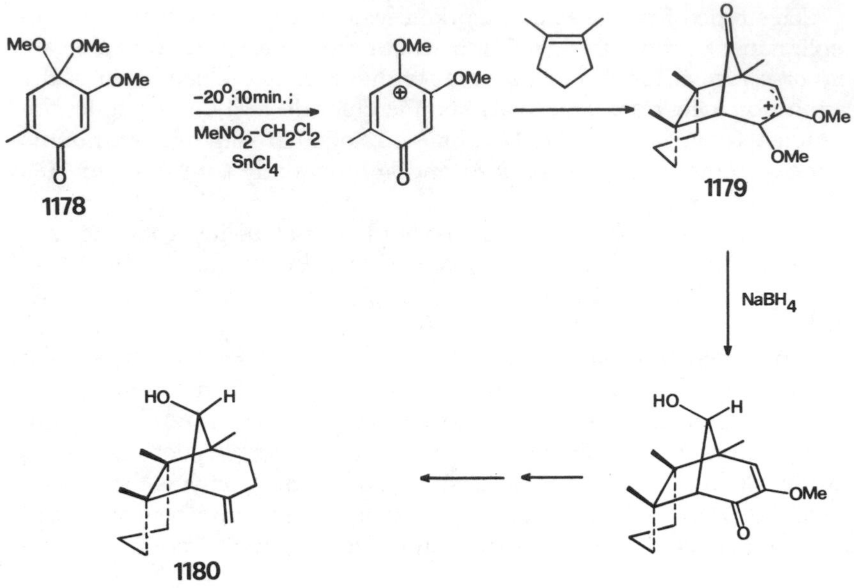

Scheme 6.13

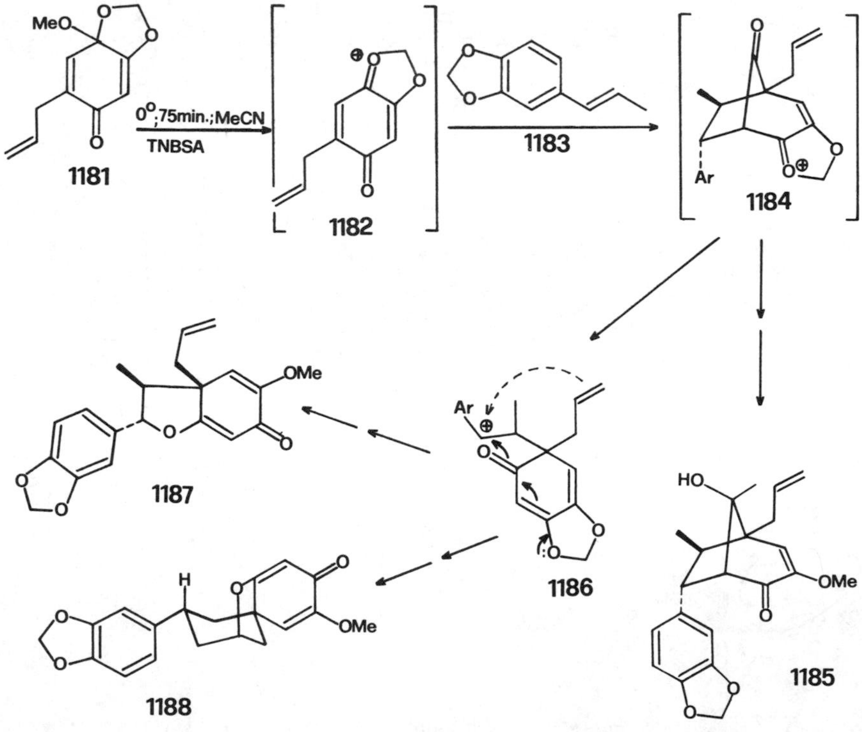

Scheme 6.14

isolated from *Perezia cuernavacana* (13). Their formation through *exo* and *endo* transition states is consistent with the well-known low preference of the *endo* vs. *exo* transition state shown by IDA when a three-atom ansa is involved (Scheme 6.15).

[2+2+2] Cycloadditions and Cycloreversions

This topic was considered in detail by Woodward and Hoffman (*14*). Two allowed thermal processes [$\pi 2s+\pi 2s+\pi 2s$] and [$\pi 2s+\pi 2a+\pi 2a$] were found to be conceivable within the conservation of orbital symmetry. The majority of these cycloadditions are assumed to involve a dipolar intermediate, and FMOs are involved in the regiochemistry of attack.

Addition of 1,1-diethoxyethene to stypandrone (**1192a**: X = H, R_1 = OH, R_2 = Ac, R_3 = Me) and 3-bromo-5-methyl-7-methoxy-1,4-naphthoquinone (**1192b**) probably proceeds through a dipolar intermediate of the type **1193**, and gives two adducts: **1194a** (R_1 = OH, R_2 =COOMe, R_3 = Me) and **1194b** (R_1 = Me, R_2 = H, R_3 = OMe) in high yields. Hydrolysis of these adducts gave the insect pigments acetylemodin (*15*) (**1195**) and deoxyerythrolaccin (*16*) (**1196**) (Scheme 6.16).

As an example of [2+2+2] cycloreversion we report an interesting synthesis of *cis*-civetone (*17*) (**1203**). Diketone **1198** obtained from the diester **1197** was converted to dihydrazone **1199** and was oxidized. We are interested in this last stage because the oxidation to acetylene **1202** (which is then hydrogenated to **1203** with molecular hydrogen and Lindlar catalyst) can proceed through **1200** and **1201.** The final step is a [2+2+2] heterocycloreversion (Scheme 6.17).

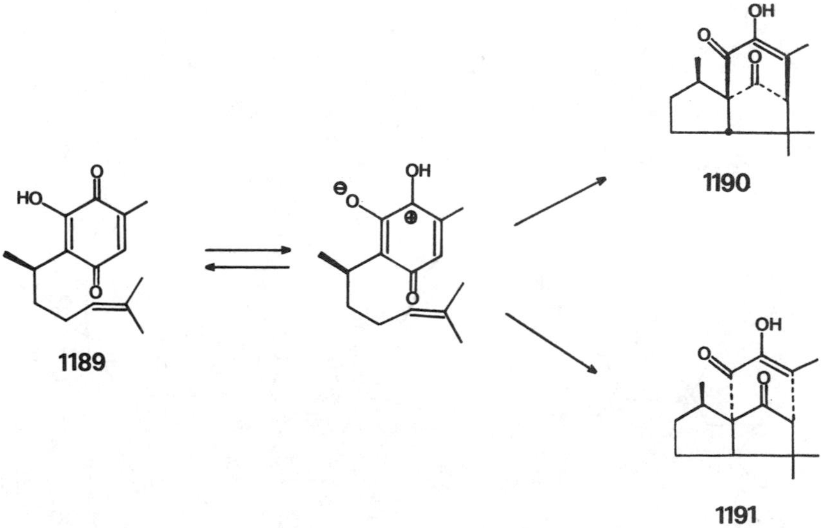

1189

1190

1191

Scheme 6.15

Scheme 6.16

Scheme 6.17

Literature Cited

1. Mukherjee, D.; Dunn, L. C.; Houk, K. N. *J. Am. Chem. Soc.* **1979**, *101*, 251.
2. Noyori, R. *Acc. Chem. Res.* **1979**, *12*, 61.
3. Hayakawa, H.; Yokoyama, K.; Noyori, R. *J. Am. Chem. Soc.* **1978**, *100*, 1791.
4. Noyori, R.; Shimizu, F.; Fukuta, K.; Takaya, H.; Hayakawa, Y. *J. Am. Chem. Soc.* **1977**, *99*, 5196.
5. Hayakawa, Y.; Shimizu, F.; Noyori, R. *Tetrahedron Lett.* **1978**, 993.
6. Noyori, R.; Nishizawa, M.; Shimizu, F.; Hayakawa, Y.; Maruoka, K.; Hashimoto, S.; Yamamoto, H.; Nozaki, H. *J. Am. Chem. Soc.* **1979**, *101*, 220.
7. Noyori, R.; Makino, S.; Okita, T.; Hayakawa, Y. *J. Org. Chem.* **1975**, *40*, 807.
8. Hayakawa, Y.; Baba, Y.; Makino, S.; Noyori, R. *J. Am. Chem. Soc.* **1978**, *100*, 1786.
9. Noyori, R.; Sato, T.; Hayakawa, Y. *J. Am. Chem. Soc.* **1978**, *100*, 2561.
10. Chidgey, R.; Hoffmann, H. M. R. *Tetrahedron Lett.* **1977**, 2633.
11. Büchi, G.; Chu, P. S. *J. Am. Chem. Soc.* **1979**, *101*, 6767.
12. Büchi, G.; Mak, C. P. *J. Am. Chem. Soc.* **1977**, *99*, 8073.
13. Walls, P.; Padilla, J.; Joseph-Nathan, P.; Giral, F.; Romo, J. *Tetrahedron Lett.* **1965**, 1577.
14. Woodward, R. B.; Hoffmann, R. *Angew. Chem., Int. Ed. Engl.* **1969**, *8*, 81.
15. Cameron, D. W.; Crossley, M. J.; Feutrill, G. I. *J. Chem. Soc., Chem. Commun.* **1976**, 275.
16. Cameron, D. W.; Crossley, M. J.; Feutrill, G. I.; Griffith, P. G. *J. Chem. Soc., Chem. Commun.* **1977**, 297.
17. Tsuji, T. *Pure Appl. Chem.* **1979**, *51*, 1235.

7

Sigmatropic Rearrangements

A sigmatropic rearrangement involves the migration of a σ bond along a conjugated π system to a new position. In a reaction of $[i,j]$ order, the new position is $(i-1)$ and $(j-1)$ away from the original position.

If the migrating group is chiral, it can retain or invert its configuration. These possibilities, together with the alternative antarafacial or suprafacial stereochemistry on the π system are dependent on the number of electrons involved in the process and on the experimental conditions. These selection rules (*1*) are summarized in Table 7.1. For chiral groups the sterically favored suprafacial migration is indicated first, but an antarafacial migration is also allowed, taking into account an opposite configuration of the migrating group.

Figure 7.1 shows two general examples of a chiral group that because of the symmetry of the orbitals involved can migrate to the position indicated. The configuration of the chiral group is retained with suprafacial migration (a), and inverted with antarafacial migration (b).

Sigmatropic rearrangements are useful tools for the construction of quaternary carbon centers, and a recent review (*2*) covers this field. We consider the $[1,m]$, $[m,m]$, and $[m,n]$ processes in this chapter.

[1,m] Sigmatropic Processes

This topic was first approached by Havinga and coworkers (*3, 4*) during his studies on vitamin D_2 chemistry.

Ergosterol (**1204**) isomerizes photolytically to precalciferol (**1205**) through an electrocyclic process that will be considered in Chapter 9. On heating, an equilibrium is established between **1205** and calciferol (**1138**) (20:80). The migration of hydrogen from the methyl group (C1) to the end of the polyene (C7) occurs with rearrangement of its bonds through a [1,7] sigmatropic process (Scheme 7.1).

If we consider the interaction between the LUMO of the π system, which acts as acceptor, and the HOMO of the σ fragment, which acts as donor (the same result is obtained with the opposite interaction), only an antarafacial migration is possible for orbital symmetry reasons

0065-7719/83/0267 $16.80/1

Table 7.1

Selection Rules for [*i,j*] Sigmatropic Rearrangements

R	*i+j*	*Thermal Conditions*	*Photochemical Conditions*
H	$4n$	antara	supra
	$4n+2$	supra	antara
Chiral	$4n$	supra-inversion	supra-retention
		antara-inversion	antara-inversion
(≠ H)	$4n+2$	supra-retention	supra-inversion
		antara-inversion	antara-retention

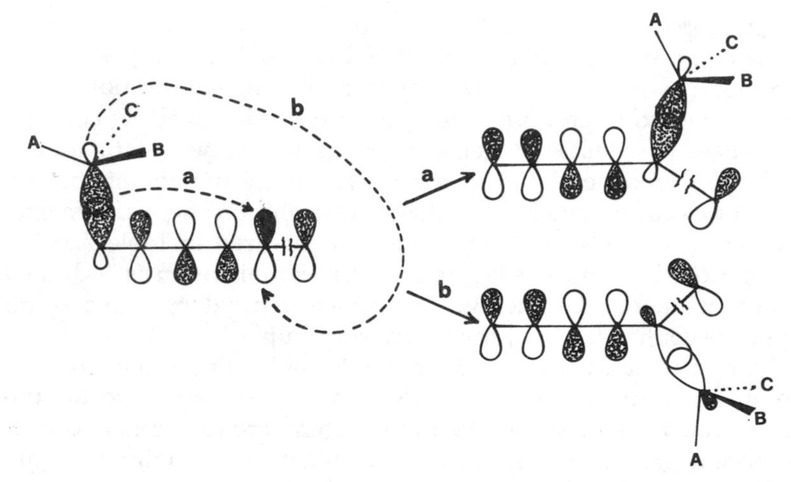

Figure 7.1 [1,5] Sigmatropic shift: suprafacial with retention of configuration (a), antarafacial with inversion of configuration (b).

(Figure 7.2). We have no direct proof that the process, involving eight electrons, is antarafacial, but a [1,7] sigmatropic hydrogen shift occurs easily in open-chain *E,E*-trienes only. These compounds can adopt a helical shape with the migrating group in a proximate position under the accepting carbon.

Barton and coworkers (*5*) found that irradiation of methyl dehydroursolate acetate (**1206a**) affords a triene **1207a** analogous to precalciferol. This triene isomerizes thermally to a further triene **1208a**, which is comparable with calciferol. Similar behavior was discovered by Jeger et al. (*6*) in 1946 for dehydro-α-amyrin acetate (**1206b**), which changes to the *lumi*-derivative **1207b** and isomerizes thermally to the so-called *pyro-lumi*-isomer **1208b** (Scheme 7.2).

All these examples, and particularly those in the vitamin D field (*7*) (the history of which is entwined with that of FMOs), constitute some of the fundamental points that led to the development of the Woodward–Hoffman theory on the conservation of orbital symmetry.

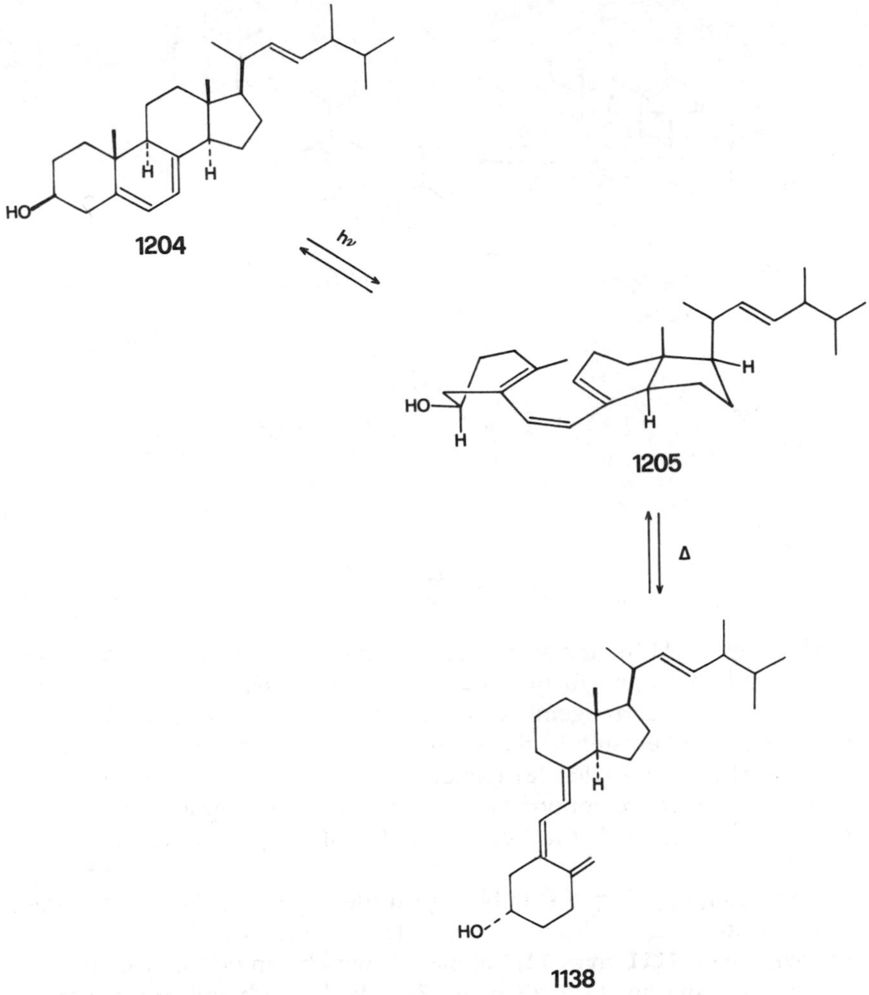

HO H H

1204

$h\nu$

HO H H H

1205

Δ

H HO

1138

Scheme 7.1

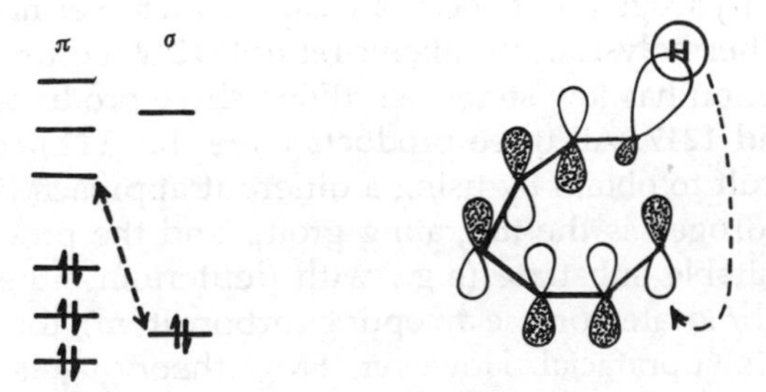

Figure 7.2. FMO interaction of thermal antarafacial [1,7] sigmatropic shift.

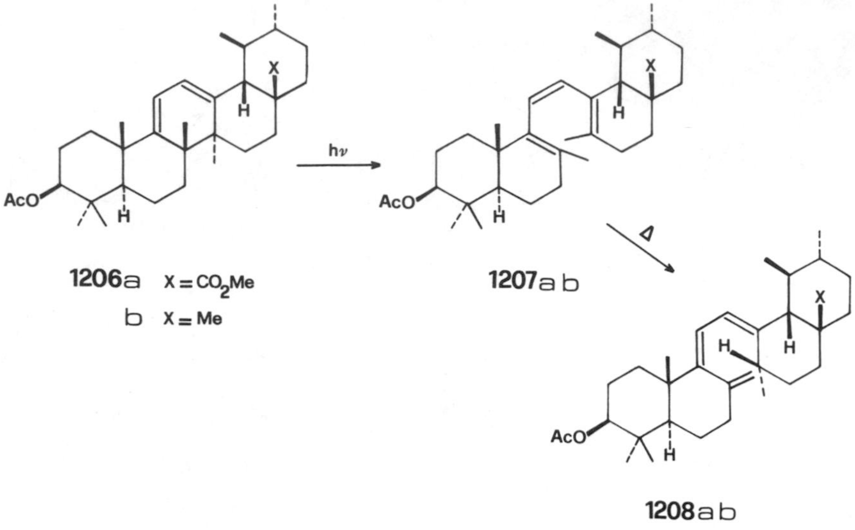

1206a X = CO_2Me
b X = Me

1207a b

1208a b

Scheme 7.2

The thermal [1,3] sigmatropic rearrangement is rarely observed and is generally assumed to proceed in a stepwise fashion. A significant example is the rearrangement of the tricyclic ketone **1209** to **1210**; this reaction is the key step in the synthesis of the plant growth promoter (±)-*cis*-sativenediol (**185**) (*8*) (Scheme 7.3). The rearrangement occurs suprafacially with retention of configuration of the migrating group and is not consistent with the MO symmetry of a process involving four electrons.

The ring opening of suitably substituted cyclopropanes sometimes can be explained in different terms. The pyrolysis of dimethyl cyclopropyl ketone **1211** gives **1212** alone (*9*), which can be easily converted to dihydrojasmone (**1213**) (Scheme 7.4). If the carbonyl group participates in the rearrangement, a homo [1,5] sigmatropic hydrogen shift can be proposed. Because six electrons are involved, the reaction occurs suprafacially.

The [1,5] hydrogen shift occurs easily under thermal conditions. For example, thermolysis of the allenic retinol (**1214**) occurs (*10*) at about 69 °C; the reaction has low stereoselectivity; three products are formed: **1215**, **1216**, and **1217**. All three products have the (11*Z*) configuration, which is difficult to obtain by using a different approach (Scheme 7.5).

When hydrogen is the migrating group and the process does not occur on a suitable substrate (e.g., with deuterium either as the migrating group or located on the accepting carbon atom), there is no proof that the shift is suprafacial. However, FMO theory takes into account the interaction between the LUMO of the π system and the HOMO of the σ fragment (Figure 7.3), and thus predicts a suprafacial reaction. An

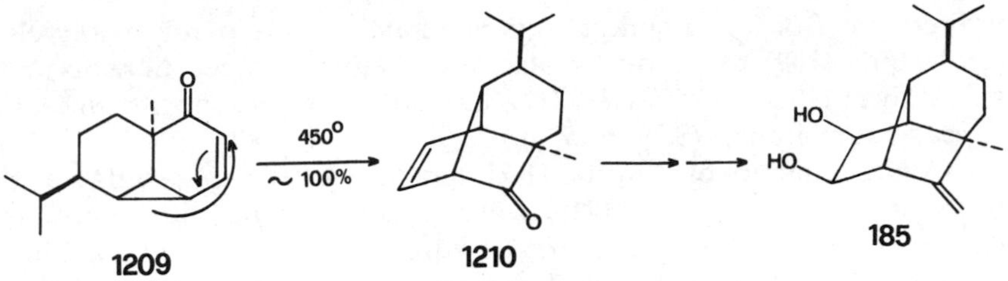

450°
~ 100%
1209
1210
185

Scheme 7.3

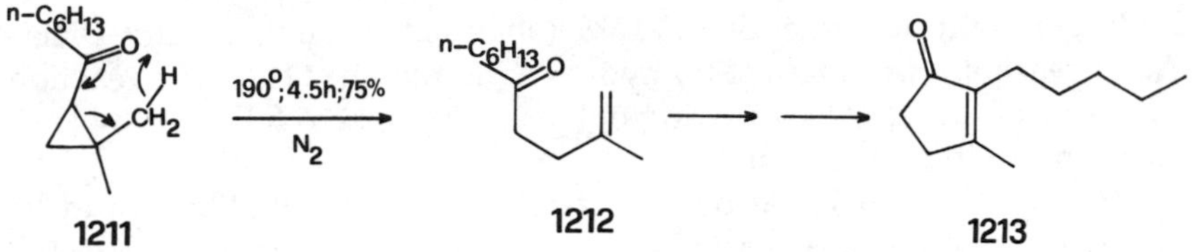

n-C_6H_{13}
190°; 4.5h; 75%
N_2
1211
1212
1213

Scheme 7.4

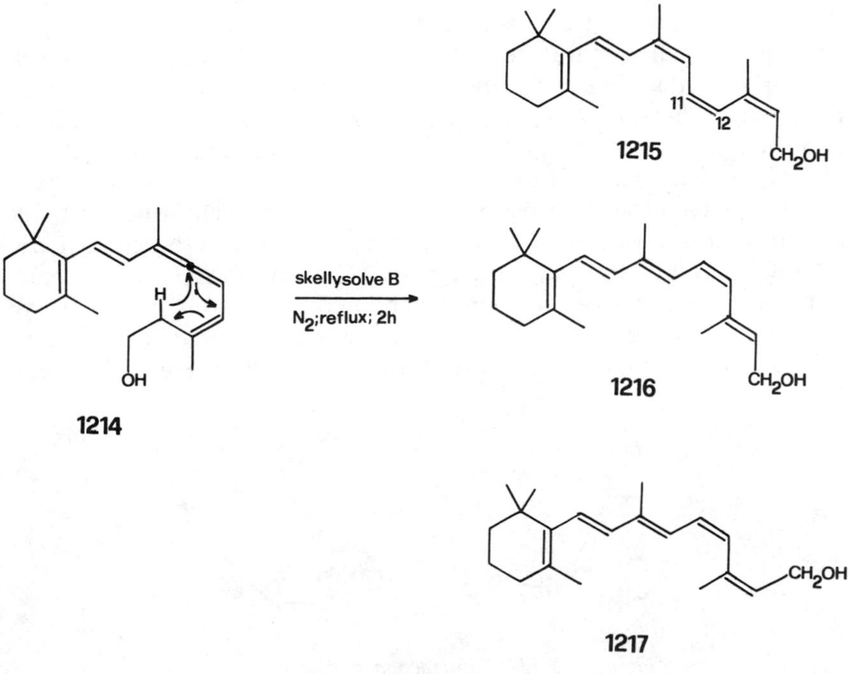

1215
CH_2OH
skellysolve B
N_2; reflux; 2h
1214
1216
1217

Scheme 7.5

indirect proof of this assumption is illustrated by the pyrolysis of vellerolactone (**108**) to pyrovellerolactone (**1218**). This reaction occurs smoothly at 110 °C in 90% yield (*11*), as would be expected with an easy suprafacial migration (Scheme 7.6).

A nice sequence of [1,5] and [1,7] sigmatropic hydrogen shifts starts from vinylallene derivative **1219**, which upon refluxing with isooctane (*12*) gives about 31% of 3-deoxy-1-hydroxy vitamin D_3 (**1220**). This product derives from a [1,5] shift; minor amounts of **1223** and **1224** (4% and 5% yield, respectively) are also formed. Those two products result from two subsequent [1,7] shifts of **1221** and **1222** (Scheme 7.7). Different yields or different isomers are obtained from stereoisomeric allenes.

The vitamin D field offers a host of examples of [1,7] sigmatropic hydrogen shifts. Irradiation of **1225** (obtained through a retro-Diels–Alder reaction), produces (22*S*)-hydroxy vitamin D_4 (**1227**). The reaction proceeds via an electrocyclic ring opening giving **1226** followed by [1,7] hydrogen shift (*13*) (Scheme 7.8).

(3*S*,10*S*)(*Z*,*Z*)-9,10-Secocholesta-5,7,14-triene-3-ol (**1229a**) and its (10*R*)-isomer **1229b** are obtained by heating a solution of *cis*-isotachysterol (**1228**). The reaction involves activation energies of about 23 kcal/mol and entropies of activation of −16 to −17 e.u., which are fully consistent with an intramolecular concerted process (*14*) (Scheme 7.9).

Examples involving longer π systems are known in the corrin field. Eschenmoser (*15*) found that the cyclization of secocorrin (**1230**) to corrin (**1232**) can be realized by using as the key step a light-induced [1,16] sigmatropic hydrogen transfer. Thus formed is the diradical **1231**, which undergoes antarafacial electrocyclic 1,15-ring closure to the desired product (*16*) (Scheme 7.10). The process occurs on the planoid complex and the nature of the metal ion is crucial—the reaction occurs with Li, Na, Mg, Ca, Zn, Cd, Pd, or Pt.

The antarafacial hydrogen shift from **1230** to **1232**, under photochemical conditions, can be explained by taking into account the 'LUMO'$_{polyene}$/LUMO$_{fragment}$ interaction. (The 'HOMO'/HOMO interaction gives the same result.) Some simple considerations of the orbital symmetry, represented in Figure 7.4, show that the process, which involves 18 electrons, can only occur antarafacially. The role of the metal

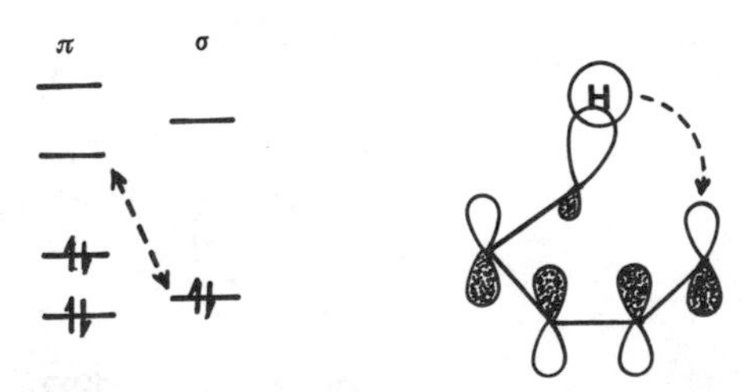

Figure 7.3 FMO interaction of thermal suprafacial [1,5] sigmatropic shift.

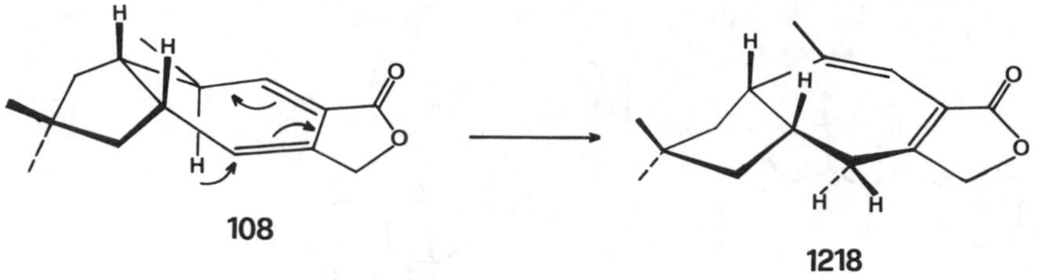

108 → 1218

Scheme 7.6

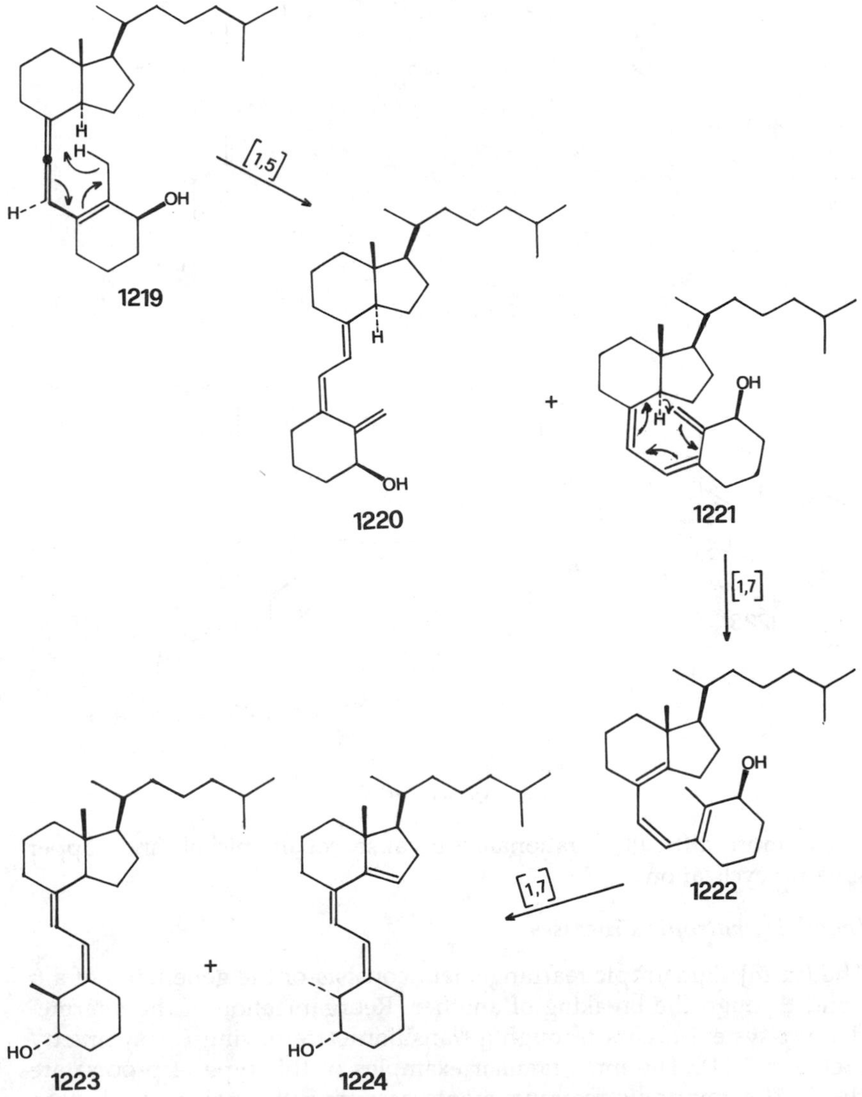

1219 —[1,5]→ 1220 + 1221

1221 —[1,7]→ 1222

1222 —[1,7]→ 1223 + 1224

Scheme 7.7

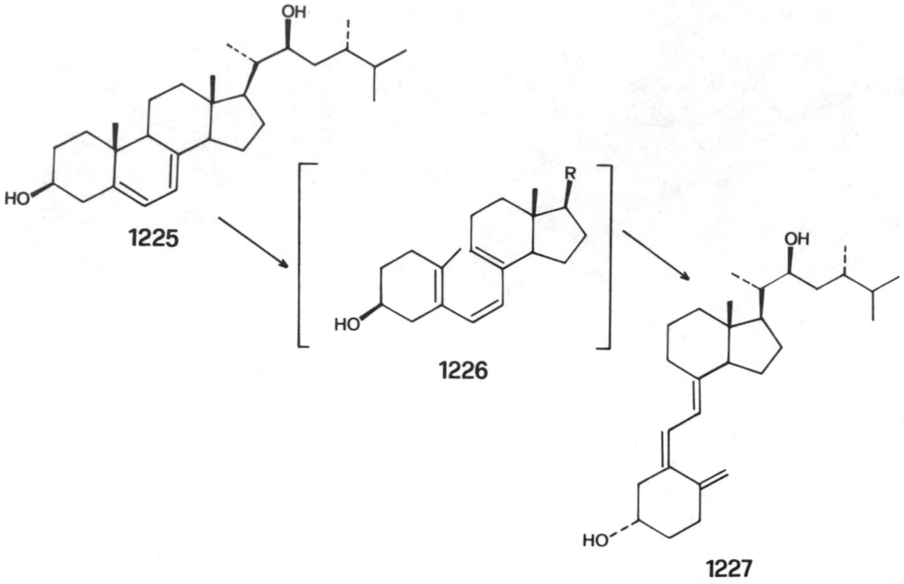

HO OH 1225 R HO 1226 OH HO 1227

Scheme 7.8

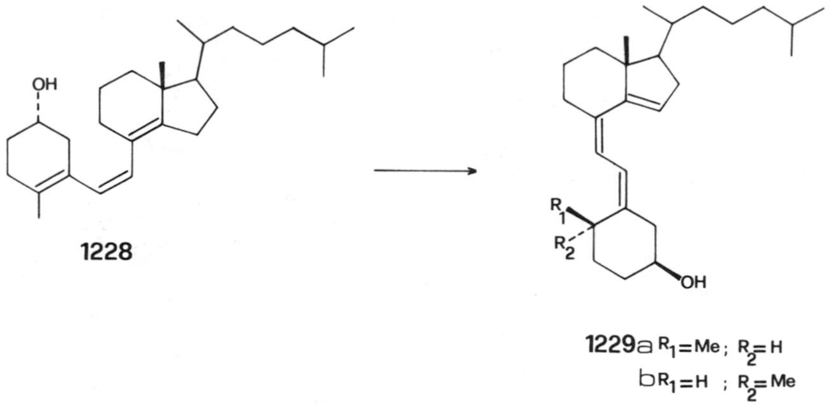

OH 1228 R1 R2 OH

1229a R_1=Me; R_2=H
b R_1=H; R_2=Me

Scheme 7.9

ion is more difficult to rationalize because cobalt, nickel, and copper give no cyclization.

[m,m] Sigmatropic Processes

The [m,m] sigmatropic rearrangement consists of the generation of a σ bond through the breaking of another. Reorganization of the intermediate π system occurs through a transition state having D_{2h} symmetry (Scheme 7.11). The most familiar examples of this type of process are the [3,3] sigmatropic rearrangements occurring through a six-electron transition state.

Cope Rearrangement. The well known Cope rearrangement is a [3,3] sigmatropic shift involving six carbon atoms (Scheme 7.12).

The interaction between the HOMO of the a,b,c (or a′,b′,c′) system with the LUMO of the a′,b′,c′ (or a,b,c) system can be used to explain the rearrangement (Figure 7.5). In principle, either a chair (Figure 7.5A) or a boat (Figure 7.5B) transition state is possible, but the former does not involve the destabilizing nonbonding interaction between b and b′ that the latter configuration does.

The same result is reached when the transition state of the Cope rearrangement is built up from the MOs of two identical allyl radicals (Figure 7.6). The interactions Ψ_1–Ψ_2' and Ψ_2–Ψ_3' (or Ψ_1'–Ψ_2 and Ψ_2'–Ψ_3) are symmetry forbidden; Ψ_1–Ψ_1' and Ψ_3–Ψ_3' give no energy gain. The interaction between Ψ_2 and Ψ_2' can give either no energy gain (if spins are parallel) or a large energy gain (if antiparallel). However, this logic is irrelevant in terms of nonbonding interactions because wave functions have nodes in b and b′ (Figure 7.6A). Hence, in spite of the large energy separation, a significant energy gain occurs from the overlap between Ψ_1 and Ψ_3' (or between Ψ_1' and Ψ_3). A suprafacial migration through a chair transition state is favored over a boat (Figure 7.6B and 7.6C, respectively) for the nonbonding repulsion between b and b′.

These findings explain the experimental observation that in the thermal rearrangement of **1233**, a chair transition state is preferred over a boat one by 5.7 kcal/mol. The ratio of the products **1234** and **1235** is 99.7:0.3 (*17*). (Scheme 7.13).

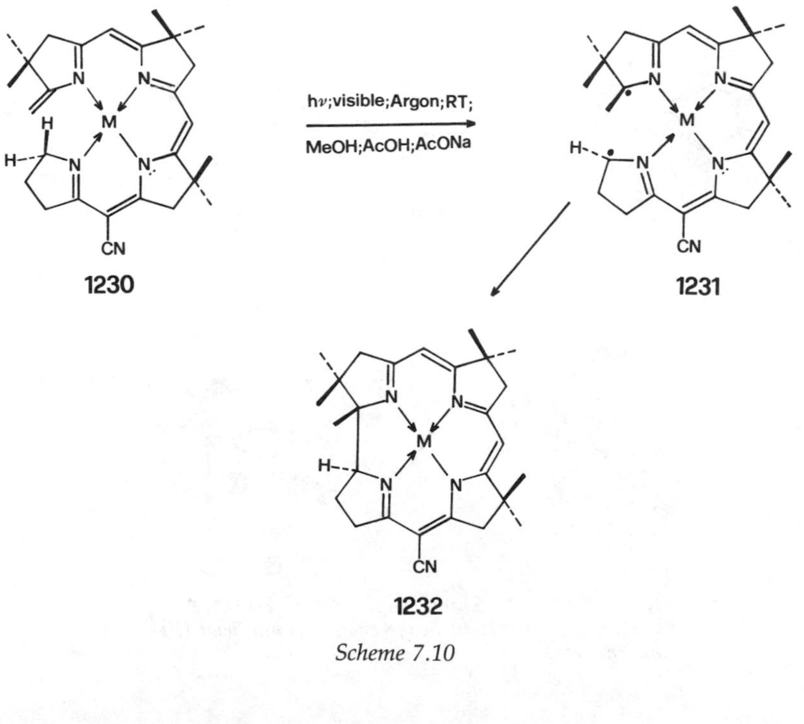

hν;visible;Argon;RT;
MeOH;AcOH;AcONa

1230 1231 1232

Scheme 7.10

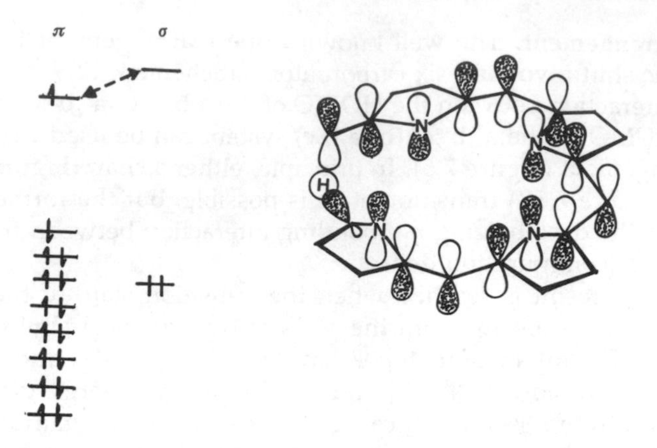

Figure 7.4 FMO interaction of photochemical antarafacial [1,16] hydrogen shift on ***1230.***

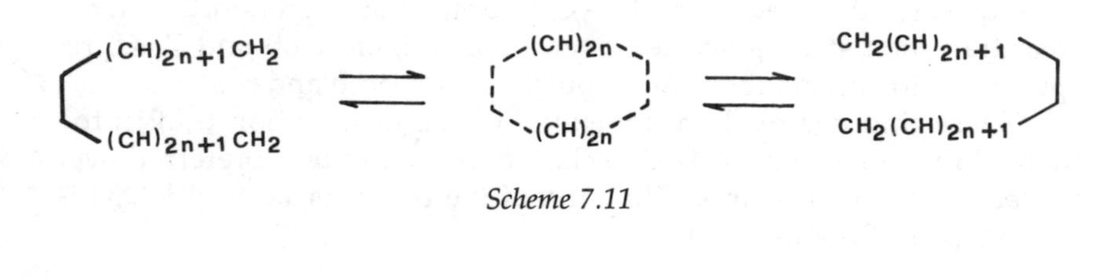

Scheme 7.11

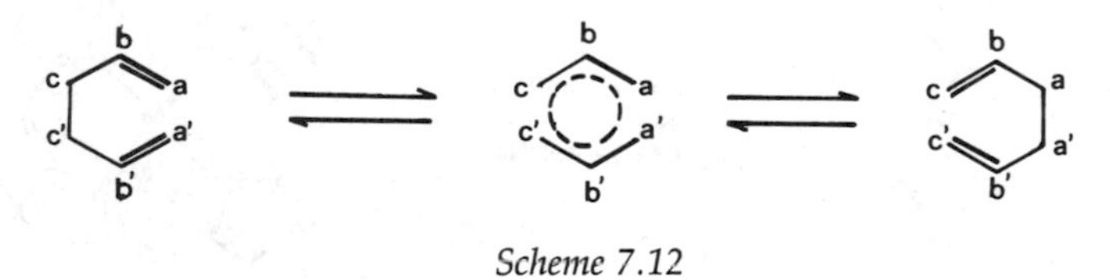

Scheme 7.12

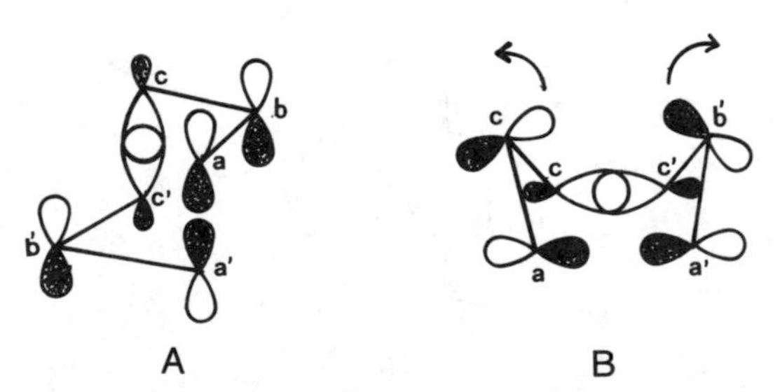

Figure 7.5 [3,3] Sigmatropic shift: $HOMO_{abc}$/$LUMO_{a'b'c'}$ interaction in the chair (A) and boat (B) conformations.

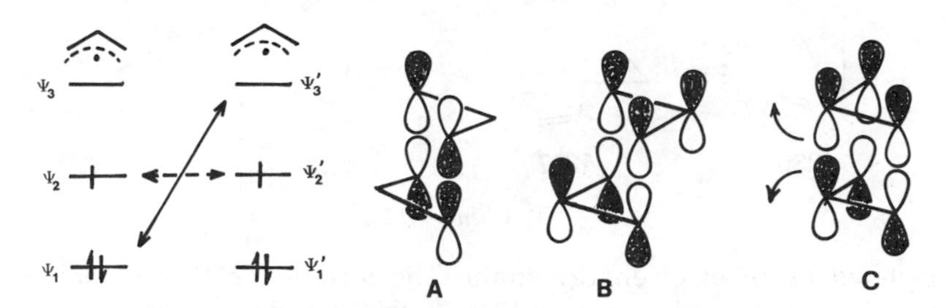

Figure 7.6. FMO interactions in the Cope rearrangement: Ψ_2-Ψ_2' *(A) and* Ψ_1-Ψ_3' *(B and C, chair and boat, respectively).*

From these MO considerations, the active orbitals involved in the Cope rearrangement are the σ orbital of the single bond to be broken and the π^*_g orbital (*18*). If we plot the experimental activation energies versus the calculated orbital energy differences, a linear correlation is obtained. The perturbation approach allows us to estimate the early part of the path leading to the transition state; therefore the relationship between calculated energy gain and experimental energy of activation follows immediately. The higher energy barrier occurs for 1,5-hexadiene (**1236**) (33.5 kcal/mol) (*19*); *cis*-1,2-divinylcyclobutane (**1237**) (23.1 kcal/mol) (*20*) and *cis*-1,2-divinylcyclopropane (**1238**) (19.4 kcal/mol) (*21*) show a significant lowering of $\Delta E^{\neq}$ due to relief of strain energy. An easier process occurs in the case of 1,5-hexadiene (**1239**) (18.2 kcal/mol) (*22*) bearing an oxyanion on C3 (the so-called oxy-Cope process) (Scheme 7.14).

Three conditions theoretically can influence the process: (1) the relief of energy strain in the substrate; (2) the presence of substituents that lower the energy gap between the interacting orbitals; and (3) the favorable overlap of the interacting orbitals [**1237** has a $\Delta E^{\neq}$ of 23.1 kcal/mol; the energy of activation of its *trans* isomer is 34.0 kcal/mol (*20*)].

The syntheses of (±)-dictyopterene C′ (**1241**), a C_{11}-hydrocarbon isolated from the essential oil of algae of the genus *Dictyopteris*, and (±)-β-himachalene (**190**) are two nice examples of Cope rearrangements fa-

1233 —Δ→ [ ] → 1234, 1235

Scheme 7.13

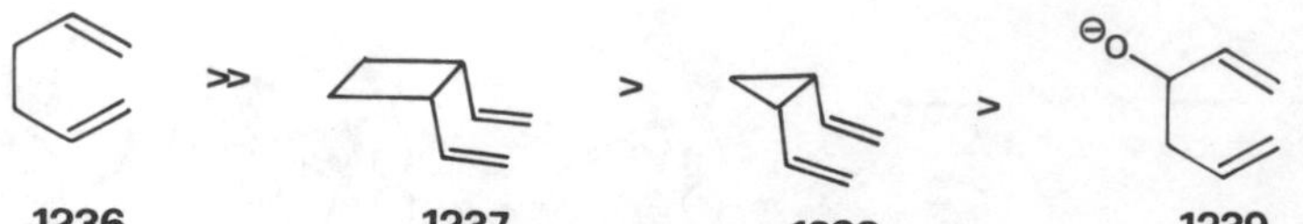

1236 ≫ 1237 > 1238 > 1239

Scheme 7.14

cilitated by relief of energy strain. The former (**1241**) is obtained by pyrolysis of *cis*-cyclopropanes **1240** [both (*E*) and (*Z*) isomers] at 80 °C (*23*). The *trans*-substituted β-cyclopropyl enone **1242** is refluxed to give **1243**, which in a few steps affords the target sesquiterpenoid (**190**) (*24*) (Scheme 7.15).

The effect of substituents is documented in the Cope rearrangement of **1244** (100–110 °C) to give **1245**, which is then transformed into **1246**. This product is a simple model compound containing the ABE ring system of hetisine-type diterpene alkaloids (*25*) (Scheme 7.16). Three electron-attracting groups, all on the same fragment (atoms a,b,c), force the fragment to behave as an acceptor, thus lowering the energy separation between $HOMO_{a'b'c'}$ and $LUMO_{abc}$.

However, the most significant substituent effect is seen in the oxy-Cope rearrangement for which the simplest substrate is **1239**.

Oxy-Cope Rearrangement. The conversion by heating of 3-oxy substituted 1,5-hexadienes (or hexaenynes) is dependent upon several conditions, all in some way related to the formation of the anion. The nature of the solvent (*26*), the presence of a base, and the eventual presence of a crown ether (22) all affect the conversion.

In the reaction product, the hydroxy group generates an enol, which tautomerizes to a carbonyl group. An example of this tautomerization is the synthesis of 3,7-dimethylpentadec-2-yl acetate (**1248**), the sex pheromone of the pine sawfly *Neodiprion lecontei*, which is obtained by pyrolyzing the dienynol **1247** (*27*) (Scheme 7.17).

The participation of a triple bond is useful in the formation of α,β-

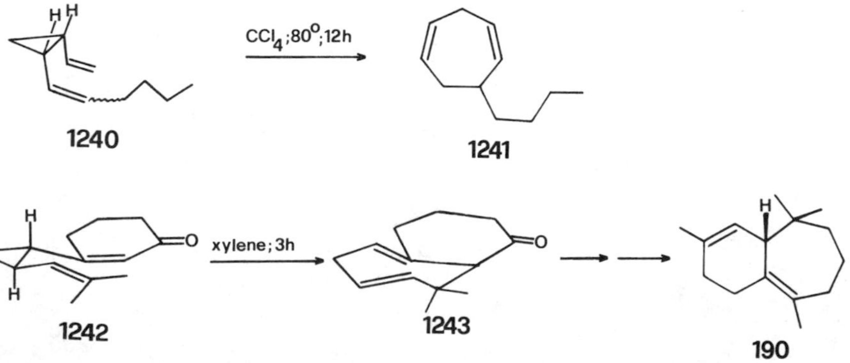

1240 —CCl_4;80°;12h→ 1241

1242 —xylene; 3h→ 1243 → → 190

Scheme 7.15

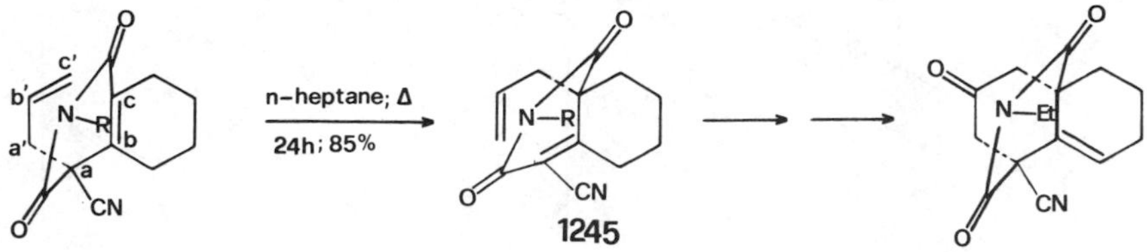

n–heptane; Δ
24h; 85%

1244 1245 1246

Scheme 7.16

unsaturated carbonyl compounds, as shown in the industrial process for the production of pseudoionone (**1251**) by oxy-Cope rearrangement of **1249** (*28*). The modification using *N*-methylpyrrolidone as solvent in the presence of a trace amount of halogen is useful because the side process involving an ene-reaction is avoided. Therefore, the double bond migration in the primary product **1250** allows the overall process to be performed in one reaction vessel (*29*) (Scheme 7.18).

Sometimes a transition state with a rigid configuration gives rise to a single specific isomer. Heating **1252** forms **1253**, the *cis* ring junction of which is retained in the final product **1254,** previously believed to be the alkaloid cannivonine (*30*). Unfortunately, the synthetic material was found to be different from the natural product (Scheme 7.19).

Similarly, the oxy-Cope rearrangement of **1255** in tetrahydrofuran in the presence of sodium hydride gives **1256**. The *cis* ring fusion of **1256** is crucial for the further transformation into coronafacic acid (**514**) (*31*) (Scheme 7.20).

Analogously, when **1257** is heated at 300 °C, the expected suprafacial [3.3] sigmatropic shift produces the octalone **1258** bearing four chiral centers. Three of the these chiral centers are necessary for the subsequent transformation into (±)-α-amorphene (**1259**) (*32*) (Scheme 7.21).

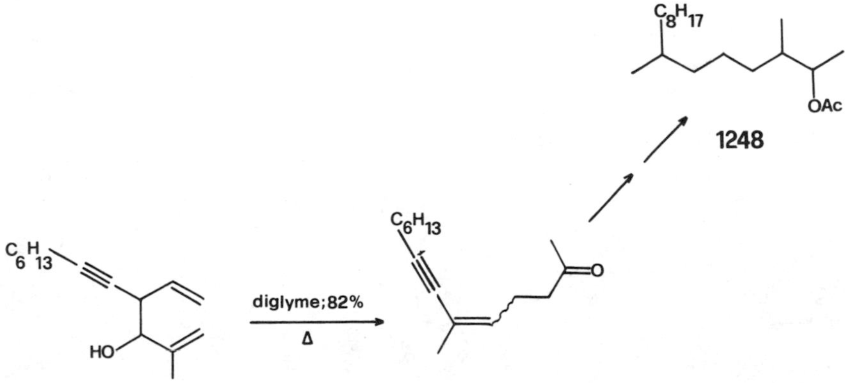

diglyme; 82%
Δ

1247 1248

Scheme 7.17

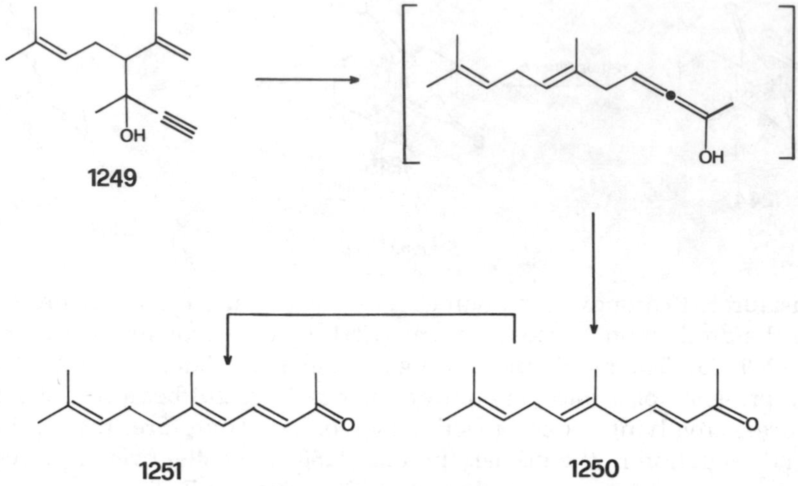

OH
1249
OH
1251
1250

Scheme 7.18

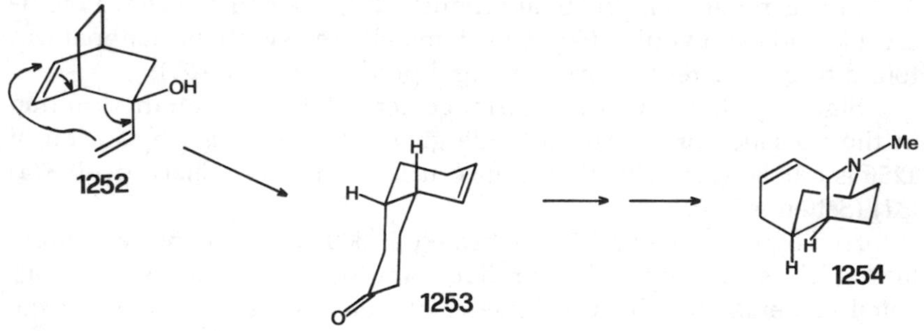

OH
1252
H
H
O
1253
Me
N
H
H
1254

Scheme 7.19

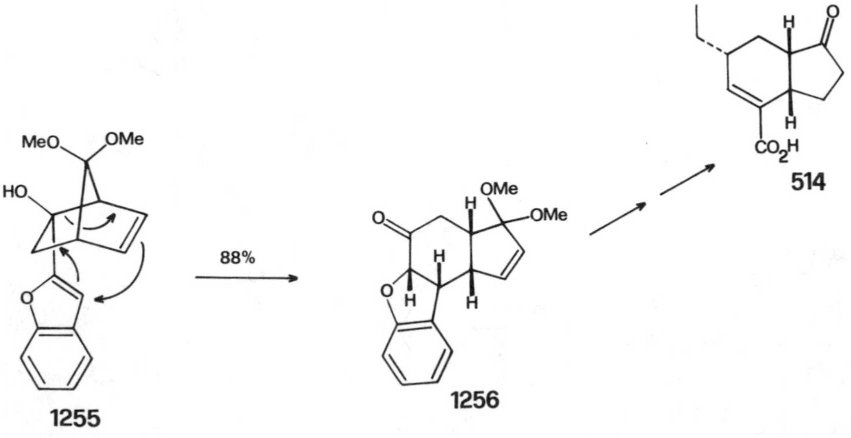

MeO OMe
HO
O
1255
88%
O
H OMe
OMe
H
O
H H
1256
H O
H
CO_2H
514

Scheme 7.20

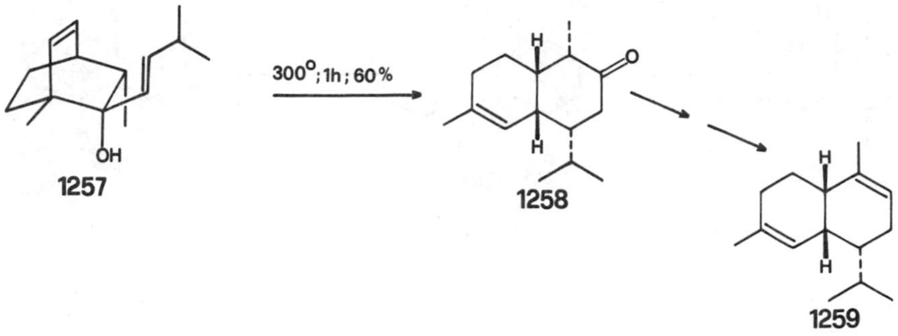

Scheme 7.21

When the transition state is not rigid, a chair-like conformation is strongly preferred, because a highly stereoselective rearrangement results. Thus **1260** (K salt from KH/THF/18-crown-6; 1 h, 70 °C) generates **1261**, retaining the (*E*)-configuration of the double bond and the α position of the isopropyl group. The product **1261** can be further elaborated to (±)-periplanone B (**1262**), the sex excitant pheromone of the American cockroach (*33*) (Scheme 7.22).

Analogously, the *erythro* configuration of **1264**, obtained from **1263** in a 91:9 ratio with its *threo* isomer through a chair-like transition state, allowed a new synthesis of (±)-*erythro*-juvabione (**1265**) (*34*), a sesquiterpene originally isolated from *Abies balsamea* (L.) Miller (Scheme 7.23).

The oxy-Cope process is important in the synthesis of prostaglandins; the Corey aldehyde (**32**) was obtained by treating *cis*-2,4,7-cyclononatrienol (**1266**) at room temperature in tetrahydrofuran in the presence of 1.2 equiv of potassium hydride (*35*). The process is also critical to the stereocontrolled introduction of the steroid side chain; a 94% yield of the β-isomer **1268** [subsequently transformed in six steps into (20*R*)-desmosterol (**1269**)] is obtained by refluxing **1267** in dioxane in the presence of potassium hydride (*36*) (Scheme 7.24).

GERMACRANE–ELEMENE REARRANGEMENT. Germacrane-type sesquiterpenes easily undergo Cope rearrangements to give elemene-type derivatives; the rearrangement is reported to proceed as follows:

However, some germacranolides undergo transanular cyclization; costunolide (**1270**) is a typical example as well as balchanolide **1271** and arctiopicrin (**1272**). Some other germacranolides do not undergo cyclization [e.g., salonitolide (**1273**), artemisiifolin (**1274**), and eupatoriopicrin (**1275**)] (*37*) (Scheme 7.25). Sometimes an equilibrium is reached, as is the case for linderalactone (**1276**)–isolinderalactone (**1277**) (ratio 2:3) and dihydrotamaulipin-A-acetate (**1278**), which gives **1279** (ratio 1:1) (*38*) (Scheme 7.26).

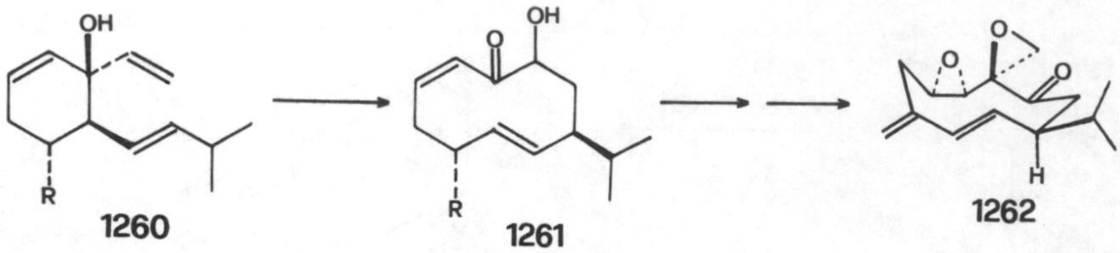

Scheme 7.22

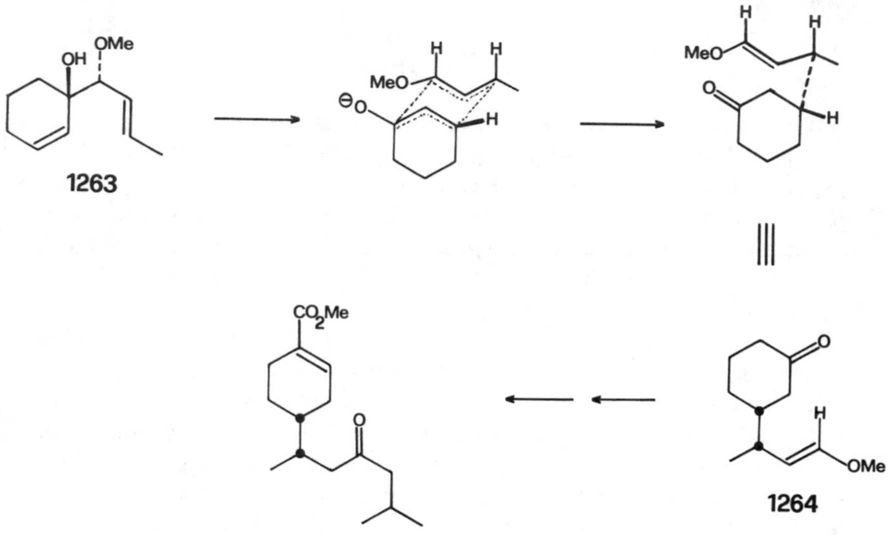

Scheme 7.23

These rearrangements demonstrate that the configuration of the substrate is critical. The stereochemical aspects as well as the conformational requirements of the transition state must be considered.

Let us look closely at the two last examples (Scheme 7.26), where the angular methyl becomes α in **1277** and β in **1279**. If we consider the conformations of the transition states **1280** and **1281**, respectively, it is not difficult to understand why the former generates an α-methyl group while the latter gives a β-isomer.

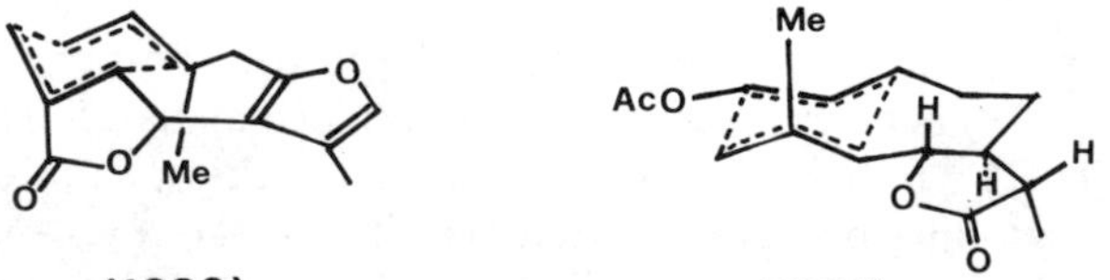

(1280) **(1281)**

The possibility of undergoing the elemene rearrangement seems to be limited to those germacranes that can adopt a TC (or parallel) or a

CC (or crossed) conformation (*39, 40*). Both configurations allow a good overlap of the interacting orbitals, the former through a boat-like transition state, and the latter through a chair-like transition state (Scheme 7.27).

Molecular mechanics calculations were carried out to evaluate the relative stabilities of the conformations of germacrene-A (**1282**) and germacrene-B (**1283**) (*41*). The former has the CC conformation more stable by 0.33 kcal/mol than the TC one; the latter has the TC one more stable

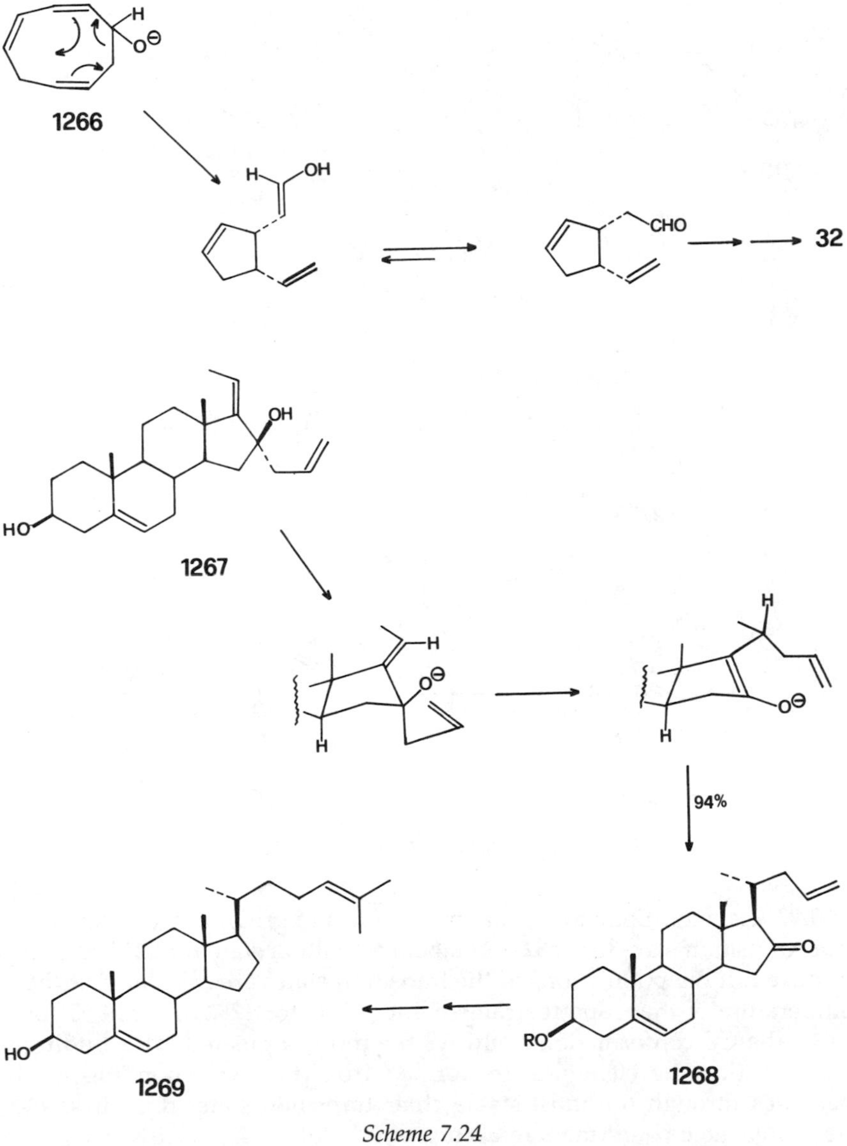

1266

CHO

32

OH

HO

1267

94%

RO

1268

HO

1269

Scheme 7.24

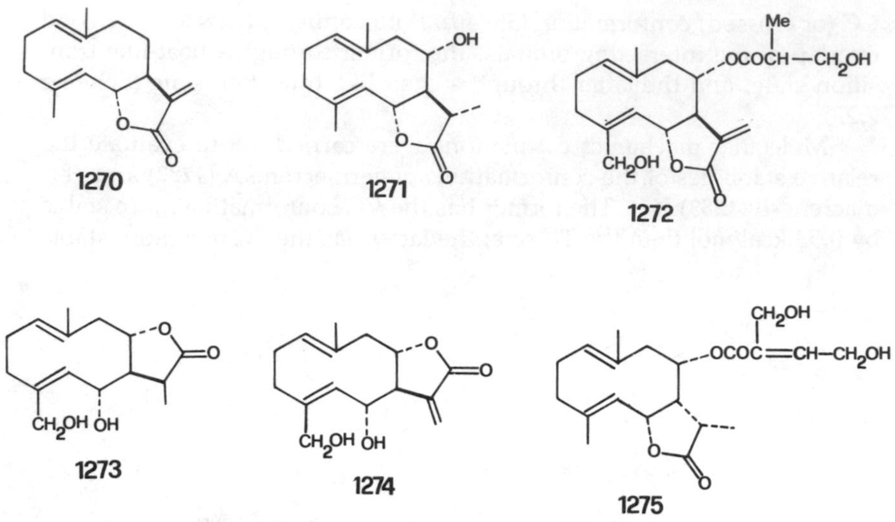
OH
Me
OCOCH—CH2OH
CH2OH
1270
1271
1272
CH2OH
OCOC=CH—CH2OH
CH2OH OH
CH2OH OH
1273
1274
1275

Scheme 7.25

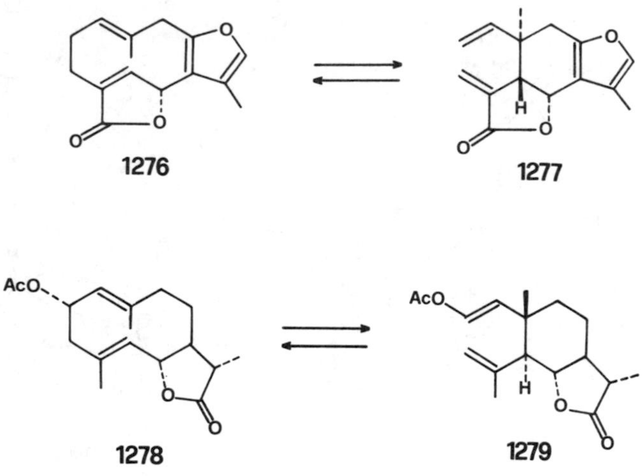
H
1276
1277
AcO
AcO
H
1278
1279

Scheme 7.26

by 0.22 kcal/mol. Both conformations are in the ground state. Hence a chair transition state for **1282** and a boat transition state for **1283** can be predicted. If the populations of the transition states are calculated at the temperature of the Cope rearrangement (25 °C for **1282** and 120 °C for **1283**), the CC conformation is always the most populated. This finding indicates that the elemenes are formed from the corresponding germacrenes through the most stable chair transition state, regardless of the more stable conformers in each ground state.

Some germacranolides do not undergo Cope rearrangement possibly because of the distortion of the medium-sized ring, which prevents overlap of the interacting orbitals. Nevertheless, it is difficult to rationalize the behavior of dihydrolaurenobiolide (**1285**), the sodium borohydride reduction product of laurenobiolide (**1284**) (*42*). This compound (**1285**) is isolated from the roots of *Laurus nobilis* L., and gives at 205 °C a reversible Cope rearrangement (**1285**:**1286** = 6:4) (Scheme 7.28). The stereochemical relationship between the substituents at C6, C7, and C8 of **1284** and those of **1273** and **1274** is evident. Furthermore, the nuclear Overhauser effect (NOE) and variable temperature ^{1}H NMR support CC (major) and TC (minor) conformations for **1284** (*43, 44*). Hence other explanations cannot be excluded.

Taking into account these points, several syntheses involving the described rearrangements will be considered. By treating **1287** with $Ni(CO)_4$ in dimethylformamide, pregeijerone (**1288**) is obtained. Rearrangement to (±)-geijerone (**1289**) (*45*), an elemene terpenoid isolated from the essential oil of *Juniperus communis* Linn., is spontaneous (Scheme 7.29).

Elemol (**1292**) was obtained by cyclizing **1290** to hedycaryol (**1291**) [in admixture with its (2Z) isomer], which was thermolyzed at 170 °C (*46*) (Scheme 7.30).

Starting from 3,5-dimethoxybenzoic acid methyl ester, after several steps, the mixture of epimeric lactones **1293a,b** was obtained. The mix-

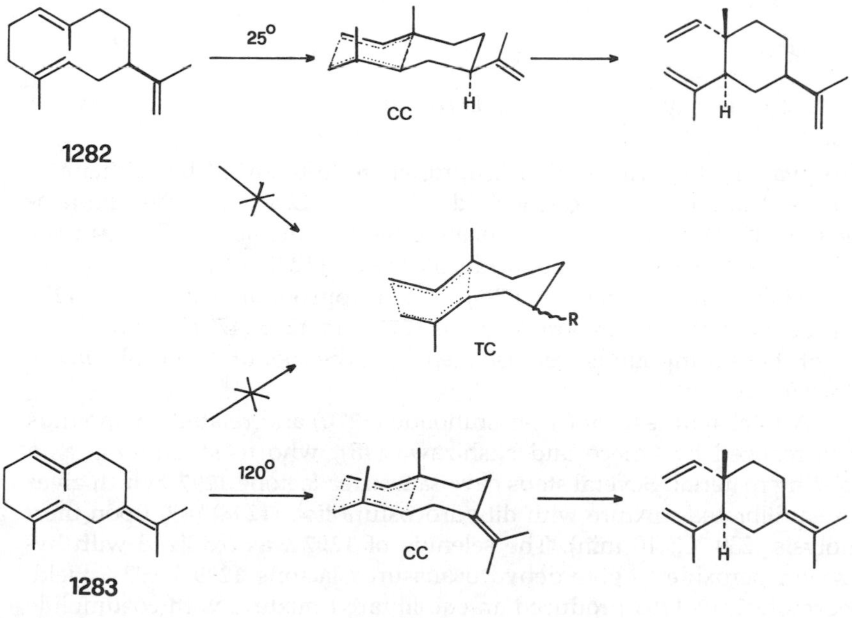

1282 25° CC H H

TC R

1283 120° CC H

Scheme 7.27

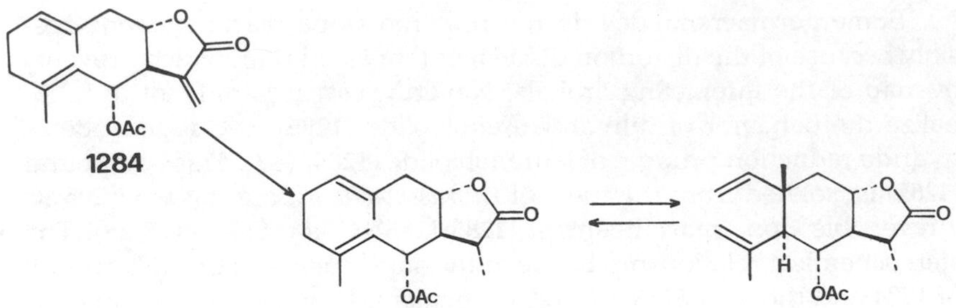

1284

1285

1286

Scheme 7.28

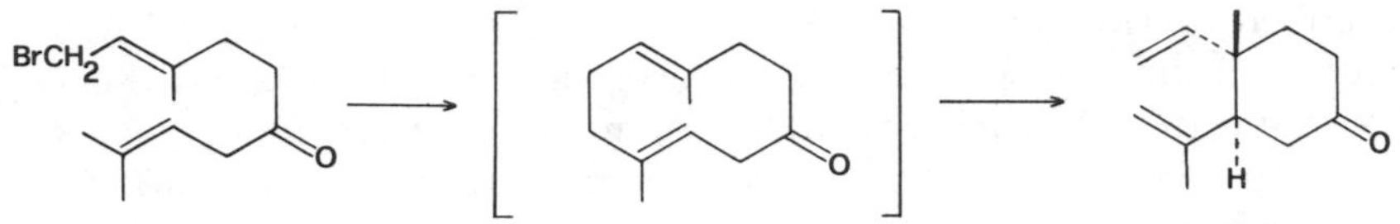

$BrCH_2$

1287

1288

1289

Scheme 7.29

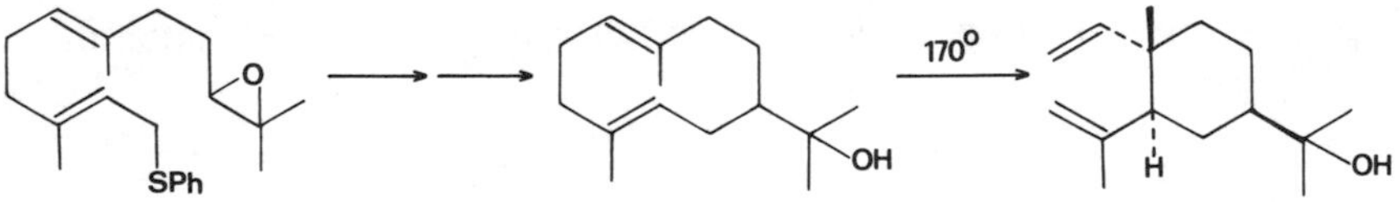

SPh

1290

1291

170°

1292

Scheme 7.30

ture was treated with lithium isopropylamide followed by Eschenmoser's salt at 20 °C to afford isolinderalactone (**1277**) and epiisolinderalactone (**1294**), respectively. By heating the mixture at 160 °C, **1294** rearranges irreversibly to neolinderolactone (**1295**); **1277** rearranges reversibly to linderalactone (**1276**). Under appropriate conditions, **1295** can be converted to an admixture of **1277** and **1276** (*47*) (Scheme 7.31). All of these compounds were isolated from the root of the shrub *Lindera strychnipolia* Vill.

A total synthesis of (+)-costunolide (**1270**) and related compounds was realized by Grieco and Nishizawa (*48*), who used santonin as a starting material. Several steps gave saussurea lactone **1297**, which gives an equilibrated mixture with dihydrocostunolide (**1298**) (*49*) upon thermolysis (230 °C, 10 min). The selenide of **1297** was oxidized with hydrogen peroxide to give dehydrosaussurea lactone **1299** in 93% yield; thermolysis of **1299** produced an equilibrated mixture with costunolide (**1270**) in the ratio 42:20 (Scheme 7.32).

Sometimes the elemenes–germacranes equilibrium is irreversibly shifted through one or more [1,5] sigmatropic rearrangements to new families of sesquiterpenes.

Coupling of isopiperitone (**1300**) with the appropriate organolithium reagent gives rise to the oxy-elemene (**1301**), which spontaneously rearranges to isoacoragermacrone (**328**) through an oxy-Cope rearrangement. Under suitable conditions, **328** isomerizes either to (±)-acoragermacrone (**329**) or formally via a [1,5] sigmatropic shift to (±)-preisocalamendiol (**1302**) (*50*). The latter compound is also obtained from (±)-shyobunone (**327**) via [3,3] and [1,5] sigmatropic rearrangements. Moreover, heating **1302** at 180 °C causes a further [1,5] shift and aromatization to give (±)-calamenene (**1303**) (*51*) (Scheme 7.33).

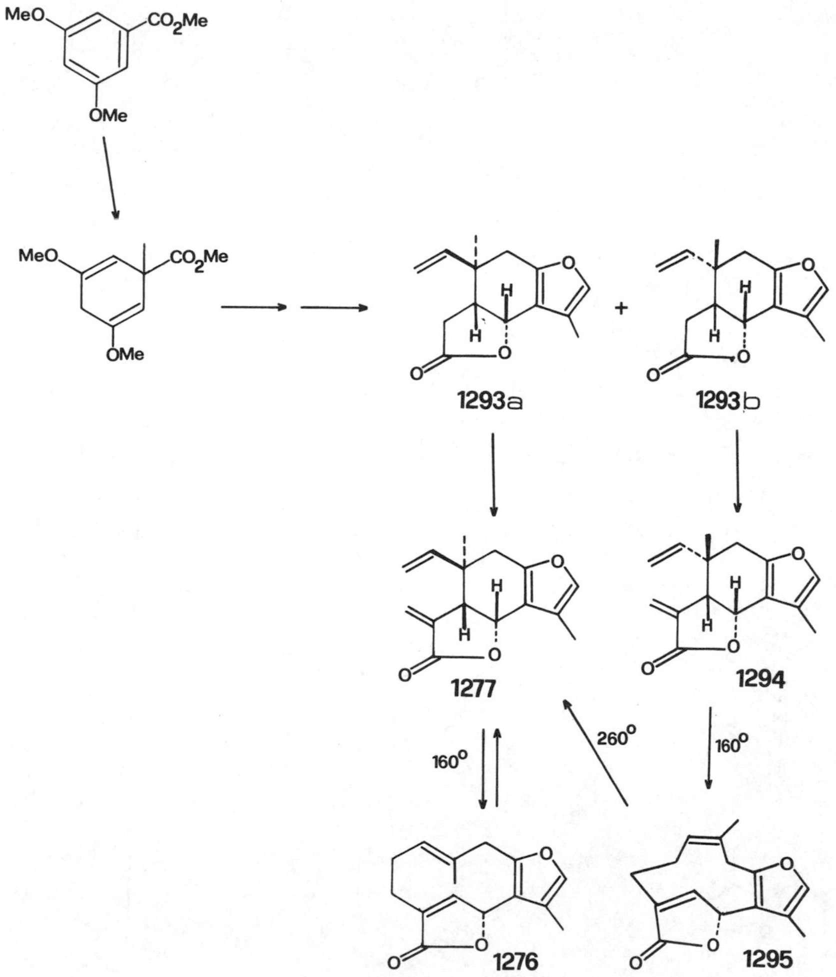

MeO CO2Me OMe

MeO CO2Me OMe

1293a 1293b

1277 1294

160° 260° 160°

1276 1295

Scheme 7.31

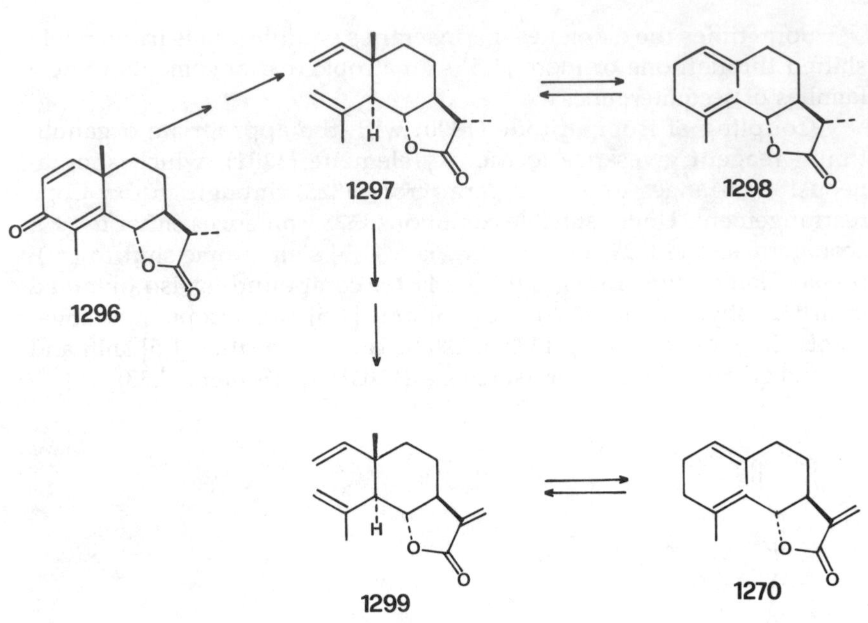

1296 1297 1298 1299 1270

Scheme 7.32

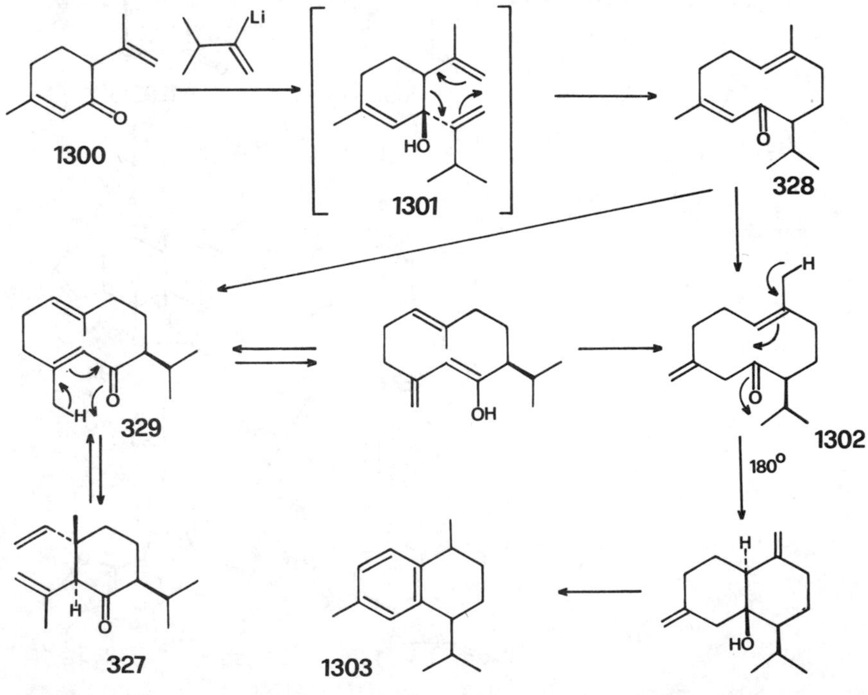

Li

1300 1301 328 329 1302 327 1303

HO OH 180°

Scheme 7.33

CATALYZED COPE REARRANGEMENT. The Cope rearrangement of 1,5-dienes bearing acyl substituents of C1 (*52*) or C2 (*53*) is strongly accelerated by protic and Lewis acids (*54*) (Scheme 7.34). A stepwise cationic-induced cyclization was proposed (*54, 55*) as an alternative to a concerted mechanism with a transition state such as that represented in Scheme 7.34.

An easy explanation can be proposed in terms of FMOs. The transition state of the catalyzed Cope rearrangement can be derived from the interaction of two allyl radicals (*see* Figure 7.6), one of them carrying a strong electron-attracting substituent (the carbocation). This fragment behaves as a strong acceptor with a decrease in the energy of its MOs. Therefore, the energy separation between Ψ_1 and Ψ'_3 is lowered, and the energy gained in the pericyclic reaction is greater (Figure 7.7).

Although this approach to the Cope rearrangement is not yet used in the natural products field, we included it in this review because with the mild conditions required it seems a promising tool.

Claisen Rearrangement. The Claisen rearrangement, like the Cope rearrangement, is a [3,3] sigmatropic shift. The two processes can be treated analogously; from a theoretical point of view, the main difference is the presence of an oxygen atom at C3 in the Claisen rearrangement.

The oxygen-including a′b′c′ fragment of **1304** can be considered as a vinyl ether, and the abc fragment as an olefin. Because conjugation with oxygen lone pairs raises the occupied orbitals (ionization potentials of vinyl ethers are lower than those of the corresponding olefins), the former fragment is the donor and the latter the acceptor.

The Claisen MO treatment does not differ greatly from that depicted in Figure 7.7 for the catalyzed Cope rearrangements. The low-lying Ψ'_1, Ψ'_2, and Ψ'_3 orbitals behave as the abc fragment of the transition

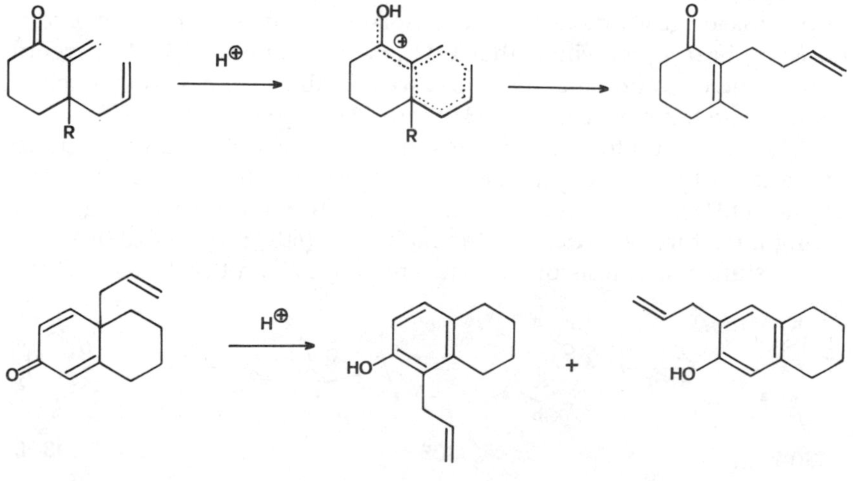

Scheme 7.34

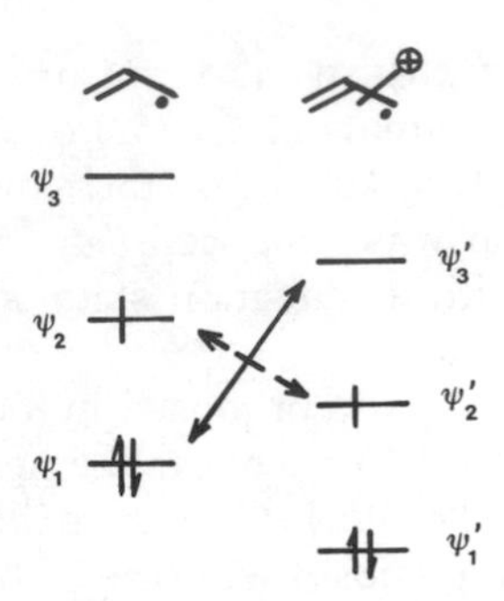

Figure 7.7. FMO interactions in the catalyzed Cope rearrangement.

state **1305**. Thus a chair-like transition state is favored for the Claisen rearrangement also (Scheme 7.35).

From a synthetic point of view, the Claisen rearrangement is a powerful tool for the preparation of γ,δ-unsaturated carbonyl compounds. Two reviews in 1977 were devoted to this reaction: Ziegler (*56*) covered natural product synthesis, and Rossi (*57*) reported the more popular routes to the synthesis of insect pheromones. Some of the already reported syntheses are outlined in this section.

To introduce the topic, we report a nice synthesis of α-sinensal (**1310**), a sesquiterpenic aldehyde isolated from the Chinese orange, *Citrus sinensis* L. The synthesis occurred through two sequential Claisen and Cope rearrangements (*58*). The vinyl ether **1307** in the more stable chair-like conformation rearranges to **1308**; a rotation around the single bond gives the more stable chair-like conformation, which rearranges to 2,6-dimethyl-10-methylenedodeca (2*Z*,6*E*)-11-trien-1-al (**1309**). This last compound finally isomerizes to α-sinensal (**1310**) as the more stable (2*E*)-isomer (Scheme 7.36).

A Claisen rearrangement was the key step in the synthesis of a C_{25}-terpene acid, gascardic acid (**1313**), isolated from *Gascardia madagascariensis*. By heating the vinyl ether **1311** in s. collidine (160 °C, 4 h, argon) a 65% yield of olefinic aldehyde **1312** was obtained. The *cis*-ring junction allowed the transformation of **1312** into **1313** (*59*) (Scheme 7.37).

As an attempt to functionalize a hydrindane system, an allyl group was attached to the ring junction. The procedure began with the diosphenol (**1314**), which retains the oxygen function on the ring, thus giving a mixture of **1315a,b** in the ratio 3.2:1 (*60*) (Scheme 7.38).

A simple synthesis of karahanaenone (**1172**), a constituent of hop

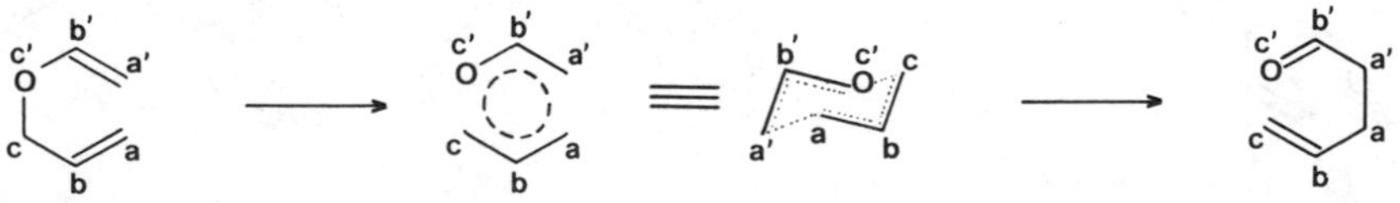

1304 **1305** **1306**

Scheme 7.35

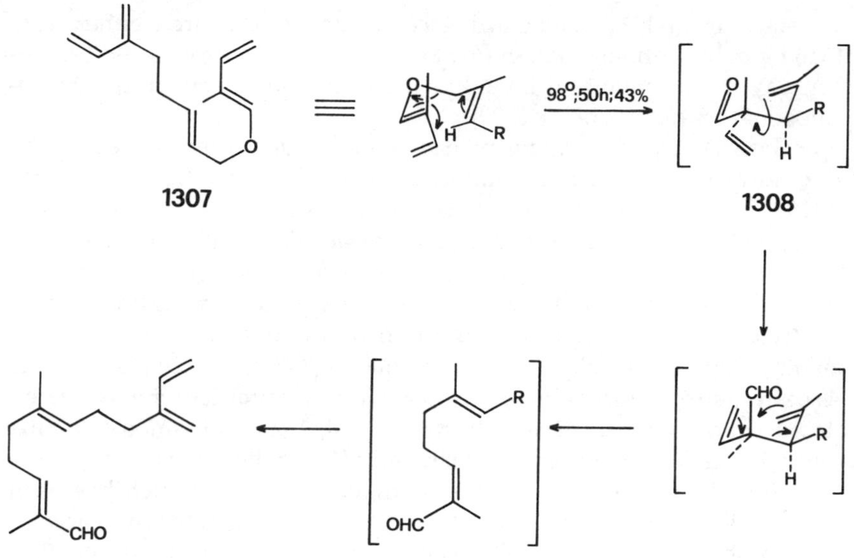

1307

98°;50h;43%

1308

1310

1309

Scheme 7.36

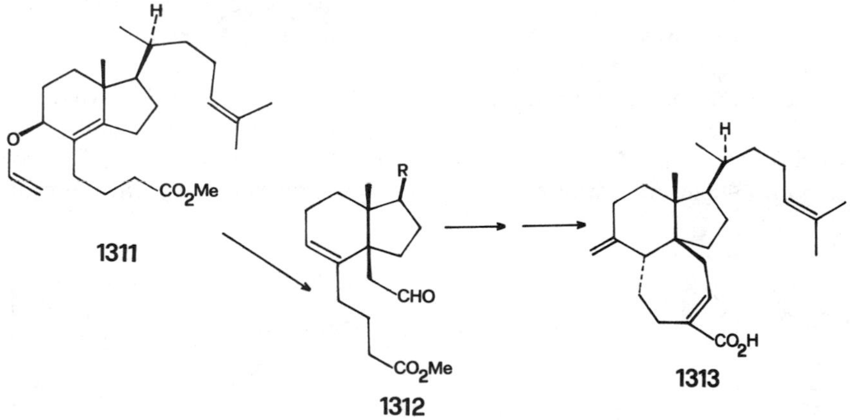

1311

1312

1313

Scheme 7.37

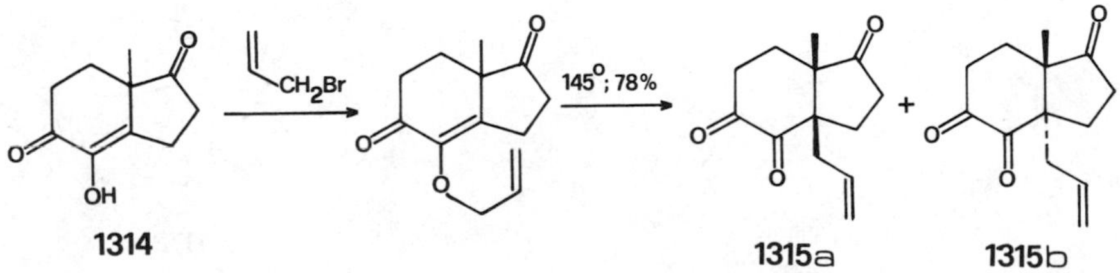

1314

145°; 78%

1315a

1315b

Scheme 7.38

oil, starts from **1317**. The starting compound is prepared either from **1316** by dehydrohalogenation (*61*) or, in lower yield, from 1,3-dibromo-3-methyl-2-butanone and 2-methylbutene by iodide ion promoted dehalogenation (*62*) (Scheme 7.39).

The Claisen rearrangement has been frequently employed in the sesquiterpenes field; good examples are the syntheses of 2,β-bisabolene (**1318**) and eremophilone (**1321**). The former substance is isolated from the essential oil of the fruits of *Schisandra chinensis* Baill, the latter from the wood oil of *Eremophila mitcheli,* a tall shrub indigenous to the drier areas of New South Wales. β_2-Bisabolene was obtained starting from 6-methylcyclohex-2-en-1-ol through two sequential Claisen rearrangements (*63*) (Scheme 7.40). (±)-Eremophilone (**1321**), a three chiral center sesquiterpene, was obtained from **1319**. The C7 chiral center was created stereorandomly via a Claisen rearrangement; **1320** is obtained as a mixture with its diastereomer in a 45:55 ratio (*64*) (Scheme 7.41).

The synthesis of the Inhoffen-Lythgoe diol **1324**, which has been converted into vitamin D_3 and into some of its metabolites, was accomplished starting from **1322**. This compound was transformed into **1323** using the flash vacuum pyrolysis technique (*65*) (Scheme 7.42).

The migration of an allyl chain along an aromatic or heteroaromatic ring is the most usual synthetic application of the Claisen rearrangement to the natural products field. We review various applications by indicating with heavy bonds the part of the final product generated by this bond reorganization process.

Several natural coumarins have been synthesized in this way: sesibiricin (**1325**) (*66*); toddaculin (**1326**) (*66*); the major metabolites of methoxsalen, known as Metabolites A and B (**1327a** and **1327b**) (*67*); and dentatin (**1328**) (*68*), a linear chromenocoumarin isolated from the root

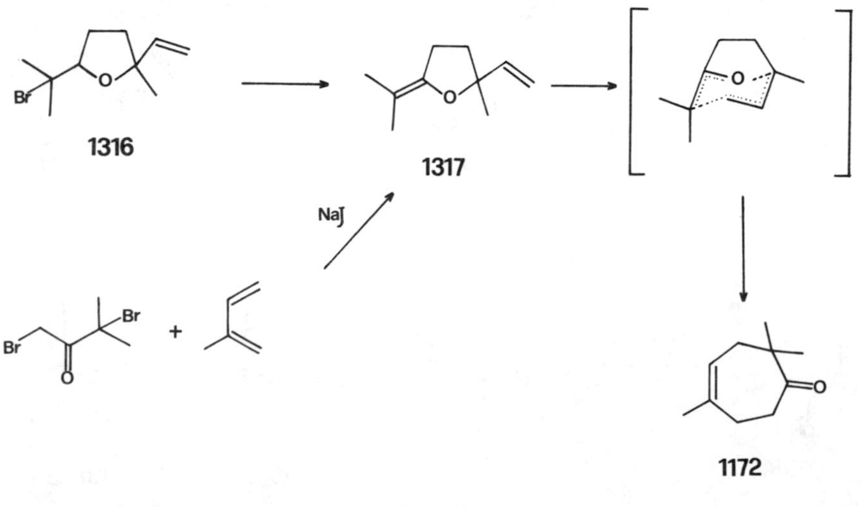

1316

1317

NaJ

1172

Scheme 7.39

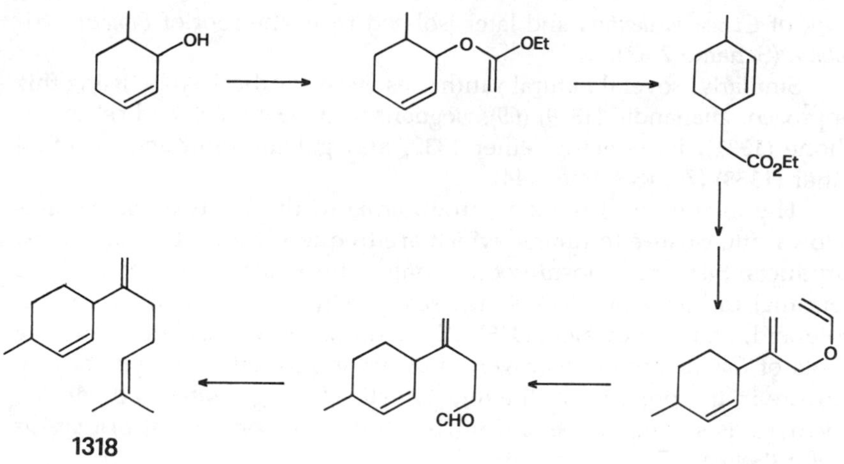

OH
OEt
CO_2Et
CHO
1318

Scheme 7.40

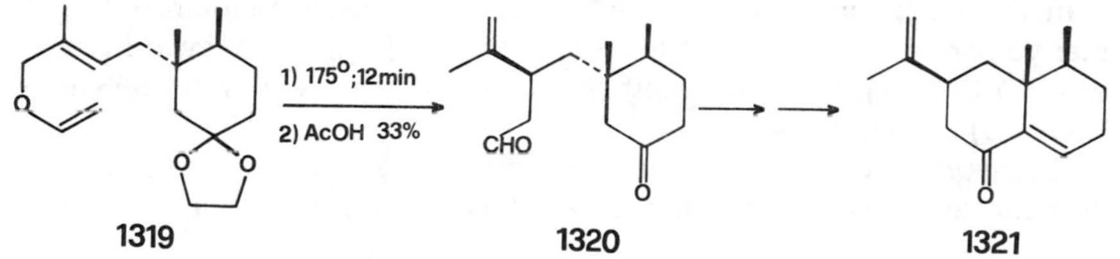

1) 175°;12min
2) AcOH 33%
CHO
1319
1320
1321

Scheme 7.41

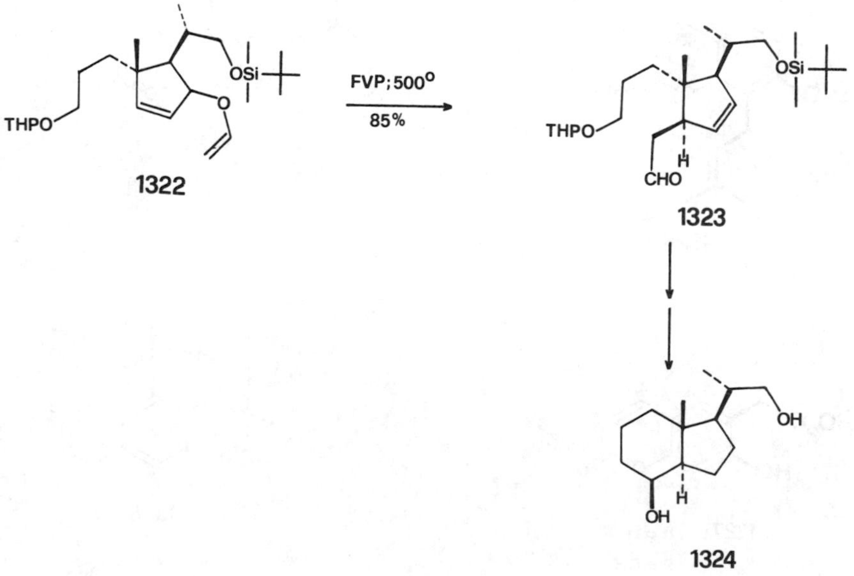

OSi
THPO
FVP; 500°
85%
THPO
H
CHO
1322
1323
OH
H
OH
1324

Scheme 7.42

bark of *Clausena dentata* and later isolated from the root of *Poncirus trifoliata* (Scheme 7.43).

Similarly, several natural xanthones were synthesized by using this approach: guanandin (**1329**) (*69*); isoguanandin (**1330**) (*69*); and celebixanthone (**1331**), its 6-methyl ether **1332**, and galvanoxanthone dimethyl ether (**1333**) (*70*) (Scheme 7.44).

The migration of the allyl group *ortho* to the hydroxyl sometimes allows ring closure to furans, which are frequently a portion of natural products. Such ring closure occurs only after oxidative removal of the terminal carbon atom. The syntheses of elliptone (**1334**) (*71*), a furan rotenoid, and (±)-cacalol (**1335**) (*72*), a sesquiterpene isolated from the roots of *Cacalia decomposita*, were thus accomplished. This type of ring closure in the presence of the intact allylic chain gives benzopyran derivatives as scabequinone (**1336**) (*73*), the major component of *Cyperus scaber* (Scheme 7.45).

Further modifications of the primary Claisen product are involved in the syntheses of several sesquiterpenes. In general, a Wittig reaction or the condensation with other organometallic reagents on the resulting aldehyde are required to complete the synthesis [e.g., the alleged humbertiol (**1337**) (*74*), bakuchiol methyl ether (**1338**) (*75*), and γ-bisabolene (**1339**) (*76*) (Scheme 7.46).

Sometimes the Claisen rearrangement occurs at the first stage of a synthetic sequence and not as the key step. We report two interesting

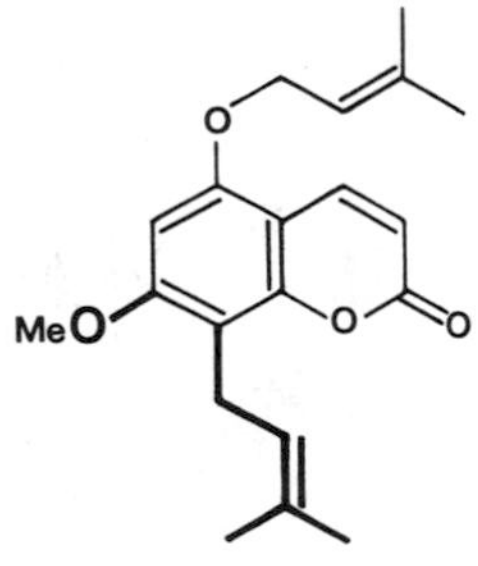

MeO O O O

1325

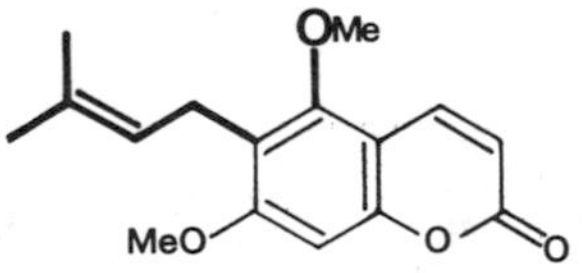

OMe MeO O O

1326

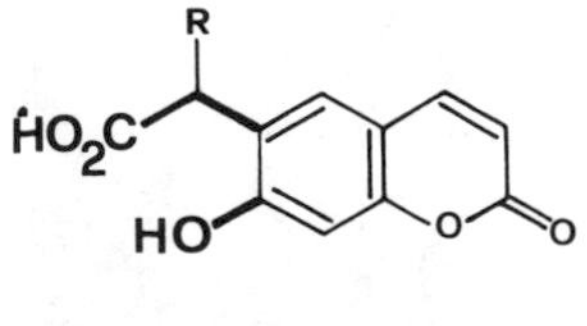

R HO₂C HO O O

1327a R = H
b R = OH

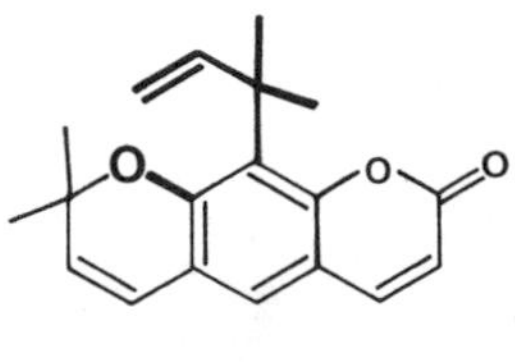

O O O

1328

Scheme 7.43

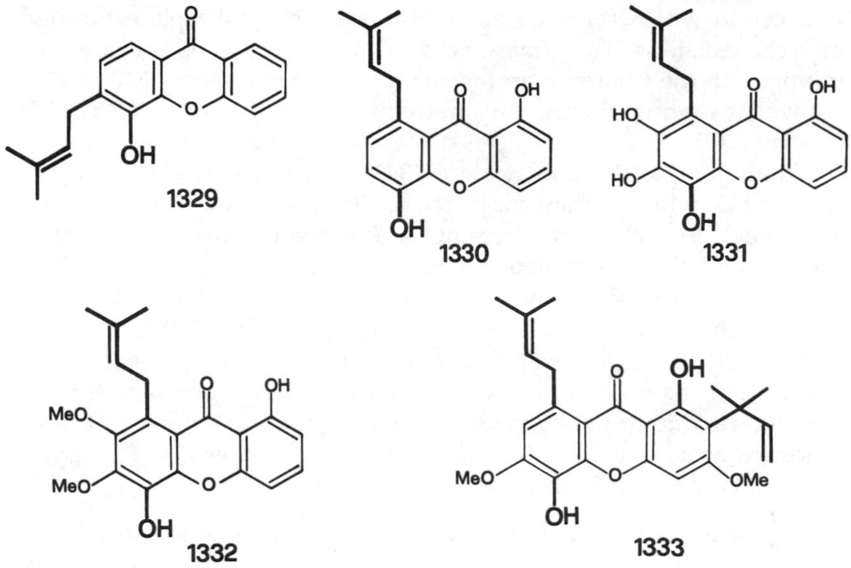

1329 1330 1331

1332 1333

Scheme 7.44

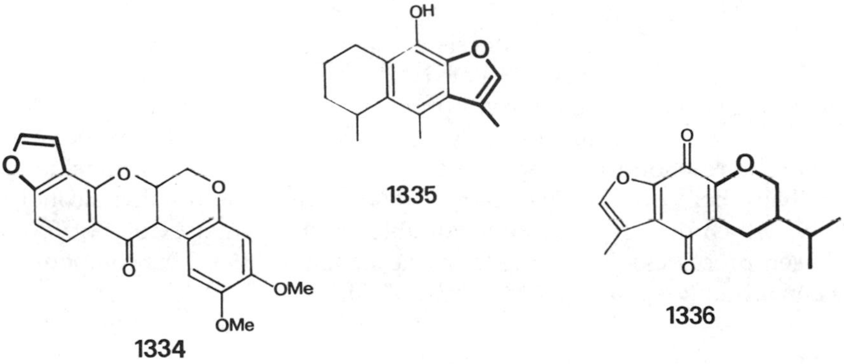

1335

1334 1336

Scheme 7.45

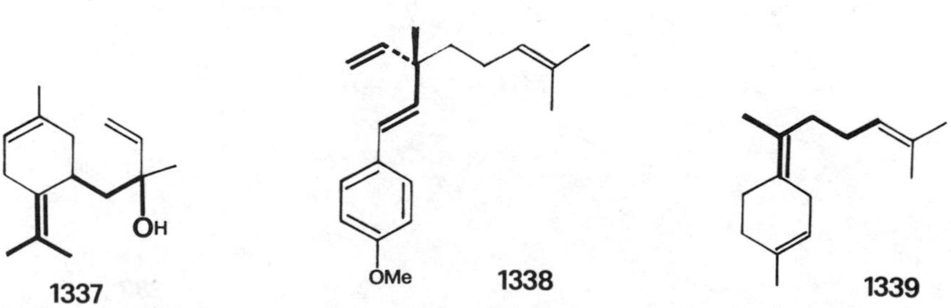

1337 1338 1339

Scheme 7.46

examples in which further steps of the synthesis take place through pericyclic reactions. Thus *trans*-deoxytaylorione (**283**) was prepared beginning with the Claisen rearrangement of the vinyl ether **1340** to **1341** followed by photochemical π methane rearrangement of **281** (*77*) (Scheme 7.47).

Analogously, the enol-acetate **1342** rearranges at 225 °C to the phenol **1343**, which is then converted to diazomalonate **1344**. An intramolecular [1,2]-cycloaddition results in the formation of a mixture of epimers **1345**, whose β isomer gives the mitosane (**1346**) (*78*). This product is suitably functionalized for conversion into the desired end-products, the naturally occurring mitomycins (Scheme 7.48).

Because of the synthetic importance of this rearrangement, many modifications have been proposed, either to facilitate access to the unsaturated residue or to obtain as rearranged products a variety of functions such as ketones, amides, carboxylic acids, and esters. These variants will be detailed in the following sections.

MARBET–SAUCY VARIANT. The Marbet–Saucy variant (*79*) was developed as a useful method for preparation of Claisen substrates. The acid-catalyzed reaction of vinyl ethers with allyl carbinols gives, after loss of alcohol and Claisen rearrangement, γ,δ-unsaturated aliphatic and alicyclic aldehydes and ketones (Scheme 7.49). The procedure takes 1–48 h at 100–180 °C, and the yields are good.

A significantly short synthesis of squalene (**1351**) provided a nice application of this variant. Bis(allyl alcohol) **1347** was treated with 5 molar equivalents of 3-methoxyisoprene in a toluene solution containing small amounts of oxalic acid (140 °C, 24 h) to produce tetraenedione **1348**. The reaction took place through a double Claisen rearrangement involving both groups. Reduction of **1348** affords the bis(allyl alcohol) **1349**, which undergoes another double rearrangement (total of four Claisen processes) giving rise to the hexaenediol **1350**. This compound is converted to squalene (*80*) (Scheme 7.50).

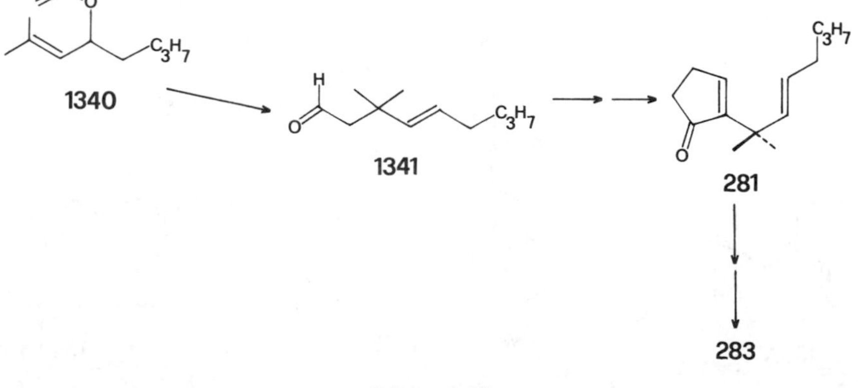

1340 → **1341** → → **281** → **283**

C_3H_7

Scheme 7.47

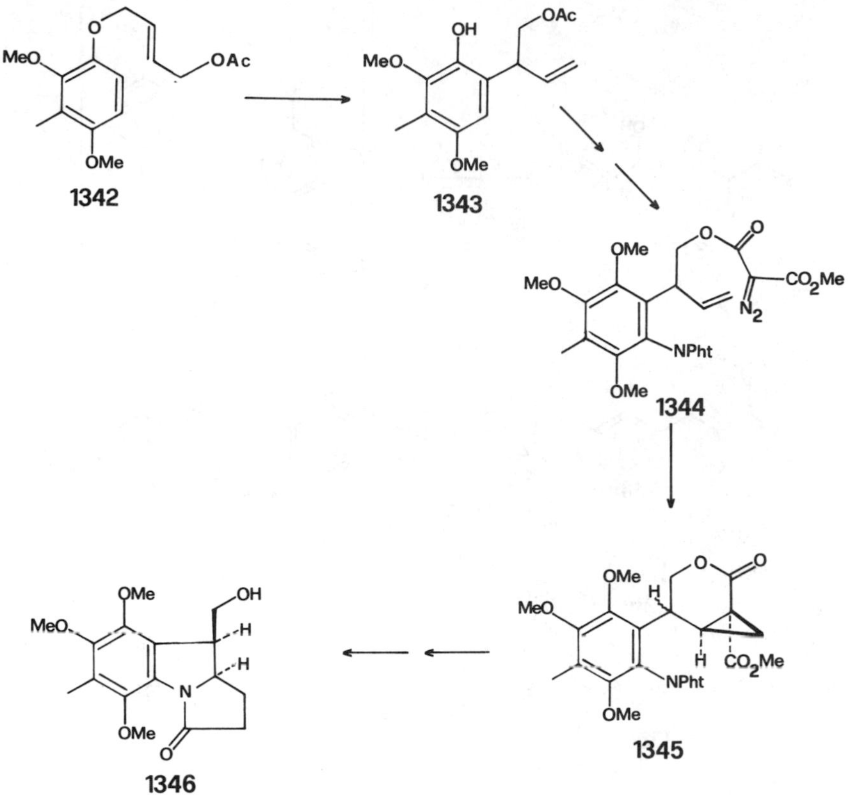

MeO OMe OAc
1342

OH MeO OMe OAc
1343

OMe MeO CO_2Me N_2 NPht OMe
1344

OMe MeO H H CO_2Me NPht OMe
1345

OMe MeO OH H H N OMe O
1346

Scheme 7.48

R R_1 O H + R_2 R_3 OR_4 → R R_1 O OR_4 R_3 R_2

R_4OH

R R_1 O R_3 R_2 → R_1 R R_3 O R_2

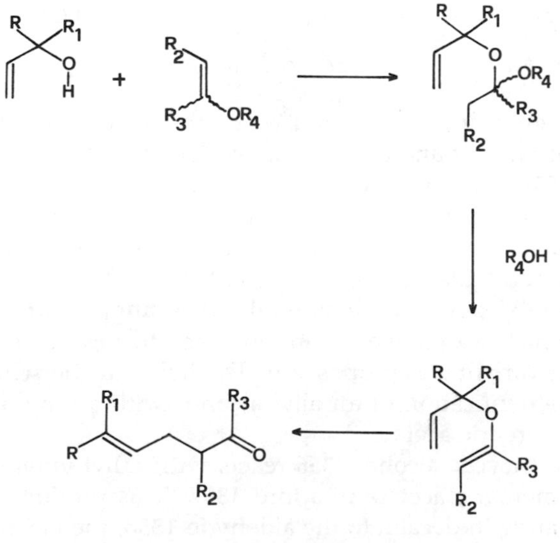

Scheme 7.49

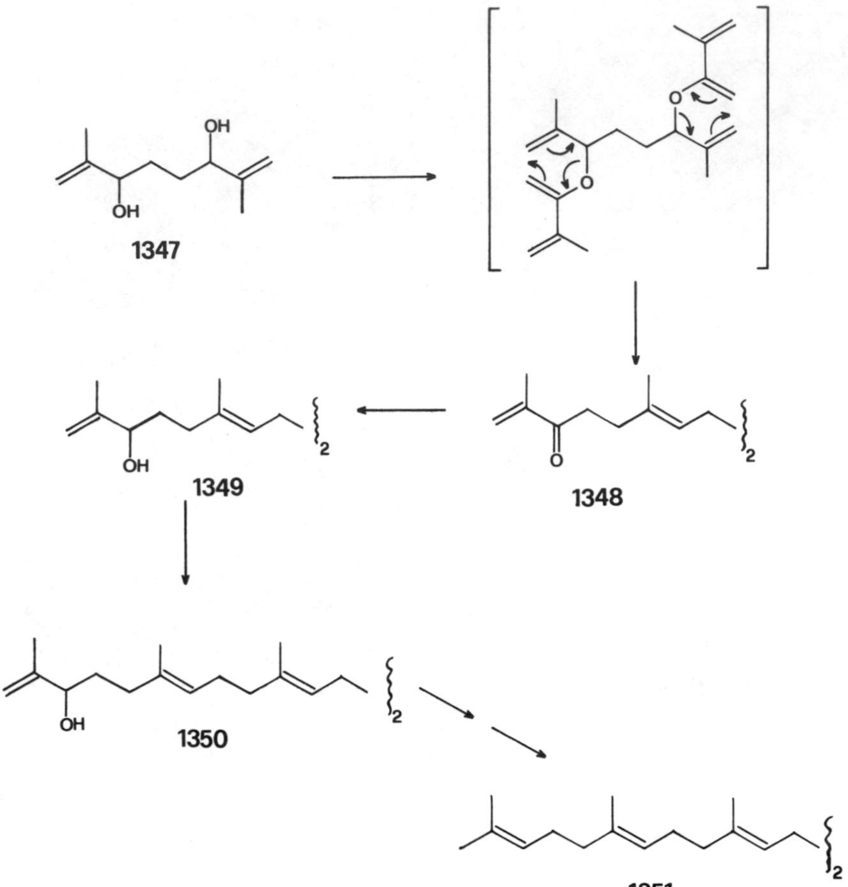

OH
OH
1347
O
O
1349
OH
2
O
1348
2
OH
1350
2
1351
2

Scheme 7.50

An analogous approach was used by the same group for the synthesis of both racemic and optically active forms of the *Cecropia* juvenile hormone (**1355**) (*80*). The reaction between methyl 2-hydroxy-3-methyl-3-butenoate and 3,3-dimethoxy-2-ethyl-1-butene (a pseudo-vinyl ether) in toluene gave ketoester **1352**, which was reduced immediately to **1353**. A second Claisen rearrangement of the latter with 2,2-dimethoxy-3-methyl-3-pentanol (eventually as a single enantiomer) gave ketol **1354**, which was converted in few steps to **1355** (Scheme 7.51).

A similar variant was proposed by Dauben and Dietsche (*81*) based on the transetherification of an allyl alcohol with a vinyl ether in the presence of mercuric acetate.

Thus the tricyclic alcohol **1356** reacts with ethyl vinyl ether in the presence of mercuric acetate to afford **1357**. This product can be converted by heating in decalin to the aldehyde **1358**, the key intermediate for a total synthesis of (±)-steviol methyl ester (**236**) (*82*) (Scheme 7.52).

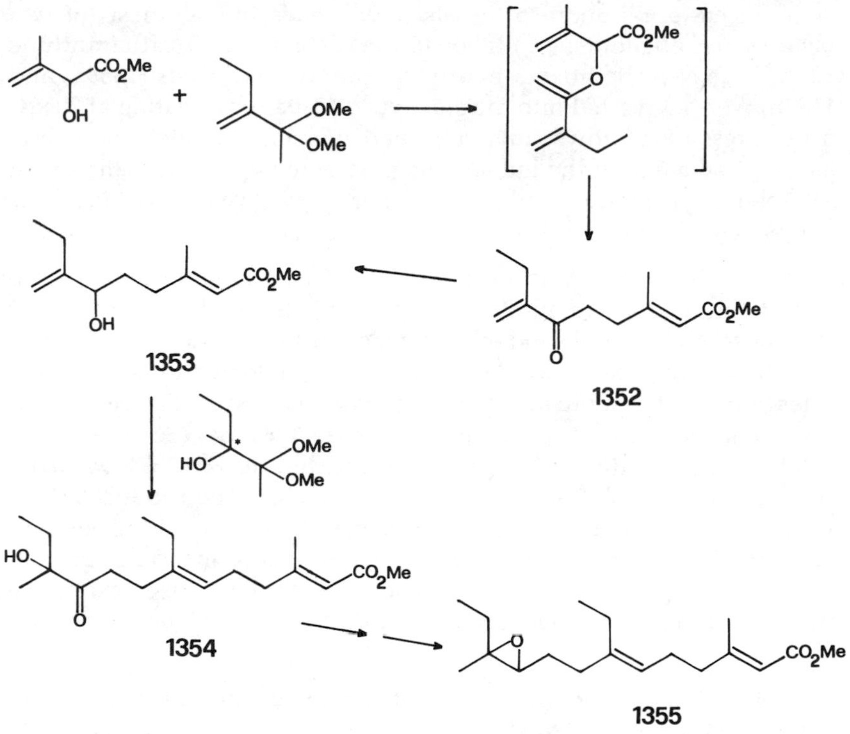

CO_2Me OH + OMe OMe → [CO_2Me O] ↓

1353 ← 1352

1354 → → 1355

Scheme 7.51

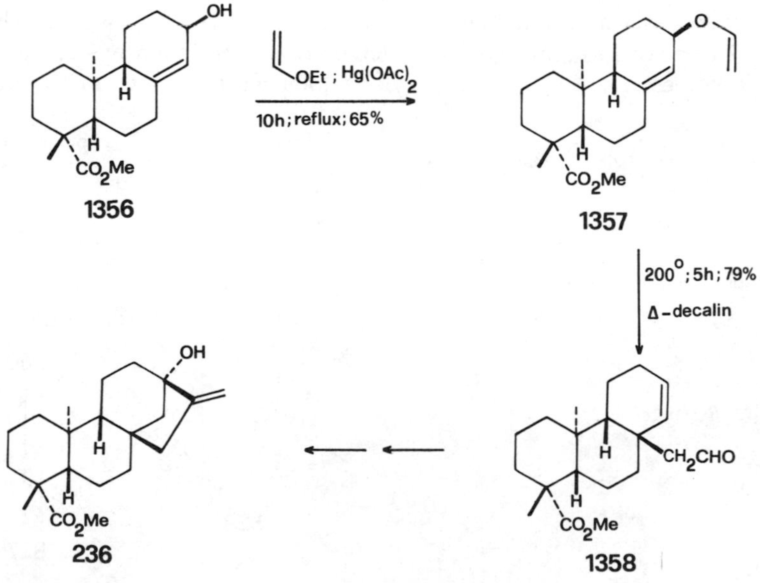

1356 —OEt ; $Hg(OAc)_2$, 10h; reflux; 65% → 1357

200°; 5h; 79%, Δ-decalin ↓

236 ← ← 1358

Scheme 7.52

A recent application to the alkaloid field is the synthesis of two *Aspidosperma* alkaloids: (±)-tabersonine (**827**) and (±)-catharanthine (**828**). Thus *N*-carbethoxy-3-hydroxytetrahydropyridines (**1359a**) and (**1359b**) were converted into the aldehydes **1360a,b** by heating at 205 °C in the presence of ethyl vinyl ether and mercuric acetate. The subsequent introduction of the indole ring, performed by condensation with β-indolylacetyl chloride, and a few further steps gave the final products (*83*) (Scheme 7.53).

SELENOXIDE VARIANT. A new version of the Claisen rearrangement consists of the reaction of an allyl carbinol with the addition product of benzeneselenyl bromide and ethyl vinyl ether to give the selenide **1361**. Oxidation to the selenoxide **1362** with $NaIO_4$, followed by loss of benzeneselenic acid in refluxing xylene, gives the ketene acetal **1363**, which can be converted into a γ,δ-unsaturated ester **1364** (*84*) (Scheme 7.54).

The total synthesis of (±)-phoracantholide J (**1367**) occurred through an intermolecular process involving **1365**. The product **1367** is the major component of the metasternal gland secretion of the eucalypt longicorn *Phoracantha synonyma*. The predominant formation of the naturally occurring (Z) isomer [in the ratio 7:2 with the (*E*) isomer] can be rationalized in terms of a less strained transition state **1366** (*85*) (Scheme 7.55).

CARROL VARIANT. The Carrol variant (*86–88*) entails the use of a β-ketoester. The conjugated-chelated tautomer of the ester rearranges to a β′-keto-γ,δ-unsaturated acid that loses carbon dioxide to give γ,δ-unsaturated ketones (Scheme 7.56).

The only application to the natural products field concerns a proposed stereocontrolled synthesis of steroid side chains (*89*). Thus 16α,17α-epoxy-20-ketopregnane (**1368**) is converted in a few steps to the allylic ketoester **1369**. In boiling xylene **1369** rearranges to **1370**, which

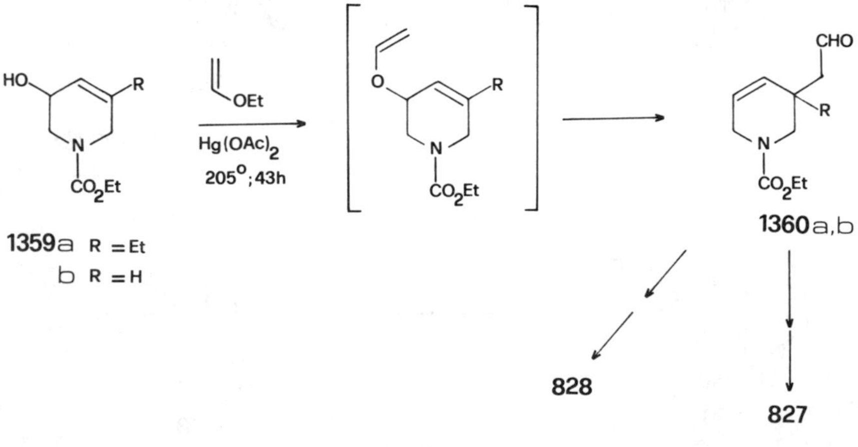

Scheme 7.53

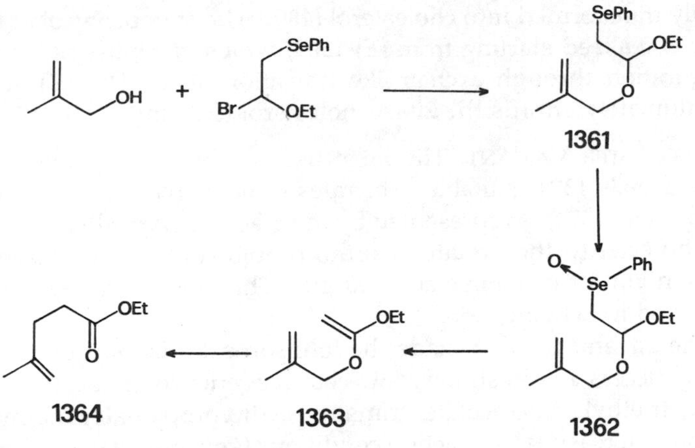

OH + SePh Br OEt → SePh OEt O 1361 → O Se Ph OEt O 1362 → OEt O 1363 → OEt O 1364

Scheme 7.54

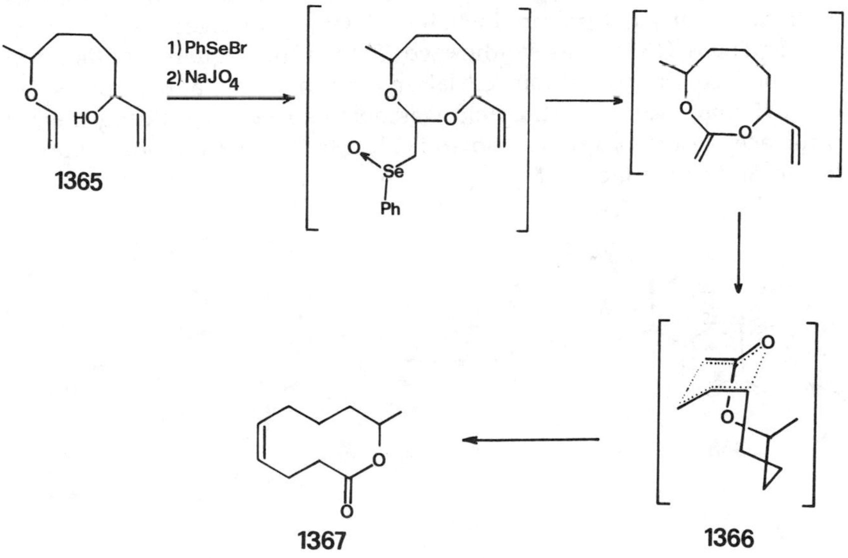

O HO 1365 — 1) PhSeBr 2) $NaJO_4$ → [O O O Se Ph] → [O O] → [O O 1366] → O O 1367

Scheme 7.55

O O O R ⇌ H O O O R → OH O O R → $-CO_2$ → O R

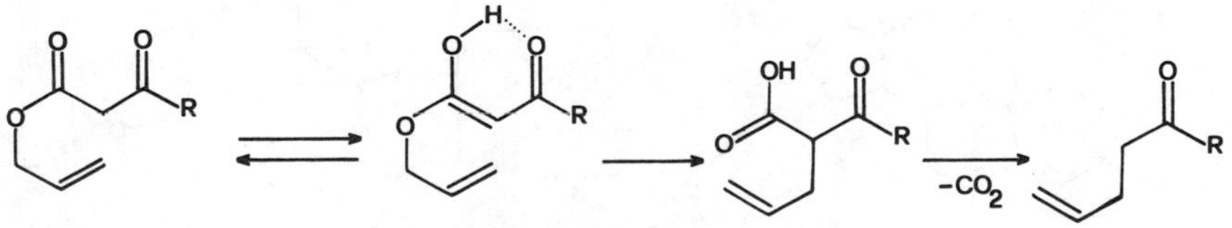

Scheme 7.56

is easily transformed into cholesterol (**480**). The stereocontrol of the C20 center is realized starting from (*E*)-**1369**, which gives the natural (20*R*)-configuration through a chair-like transition state. The (*Z*)-isomer of **1369** ultimately affords the 20-isocholesterol (Scheme 7.57).

JOHNSON-ESTER VARIANT. The most useful method of generating allyl ketene acetals **1371**, suitable substrates to be converted into γ,δ-unsaturated esters **1372**, is represented by the reaction of an allyl alcohol with an ortho ester (ortho acetate or ortho propionate) in the presence of a small amount of propionic acid (*90, 91*). The overall reaction sequence is reported in Scheme 7.58.

The advantages that made the Johnson-ester variant popular were an easy access to substrates, low-cost reagents (e.g., trimethyl ortho acetate, triethyl ortho acetate, trimethyl ortho propionate, triethyl ortho propionate), reasonable reaction conditions (generally a few hours, 130–140 °C), and excellent yields of esters **1372**. Because the ionization potentials of ketene acetals are lower than those of the corresponding vinyl ethers, the ethoxy group in **1371** increases the donor character of the O–C–C fragment. We discuss this point in detail later. By using this approach, many natural products have been synthesized.

Squalene (**1351**) was synthesized (*90*) via the sequence outlined in Scheme 7.50, through a double Claisen rearrangement at the same stage on **1347** and **1349**. A better yield was obtained by substituting triethyl ortho acetate and propionic acid and (3 h, 138 °C) for 3-methoxyisoprene in the first rearrangement.

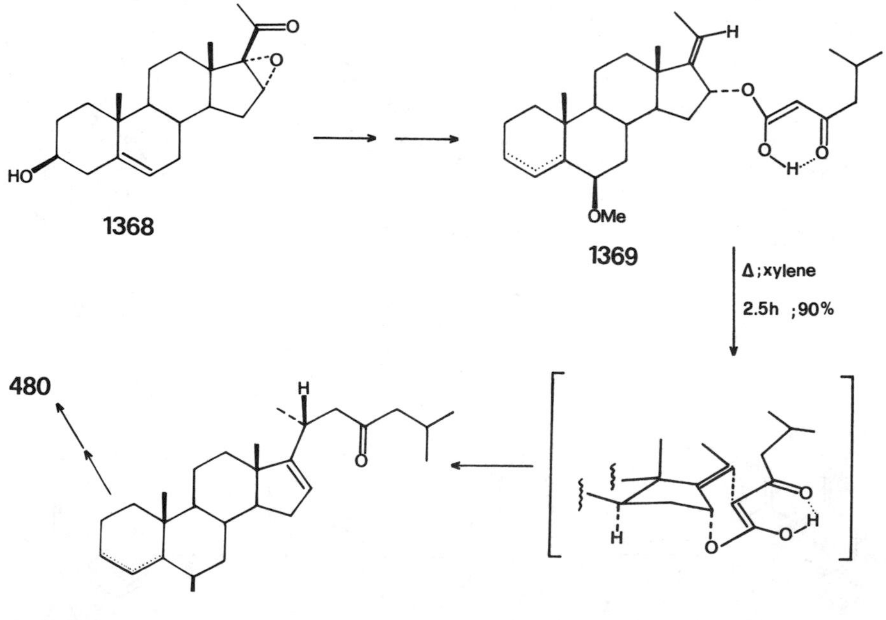

Scheme 7.57

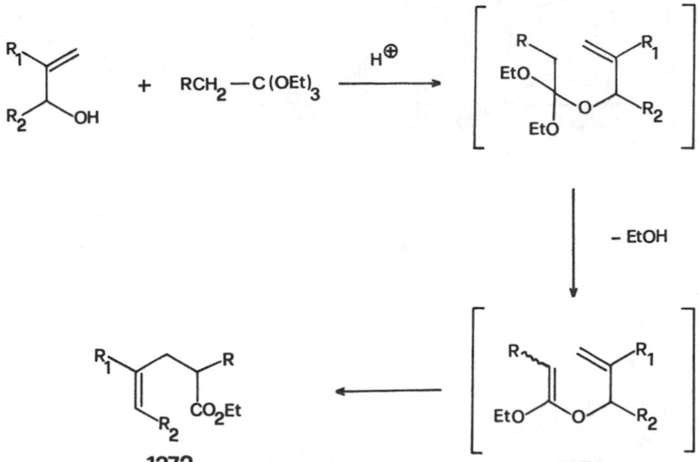

Scheme 7.58

A synthesis of (±)-progesterone (**368**) was realized by Johnson et al. (*92*) through an olefinic-acetylenic cyclization of **1376**. The key step was a Wittig condensation of the aldehyde **1374** with the ylide **1375**. The aldehyde was obtained in good yields starting from enynol **1373** by action of triethyl ortho acetate (Scheme 7.59).

Stork's efforts in the development of prostaglandins chemistry are of leading importance to the progress of the field. We report here two beautiful examples.

A chiral synthesis of (±)-(15*S*) prostaglandin A_2 (**1383**) (*93*) starts from 2,3-isopropylidene-L-erythrose (**1377**), which was transformed into **1378**. This product then underwent two Claisen rearrangements, first to **1379** and then **1380**. By using the ortho ester **1381**, **1382** was produced from **1380**. The desired product was formed from **1382** (Scheme 7.60).

Prostaglandin $F_{2\alpha}$ (**53**) was synthesized with transfer of chirality starting from D-glucose (*94*). This sugar was converted to the allylic alcohol **1384**, and the proper chirality of the eventual C12 center was secured by the Claisen ortho ester method to give 80% yield of **1385**. Several steps, including the introduction of the second chain, allowed the completion of the synthesis (Scheme 7.61).

L(+)-Cysteine was the starting material for a stereospecific total synthesis (*95*) of (+)-biotin (**128**). The synthesis featured a Claisen step from **1386** to **1387** instead of a cycloaddition approach (previously outlined in Chapters 3 and 4) (Scheme 7.62).

The Johnson-ester method is very useful for the delivery of a chain with a complete transfer of chirality from the allylic alcohol to the optically active products, as illustrated by the following three examples. The stereospecific synthesis of the side chain of oogoniol (**1390**), a sex hormone of the water mold *Achlya*, proved that the stereochemistry at

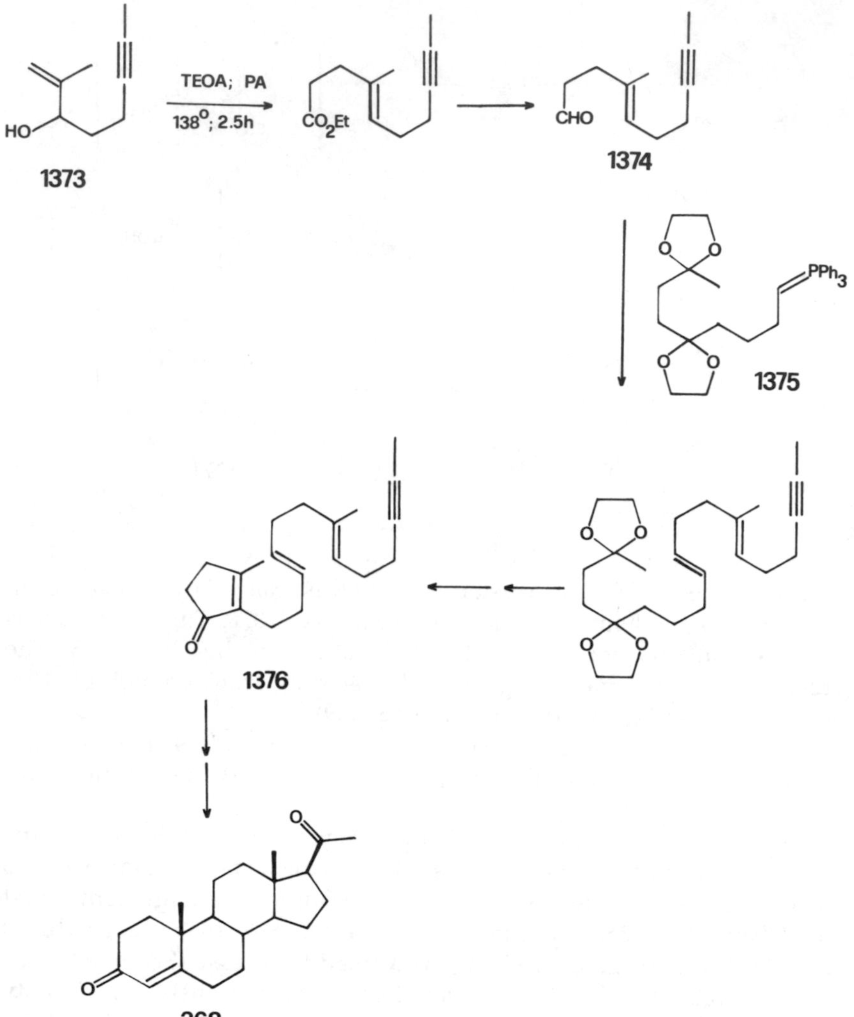

TEOA; PA

138°; 2.5h

CO_2Et

CHO

1373

1374

PPh_3

1375

1376

368

Scheme 7.59

the C24 center has an (*R*)-configuration. This center was realized by rearranging the (*Z*)(22*S*)-allylic alcohol **1388** to produce the (*E*)(24*R*)-ester **1389** through a chair-like transition state (*96*) (Scheme 7.63).

The C24 configuration of ficisterol (**1393**), isolated from the marine sponge *Petrosia ficiformis,* was demonstrated by synthesizing all four possible C23,C24 stereoisomers via the Claisen-ester rearrangement of **1391** to **1392** (*97*) (Scheme 7.64).

The synthesis of (2*R*,4′*R*,8′*R*)-α-tocopheryl acetate (**1396**), the vitamin E acetate, was achieved by Claisen-ester rearrangement of (*R*,*E*)- and (*S*,*Z*)-allylic alcohols (**1394a** and **1394b**, respectively). Both alcohols gave the same (*R*,*E*)-unsaturated ester **1395** with complete transmission

Scheme 7.60

Scheme 7.61

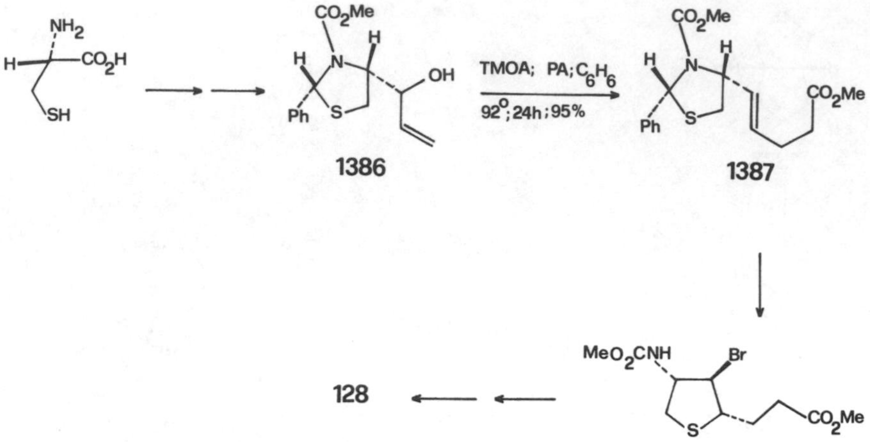

Scheme 7.62

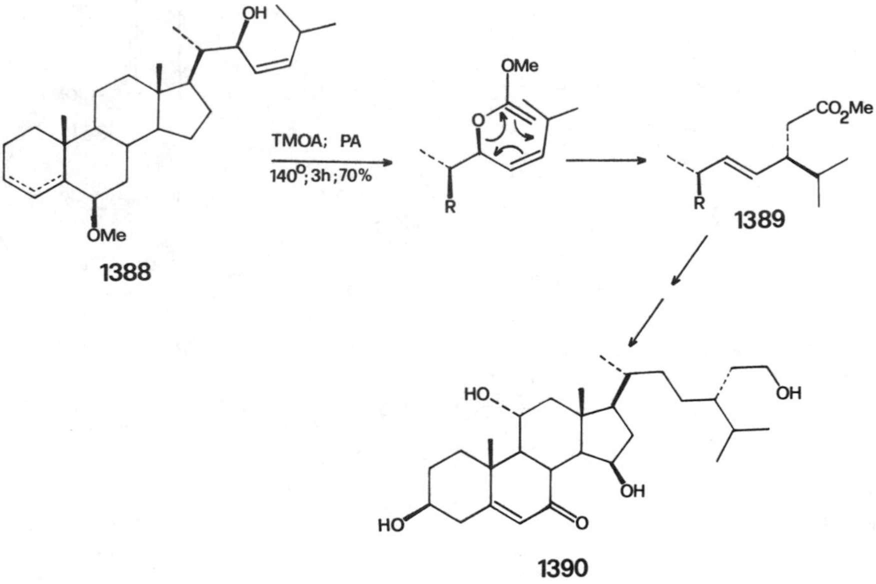

Scheme 7.63

of chirality. Coupling with an active nine-atom synthon and a few standard transformations gave **1396** (*98*), thus demonstrating the wide potential applicability of the Claisen process to the synthesis of optically active substances (Scheme 7.65).

Some applications have been made in the terpene field. The synthesis of (±)-widdrol (**1399**) involves in the early stage the rearrangement of allylic alcohol **1397** to ester **1398** (*99*). Treatment of allylic alcohol **1400** in the presence of triethyl ortho acetate and propionic acid to furnish

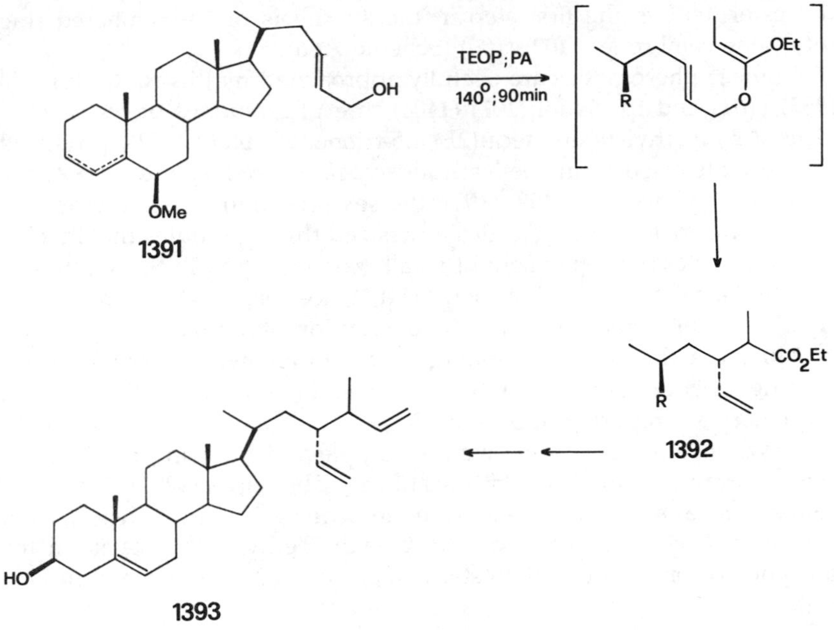

1391

TEOP; PA

140°; 90min

OEt

R

1392

CO_2Et

1393

HO

OMe

Scheme 7.64

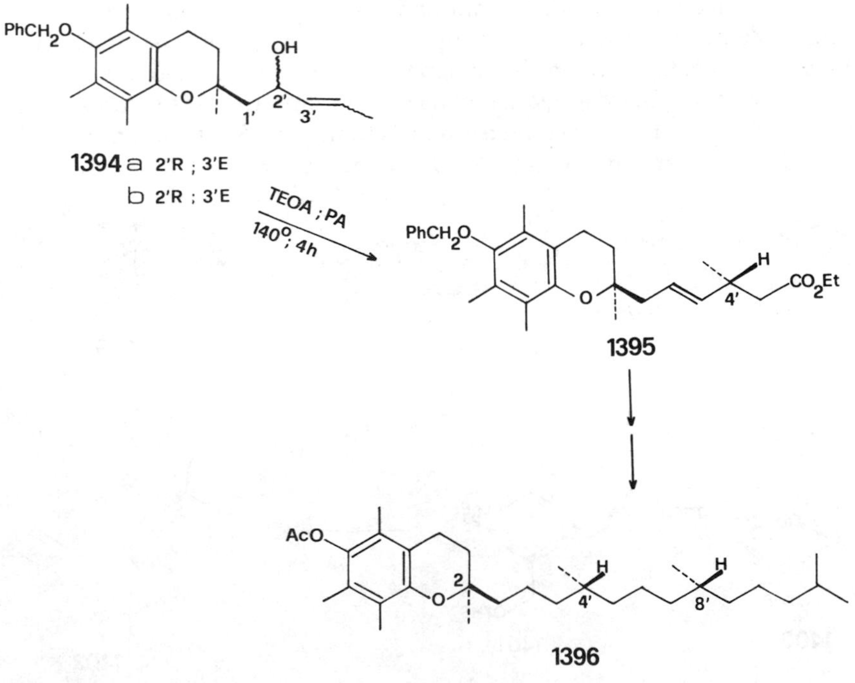

$PhCH_2O$

OH

1'

2'

3'

1394 a 2'R ; 3'E

b 2'R ; 3'E

TEOA ; PA

140°; 4h

1395

4'

CO_2Et

AcO

2

4'

8'

H

1396

Scheme 7.65

the ester **1401** is the first step in the synthesis of 14-membered ring diterpene cembrene (**1402**) (*100*) (Scheme 7.66).

Linear pheromones are usefully approached by this route. Ipsenol (**953**) (*101*) and ipsdienol (*101*) (**1404**), the pheromones of *Ips paraconfusus;* (±)-methyl *n*-tetradeca-(2*E*),4,5-trienoate (**1406**) (*102*), produced by the male dried bean beetles *Acanthoscelides obtectus;* and (7*E*,9*Z*)-dodecatrien-1-yl acetate (**1409**) (*103*), the sex pheromone of the virgin female of *Lobesia botrana,* were all synthesized through routes that involve a Claisen-ester rearrangement of an allyl alcohol early in the syntheses. The allyl alcohols used were allenic (**1403**), acetylenic (**1405**), and enynic (**1407**) alcohol, respectively. In the competition of double vs. triple bond of **1407**, the double bond undergoes rearrangement, possibly by assuming a better configuration in the transition state **1408**. The reaction sequences are reported in Scheme 7.67.

Two alkaloid syntheses were accomplished by using the Claisen-ester rearrangement. The synthesis of (±)-tabersonine (**827**) (*104*) is essentially a variant of the one reported in Scheme 7.53. An allylic alcohol similar to **1359a** was heated at 140 °C with triethyl ortho acetate in the presence of pivalic acid; the ester analog of **1360a** was formed in 74% yield.

The synthesis of (±)-camptothecin (**1413**), an alkaloid originally isolated from *Camptotheca acuminata* (Nyssaceae) is more interesting. The introduction of the required residue via a Claisen-process ester (*105*), first investigated on model compounds with the pyridone-lactone DE ring system, was extended to camptothecin. The piperitone derivative **1410** was transformed into methylene lactam **1411**. A Friedlander condensation gave the required condensed quinoline ring **1412**, containing all substituents. Further conversion of **1412** into **1413** was accomplished by using standard methods (*106, 107*) (Scheme 7.68).

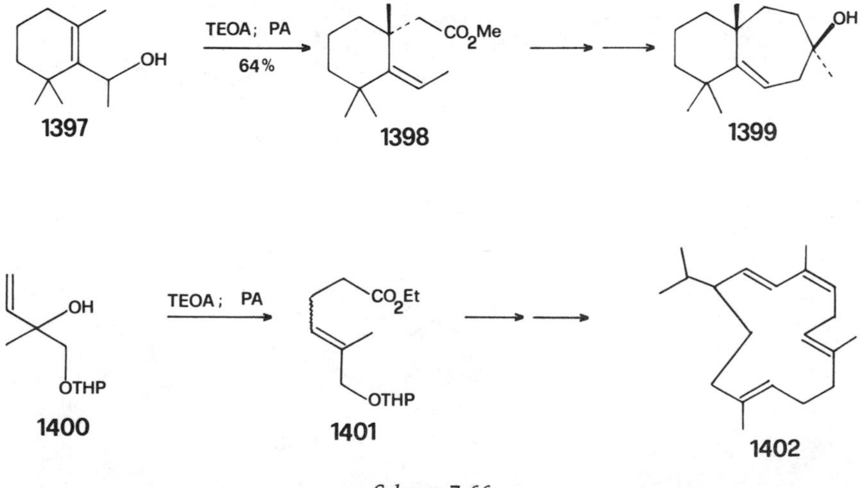

Scheme 7.66

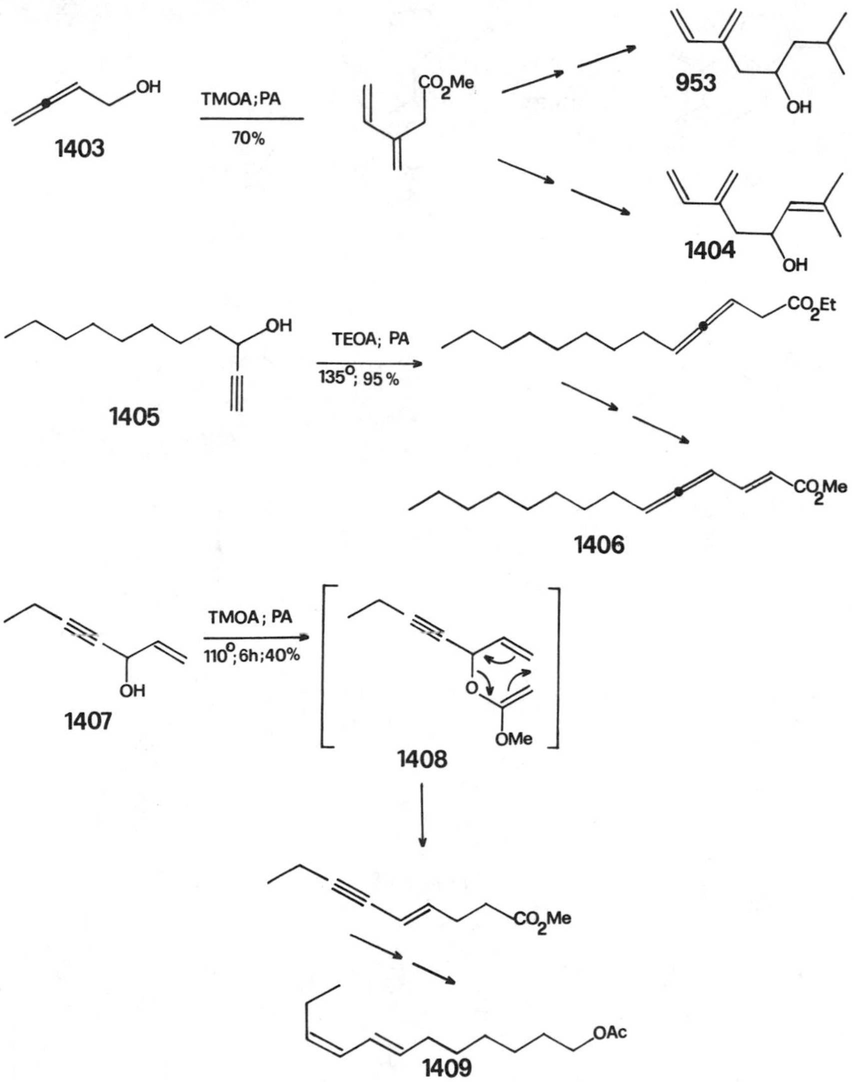

TMOA;PA 70%

1403 953 1404

TEOA; PA 135°; 95 %

1405 1406

TMOA; PA 110°;6h;40%

1407 1408 1409

Scheme 7.67

ESCHENMOSER VARIANT. On heating an allylic alcohol in the presence of *N,N*-dimethylacetamide dimethylacetal, a Claisen rearrangement analogous to that described for ortho esters takes place. The products are γ,δ-unsaturated amides (*108*) instead of the corresponding esters (Scheme 7.69).

This Eschenmoser variant has two main advantages: a nitrogen atom at the reaction site and a lower energy of activation of the process due to the higher donor character of the oxygen-including fragment. The reagent behaves like a ketene amino acetal, which has an ionization

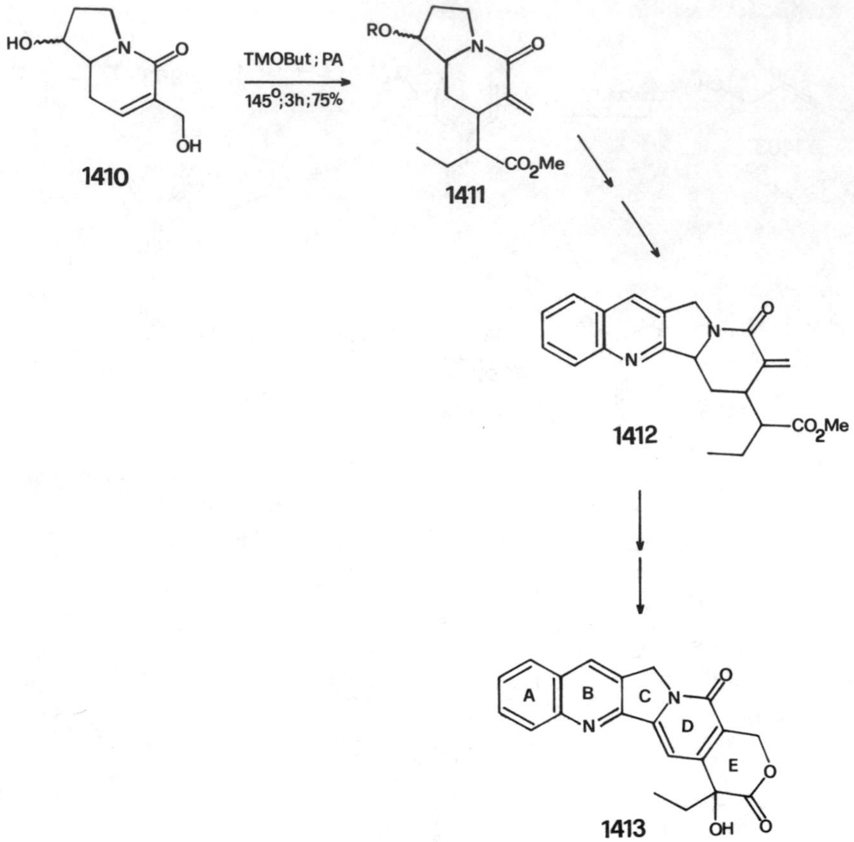

HO ... 1410 — TMOBut; PA, 145°; 3h; 75% → RO ... CO_2Me 1411 → 1412 (CO_2Me) → → 1413 (A B C D E, OH)

Scheme 7.68

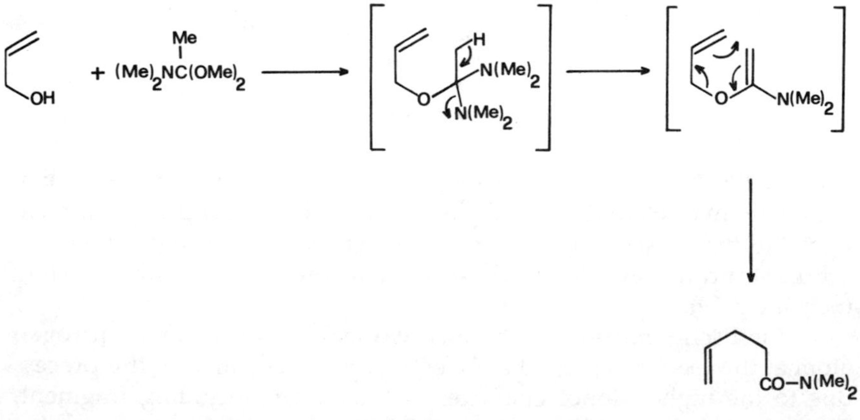

allyl alcohol (OH) + $(Me)_2NC(OMe)_2$ with Me → [O, H, $N(Me)_2$, $N(Me)_2$] → [O, $N(Me)_2$] → $CO{-}N(Me)_2$

Scheme 7.69

potential lower than that of the corresponding ketene acetal. Hence this should be the easiest Claisen variant of those described thus far.

The advantage given by a nitrogen atom suitably placed in the molecular frame can be appreciated in a recent total synthesis of a simple alkaloid: *O*-methyljoubertiamine (**1416**). 3-Aryl-2-cyclohexen-1-ol (**1414**) was heated with *N,N*-dimethylacetamide dimethylacetal (2h, 100–120 °C) to afford a 71% yield of cyclohexene acetamide **1415**, which was easily converted by standard method to **1416** (*109*) (Scheme 7.70).

The Corey synthesis of thromboxane B_2 (**1420**) (*110*) starts from D-glucose and involves the rearrangement of unsaturated sugar derivative **1417** to **1418**. This product can be subsequently converted to lactone **1419**, a useful synthon to thromboxane (Scheme 7.71). Although the Johnson-ester variant, at a first glance the preferred one, gives low yields only, the Eschenmoser variant furnishes **1418** in 75% yield, thus showing the advantage of the latter method over the former one.

The last example is an interesting total synthesis of (±)-estrone (**864a**) as its methyl ether. The Claisen–Eschenmoser rearrangement from **1421** to **1422** allows the introduction of a substituent suitable for the anellation of the C ring (*111*) (Scheme 7.72).

IRELAND-ENOLATE VARIANT. In 1972 Arnold and Hoffmann (*112*) found that allylic esters (e.g., **1423**) undergo an easy Claisen rearrangement if the enolate **1424** is generated by action of a strong base [e.g., lithium diisopropylamide (LDA)]. The reaction generally occurs at room temperature, and stereochemical control is exerted by the solvent through a stereoselective formation of the enolate (*113*). For example, (*E*)-crotyl propanoate (**1423**) leads to *erythro* acid **1427** when the enolization is carried out in tetrahydrofuran, but the *threo* isomer **1428** is formed when hexamethylphosphoramide is present in the solvent (Scheme 7.73).

The coordinating character of the solvent seems to account for the stereoselective enolate formation. The presence of hexamethylphos-

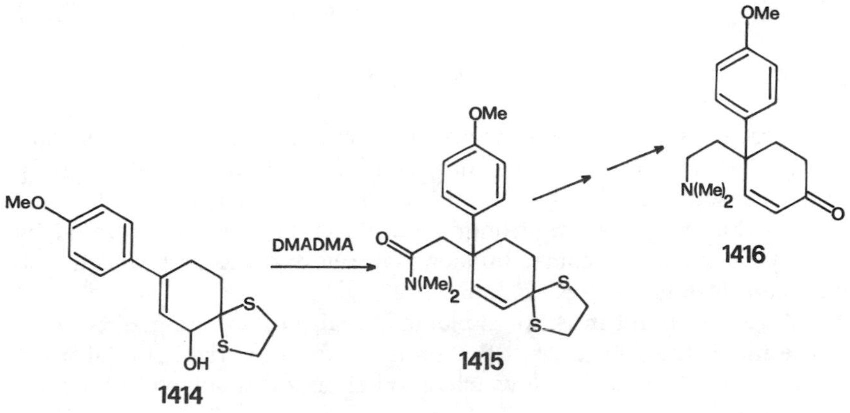

MeO OH S S **1414** DMADMA O N(Me)$_2$ OMe S S **1415** OMe N(Me)$_2$ O **1416**

Scheme 7.70

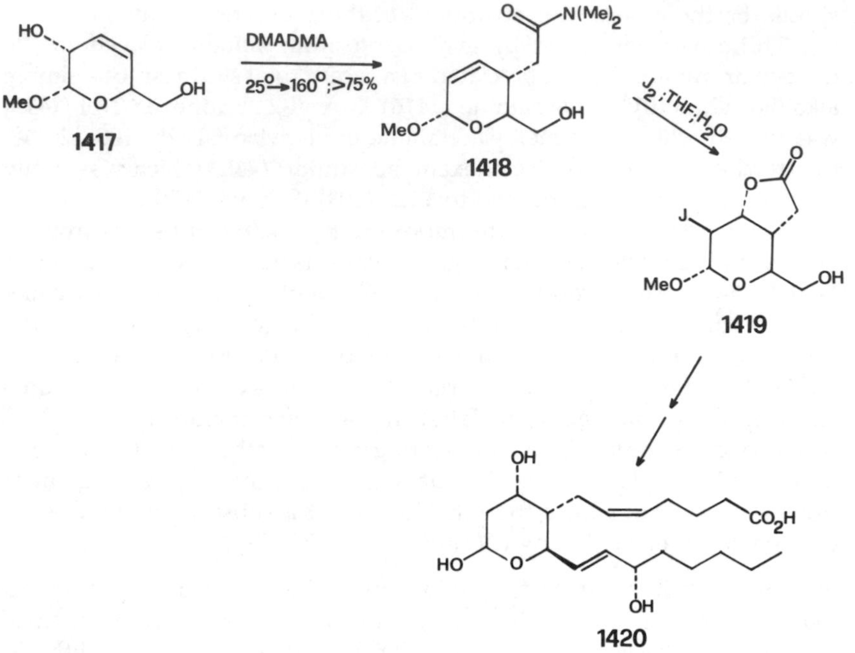

Scheme 7.71

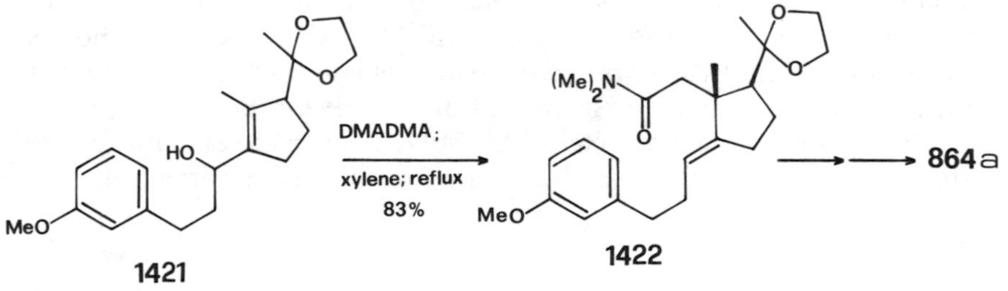

Scheme 7.72

phoramide results in a high degree of solvation of Li^+, hence its coordination with oxygen is weak. These conditions allow the methyl group to be eclipsed with C–O **1425**, and the *threo* isomer is obtained. However, if the oxygen atom strongly complexes Li^+, poorly solvated by tetrahydrofuran, it becomes bulkier than OR and the preferred transition state **1426** gives the *erythro* acid **1427**.

A further point must be explained. Claisen rearrangements occur in the range 120–200 °C , but the enolate variant works easily at room temperature. Therefore, a low activation energy is involved in the variant process.

We can reconsider the MO interactions occurring in the transition state of the [3,3] sigmatropic rearrangement as arising from the overlap of two allylic radicals. By comparing Cope, Claisen, Johnson–Claisen, Eschenmoser–Claisen, and Ireland–Claisen rearrangements (Figure 7.8), it seems evident from a theoretical point of view that the order of reactivity should increase along the above sequence, if the interaction between Ψ_1' and Ψ_3 is taken as dominant. Because the Ψ_2/Ψ_2' interaction is more favorable for the Cope rearrangement, spin polarization could be the reason for its low influence.

The enolate variant was usefully adopted in the total synthesis of the iridoid monoterpene (±)-sarracenin (**1432**). Treatment of α-methoxyacetate **1429** with lithium diethylamide in tetrahydrofuran produces the enolate **1430**, which rearranges to **1431**. This last compound can be converted in few steps to **1432** (*114*) (Scheme 7.74).

Shortly after the first observation that enolates undergo Claisen rearrangements, Ireland and Mueller (*115*) found a simple and useful modification quenching the lithium enolate at −78 °C with trimethyl chlorosilane. The resulting trimethylsilyl ketene acetal rearranges quite rapidly at room temperature without side reactions (Scheme 7.75).

In general, a Claisen rearrangement occurs through a chair-like transition state; in heterocyclic systems, a boat-like transition state is sometimes preferred. An example of the latter is shown in the total synthesis of (−)- and (+)-nonactic acids [(7*R*) and (8*S*)], which are the components of the macrotetrolide nonactin. The synthesis of the (7*R*)-component (*116*) **1437** begins with the transformation of D-mannose (**1433**) into the propionic ester of the furanoid glycal **1434**. In the presence of lithium diisopropylamide/tetrahydrofuran/trimethylsilyl chloride, **1434** gives **1436** through the boat transition state **1435**; **1436** can be

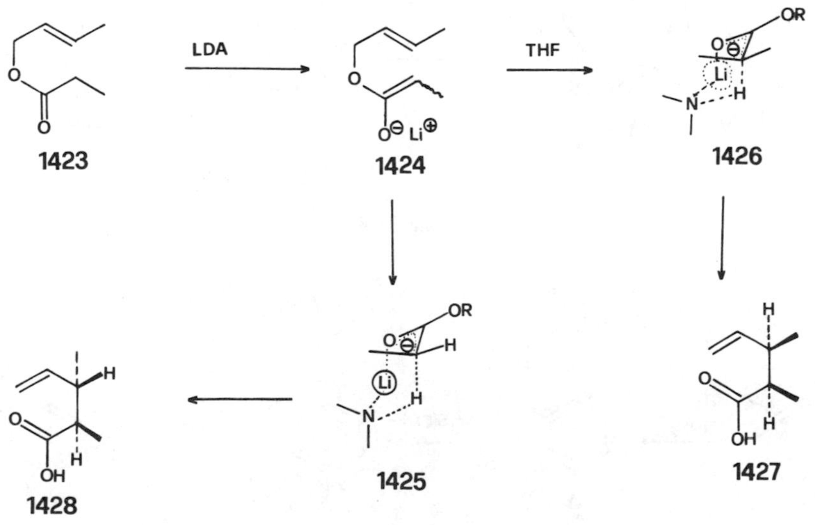

Scheme 7.73

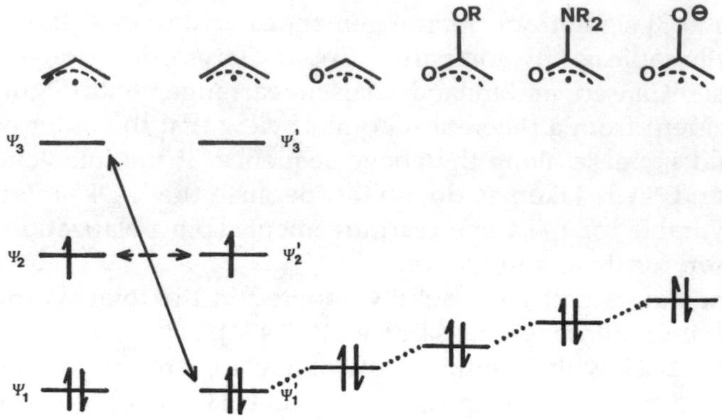

Figure 7.8. FMO interactions in the Cope and Claisen rearrangements.

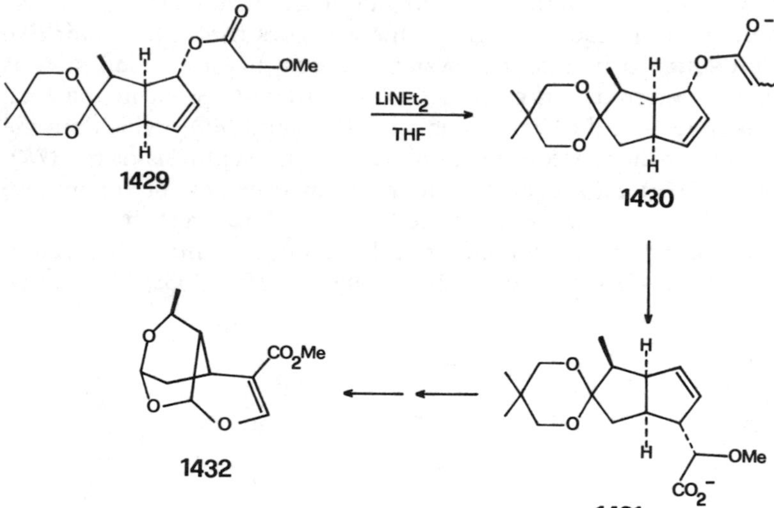

Scheme 7.74

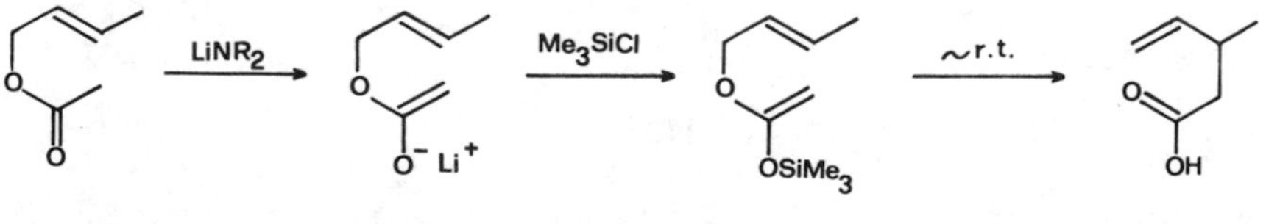

Scheme 7.75

further converted into **1437**. The (+)-acid was synthesized analogously starting from D-glucono-γ-lactone (*116*) (Scheme 7.76).

Ireland applied his variant to the synthesis of several natural products. Dihydrojasmone (**1213**) (*117*) was obtained from 2-ethoxy-2-octenyl propionate (**1438**), through **1439**, which in three steps gave the final product. Analogously, the prostaglandin skeleton (*117*) was built up starting from the ester **1440**, converted by lithium diisopropylamide/tetrahydrofuran/dimethylbutylsilyl chloride into **1441**. This product rearranges in refluxing methanol to **1442**, which is easily transformed by standard reactions into **1443** (Scheme 7.77).

The third and more fascinating target synthesized in this way by Ireland et al. (*118*) was the antibiotic lasalocid A (**982**), a polyether possessing a potent physiological activity as cation carrier. This compound undergoes reverse aldol-type cleavage to give the aldehyde **981** [this synthesis was outlined in Chapter 5 in the section entitled "Elimination of the Dienophile" (Scheme 5.130)] and the actual target. Treatment of the furanoid glycal **1444** with lithium diisopropylamide/hexamethylphosphoramide–tetrahydrofuran/trimethylsilyl chloride at room temperature followed by several standard steps gave **1445**, which was condensed with glycal **1446** to give **1447**. This last compound was transformed into **1448** in four steps (*118*) (Scheme 7.78).

The simpler antibiotic and antileukemic agent botryodiplodin (**1451**) was synthesized within three steps from (*Z*)-crotyl senecioate (**1449**). This ester rearranges to the (2*S*,3*R*)-acid **1450** (yield 88%, ratio with

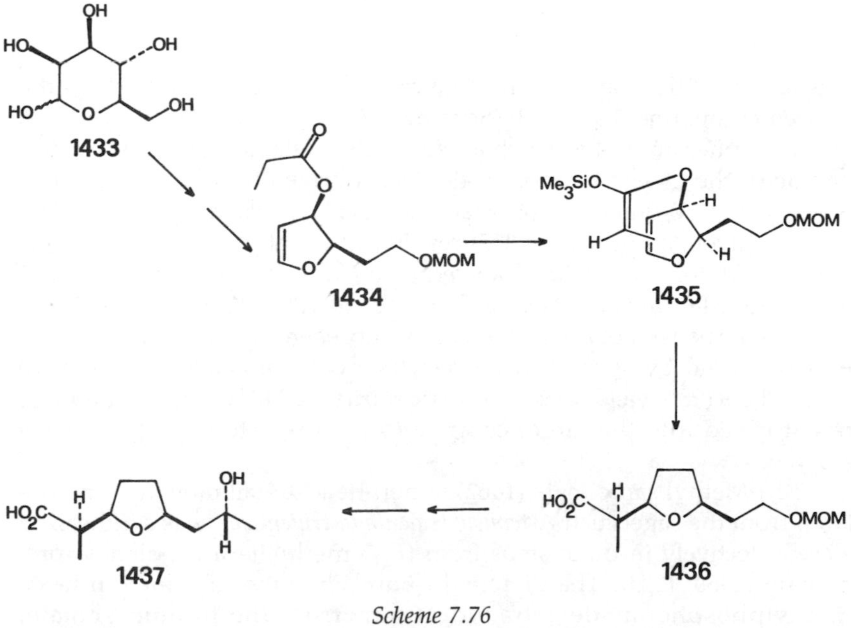

HO, OH, OH, HO, O, OH
1433
O, O, O, OMOM
1434
Me_3SiO, O, H, OMOM, H, O, H
1435
HO_2C, H, O, OMOM
1436
OH, HO_2C, H, O, H
1437

Scheme 7.76

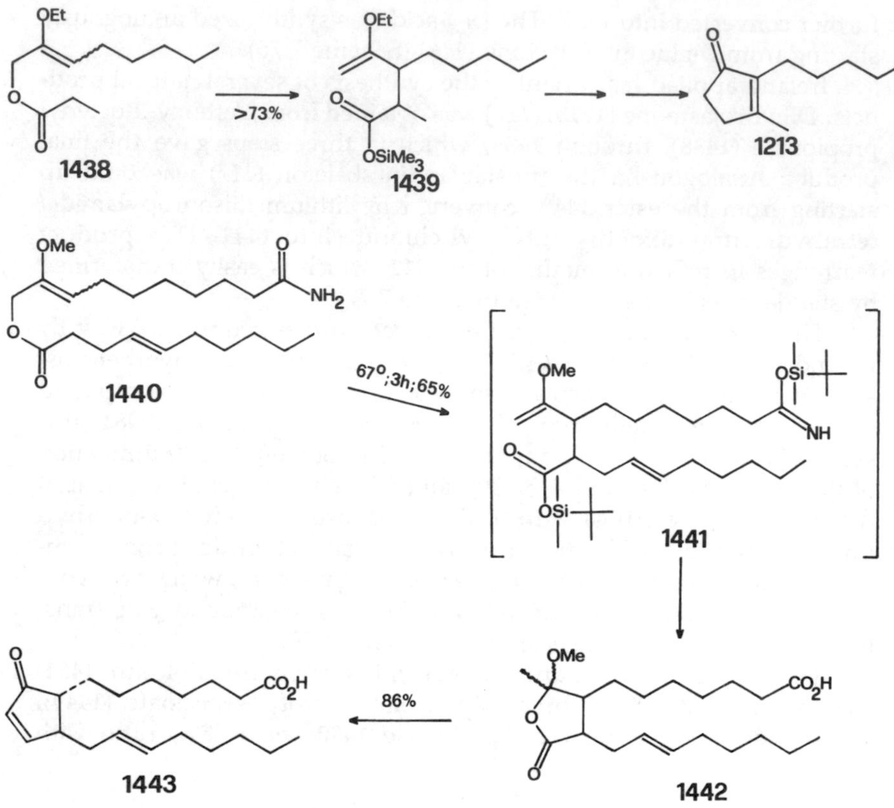

Scheme 7.77

diastereomer 91:9), and can be converted by standard methods to the desired compound **1451** (*119*) (Scheme 7.79).

Two pheromones were obtained by the Ireland variant. The pheromone of the Queen butterfly (**1454**) was derived from (*E*)-3-acetoxy-8-mesitoyloxy-2,6-dimethyl-1,6-octadiene (**1452**) via acid **1453** (*120*). The *Cecropia* juvenile hormone (**1355**) resulted through a Claisen rearrangement of (2*E*,5*Z*)-ester **1455**, followed by a Cope rearrangement of **1456**, which afforded the acid **1457**, a precursor of **1355** (*121*) (Scheme 7.80).

The terpenes field is conveniently represented. Shyobunone (**327**) was prepared by starting from neryl senecioate (**1458**), which gave *threoid*-**1459** (98% yield, diastereomeric purity 82%). This compound was transformed into the target compound in three steps (*122*) (Scheme 7.81).

(±)-Methyl santolinate (**1462**), a non-head-to-tail monoterpene isolated from the sagebrush *Artemisia tridentada tridentada*, was synthesized stereoselectively in three steps from (*E*)-5-methylhexa-2,4-dienoyl propionate (**1460**) (*123*). The dimethyl *t*-butylsilyl ether obtained in hexamethylphosphoramide–tetrahydrofuran from the lithium enolate,

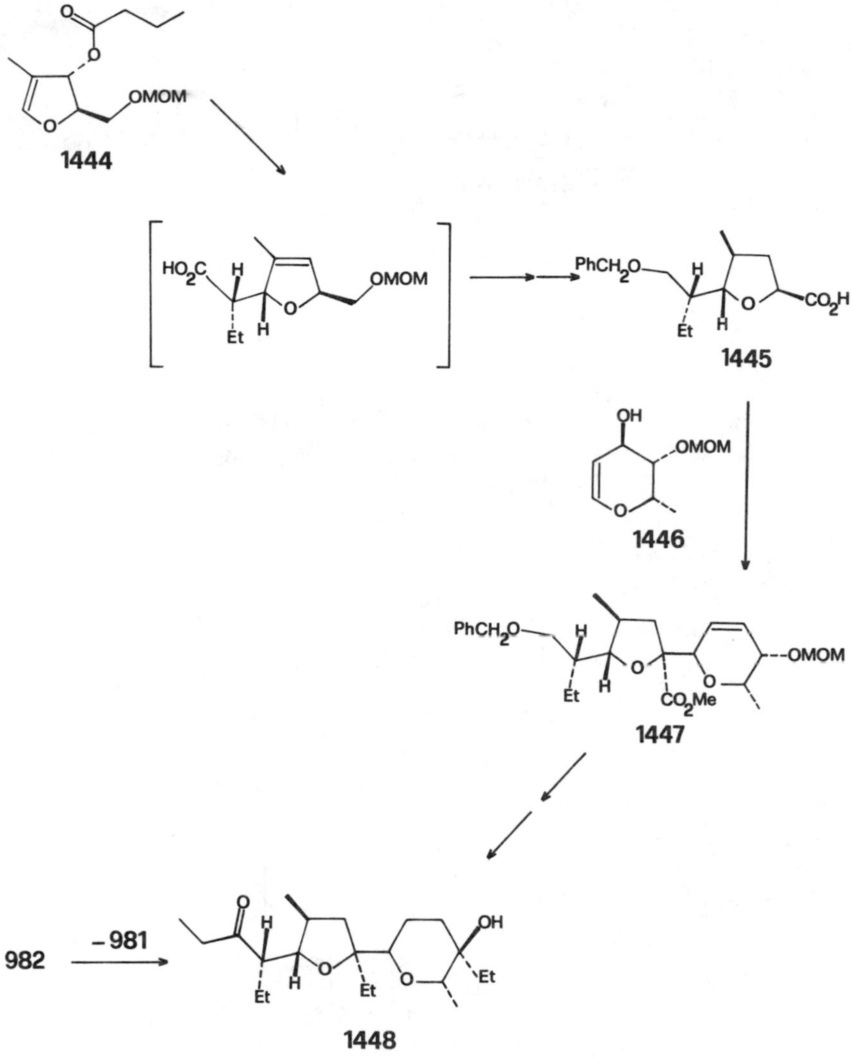

1444

1445

1446

1447

982 — 981 →

1448

Scheme 7.78

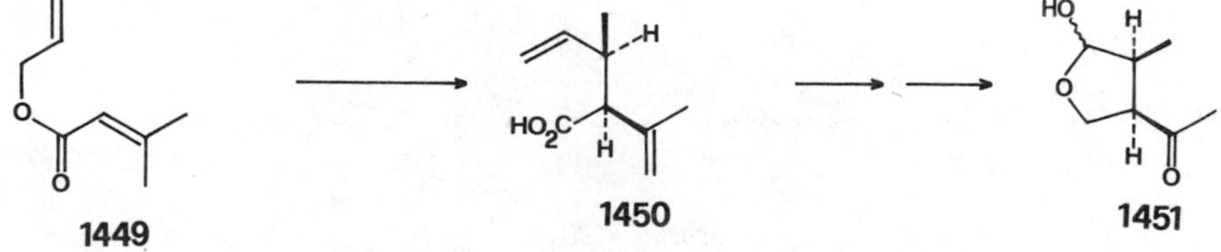

1449

1450

1451

Scheme 7.79

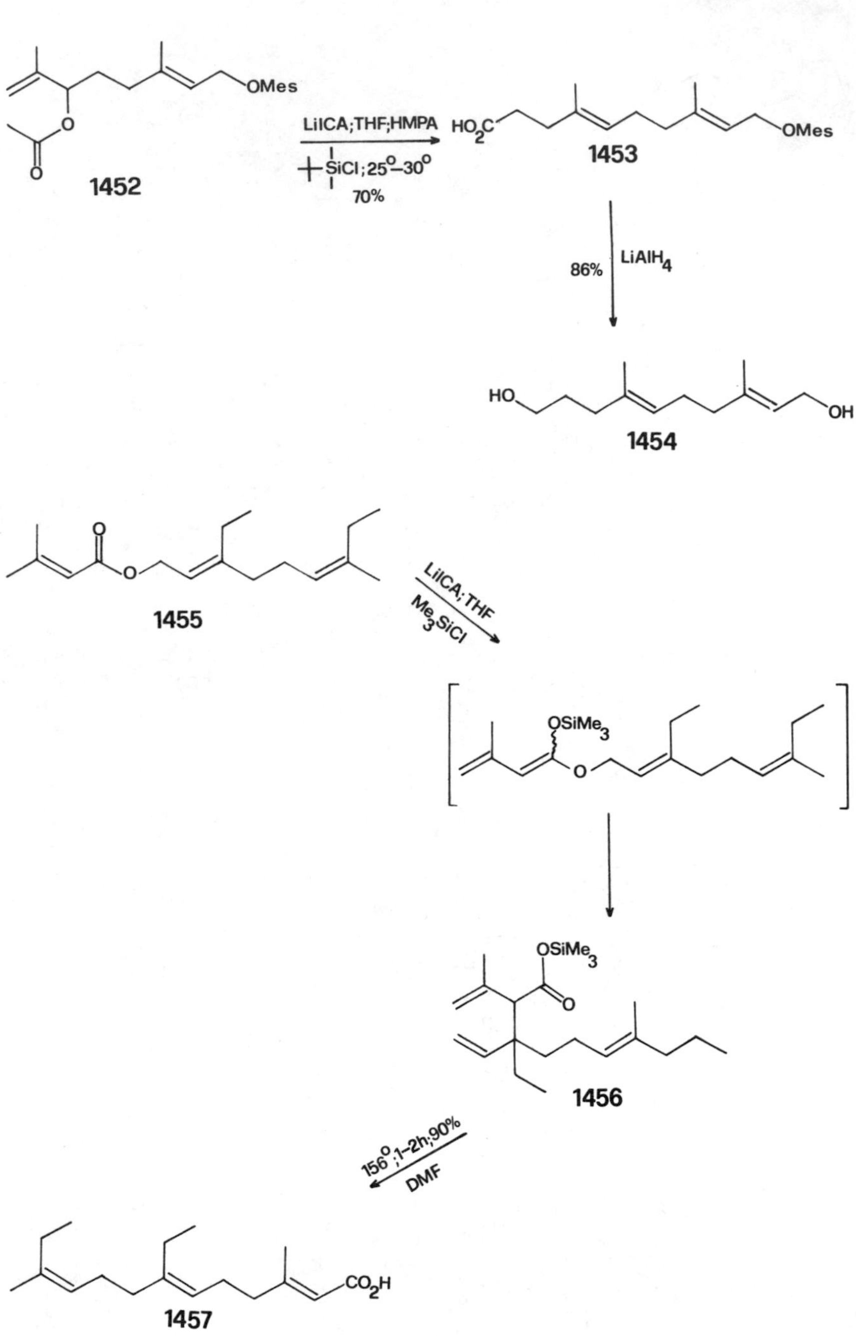

OMes

1452

LiICA;THF;HMPA

$+SiCl$; 25°–30°

70%

HO_2C

OMes

1453

86%

$LiAlH_4$

HO

OH

1454

O

O

1455

LiICA; THF

Me_3SiCl

$OSiMe_3$

O

$OSiMe_3$

O

1456

156°; 1–2h; 90%

DMF

CO_2H

1457

Scheme 7.80

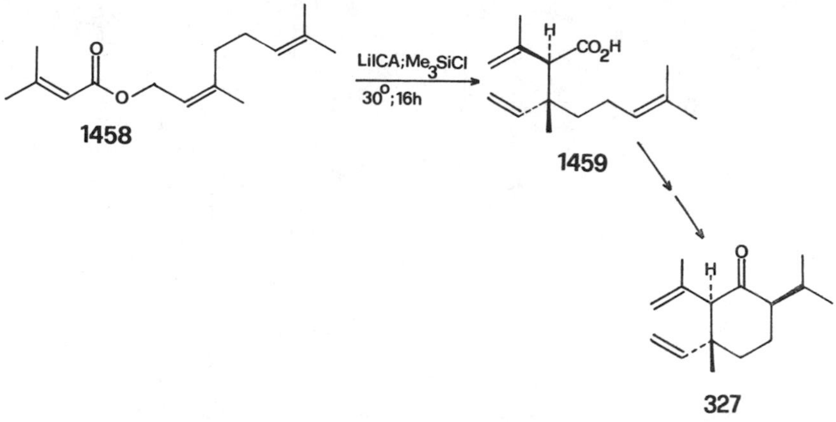

1458

LiICA;Me_3SiCl
30°;16h

1459

327

Scheme 7.81

which should be almost exclusively the (*E,E*)-isomer **1461**, was heated at 65 °C for 3 h. After hydrolysis and methylation the required *erythro* isomer **1462** was the major product (ratio *erythro:threo,* 8:1) (Scheme 7.82).

A synthesis of the allergenic sesquiterpene (±)-frullanolide (**1465**) involved as a key step the conversion of the β-pyrrolidinopropionate **1463** to the acrylic ester **1464**. This process was realized by transforming the former into the corresponding silylketene acetal by treatment with lithium diisopropylamide/trimethylsilyl chloride followed by rearrangement in refluxing toluene (20 min) and elimination with dimethyl sulfoxide/methanol/potassium carbonate. Two standard steps gave the final product **1465** (*124*) (Scheme 7.83).

Hetero-Claisen Rearrangements. The Claisen rearrangement applied to allyl vinyl ether moieties with a different heteroatom substituted for the oxygen atom allows the approach to new functionalized compounds. The most common substitution occurs with sulfur.

THIO-CLAISEN REARRANGEMENTS. Suitably substituted allyl vinyl sulfides **1468** derived from **1466** and alkylated via anion **1467** decompose stereospecifically to (*E*)-γ,δ-unsaturated thioaldehydes **1469** (in general, aldehydes under hydrolytic conditions). The decomposition occurs through the more stable chair-like transition state **1468** having the substituent in the equatorial position (*125*) (Scheme 7.84).

Thus propylure (**1472**), the sex-attractant of the pink bollworm moth *Pectinophora gossypiella* Saunders, was obtained from **1471**, which was derived by decomposition of **1470** in the presence of calcium carbonate (*125*) (Scheme 7.85).

Reaction of the tetracyclic thiolactam **1473a,b** with 2-bromomethyl-1-butene or allyl bromide in the presence of bases produced **1474a,b**, which underwent a thio-Claisen rearrangement at room tem-

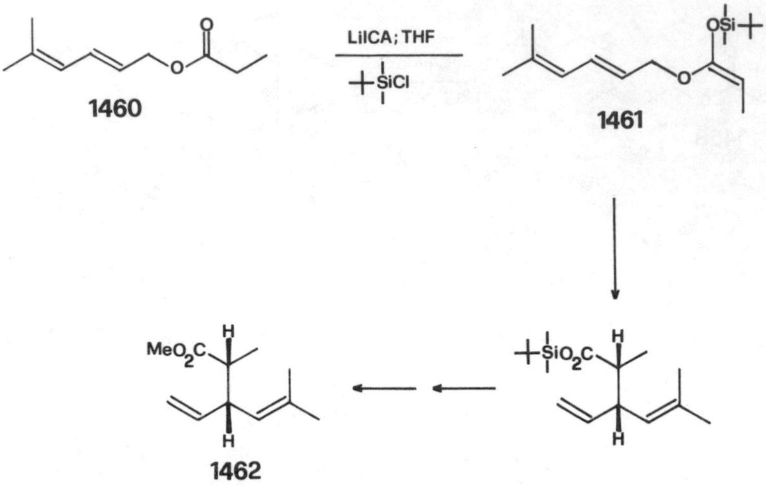

1460 LiICA; THF / SiCl 1461

1462

Scheme 7.82

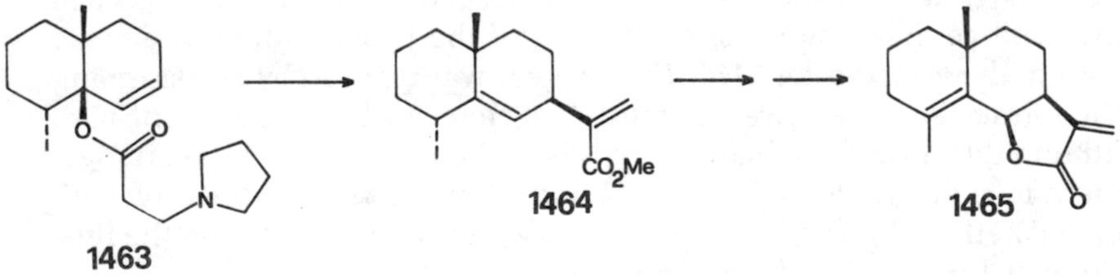

1463 1464 1465

Scheme 7.83

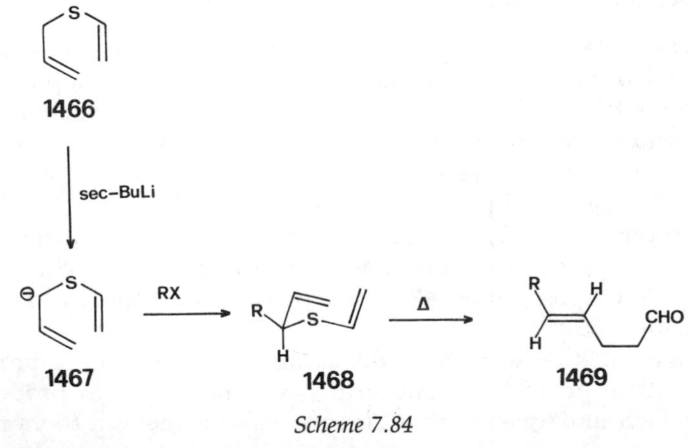

1466 sec-BuLi 1467 RX 1468 Δ 1469 CHO

Scheme 7.84

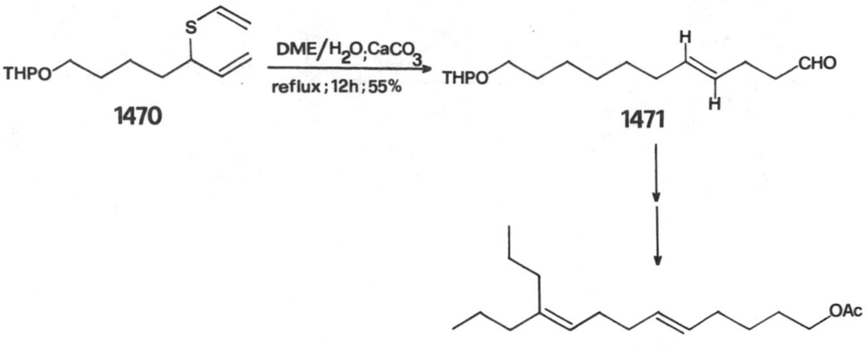

THPO 1470 → DME/H_2O;$CaCO_3$ reflux;12h;55% → THPO 1471 CHO → 1472 OAc

Scheme 7.85

perature to afford **1475a,b** (*126*). These compounds are the key intermediates for, respectively, 4α-dihydrocleavamine (**1476a**), a degradation product of the alkaloid catharanthine, and (±)-quebrachamine (**1476b**), a widely diffused alkaloid of the aspidospermine group (Scheme 7.86).

A nice sequence of [3,3] sigmatropic shifts starts with the thio-Claisen rearrangement of the sulfide **1477**, in turn obtained from 2-methyl-2-butenoylpyrrolidine and 9-bromolimonene. The rearrangement afforded a 3:2 diastereomeric mixture of thioamides **1478**. A Cope rearrangement in refluxing decaline and conversion of thioamide to the ester gave ethyl (*E*)-lanceolate (**1479**) (*127*) in a 2:1 mixture with its (*Z*)-isomer, the precursor (*128*) of sesquiterpene (*E*)-lanceol (**1480**) (Scheme 7.87).

MISCELLANEOUS HETERO-CLAISEN REARRANGEMENTS. In this section we include several [3,3] sigmatropic rearrangements related, from the formal point of view, to the Claisen reaction. In each case, a different heteroatom, generally sulfur or nitrogen, is substituted for the oxygen and/or a carbon of the allyl vinyl ether moiety.

Allyl thiocarbamate **1481** (prepared in situ from the suitable allyl alcohol and *N,N*-dimethylthiocarbamoyl chloride) rearranges to (*E*)-allyl thiocarbamate (**1482**) without detectable traces of other isomers. This compound can be cleanly transformed either into (*E*)-2,4-dimethyl-2-hexenoic acid (**1483**) or into the so-called manicone (**1484**), two alarm pheromones of ants (*129*) (Scheme 7.88).

Allylic dithiocarbamates **1485** can undergo rearrangement in refluxing chloroform, and, taking advantage of the acidity of the protons α to sulfur before and after the rearrangement, a variety of pheromones can be synthesized. (*E*)-6-Nonen-1-ol (**302**), (*E*)-7-dodecen-1-ol (**1486a**), and (*E*)-7-tetradecen-1-ol (**1486b**), which are all sex pheromones or attractants of lepidopterous insects, were thus synthesized (*130*) (Scheme 7.89).

The imidate–amide (**1487–1488**) rearrangement, occurring under the mildest conditions because of the exothermicity of the reaction, was

Scheme 7.86

Scheme 7.87

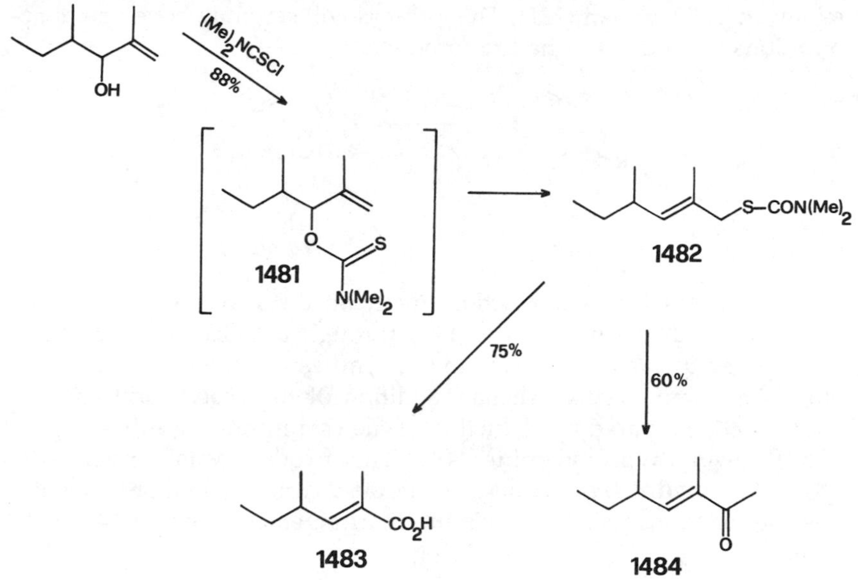
(Me)2NCSCl
88%
OH
O
S
1481
N(Me)2
S-CON(Me)2
1482
75%
60%
CO2H
1483
O
1484

Scheme 7.88

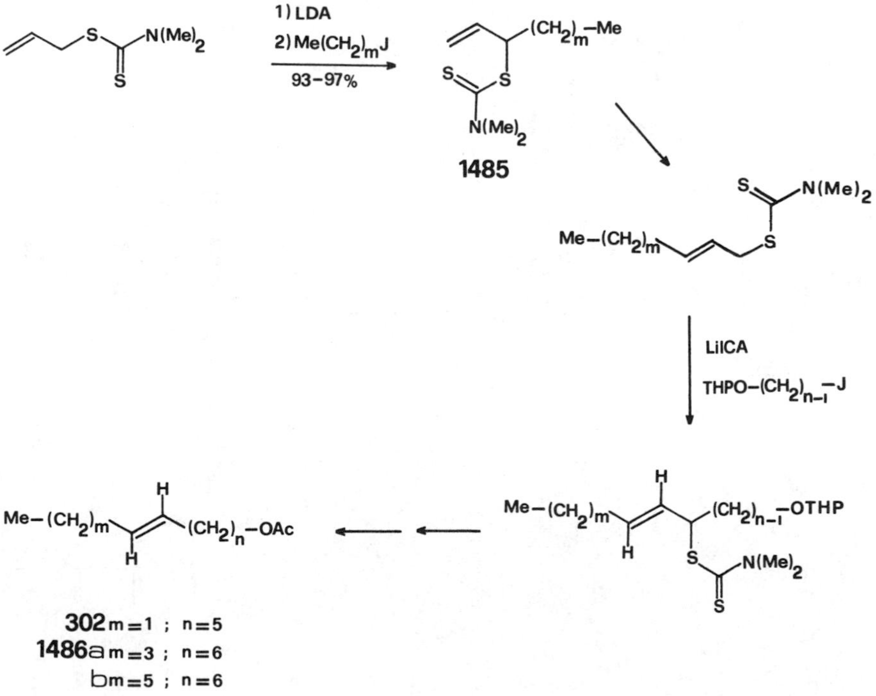
S
N(Me)2
S
1) LDA
2) Me(CH2)mJ
93-97%
(CH2)m-Me
S
S
N(Me)2
1485
S
N(Me)2
Me-(CH2)m
S
LiICA
THPO-(CH2)n-1-J
H
Me-(CH2)m
(CH2)n-1-OTHP
H
S
N(Me)2
S
H
Me-(CH2)m
(CH2)n-OAc
H
302 m=1 ; n=5
1486a m=3 ; n=6
b m=5 ; n=6

Scheme 7.89

reviewed by Overman (*131*). This process will certainly have useful applications in the field of natural products.

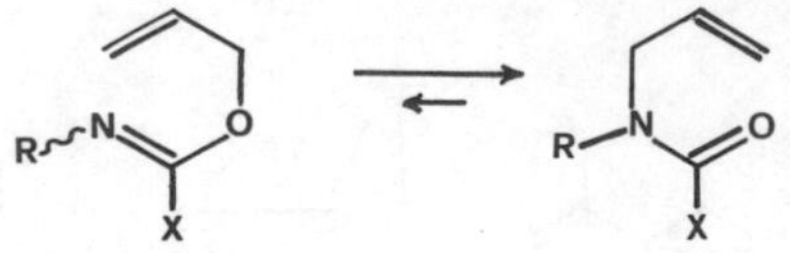

1487 1488

A somewhat similar reaction involving the rearrangement of hydroxylamine derivative **1489** to **1490** through a 1-aza-1′-oxa [3,3] sigmatropic shift was the key step in the synthesis of the mitomycin skeleton. An intramolecular Michael addition of the acetylenic hydroxylamine **1491**, prepared by reduction of the corresponding nitrone, gave *N*-aryl-*O*-vinylhydroxylamine **1492**. This product was converted to ethyl 2,3-dihydro-1*H*-pyrrolo[1,2-*a*]indole-9-carboxylate (**1494**), which has the mitomycin skeleton, via the rearrangement product **1493** (*132*) (Scheme 7.90).

Polar [3,3] Rearrangements. We have discussed [3,3] sigmatropic rearrangements involving six neutral atoms. When one of the atoms, generally a heteroatom, is charged, polar [3,3] rearrangements take place.

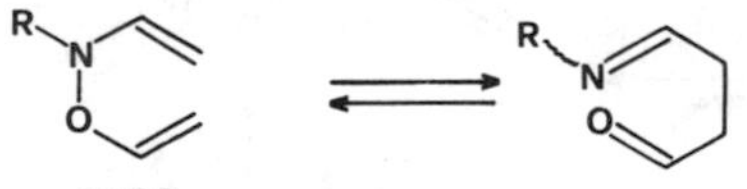

1489 1490

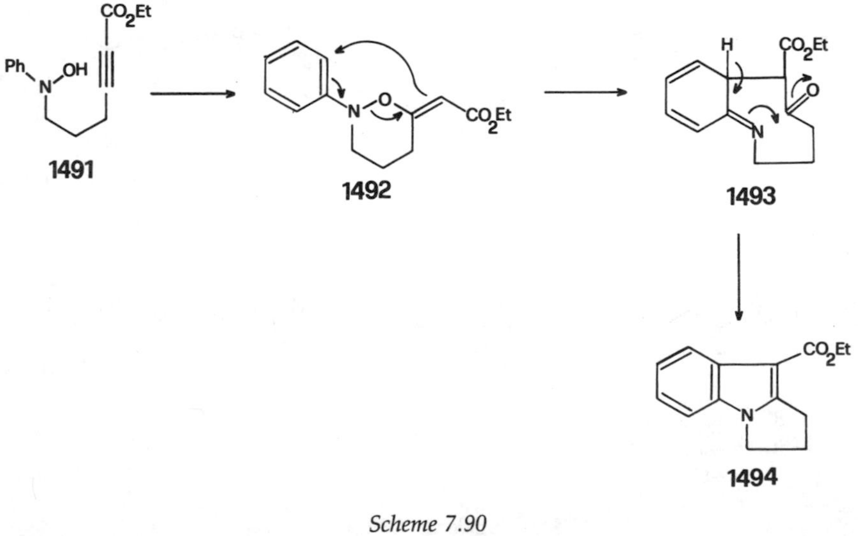

1491 1492 1493

1494

Scheme 7.90

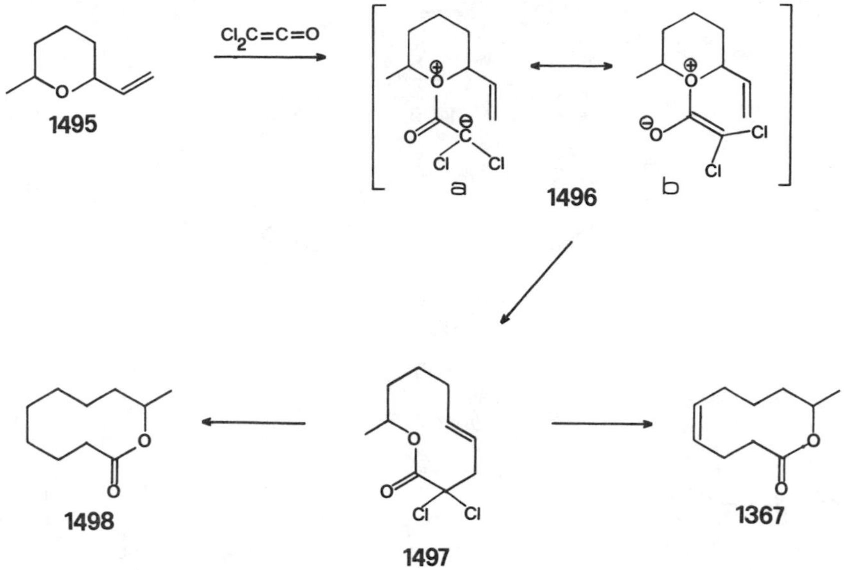

Scheme 7.91

A rather unusual example (because the second double bond involved is polarized to a negative charge if resonance structure **1496b** is not taken into account) is offered by the syntheses of two naturally occurring macrolides: (±)-phoracantholide I (**1498**) and (±)-phoracantholide J (**1367**). Thus the tetrahydropyran **1495**, derived from the heterodiene dimer of methyl vinyl ketone, reacted with dichloroketene to give a dipolar intermediate **1496**. This intermediate undergoes a dipolar-Claisen rearrangement to give **1497**, which is then easily converted to the desired macrolides (*133*) (Scheme 7.91).

A very useful polar [3,3] sigmatropic rearrangement is the so-called 2-azonia process, which is essentially a Cope rearrangement where a positively charged nitrogen is substituted for a C2 carbon. The reaction conditions are mild (generally, below 100 °C) (*134*). The process has been proposed to occur either via a concerted mechanism, through transition state **1499**, or via a two-step mechanism involving an intermediate carbonium ion **1500** (*135*) (Scheme 7.92).

Useful application of the 2-azonia rearrangement in the natural products field is illustrated by the transformation of 1-allylnorharman derivative **1501** into indoloquinolizidine derivative **1502** (*136*). This product has the basic skeleton of corynantheine alkaloids. The reaction proceeds in the presence of formaldehyde and after stereoselective cyclization promoted by the solvent (Scheme 7.93).

Further development of this aproach provided a simple access to the yohimbane skeleton. This synthesis started from β-carboline **1503**, which was reduced to **1504** and treated with formaldehyde at room

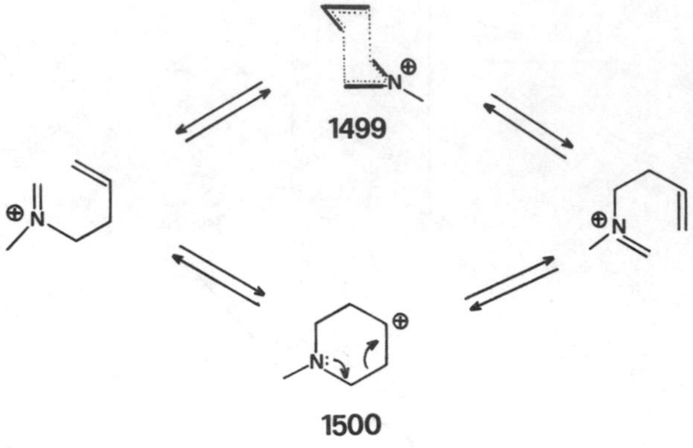

1499

1500

Scheme 7.92

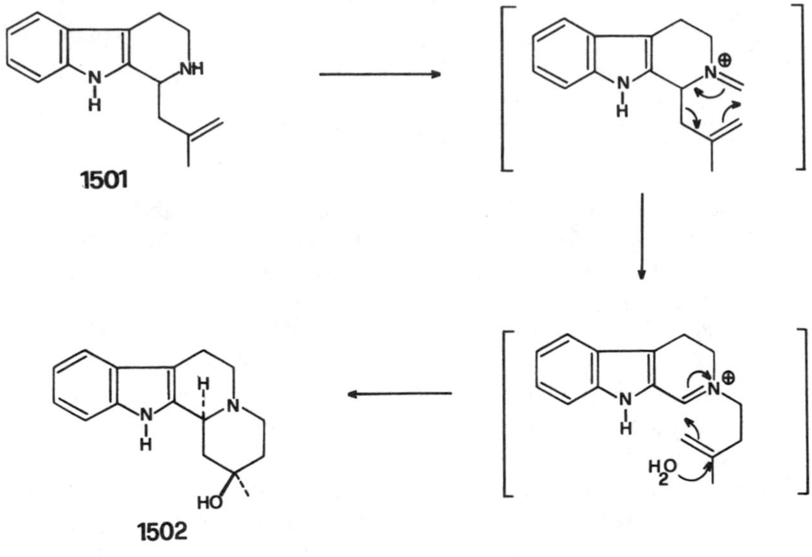

1501

1502

Scheme 7.93

temperature to yield 96% of 15β-methoxy-yohimban-17-ol (**1505**) (*137*) (Scheme 7.94).

A more recent application of the 2-azonia rearrangement is the synthesis of (±)-perhydrogephyrotoxin (**558**) (*138*), mentioned in Chapter 5 because the starting material was obtained through a regiospecific Diels–Alder reaction. By heating the homoallylic amine **1506** in benzene in the presence of *p*-toluenesulfonic acid, a cyclization to iminium ion **1507** took place. This reaction was followed by a 2-azonia rearrangement that proceeds stereoselectively across the convex face of **1507** to give **1508**. A ring closure to the perhydropyrrolo[1,2-*a*]quinoline (**1509**) and its conversion to **558** completed the synthesis (Scheme 7.95).

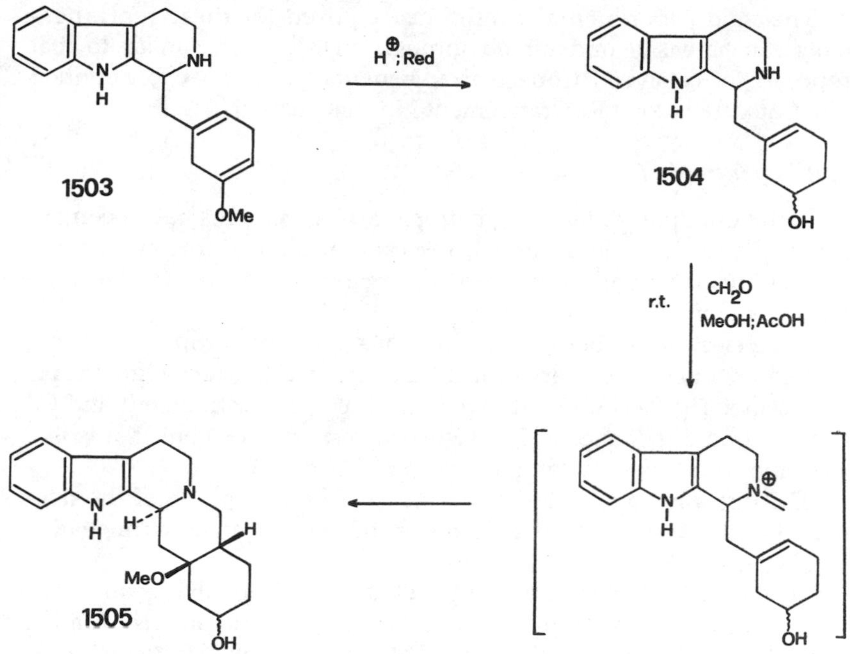

1503 → ($H^{\oplus}$; Red) → 1504 → (CH_2O, MeOH; AcOH, r.t.) → [intermediate] → 1505

Scheme 7.94

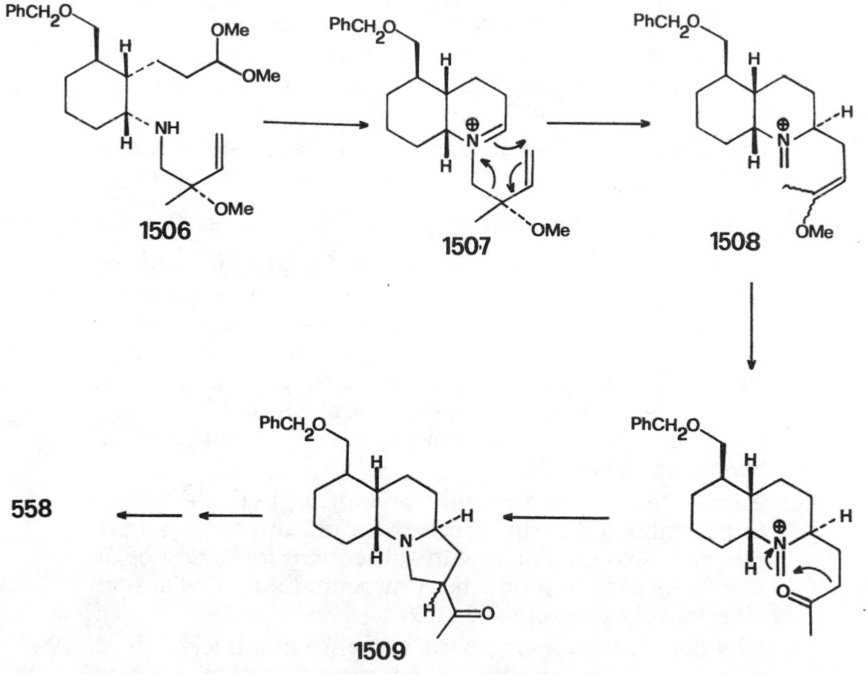

1506 → 1507 → 1508 → → 1509 → → 558

Scheme 7.95

The mild experimental conditions required for these rearrangements can be easily understood through an approach similar to that proposed for catalyzed Cope rearrangements (*see* the section entitled "The Catalyzed Cope Rearrangement" in this chapter, Figure 7.7).

[m,n] Sigmatropic Processes

Many examples of [*m*,*n*] sigmatropic rearrangements have been described, for example, [3,4] sigmatropic cationic shifts (*1*). However, the most fruitful reactions in the field of organic syntheses are the [2,3] rearrangements.

If the electrons of one double bond of a [3,3] sigmatropic rearrangement lie as a negative charge on the carbon at C2, a new sigmatropic shift occurs. This second shift involves C2 of the first fragment and C3 of the second one; hence a [2,3] sigmatropic rearrangement that is isoelectronic with the [3,3] one takes place.

The difficult stage in these processes is the generation of the negative charge, particularly if it is not stabilized on an electronegative heteroatom.

Scheme 7.96 includes examples of processes involving both formally neutral species (as ylides or carbenes) and true anions. Among the routes involving neutral intermediates are the following reactions:

a) The rearrangement of sulfoxides to sulfenate esters, which are easily converted to allylic alcohols in refluxing methanol in the presence of nonahydrated sodium sulfide.
b) The conversion of sulfur methylides (prepared from sulfonium salts deprotonated by strong bases (e.g., *n*-butyllithium) into butenylic thioethers.
c) The conversion of allylic ammonium ylides (stabilized by electron-attracting substituents) to homoallylic amines.

The anions can be formed through the following reactions:

d) Allylic alcohols can be alkylated with iodomethyltributyltin to allyl stannylmethyl ethers which, in the presence of *n*-butyllithium, give a tin–lithium exchange. The resulting anion rearranges smoothly to the homoallylic alcohols. An alternative to this method is heating allylic alcohols with dimethylformamide acetal at 160 °C for several hours to give allylic amides, possibly through a carbenoid species (*139*).
e) Carbazates, in the presence of sodium hydride, give a fairly stabilized carbanion that rapidly undergoes rearrangement to thioanion with subsequent formation of dithioesters as final products. A carbenoid intermediate can alternatively be proposed (*139*).
f) Allyl benzyl thioethers, in the presence of *n*-butyllithium, give benzylic carbanions, which rapidly undergo a rearrangement to thioanions (*140*). The fate of the thioanions

depends on the reaction conditions; for example, they can be easily alkylated.

The five-membered ring transition state is of the "folded envelope" type; adoption of the preferred conformation with the substituent in the pseudo-equatorial **1510a** (*139*) or pseudo-axial position **1510b** (*141*) seems to depend on the steric interactions involved (Scheme 7.97).

The rearrangement of allylic sulfoxides to allylic alcohols is the key step in the approach to the hasubanan alkaloid skeleton **1512** and to verrucarol (**515**). The former compound was obtained from tetracyclic sulfoxide **1511**, which was constructed by a Diels–Alder reaction (*142*).

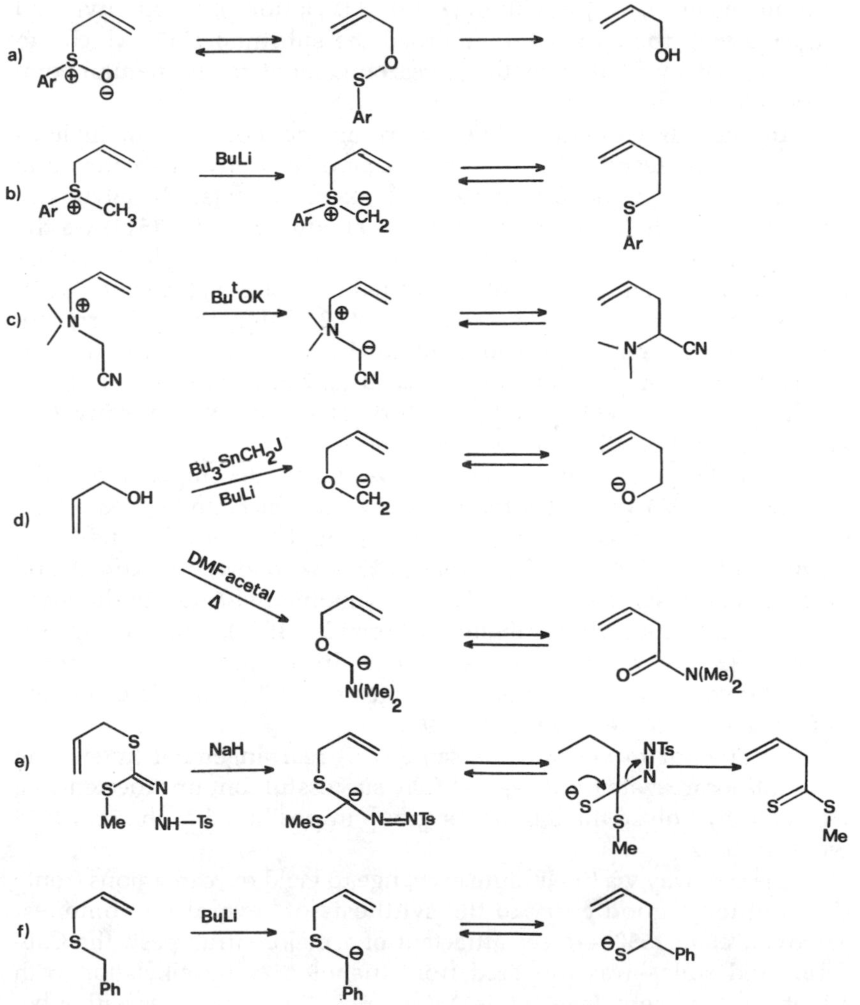

a) Ar ⊕ S O ⊖ ⇌ S O Ar → OH

b) Ar ⊕ S CH_3 —BuLi→ Ar ⊕ S ⊖ CH_2 ⇌ S Ar

c) ⊕ N CN —Bu^tOK→ ⊕ N ⊖ CN ⇌ N CN

d) OH —Bu_3SnCH_2J, BuLi→ O ⊖ CH_2 ⇌ ⊖O

—DMF acetal, Δ→ O ⊖ $N(Me)_2$ ⇌ O $N(Me)_2$

e) S S N Me NH—Ts —NaH→ S ⊖ MeS N═NTs ⇌ ⊖S NTs N S Me → S S Me

f) S Ph —BuLi→ S ⊖ Ph ⇌ ⊖S Ph

Scheme 7.96

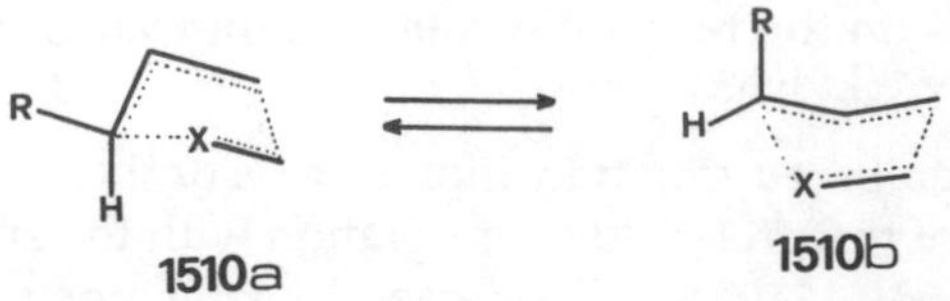

1510a **1510b**

Scheme 7.97

Verrucarol was formed by rearrangement of the benzofuran sulfoxide **1513** to the alcohol derivative **1514**; ring opening to tetrahydrochromanone (*143*) (**1515**) allows the conversion to **515**, the desired product (Scheme 7.98).

A nice application of sulfoxide rearrangement is the synthesis of sesquiterpene (±)-(*E*)-nuciferol (**1518**). Alkylation of isobutenyl tolyl sulfoxide with the iodide **1516** afforded the substrate **1517**, which can be converted by [2,3] sigmatropic rearrangement to the natural compound (*144*) (Scheme 7.99).

An interesting example of [2,3] rearrangement of sulfur methylides (route b) was described in the synthesis of isodehydroabietenolide (**1522**), used as a model for further antileukemic targets. The allylic sulfonium methylide **1520**, prepared from dehydroabietone (**1519**) via the corresponding sulfonium salt, rearranged to the butenylic thioether **1521**. This thioether can be easily converted to **1522** (*145*) (Scheme 7.100).

We discuss in detail the application of [2,3] sigmatropic rearrangement of allylic ammonium ylides (route c) because it enables the synthesis of (±)-14-norhelminthosporic acid (**1528**). This compound is a synthetic gibberellin-like plant growth regulator; its synthesis involves other pericyclic reactions as well (Scheme 7.101).

In the first stages of the synthetic sequence, decomposition of the diazoketone **1523** generates the ketocarbene, which undergoes a [1,2] intramolecular cycloaddition to the cyclopropyl ketone **1524**. Its transformation to the bicyclo[3.2.1]ketone **1525** was followed by conversion of the latter by standard methods to *N*-cyanomethyl-*N*-(13,14-dinorhelminthospor-6-en-12-yl)pyrrolidinium bromide (**1526**), the starting material for [2,3] sigmatropic rearrangement. In the presence of a strong base, the corresponding anion rearranges to **1527**, which is converted in few steps into the target **1528** (*146*).

An attempt to perform the same [2,3] rearrangement by starting from dithiocarbazate (route e) was fully successful, but unexpectedly all efforts to hydrolyze the dithioester group in position 6 to the acid **1528** were fruitless.

The pathway via tin–lithium exchange to yield oxycarbanions (route d) provided a good entry to the synthesis of several pheromones. Trienyl acetate **1530b**—a sex-attractant of a major citrus pest, the California red scale—was prepared from trienol **1529** by alkylation with tributylstannylmethylene iodide. Its stannyl ether exchanges with *n*-butylithium to generate an oxycarbanion, which rearranges to homoallyic alcohol **1530a**. This alcohol was acylated with acetic anhydride in pyridine to **1530b** (*141*). The *Cecropia* juvenile hormone **1355** was obtained through a similar, but more sophisticated, rearrangement. The trisallylic alcohol derivative **1531** was treated with tributylstannylmethylene iodide to give **1532**. In the presence of *n*-butyllithium, **1532** gave a double [2,3] sigmatropic rearrangement of the oxycarbondianion **1533** to yield bis-homofarnesyl derivative **1534** (*147*), which can be converted into **1535** (Scheme 7.102).

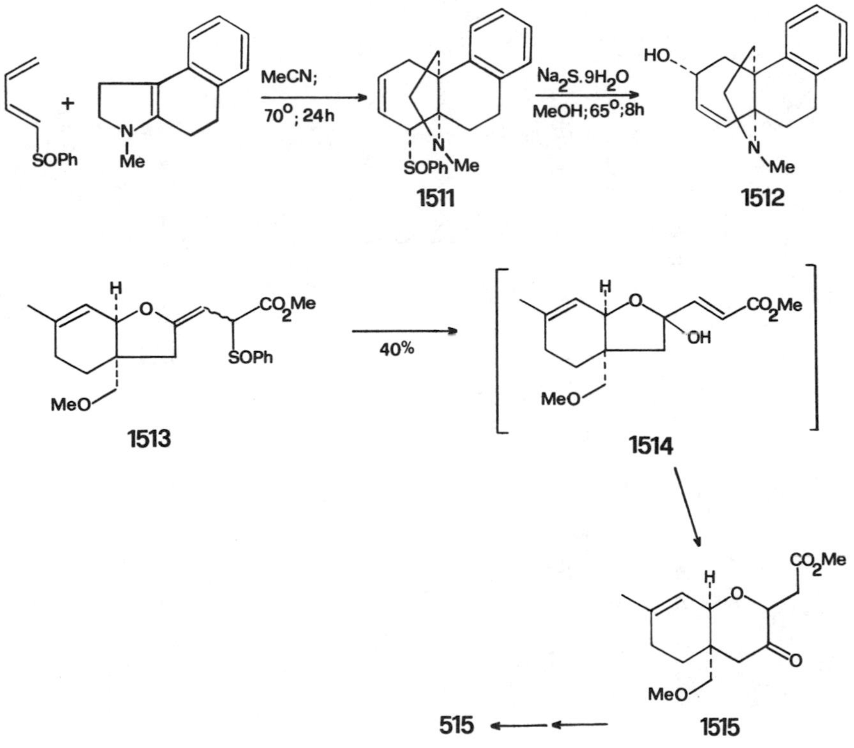

SOPh + N Me — MeCN; 70°; 24h → **1511** (SOPh, N, Me) — $Na_2S.9H_2O$, MeOH; 65°; 8h → **1512** (HO, N, Me)

1513 (H, O, CO_2Me, SOPh, MeO) — 40% → [**1514** (H, O, OH, CO_2Me, MeO)] → **1515** (H, O, CO_2Me, O, MeO) → → **515**

Scheme 7.98

1516 (H, CH_2J) — S(O)Ar; BuLi; THF; −50° → **1517** (H, S, Ar, O) — 58% → **1518** (H, HO)

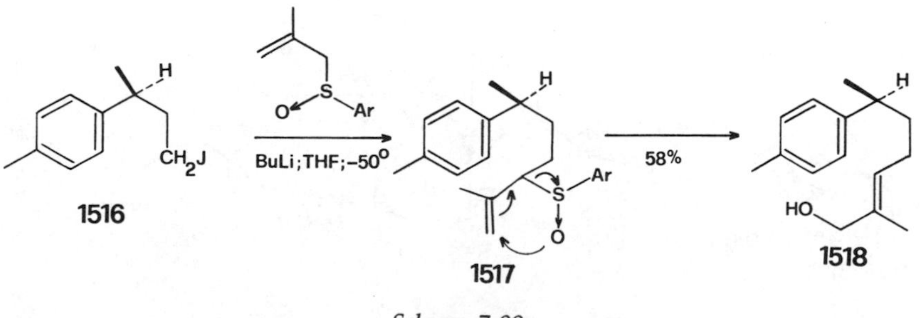

Scheme 7.99

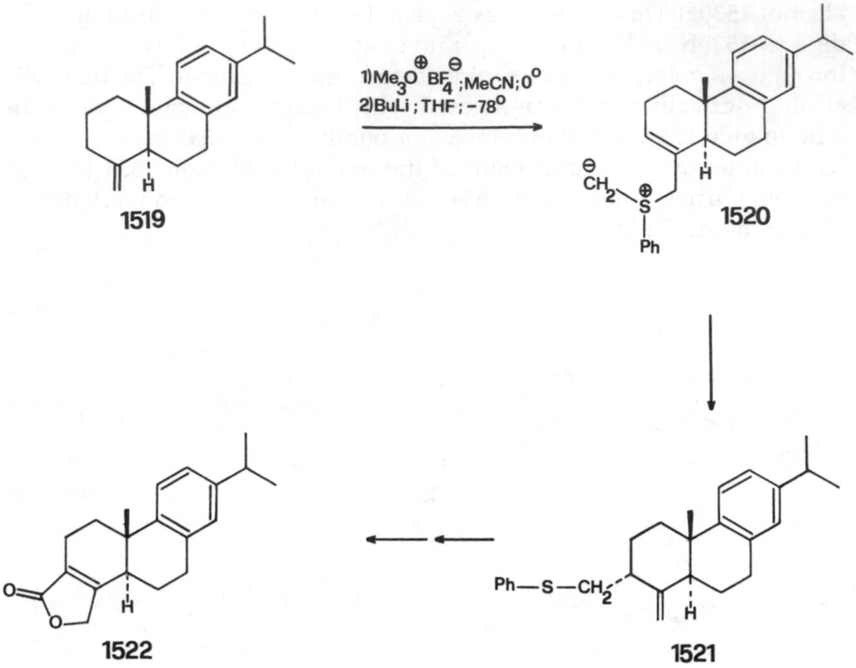

1519

1) $Me_3O^{\oplus}BF_4^{\ominus}$; MeCN; 0°
2) BuLi; THF; −78°

1520

1521

1522

Scheme 7.100

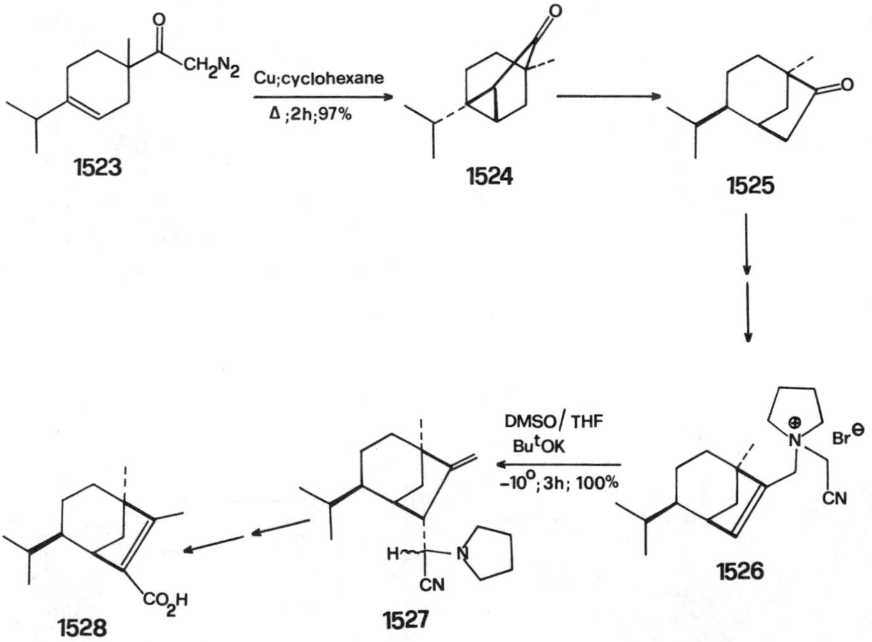

1523

Cu; cyclohexane
Δ; 2h; 97%

1524

1525

1526

DMSO/THF
Bu^tOK
−10°; 3h; 100%

1527

1528

Scheme 7.101

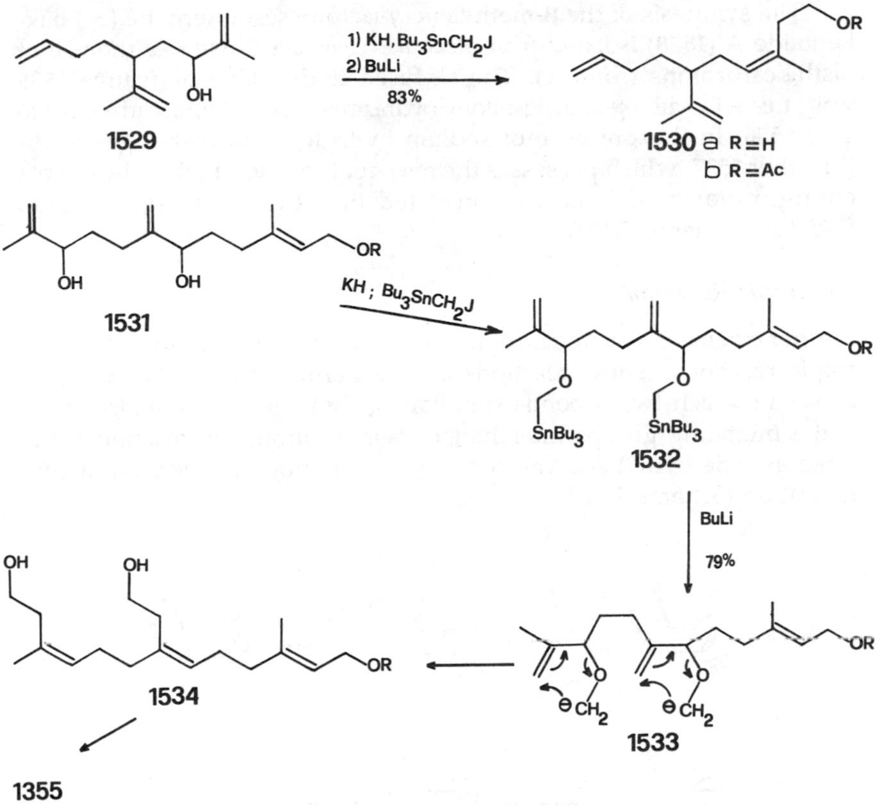

1) KH, Bu_3SnCH_2J
2) BuLi
83%

1529

1530 a R=H
b R=Ac

1531

KH ; Bu_3SnCH_2J

1532

BuLi
79%

1533

1534

1355

Scheme 7.102

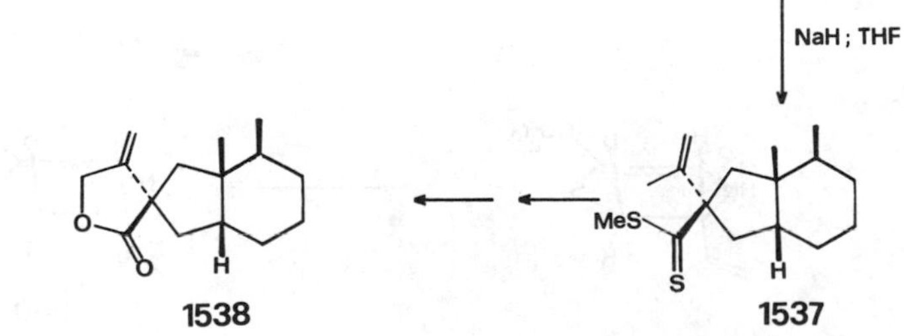

Ts—NH—NH—CS_2Me

1535

1536

NaH ; THF

1537

1538

Scheme 7.103

The synthesis of the β-methylene-γ-lactone sesquiterpene (±)-bakkenolide A (**1538**) isolated from *Petasides japonicus* illustrates the use of bisthiocarbanions (route c). The *cis*-fused hydrindane derivative **1535** was treated with *p*-toluenesulfonylhydrazine-*S*-methyldithiocarbazate to give **1536**. In the presence of sodium hydride, **1536** rearranges to dithioether **1537**, which possesses the new quaternary center in the correct configuration and is easily converted into the final product **1538** (*148,149*) (Scheme 7.103).

Dyotropic Reactions

Further extensions of sigmatropic rearrangements are the so-called dyotropic reactions. These reactions can be defined as uncatalyzed processes in which two σ-bonds simultaneously migrate intramolecularly. If the migrating groups interchange their position, the reaction is defined as type I (*150*) and can occur with retention or inversion of configuration (Scheme 7.104).

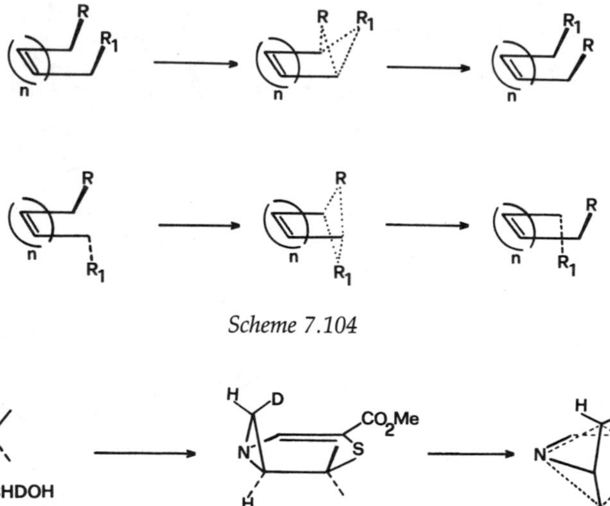

Scheme 7.104

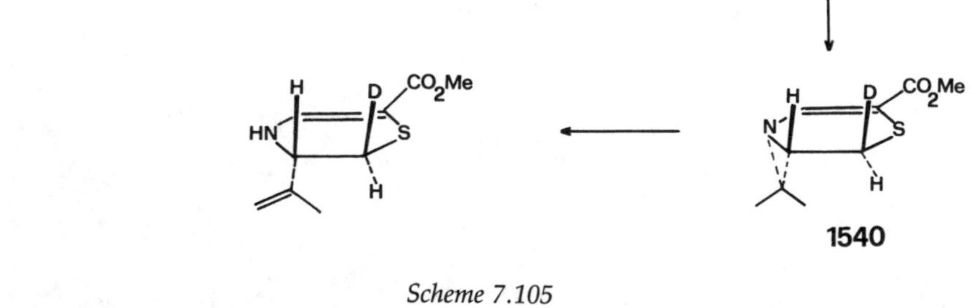

Scheme 7.105

If the migrating groups do not interchange their position, the reaction is defined as type II (*151*).

Although these reactions have not yet been used to synthesize natural products, we report here an example of future developments. In a field related to penicillanoyl derivatives, a type I [4,4] dyotropic shift occurs when **1539** rearranges to **1540**. The groups interchange with retention of configuration (*152*) (Scheme 7.105).

Literature Cited

1. Woodward, R. B.; Hoffmann, R. *Angew. Chem., Int. Ed. Engl.* **1969**, *8*, 781.
2. Martin, S. F. *Tetrahedron* **1980**, *36*, 419.
3. Rappold, M. P.; Havinga, E. *Recl. Trav. Chim. Pays–Bas* **1960**, *79*, 369.
4. Havinga, E.; Schlatmann, J. L. M. A. *Tetrahedron* **1961**, *16*, 146.
5. Autrey, R. L.; Barton, D. H. R.; Canguly, A. K.; Reusch, W. H. *J. Chem. Soc.* **1961**, 3313.
6. Jeger, O.; Redel, J.; Nowak, R. *Helv. Chim. Acta* **1946**, *29*, 1241.
7. Jacobs, H. J. C.; Havinga, E. *Adv. Photochem.* **1979**, *11*, 305.
8. McMurry, J. E.; Silvestri, M. G. *J. Org. Chem.* **1976**, *41*, 3953.
9. Ho, T. L. *Synth. Commun.* **1977**, *7*, 351.
10. Knudsen, C. G.; Carey, S. C.; Okamura, W. H. *J. Am. Chem. Soc.* **1980**, *102*, 6355.
11. Froborg, J.; Magnusson, G. *J. Am. Chem. Soc.* **1978**, *100*, 6728.
12. Hammond, M. L.; Mourino, A.; Okamura, W. H. *J. Am. Chem. Soc.* **1978**, *100*, 4907.
13. Crump, D. R.; Williams, D. H.; Pelc, B. *J. Chem. Soc., Perkin Trans. 1* **1973**, 2731.
14. Onisko, B. L.; Schnoes, H. K.; DeLuca, H. F. *J. Org. Chem.* **1978**, *43*, 3441.
15. Eschenmoser, A. *Naturwissenschaften* **1974**, *61*, 513.
16. Yamada, Y.; Miljkovic, D.; Wehrli, P.; Golding, B.; Löliger, P.; Keese, R.; Müller, K.; Eschenmoser, A. *Angew. Chem., Int. Ed. Engl.* **1969**, *8*, 343.
17. VonDoering, W. E.; Roth, W. R. *Tetrahedron* **1962**, *18*, 67.
18. Ahlgren, G. *Tetrahedron Lett.* **1979**, 915.
19. VonDoering, W. E.; Toscano, V. G.; Beasley, G. H. *Tetrahedron* **1971**, *27*, 5299.
20. Hammond, G. S.; Deboer, C. D. *J. Am. Chem. Soc.* **1964**, *86*, 899.
21. Brown, J. M.; Golding, B. T.; Stofko, J. J., Jr., *J. Chem. Soc., Chem. Commun.* **1973**, 319.
22. Evans, D. A.; Golob, A. M. *J. Am. Chem. Soc.* **1975**, *97*, 4765.
23. Billups, W. E.; Chow, W. Y.; Cross, J. H. *J. Chem. Soc., Chem. Commun.* **1974**, 252.
24. Piers, E.; Ruediger, E. H. *J. Chem. Soc., Chem. Commun.* **1979**, 168.
25. Van der Baan, J. L.; Bicklhaupt, F. *Recl. Trav. Chim. Pays–Bas* **1975**, *94*, 109.
26. Onishi, T.; Fujita, Y.; Nishida, T. *J. Chem. Soc., Chem. Commun.* **1978**, 651.
27. Place, P.; Roumestant, M. L.; Gore, J. *J. Org. Chem.* **1978**, *43*, 1001.
28. Onishi, T.; Fujita, Y.; Nishida, T. *Synthesis*, in press.
29. Fujita, Y.; Onishi, T.; Hino, K.; Nishida, T. *Tetrahedron Lett.* **1980**, 1347.
30. Evans, D. A.; Golob, A. M.; Mandel, N. S.; Mandel, G. S. *J. Am. Chem. Soc.* **1978**, *100*, 8170.
31. Jung, M. E.; Hudspeth, J. P. *J. Am. Chem. Soc.* **1980**, *102*, 2463.
32. Gregson, R. P.; Mirrington, R. N. *J. Chem. Soc., Chem. Commun.* **1973**, 598.
33. Still, W. C. *J. Am. Chem. Soc.* **1979**, *101*, 2493.
34. Evans, D. A.; Nelson, J. V. *J. Am. Chem. Soc.* **1980**, *102*, 774.
35. Paquette, L. A.; Crouse, G. D.; Sharma, A. K. *J. Am. Chem. Soc.* **1980**, *102*, 3972.
36. Koreeda, M.; Tanaka, Y.; Schwartz, A. *J. Org. Chem.* **1980**, *45*, 1172.
37. Sôrm, F. *Pure Appl. Chem.* **1970**, *21*, 263.
38. Takeda, K. *Pure Appl. Chem.* **1970**, *21*, 181.
39. Takeda, K.; Horibe, I.; Minato, M. *J. Chem. Soc. C* **1970**, 1142.

40. Takeda, K.; Tori, K.; Horibe, I.; Ohtsuru, M.; Minato, H. *J. Chem. Soc. C* **1970**, 2697.
41. Terada, Y.; Yamamura, S. *Tetrahedron Lett.* **1979**, 3303.
42. Tada, H.; Takeda, K. *J. Chem. Soc., Chem. Commun.* **1971**, 1391.
43. Tori, K.; Horibe, I.; Kuriyama, K.; Tada, H.; Takeda, K. *J. Chem. Soc., Chem. Commun.* **1971**, 1393.
44. Takeda, K. *Tetrahedron* **1974**, *30*, 1525.
45. Vig, O. P.; Bari, S. S.; Sood, O. P.; Sharma, S. D. *Indian J. Chem.* **1976**, *14B*, 564.
46. Kodama, M.; Matsuki, Y.; Itô, S. *Tetrahedron Lett.* **1976**, 1121.
47. Gopalan, A.; Magnus, P. *J. Am. Chem. Soc.* **1980**, *102*, 1756.
48. Grieco, P. A.; Nishizawa, M. *J. Org. Chem.* **1977**, 42, 1717.
49. Jain, T. C.; Banks, C. M.; McCloskey, J. E. *Tetrahedron Lett.* **1970**, 841.
50. Still, W. C. *J. Am. Chem. Soc.* **1977**, *99*, 4186.
51. Iguchi, M.; Nishiyama, A.; Yamamura, S.; Hirata, Y. *Tetrahedron Lett.* **1969**, 4295.
52. Windmer, U.; Zsindely, J.; Hansen, H. J.; Schmid, H. *Helv. Chim. Acta* **1973**, *56*, 75.
53. Dauben, W. G.; Chollet, A. *Tetrahedron Lett.* **1981**, 1583.
54. Overman, L. F.; Knoll, F. M. *J. Am. Chem. Soc.* **1980**, *102*, 865, and references therein.
55. Lutz, R. P.; Berg, H. A.; Wang, P. J. *J. Org. Chem.* **1976**, *41*, 2048.
56. Ziegler, F. E. *Acc. Chem. Res.* **1977**, *10*, 227.
57. Rossi, R. *Synthesis* **1977**, 817.
58. Thomas, A. F. *J. Chem. Soc., Chem. Commun.* **1967**, 947.
59. Boeckman, R. K., Jr.; Blum, D. M.; Arthur, S. D. *J. Am. Chem. Soc.* **1979**, *101*, 5060.
60. Dauben, W. G.; Ponaras, A. A.; Chollet, A. *J. Org. Chem.* **1980**, *45*, 4413.
61. Demole, E.; Enggist, P. *Helv. Chim. Acta* **1971**, *54*, 456.
62. Chidgey, R.; Hoffmann, H. M. R. *Tetrahedron Lett.* **1978**, 1001.
63. Vig, O. P.; Sharma, S. D.; Kad, G. L.; Sharma, M. L. *Indian J. Chem.* **1975**, *13*, 439.
64. Ziegler, F. E.; Reid, G. R.; Studt, W. L.; Wender, P. A. *J. Org. Chem.* **1977**, *42*, 1991.
65. Trost, B. M.; Bernstein, P. R.; Funfschilling, P. C. *J. Am. Chem. Soc.* **1979**, *101*, 4378.
66. Murray, R. D. H.; Ballantyne, M. M.; Hogg, T. C.; McCare, P. H. *Tetrahedron* **1975**, *31*, 2960.
67. Confalone, P. N.; Confalone, D. L. *J. Org. Chem.* **1980**, *45*, 1470.
68. Mowat, D.; Murray, R. D. H. *Tetrahedron* **1973**, *29*, 2943.
69. Quillinan, A. F.; Scheinmann, F. *J. Chem. Soc., Perkin Trans. 1* **1972**, 1382.
70. Ibid., **1975**, 241.
71. Chandrashekar, V.; Krishnamurti, M.; Seshadri, T. R. *Tetrahedron* **1967**, *23*, 2505.
72. Yuste, F.; Walls, F. *Aust. J. Chem.* **1976**, *29*, 2333.
73. Macleod, J. K.; Worth, B. R.; Wells, R. J. *J. Chem. Soc., Chem. Commun.* **1973**, 718.
74. Hoppmann, A.; Weyerstahl, P. *Chem. Ber.* **1974**, *107*, 1102.
75. Damodaran, N. P.; Dev., S. *Tetrahedron* **1973**, *29*, 1209.
76. Faulkner, D. J.; Wolinsky, L. F. *J. Org. Chem.* **1975**, *40*, 389.
77. Pattenden, G.; Whybrow, D. *J. Chem. Soc., Perkin Trans. 1* **1981**, 1046.
78. Danishefsky, S.; Doehner, R. *Tetrahedron Lett.* **1977**, 3031.
79. Marbet, R.; Saucy, G. *Helv. Chim. Acta* **1967**, *50*, 2091, 2095.
80. Faulkner, D. J.; Petersen, M. R. *J. Am. Chem. Soc.* **1973**, *95*, 553.
81. Dauben, W. G.; Dietsche, T. J. *J. Org. Chem.* **1972**, *37*, 1212.
82. Nakahara, Y.; Mori, K.; Matsui, M. *Agri. Biol. Chem.* **1971**, *35*, 918.
83. Imanishi, T.; Shin, H.; Yagi, N.; Hanaoka, M. *Tetrahedron Lett.* **1980**, 3285.
84. Petrzilka, M. *Helv. Chim. Acta* **1978**, *61*, 2286.
85. Ibid., 3075.
86. Carrol, M. F. *J. Chem. Soc.* **1940**, 704.
87. Ibid., 1266.
88. Ibid., **1941**, 507.

89. Tanabe, M.; Hayashi, K. *J. Am. Chem. Soc.* **1980**, *102*, 862.
90. Johnson, W. S.; Werthmann, L.; Bartlett, W. R.; Brocksom, T. J.; Li, T. T.; Faulkner, D. J.; Petersen, M. R. *J. Am. Chem. Soc.* **1970**, *92*, 741.
91. Johnson, W. S.; Gravenstock, M. B.; Parry, R. J.; Myers, R. F.; Bryson, T. A.; Howard Miles, D. *J. Am. Chem. Soc.* **1971**, *93*, 4330.
92. Johnson, W. S.; Gravestock, M. B.; McCarry, B. E. *J. Am. Chem. Soc.* **1971**, *93*, 4332.
93. Stork, G.; Raucher, S. *J. Am. Chem. Soc.* **1976**, *98*, 1583.
94. Stork, G.; Takahashi, T.; Kawamoto, I.; Suzuki, T. *J. Am. Chem. Soc.* **1978**, *100*, 8272.
95. Confalone, P. N.; Pizzolato, G.; Baggiolini, E. G.; Lollar, D.; Uskokovic, M. R. *J. Am. Chem. Soc.* **1977**, *99*, 7020.
96. Wiersig, J. R.; Waespe-Sarcevic, N.; Djerassi, C. *J. Org. Chem.* **1979**, *44*, 3374.
97. Karpf, M.; Djerassi, C. *Tetrahedron Lett.* **1980**, 1603.
98. Chan, K. K.; Specian, A. C., Jr.; Saucy, G. *J. Org. Chem.* **1978**, *43*, 3435.
99. Danishefsky, S.; Tzuzuki, K. *J. Am. Chem. Soc.* **1980**, *102*, 6892.
100. Dauben, W. G.; Beasley, G. H.; Broadhurst, M. D.; Muller, B.; Peppard, D. J.; Pesnelle, P.; Suter, C. *J. Am. Chem. Soc.* **1974**, *96*, 4724.
101. Bertrand, M.; Viala, J. *Tetrahedron Lett.* **1978**, 2575.
102. Kocienski, P. J.; Cernigliaro, G.; Feldstein, G. *J. Org. Chem.* **1977**, *42*, 353.
103. Labovitz, J. N.; Henrich, C. A.; Corbin, V. L. *Tetrahedron Lett.* **1975**, 4209.
104. Ziegler, F. E.; Bennett, G. B. *J. Am. Chem. Soc.* **1973**, *95*, 7458.
105. Plattner, J. J.; Gless, R. D.; Rapoport, H. *J. Am. Chem. Soc.* **1972**, *94*, 8613.
106. Tang, C.; Rapoport, H. *J. Am. Chem. Soc.* **1972**, *94*, 8615.
107. Tang, C.; Morrow, C. J.; Rapoport, H. *J. Am. Chem. Soc.* **1975**, *97*, 159.
108. Felix, D.; Gschwend-Steen, K.; Wick, A. E.; Eschenmoser, A. *Helv. Chim. Acta* **1969**, *52*, 1030.
109. Strauss, H. F.; Wiechers, A. *Tetrahedron* **1978**, *34*, 127.
110. Corey, E. J.; Shibasaki, M.; Knolle, J. *Tetrahedron Lett.* **1977**, 1625.
111. Bryson, T. A.; Reichel, C. J. *Tetrahedron Lett.* **1980**, 2381.
112. Arnold, R. T.; Hoffmann, C. *Synth. Commun.* **1972**, 2, 27.
113. Ireland, R. E.; Mueller, R. H.; Willard, A. K. *J. Am. Chem. Soc.* **1976**, *98*, 2868.
114. Whitesell, J. K.; Matthews, R. S.; Helbling, A.M. *J. Org. Chem.* **1978**, *43*, 784.
115. Ireland, R. E.; Mueller, R. H. *J. Am. Chem. Soc.* **1972**, *94*, 5897.
116. Ireland, R. E.; Vevert, J. P. *J. Org. Chem.* **1980**, *45*, 4259.
117. Ireland, R. E.; Mueller, R. H.; Willard, A. K. *J. Org. Chem.* **1976**, *41*, 986.
118. Ireland, R. E.; Thaisrivangs, S.; Wilcox, C. S. *J. Am. Chem. Soc.* **1980**, *102*, 1155.
119. Wilson, S. R.; Myers, R. S. *J. Org. Chem.* **1975**, *40*, 3309.
120. Katzenellenbogen, J. A.; Christy, K. J. *J. Org. Chem.* **1974**, *39*, 3315.
121. Frater, G. *Helv. Chim. Acta* **1975**, *58*, 442.
122. Frater, G. *Chimia* **1975**, *29*, 528.
123. Boyd, J.; Epstein, W.; Frater, G. *J. Chem. Soc., Chem. Commun.* **1976**, 380.
124. Still, W. C.; Schneider, M. J. *J. Am. Chem. Soc.* **1977**, *99*, 948.
125. Oshima, K.; Takahashi, H.; Yamamoto, H.; Nozaki, H. *J. Am. Chem. Soc.* **1973**, *95*, 2693.
126. Takano, S.; Hirama, M.; Araki, T.; Ogasawara, K. *J. Am. Chem. Soc.* **1976**, *98*, 7084.
127. Tamaru, Y.; Harada, T.; Yoshida, Z. *J. Am. Chem. Soc.* **1980**, *102*, 2392.
128. Crawford, R. J.; Erman, W. F.; Broaddus, C. D. *J. Am. Chem. Soc.* **1972**, *94*, 4298.
129. Nakai, T.; Mimura, T.; Kurokawa, T. *Tetrahedron Lett.* **1978**, 2895.
130. Hayashi, T.; Midorikowa, H. *Synthesis* **1975**, 100.
131. Overman, L. E. *Acc. Chem. Res.* **1980**, *13*, 218.
132. Coates, R. M.; Hutchins, C. W. *J. Org. Chem.* **1979**, *44*, 4742.
133. Malherbe, R.; Bellus, D. *Helv. Chim. Acta* **1978**, *61*, 3096.
134. Overman, L. E.; Kakimoto, M. *J. Am. Chem. Soc.* **1979**, *101*, 1310.
135. Kinsman, R. G.; Dyke, S. F. *Tetrahedron Lett.* **1975**, 2231.
136. Rischke, H.; Wilcock, J. D.; Winterfeldt, E. *Chem. Ber.* **1973**, *106*, 3106.

137. Ahmad, V. U.; Feuerherd, K. H.; Winterfeldt, E. *Chem. Ber.* **1977**, *110*, 3624.
138. Overman, L. E.; Fukaya, C. *J. Am. Chem. Soc.* **1980**, *102*, 1454.
139. Nakai, T.; Mikami, K. *Chem. Lett.* **1979**, 1081.
140. Biellman, J. F.; Ducep, J. B.; Schrlin, D. *Tetrahedron* **1980**, *36*, 1249.
141. Still, W. C.; Mitra, A. *J. Am. Chem. Soc.* **1978**, *100*, 1927.
142. Evans, D. A.; Bryan, C. A.; Sims, C. L. *J. Am. Chem. Soc.* **1972**, *94*, 2891.
143. Trost, B. M.; Rigby, J. H. *J. Org. Chem.* **1978**, *43*, 2938.
144. Grieco, P. A.; Finkelhor, R. S. *J. Org. Chem.* **1973**, *38*, 2245.
145. Van Tamelen, E. E.; Taylor, E. G.; Leiden, T. M.; Kreft, A. F., III, *J. Am. Chem. Soc.* **1979**, *101*, 7423.
146. Mander, L. N.; Turner, J. V.; Coombe, B. G. *Aust. J. Chem.* **1974**, *27*, 1985.
147. Still, W. C.; McDonald, J. H.; Collum, D. B.; Mitra, A. *Tetrahedron Lett.* **1979**, 593.
148. Evans, D. A.; Sims, C. L. *Tetrahedron Lett.* **1973**, 4691.
149. Evans, D. A.; Sims, C. L.; Andrews, G. C. *J. Am. Chem. Soc.* **1977**, *99*, 5453.
150. Reetz, M. T. *Angew. Chem., Int. Ed. Engl.* **1972**, *11*, 129.
151. Ibid., 130.
152. Kitchin, J.; Stoodley, R. J. *J. Am. Chem. Soc.* **1973**, *95*, 3439.

Ene Reactions

The ene reaction consists of a thermal intermolecular or intramolecular process between an olefin carrying an allylic hydrogen, which acts as a donor (ene), and a double or triple bond (enophile), which acts as acceptor. The ene is usually a three-carbon fragment and sometimes an allylic alcohol; however, the enophile can involve a large variety of groups: olefins, acetylenes, carbonyls, thiocarbonyls, nitrosoalkanes, azoalkanes, and singlet oxygen.

The MO interactions between the HOMO of ene and the LUMO of enophile (Figure 8.1) explain the symmetry-allowed concertedness of the reaction; a suprafacial use of both ene and enophile; and the a priori possibility of both *exo* and *endo* transition states, a choice that becomes of great interest because of the stereochemical consequences in the intramolecular version. We thus consider both the intermolecular and intramolecular processes.

Intermolecular Ene Reactions

Despite its popularity, this reaction has not been widely applied to the natural product field, except when singlet oxygen is used as the enophile. From the few examples known we report a two-step synthesis of aristolene (**1544**) from its isomer calarene (**1541**). The process occurs via an ene reaction of **1541** with *N*-sulfinylbenzene sulfonamide (**1542**), whose nitrogen–sulfur double bonds act as enophile, followed by desulfurization of the adduct, **1543**, with Raney nickel/hydrogen (*1*) (Scheme 8.1). The overall process is a clean and stereospecific olefin isomerization of great synthetic value.

The most investigated intermolecular ene process involves singlet oxygen as enophile. (The generation and MO description of singlet oxygen were reported in Chapter 5 in the section "Diels–Alder Reactions with Singlet Oxygen.") Because a survey on this argument is available (2), the topic does not need detailed consideration here.

The process consists of the formation of an allylic hydroperoxide, **1545**, from an olefin (having an allylic hydrogen) and singlet oxygen.

0065-7719/83/0180-0339 $06.25/1

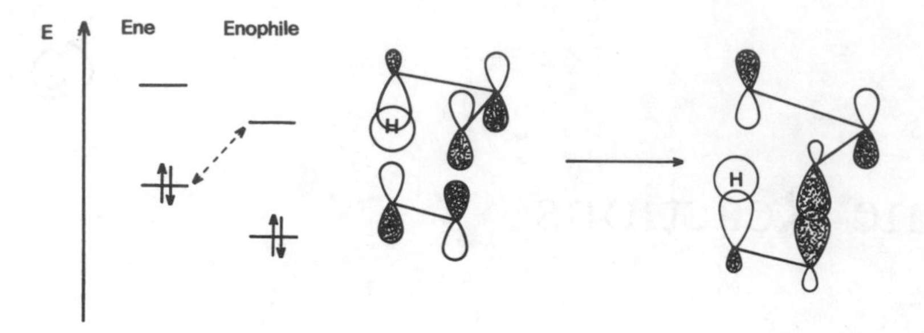

Figure 8.1. FMO interaction in the ene reaction.

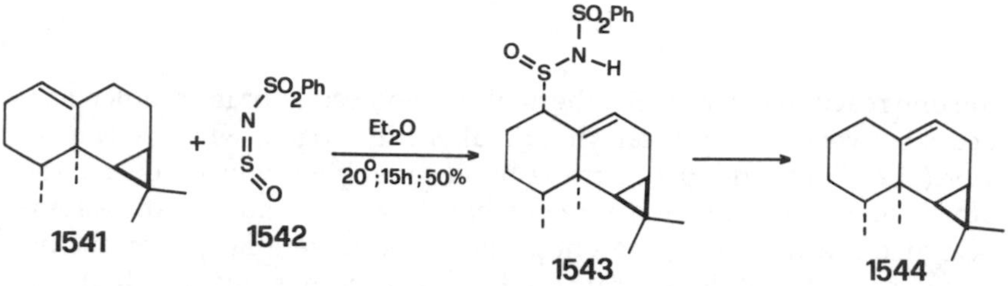

1541 + 1542 $\xrightarrow[20^{\circ};\ 15h;\ 50\%]{Et_2O}$ 1543 ⟶ 1544

Scheme 8.1

Compounds such as **1545** can be either used directly or suitably modified; they can be reduced, generally by lithium aluminum hydride, cleaved to carbonyls by acidic hydrolysis (Hock-type fragmentation), or oxidized to α, β-unsaturated carbonyl derivatives (Scheme 8.2).

From a theoretical point of view, the strong electrophilic character of singlet oxygen suggests that the dominant interaction is $HOMO_{olefin}/LUMO_{oxygen}$. By adopting this theory, we can explain the following experimental facts (*3*): The reaction is stereospecific and occurs suprafacially; the deuterium isotope effect shows that groups with a (*Z*) relationship are competitive but those that are *gem* or (*E*) are not; and the preferred hydrogen abstraction occurs on the disubstituted side of a trisubstituted olefin. These explanations presuppose that a complex is first involved where the oxygen π^* interacts with the π-HOMO of the olefin that has allylic hydrogens antibonding with the π system (hence, the olefin (*Z*) configuration is preferred) (Figure 8.2.)

If we consider that chlorophyll is a good sensitizer of the photooxygenation process, it is not astonishing that hydroperoxides are naturally occurring species. Hydroperoxides are generally isolated from leaves of several arboreal species; peroxycostunolide (**1546**) and peroxyparthenolide (**1548**) are two cytotoxic germacranolide hydroperoxides isolated from the leaves of *Magnolia grandiflora*, neoconcindiol hydroperoxide from red alga *Laurencia snyderiae*, and 3α,22α-dihydroxy-7α-hydroperoxy-Δ^5-stigmastene from the leaves of *Aesculus hippocastanum*. The first two compounds were easily synthesized starting from costunolide (**1270**) and parthenolide (**1547**), respectively, by methylene blue-sensitized photooxygenation (*4*) (Scheme 8.3).

To synthesize allylic alcohols, the hydroperoxide must be reduced. Several alkaloids and terpenes have been prepared by this route.

The approach to garrya and atisine alkaloids was realized through the syntheses of C/D ring systems **1550** and **1552**, starting from isokaurene (**1549**) and the analogous olefin **1551**, respectively (*5*). In a more sophisticated preparation, the first synthesis of garriyne (**1555**) occurred through a hematoporphyrin-sensitized photooxygenation of **1553** to give **1554**, which was then converted into the target (*6*) (Scheme 8.4).

We report four examples from the terpene field that illustrate the variety of strategic approaches to these compounds.

In the previously mentioned Büchi (*7*) synthesis of (±)-khusimone (**790**), the contrathermodynamic isomerization of isokhusimone (**1556**) to **1557** was accomplished by photooxygenation. This process was followed by work-up with triethyl phosphite and reduction of the allylic alcohol **1557** to racemic **790** (in admixture with its epimer, ratio 7:3) (Scheme 8.5).

The key step of a synthesis of α- and β-himachalenes (**942** and **190**, respectively) was an intramolecular Diels–Alder reaction, but the critical **1559** was in turn obtained by photooxygenation of **1558**. Upon dehydration, the resultant mixture of allylic alcohols gave **1559**, which was easily transformed into the required product (**1558**) (*8*) (Scheme 8.6).

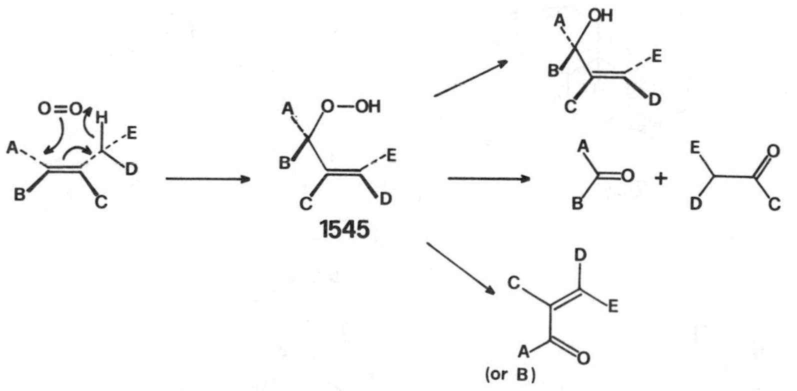

Scheme 8.2

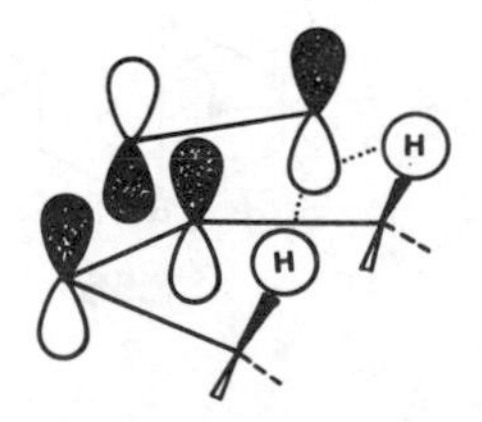

Figure 8.2. FMO interaction between LUMO of singlet oxygen and HOMO of a (Z)-olefin.

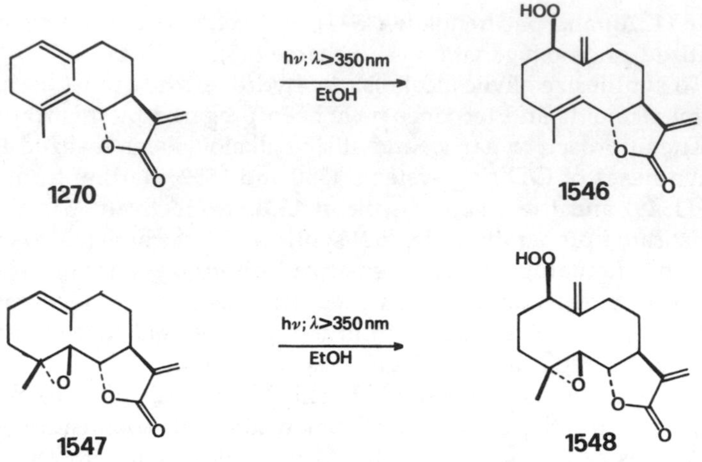

1270 → (hν; λ>350 nm, EtOH) → 1546

1547 → (hν; λ>350 nm, EtOH) → 1548

Scheme 8.3

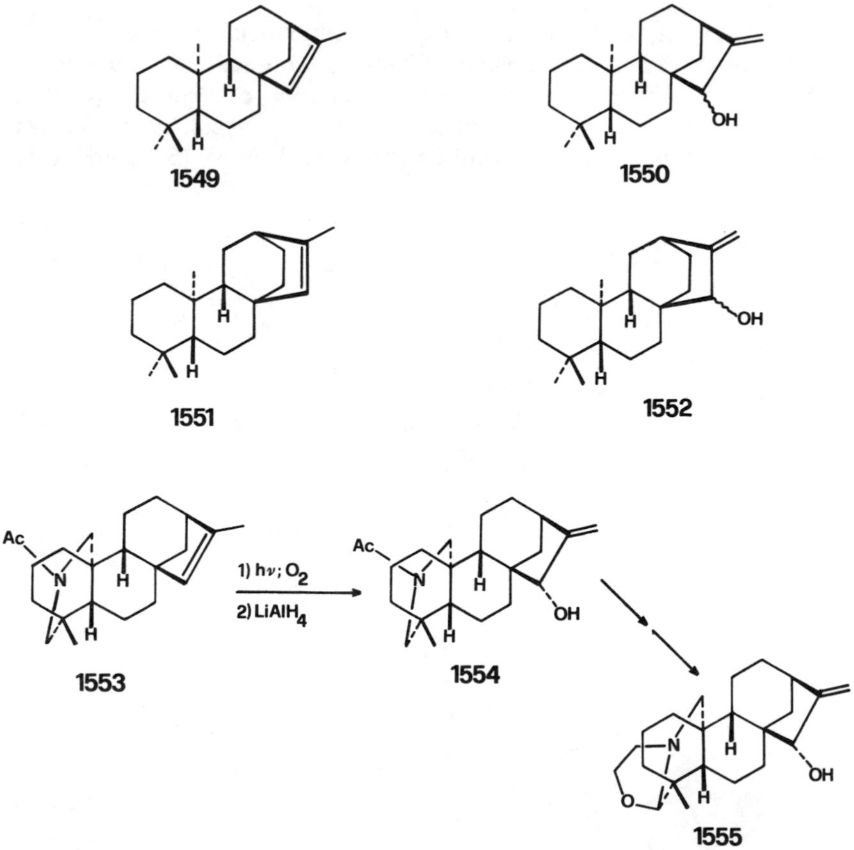

1549

1550

1551

1552

1553 → (1) hν; O_2 2) $LiAlH_4$) → 1554 → 1555

Scheme 8.4

The synthesis of the unsymmetrical pentacyclic triterpene *dl*-germanicol (**1564**) requires the functionalization of the tricyclic derivative **1561** to build up the new two rings. This goal was realized by photooxygenation of **1561** to **1562**, from which **1563** was obtained. This product was converted to **1564** (*9*) (Scheme 8.7).

The hematoporphyrin-photosensitized oxygenation of unsaturated lactone, **1565**, afforded **1566** as the major product. This compound's oxygenated function can be retained for its conversion to (±)-telekin (**1567**) or eliminated to give the double bond required for its conversion into (±)-alantolactone (**1568**) (*10*). Both compounds are lactones of the eudesmane class of sesquiterpenes (Scheme 8.8).

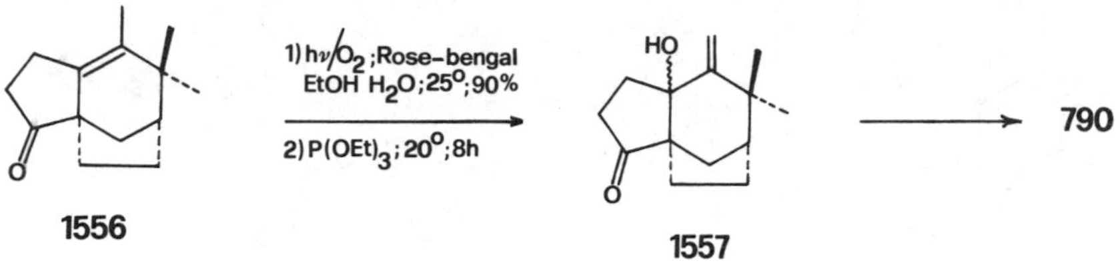

1556 — 1) hν/O_2; Rose-bengal EtOH H_2O; 25°; 90% 2) $P(OEt)_3$; 20°; 8h → 1557 (HO) → 790

Scheme 8.5

1558 (CO_2Et) — 360nm; O_2 hematoporphyrin pyridine; 50°; 16h → (OH, CO_2Et) → 1559 (CO_2Et) → 1560 (O) → 942 + 190

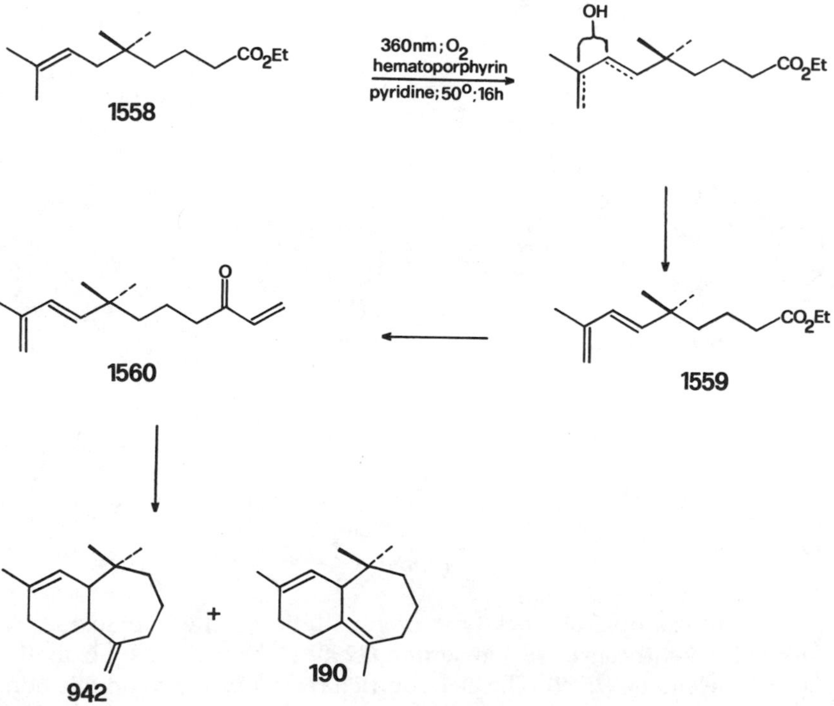

Scheme 8.6

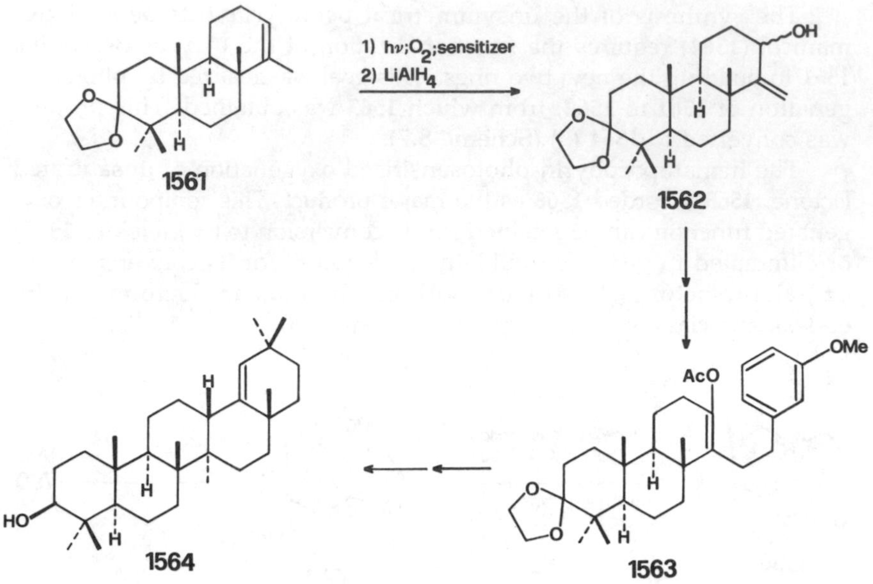

1) hν; O_2; sensitizer
2) $LiAlH_4$

1561 1562 1563 1564

Scheme 8.7

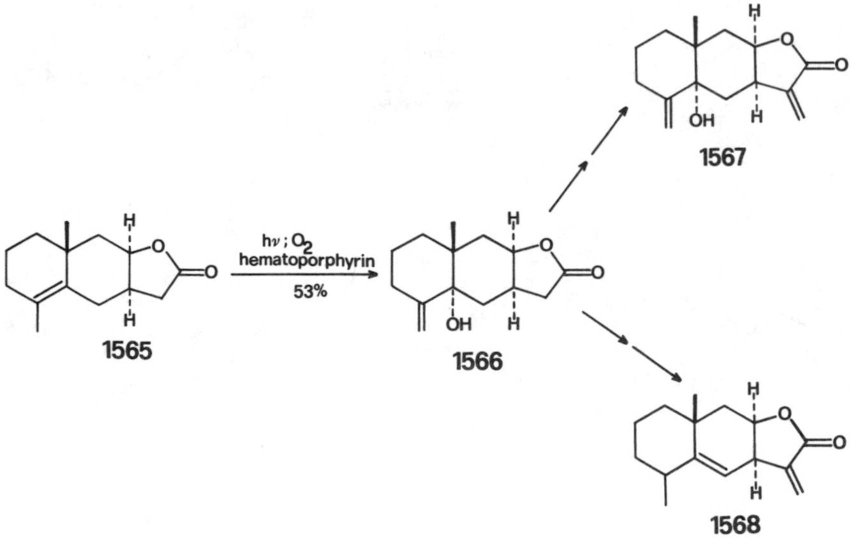

hν; O_2
hematoporphyrin
53%

1565 1566 1567 1568

Scheme 8.8

As an example of Hock-type fragmentation of hydroperoxide, we report the synthesis of (−)-geijerone (**1290**) (*11*) obtained by oxidation of (+)-γ-elemene (**1569**). The electron-rich tetrasubstituted double bond is attacked by singlet oxygen to produce a mixture of allylic peroxides, which are cleaved to give a single product (Scheme 8.9).

Intramolecular Ene Reactions

This variant has been extensively applied to the synthesis of natural products; the topic was reviewed by Oppolzer and Snieckus (*12*). The Oppolzer group is deeply involved in the field.

The main difference between the intramolecular and the intermolecular processes is evident in the synthesis of cyclonerodiol (**1572**), a sesquiterpenoid isolated from the cultured broth of a strain of *Trichothecium* species of fungi. The starting linalool (or its silyl ether) **1570** was pyrolyzed at 240–290 °C to afford a mixture of cyclopentanols; the 2,3-*cis* isomers were the main products. The β-isomer **1571** was then converted in few steps into **1572** (*13*) (Scheme 8.10). A careful kinetic investigation (*14*) showed a $\Delta H^{\neq}$ value of 31.5 kcal/mol, which is not far from values of analogous intermolecular processes. The $\Delta S^{\neq}$ value is −19 e.u.; however, intermolecular reactions have entropy values in the range of −30 to −45 e.u. (*15*).

Similar entropy values ($\Delta S^{\neq}$ = −13.7 e.u.) were found for the conversion of dehydrolinalool (**1573**) into **1574**. This latter compound is the starting material for a three-step synthesis of the spirosesquiterpene β-acoratriene (**1575**) (*16*) (Scheme 8.11).

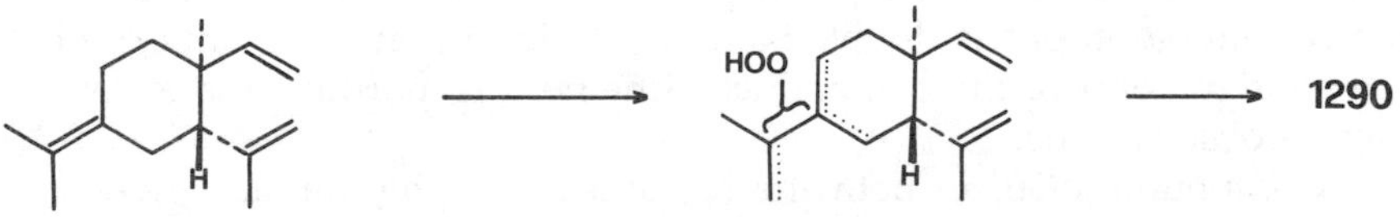

1569 → HOO intermediate → **1290**

Scheme 8.9

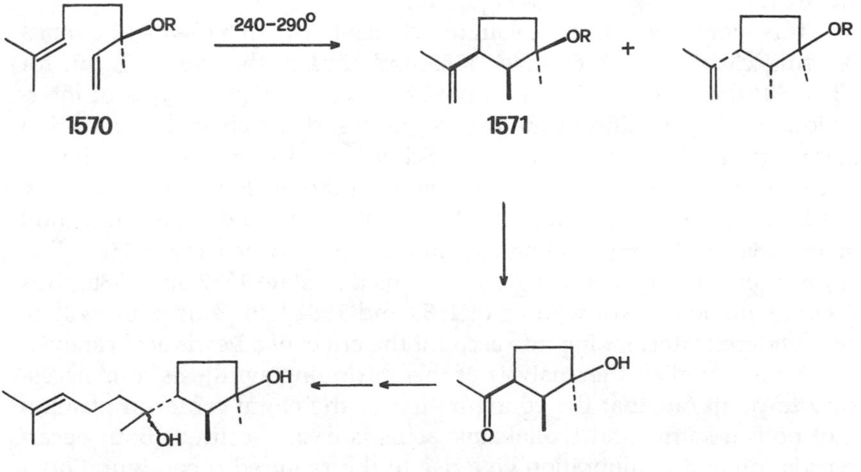

1570 (OR) —240–290°→ **1571** (OR) + isomer (OR)

→ ketone (OH) → **1572** (OH, OH)

Scheme 8.10

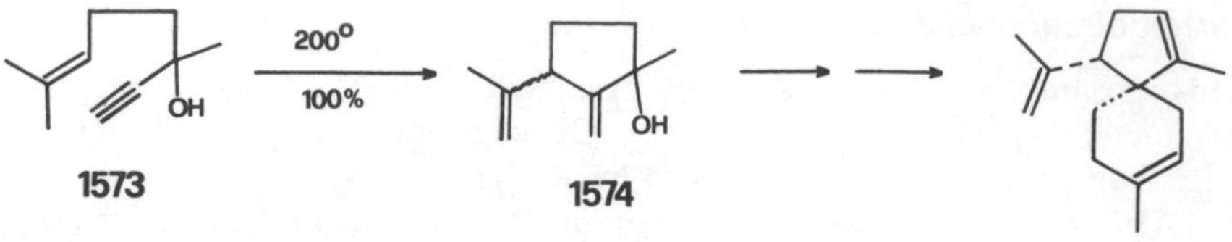

Scheme 8.11

Stork and Kraus (*17*) utilized a remarkably similar sequence to prepare the 2-methylenecyclopentanone derivative **1578**, an important intermediate in an elegant approach to prostaglandins. The facile thermal cyclization of **1576** gives **1577**, which can be converted to **1578** in a few steps (Scheme 8.12).

The α-allokainic (*18*) and α-kainic (*19*) acids (**1585** and **1582**, respectively) are two isomeric amino acids possessing potent neurobiological activity isolated from the marine algae *Digenea simplex* Ag. Their syntheses, from which we can learn about the transition state of intramolecular ene reactions, are simple. 4-Aza-1,6-dienes **1579** and **1580** cyclize nicely; the required temperature is strictly dependent on the presence of electron-attracting groups at C5 and partially dependent on the configuration of the double bond at C6. Both points are critical for the configuration of the ene adduct. Scheme 8.13 portrays the experimental observations.

Compound **1580**, in both the (*Z*) and (*E*) configurations, gives a ring closure easier than does **1579**. This finding is clearly understood in terms of the electron-attracting effect of substituents at C5 that lower the LUMO of the enophile. Thus, the energy separation of $HOMO_{ene}$/$LUMO_{enophile}$ is lowered (*see* Figure 8.1)

It is more difficult to explain the choice between a *cis*-3,4 or a *trans*-3,4 ring closure; (*Z*)-**1580** gives **1583** and **1584** in the ratio 1:3, but for (*E*)-**1580**, the ratio is 1:1 (*20*). A possible explanation in terms of interactions in the transition state can be proposed if a chair-like transition state different from that shown in Scheme 8.14 is postulated. The required transition state has the hydrogen migrating along the axis joining the termini of ene and enophile. The (Z)-isomer could prefer transition state **1586**, free from nonbonding interactions, which gives **1584**. The (*E*)-isomer could react through both transition state **1587** and **1588**, thus giving equimolecular amounts of **1583** and **1584** (*20*). This point will be reconsidered later, taking into account the effect of a Lewis acid catalyst.

To accomplish the analysis of this intriguing synthesis, it is necessary to point out that the configuration of the chiral center in position 2 of both α-kainic and α-allokainic acids is easily secured. Both decarboxylation and equilibration give rise to the required α configuration.

The choice of the *exo* versus *endo* transition state in this kind of reaction is generally not a matter of MO interactions, but is simply the result of an attempt to minimize the steric interactions in the transition state. This result is evident in the pyrolysis of (2*E*, 7*Z*)- and (2*E*, 7*E*)-methylnonadienoates (**1589** and **1590**, respectively) at 300 °C to give solely *cis*-substituted cyclopentane **1593**. The first isomer reacts via transition state **1591** with an *exo*-oriented carbomethoxy group, the second via transition state **1592** with the same group *endo*-oriented (*21*) (Scheme 8.15)

The syntheses of two unusual tricyclic sesquiterpenes further illustrate this concept. Isocomene (**261**), isolated from the toxic plant *Isocoma wrightii*, has four chiral centers in the 1, 4, 7, and 8 positions. It was

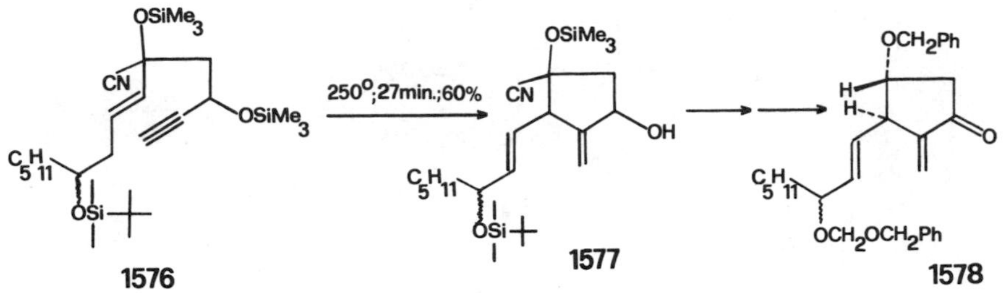

Scheme 8.12

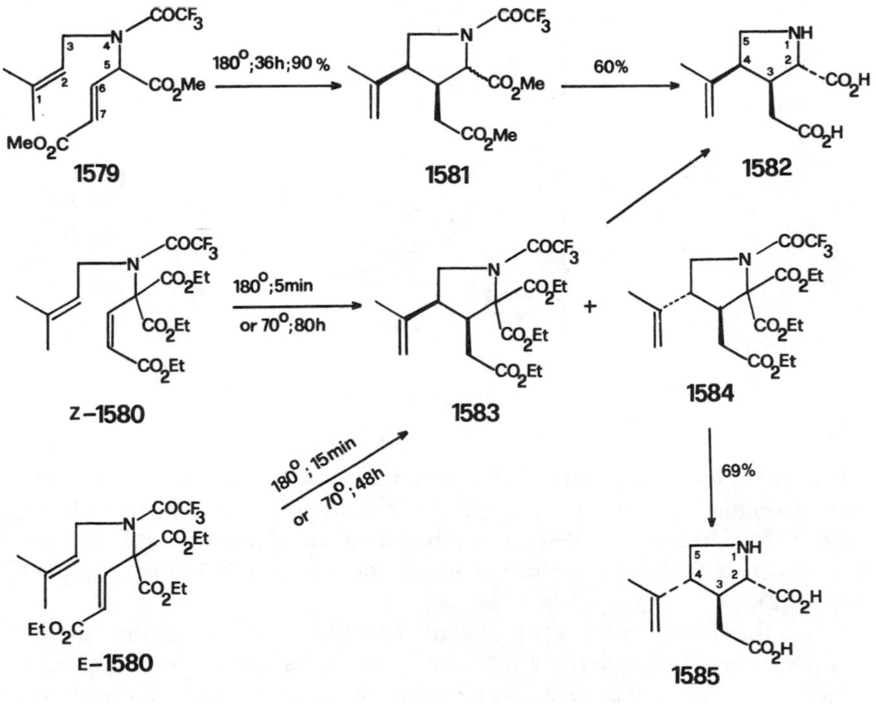

Scheme 8.13

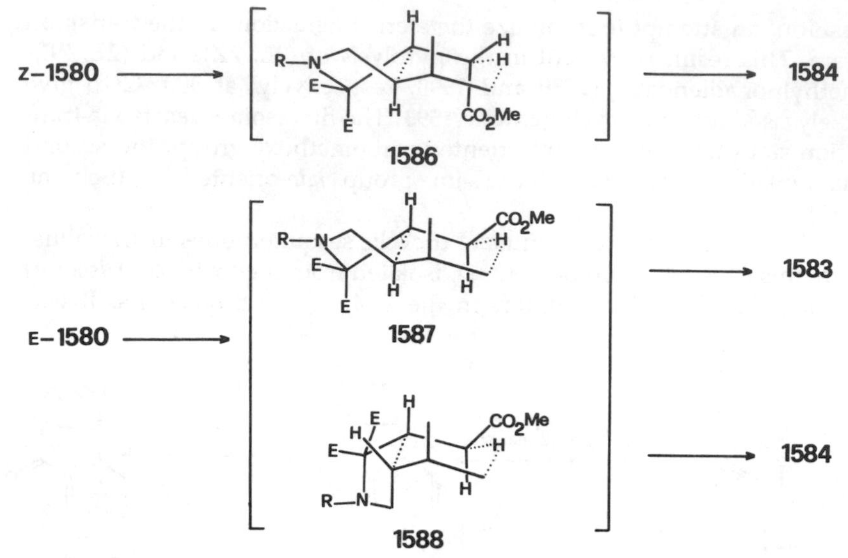

Z–1580 → 1586 → 1584

E–1580 → 1587 → 1583; 1588 → 1584

Scheme 8.14

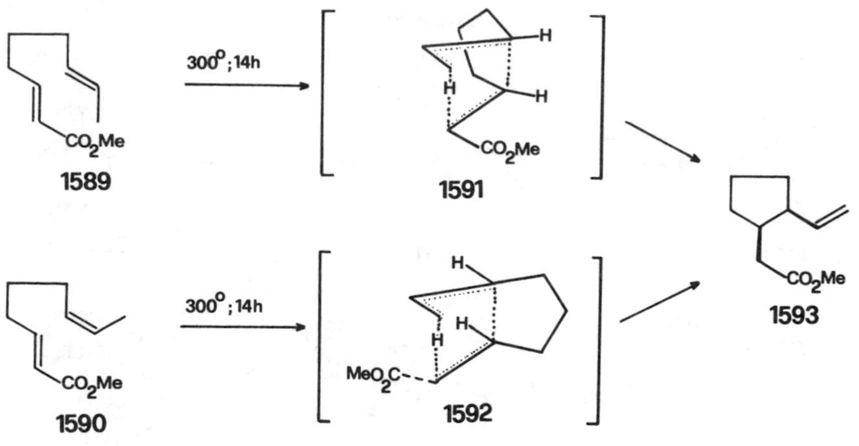

1589 —300°; 14h→ 1591 → 1593

1590 —300°; 14h→ 1592 → 1593

Scheme 8.15

synthesized by starting from **1594**, which has two chiral centers, by an intramolecular ene reaction through the less encumbered *endo* transition state, **1595**. The product **1596**, which has the four required chiral centers in the correct configuration, can be easily converted to **261** in five simple steps (22).

Modhephene (**1600**), a propellanic sesquiterpene containing three chiral centers, isolated from the same source as isocomene, was obtained by a similar approach. However, because of the steric disposition of the starting substrate, the intramolecular ene reaction of **1597** occurs exclusively through *exo* transition state **1598** to give **1599** (*23*) (Scheme 8.16).

These examples show that the configuration of the ene adduct can be predicted; it will be the same as that of the less sterically hindered transition state.

Hetero-Ene Reactions

Sometimes a carbon atom of either the ene or enophile is substituted by oxygen. An example of an oxygen-containing ene is given by the thermal conversion of (+)-dihydrocarvone (**1601**) into (+)-camphor (**1603**). This process proceeds in 55% yield and 90% optical purity through the enol **1602** (*24*) (Scheme 8.17). An example with an oxygenated enophile is the previously described rearrangement at 180–220 °C of shyobunone (**327**), which affords several products. The plausible mech-

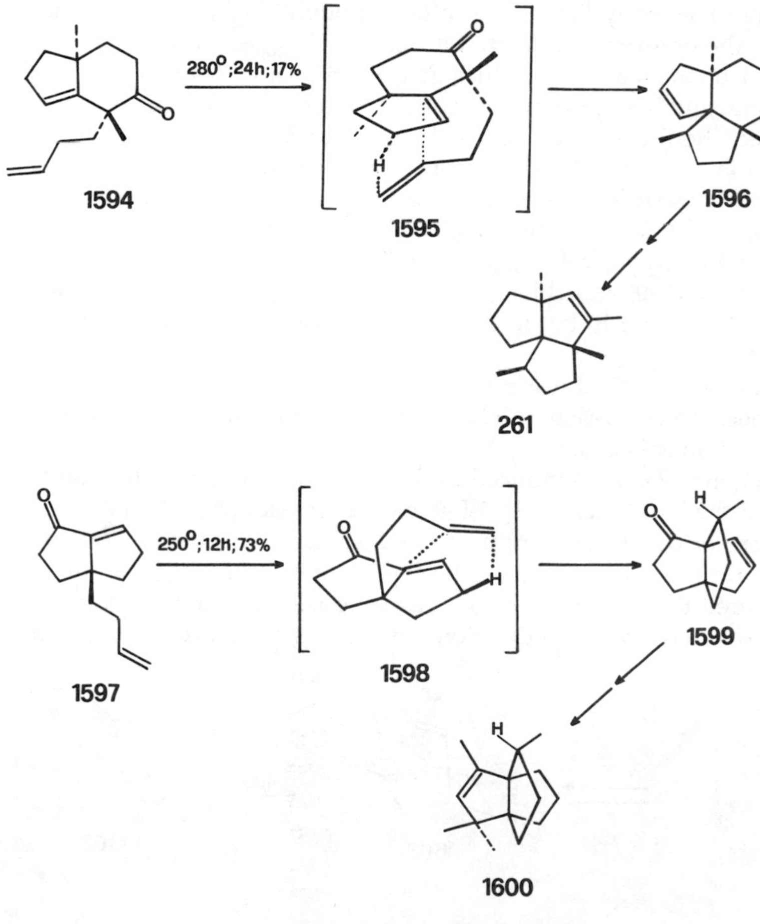

Scheme 8.16

anism outlined in Scheme 7.33 explains the formation of some of the products through a Cope rearrangement, a [1,5] heterosigmatropic shift, and a final intramolecular oxyene reaction of **1302** (*25, 26*).

A magnificent 12-electron oxyene reaction was proposed in the biogenetic pathway of the alkaloid ochotensimine (**1608**) from the precursor **1604**. Two electrocyclic processes are proposed; the first is an oxyene reaction from **1605** to **1606** (12 electrons are involved, and hence, a $4n$-electron transition state occurring in a supra–antara fashion), and the second a [1,7] sigmatropic shift giving **1607**. These proposed reactions are supported by a 75% yield conversion of protoberberinium methyl iodide (**1609**) to **1610** through steps that parallel previous hypotheses (*27*) (Scheme 8.18).

Lewis Acid Catalyzed Ene Reactions

The severe experimental conditions (a temperature higher than 250 °C is common) severely limit the synthetic potential of the ene reaction. Because the ene reaction is $HOMO_{ene}/LUMO_{enophile}$ controlled, complexation of the enophile with a Lewis acid should lower the FMO separation, and hence lower the activation energy of the process as well. This effect should be particularly evident for carbonyl enophiles because coordination with Lewis acids depresses the high energy LUMOs. For example, the vertical electron affinities of acetaldehyde and acrolein are -1.19 and ~ 0 eV, respectively (*28*); the effect of complexation lowers the LUMO energy level of the latter (*29*). Thus the reaction between (−)-β-pinene (**240**) and chloral in the presence of aluminum trichloride occurs at 20 °C in 2 h, but in the absence of catalyst, 95 °C and 4 h are required (*30*).

Although this field was reviewed (*31*), we wish to emphasize some points because this reaction has the potential to instrument future developments in this area.

Asymmetry can be induced in the ene reaction, depending on the nature of the Lewis acid. (−)-β-Pinene (**240**) reacts with chloral to give a mixture of diastereomers with **1612** as the main isomer (ratio of **1612:1614**, 83:17). This result can be explained by assuming that the less encumbered transition state **1611**, with the trichloromethyl group *exo*, is the preferred one. Coordination with Lewis acids occurs on the ox-

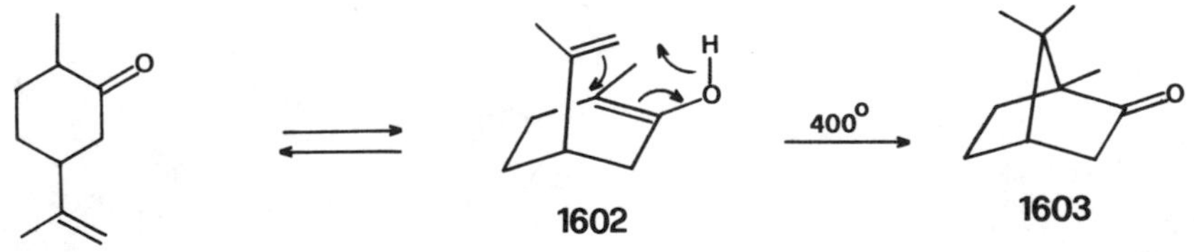

1601 ⇌ 1602 —400°→ 1603

Scheme 8.17

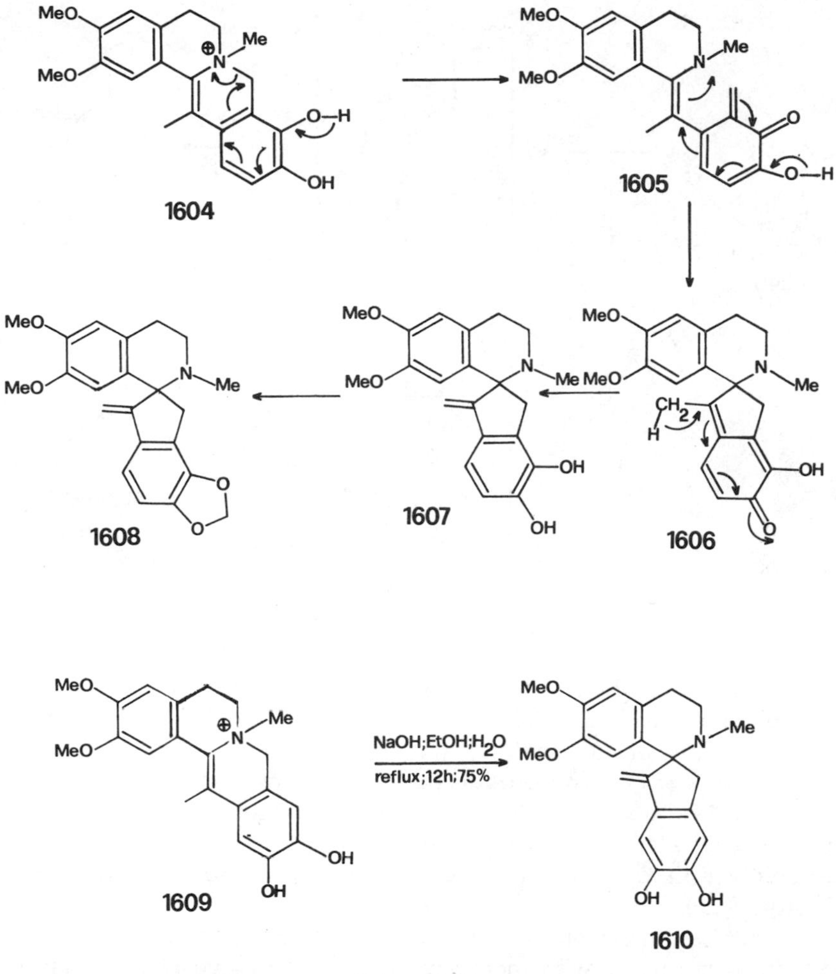

MeO MeO Me N OH OH 1604 1605 1606 1607 1608 1609 1610

NaOH;EtOH;H_2O reflux;12h;75%

Scheme 8.18

ygen atom *anti* to the bulky trichloromethyl group; therefore, when the Lewis acid is bulky, transition state **1613**, with this group *endo*, is more favored. This relationship is proportionate. Thus the ratio **1612:1614** becomes 1:1 with boron trichloride, 1:3 with aluminum trichloride, 1:9 with stannic chloride, 3:97 with ferric chloride; titanium tetrachloride gives **1614** as the only reaction product (*32*) (Scheme 8.19).

A useful application of these concepts was realized by Oppolzer and Robbiani (*20*) as a development of the synthesis of α-kainic (**1582**) and α-allokainic (**1585**) acids (*see* Scheme 8.13). They found that (*Z*)-**1580**, in the presence of diethylaluminum chloride at −78 °C, gives **1584** as the only product, instead of the predicted 1:3 mixture of **1583** and **1584**. Also, (*E*)-**1580** gives the expected products, but in a ratio of 1:9 instead of 1:1. They accomplished an enantiomeric synthesis of the nat-

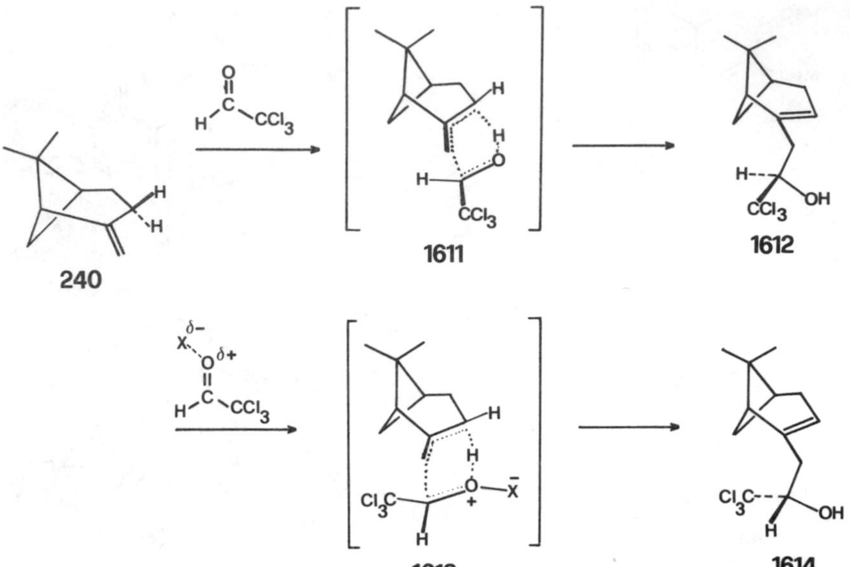

Scheme 8.19

ural isomer (+)-α-allokainic acid (**1585**) by ring closure of (Z)-enophile **1615** [obtained from *cis*-β-chloroacrylic acid and (−)-8-phenylmenthol] in the presence of diethylaluminum chloride (or dimethylaluminum chloride) at −35 °C. The products are *trans*-pyrrolidines; the (3*S*,4*R*)-isomer **1617** is obtained in 95% optical purity, and is subsequently hydrolyzed to the target. An explanation of the high asymmetric induction can be given in terms of the less encumbered transition state **1616** (*33*) (Scheme 8.20).

Several examples of intermolecular processes have been reported. (+)-Limonene reacts with methyl vinyl ketone to give **1618**, which is easily converted to β-bisabolene (**1619**) (*34*). (−)-Carvone (**1620**) reacts with the same enophile to give **1621**, transformed in two steps into (−)-cryptomerion (**1622**) (*34*). These processes occur at room temperature in the presence of aluminum trichloride. To emphasize the catalytic effect, consider that the reaction of limonene with methyl propiolate to produce **1623**, a suitable intermediate for the synthesis of (+)-β- and (+)-α-atlantones (**922** and **505**, respectively), occurs only at 160 °C (*34*) (Scheme 8.21).

A nice one-step synthesis of ipsenol (**953**) was realized from isovaleraldehyde and isoprene in the presence of dimethylaluminum chloride; aluminum trichloride cannot be used because **953** is acid sensitive (*35*). The yield is low (16%) because the major product (60%) is the Diels–Alder adduct **1624**.

Another pheromone, the AI component **1627** produced by the California Red Scale, *Aonidiella aurantii*, was synthesized by reacting citro-

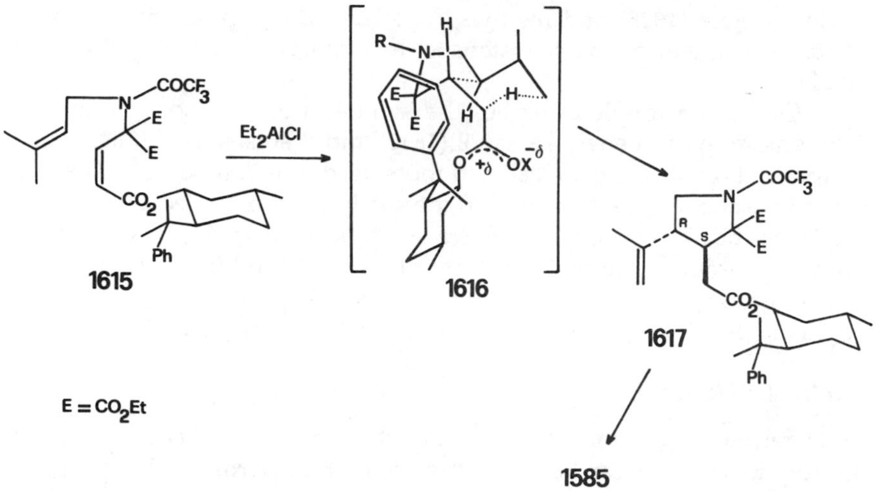

Scheme 8.20

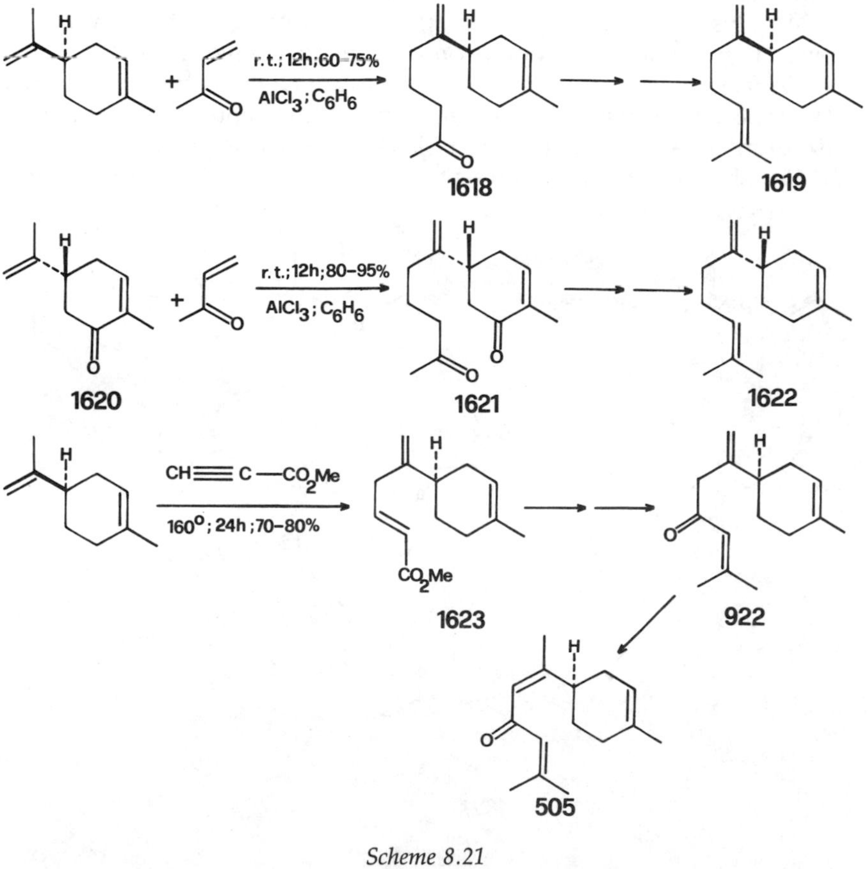

Scheme 8.21

nellyl acetate (**1625**) and methyl propiolate. The regiospecific product **1626**, was converted in a few steps to the required product (*36*) (Scheme 8.22).

One last example describes the synthesis of two isomeric hydroazulenic sesquiterpenes, kessanol (**1631**) and 8-epikessanol (**1630**), two important constituents of Valerian roots used in folk medicine. The key step is centered on a stannic chloride-catalyzed intramolecular reaction of **1628**; the aldehydic group acts as the enophile, giving rise to the 8-hydroxy group of **1629**. Even through several modifications, this hydroxy group is retained at the end of the steps required to give first the epimer (β-OH) and then kessanol itself (α-OH) (*37*) (Scheme 8.23).

Retro-Ene Reactions

The retro-ene reaction (*38*) is a very diffuse process because several pyrolytic cleavages can be defined as such. Thus thermolysis of acetates, β-hydroxyolefins, xanthates (the Chugaev reaction), keto acids, and allylic esters gives rise to an olefin as the main product, which sometimes is the ene, sometimes the enophile, as depicted in the following Scheme (8.24). This is not an easy process; the equilibrium is shifted toward the generation of ene and enophile only under precise conditions.

A factor that can favor the reaction is the relief of strain. Thus **1632** decomposed thermally to generate 5-methyl-Δ^4-heptenal (**1633**) (*39*), a useful intermediate for the synthesis (*40*) of *Cecropia* juvenile hormone (**1355**). Similarly, the thermolysis of **1634** gives isomerization and retro-

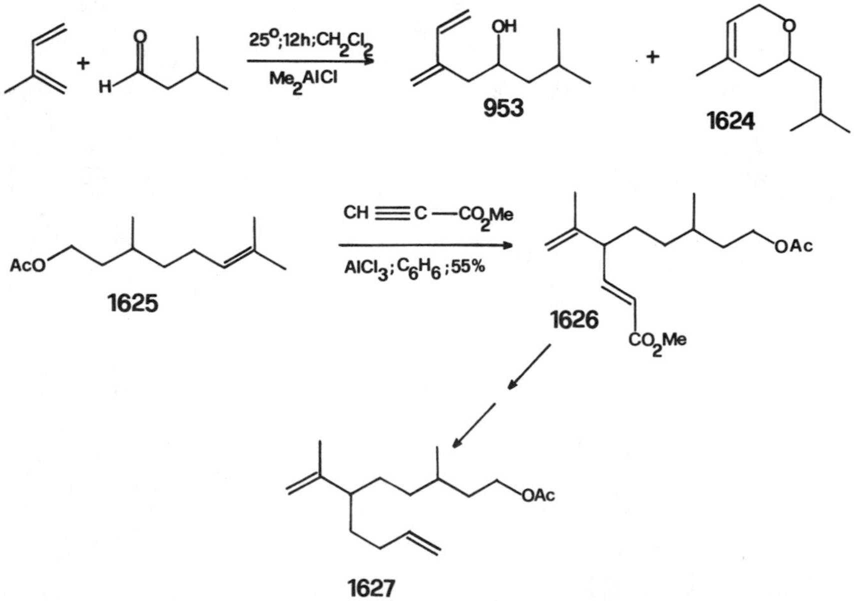

25°;12h;CH_2Cl_2
Me_2AlCl
OH
953
1624

$CH \equiv C - CO_2Me$
$AlCl_3$;C_6H_6;55%
AcO
1625
OAc
1626
CO_2Me
OAc
1627

Scheme 8.22

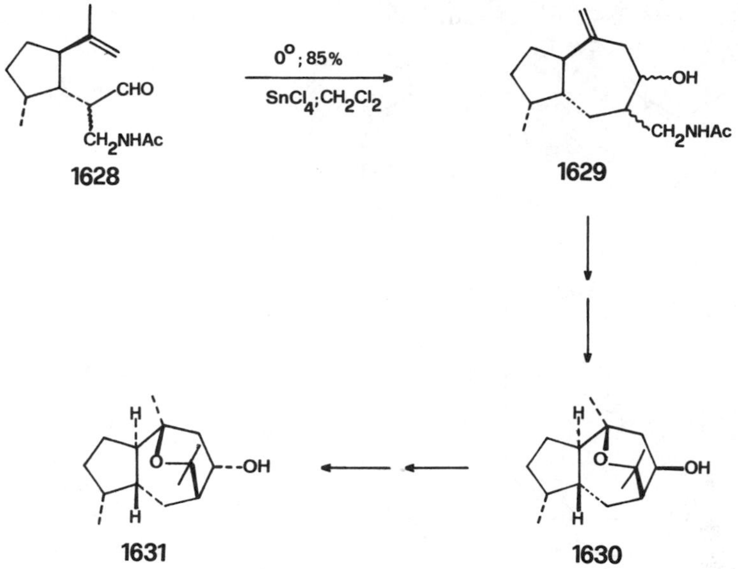

Scheme 8.23

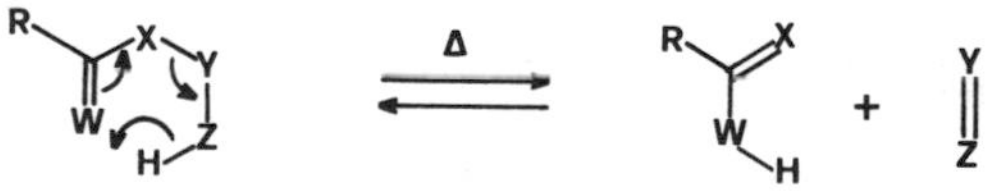

Scheme 8.24

ene ring opening to **1635**, which is easily transformed into *cis*-jasmone (**8**) (*41*). The thermal conversion of (+)-3-hydroxymethyl-Δ^4-carene (**1636**) into (+)-6-hydroxymethyl-$\Delta^{2,8}$-*p*-menthadien (**1637**) (*42*) shows that the retro-ene process (as well as the ene reaction) occurs via suprafacial migration of hydrogen (Scheme 8.25).

A second condition favoring a retro-ene process is the loss of a small volatile fragment. The equilibrium shown in Scheme 8.24 thus would be displaced. A nice example of a retro-ene reaction proceeding with loss of nitrogen (as enophile) is the synthesis of two marine sesquiterpenes present in *Dictyota* algae: (+)-pachydictyol A (**1641**) and (−)-dictyolene (**1644**). Reduction of the enone **1638** to the *trans*-fused product is prevented by the sterically hindered approach to the β face of the molecule. The *cis*-fused compound is obtained under a variety of conditions. The correct configuration was secured by decomposing β-diazene **1639**, which was derived from reduction of the tosylhydrazone with catecholborane; this decomposition obviously occurs on the α face of molecule and causes hydrogen to migrate to the more hindered β face, thus producing **1640**. This compound can be converted to **1641**. Analogously, 6-epi-α-santonin (**1642**) is stereospecifically transformed

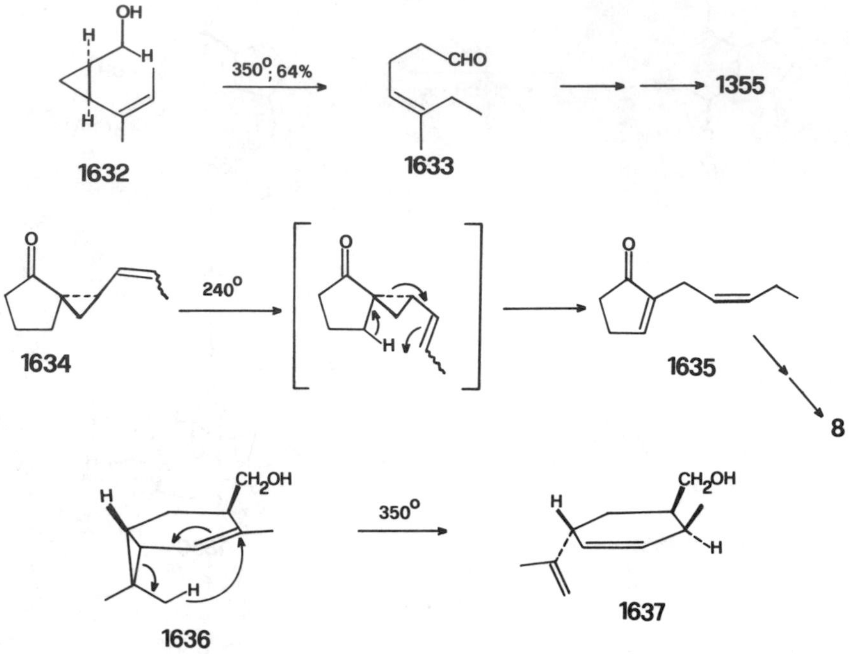

OH H H H 350°; 64% CHO 1355 1632 1633

O 240° O O 1634 H 1635 8

H CH_2OH 350° CH_2OH H H H 1636 1637

Scheme 8.25

into **1643**, which can be converted to dictyolene (**1644**) (*43*) (Scheme 8.26).

A similar explanation can be advanced for the dehydration with dicyclohexylcarbodiimide (DCC) of β-hydroxycarbonyl compound **1645** to the enone **1647** in the Corey synthesis of prostaglandins (*44*). Although the other product, dicyclohexylurea, is not a volative compound, the same explanation can be proposed. The carbammate intermediate **1646** undergoes a retro-ene fragmentation under neutral conditions. (Scheme 8.27).

We end this section by outlining the synthesis of some spirosesquiterpenes described by Oppolzer and his associates (*45*). The key step is an intramolecular thermal ene reaction of ethyl 2-(1-cyclohexenyl)-5-hexenoate (**1648**); the process proceeds through *endo* transition state **1649** to give a mixture of epimers **1650a** and **1650b**. The former epimer was converted to acorenone B (**36**) and acorenone (**1651**) by standard methods, and the latter was transformed into the acetal **1652**. By heating this compound at 280 °C, a retro-ene reaction took place and produced the ester **1653**, an important intermediate for an easy conversion into β-acorenol (**1654**) and β-acoradiene (**1655**) (*45*) (Scheme 8.28). The overall process generates three chiral centers through the ene-reaction, and isomerizes a double bond through the retro-ene fragmentation.

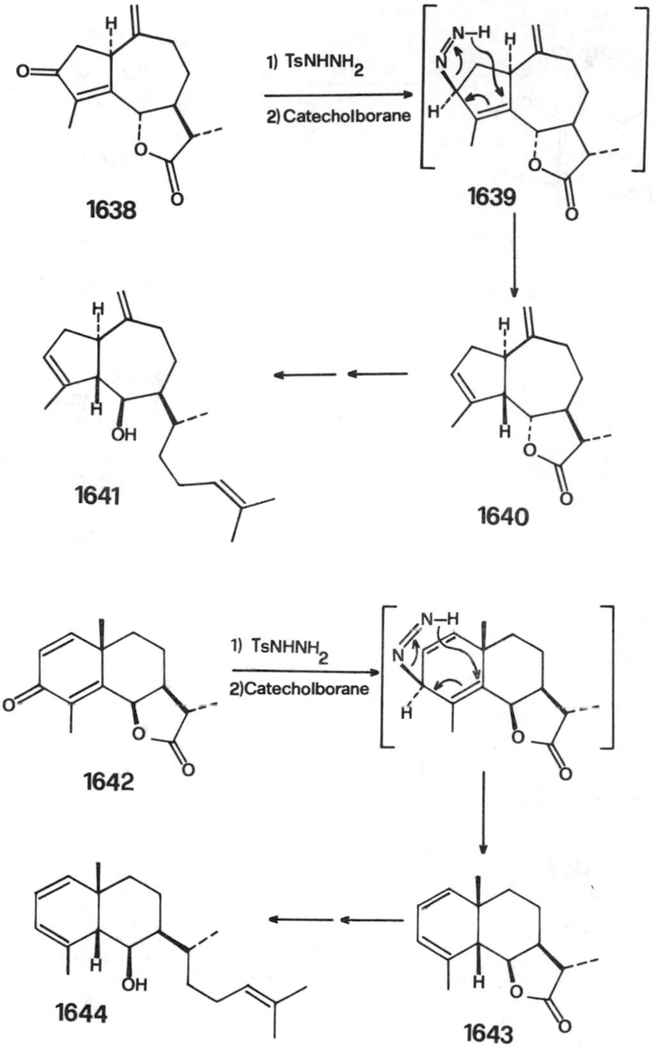

Scheme 8.26

The ene reaction, thought to be a limited process suitable for simple molecules only, is a fruitful approach to complex natural products. These products can be synthesized stereospecifically; as many as three chiral centers are generated in the pericyclic step. The synthetic potential of the reaction is somewhat limited by the high temperatures that are required, but if the LUMO of the enophile can be lowered by complexation with a Lewis acid, the overall process can also occur at room temperature.

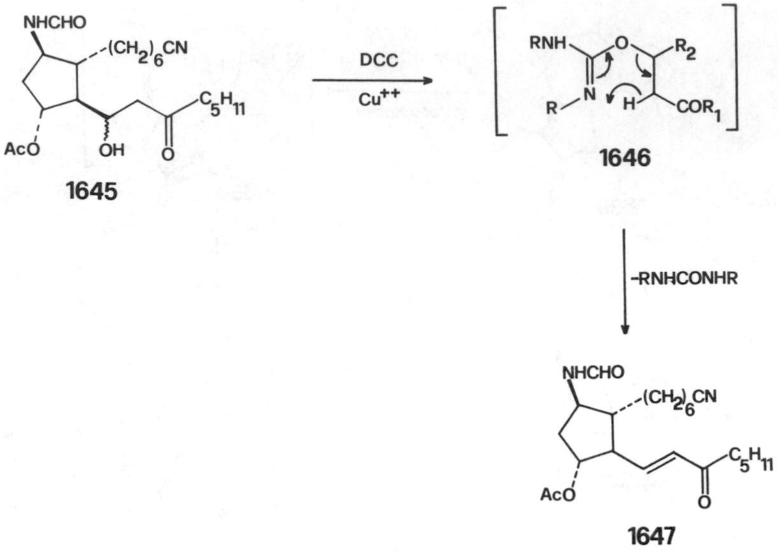

NHCHO
$(CH_2)_6CN$
C_5H_{11}
AcO OH O
1645

DCC
Cu^{++}

RNH O R_2
R N H COR_1
1646

$-RNHCONHR$

NHCHO
$(CH_2)_6CN$
C_5H_{11}
AcO O
1647

Scheme 8.27

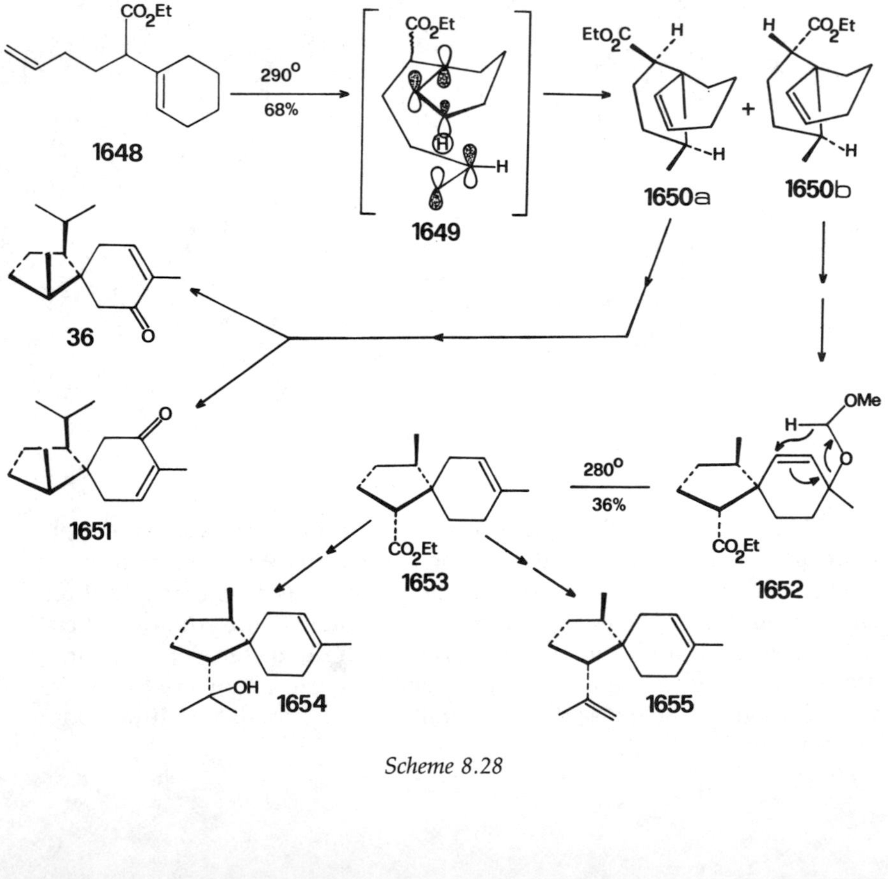

CO_2Et
1648

290°
68%

CO_2Et
H
H
1649

EtO_2C H
H
1650a

H CO_2Et
H
1650b

36

1651

OMe
H
O
CO_2Et
1652

280°
36%

CO_2Et
1653

OH
1654

1655

Scheme 8.28

Literature Cited

1. Deleris, G.; Dunogues, J.; Calas, R. *Tetrahedron Lett.* **1979**, 4835.
2. Barrett, H. C.; Büchi, G. *J. Am. Chem. Soc.* **1967**, *89*, 5665.
3. Stephenson, L. M. *Tetrahedron Lett.* **1980**, 1005.
4. El-Feraly, F. S.; Chan, J. M.; Fairchild, E.; Doskotch, R. W. *Tetrahedron Lett.* **1977**, 1973.
5. Bell, R. A.; Ireland, R. E.; Mander, L. N. *J. Org. Chem.* **1966**, *31*, 2536.
6. Masamune, S. *J. Am. Chem. Soc.* **1964**, *86*, 290.
7. Büchi, G.; Hauser, A.; Limacher, J. *J. Org. Chem.* **1977**, *42*, 3323.
8. Wenkert, E.; Naemura, M. *Synth. Commun.* **1973**, *3*, 45.
9. Ireland, R. E.; Baldwin, S. W.; Dawson, D. J.; Dawson, M. I.; Dolfini, J. E.; Nembould, J.; Johnson, W. S.; Brown, M.; Crawford, R. J.; Hudrlik, R. F.; Rasmussen, G. H.; Schmiegel, K. K. *J. Am. Chem. Soc.* **1970**, *92*, 5743.
10. Marshall, J. A.; Cohen, N.; Hochstetler, A. R. *J. Am. Chem. Soc.* **1966**, *88*, 3408.
11. Thomas, A. F. *Helv. Chim. Acta* **1972**, *55*, 2429.
12. Oppolzer, W.; Snieckus, V. *Angew. Chem., Int. Ed. Engl.* **1978**, *17*, 476.
13. Nozoe, S.; Morisaki, N.; Goi, M. *Tetrahedron Lett.* **1971**, 3701.
14. Pickenhagen, W.; Ohloff, G.; Russel, R. K.; Roth, W. R. Unpublished results (reported in Ref. 12).
15. Hoffmann, H. M. R. *Angew. Chem., Int. Ed. Engl.* **1969**, *8*, 566.
16. Naegeli, P.; Kaiser, R. *Tetrahedron Lett.* **1972**, 2013.
17. Stork, G.; Kraus, G. *J. Am. Chem. Soc.* **1976**, *98*, 6747.
18. Oppolzer, W.; Andres, H. *Tetrahedron Lett.* **1978**, 3397.
19. Oppolzer, W.; Andres, H. *Helv. Chim. Acta* **1979**, *62*, 2282.
20. Oppolzer, W.; Robbiani, C. *Helv. Chim. Acta* **1980**, *63*, 2010.
21. Oppolzer, W.; Sarkar, T. Unpublished results (reported in Ref. 12).
22. Oppolzer, W.; Bätig, K.; Hudlicky, T. *Helv. Chim. Acta* **1979**, *62*, 1493.
23. Oppolzer, W.; Marazza, F. *Helv. Chim. Acta* **1981**, *64*, 1575.
24. Conia, J. M.; Lange, G. L. *J. Org. Chem.* **1978**, *43*, 564.
25. Iguchi, M.; Nishiyama, A.; Yamamura, S.; Hirata, Y. *Tetrahedron Lett.* **1969**, 4295.
26. Yamamura, S.; Iguchi, M.; Nishiyama, A.; Niwa, M.; Koyama, H.; Hirata, Y. *Tetrahedron* **1971**, *27*, 5419.
27. Shamma, M.; Jones, C. D. *J. Am. Chem. Soc.* **1969**, *91*, 4009.
28. Jordan, K. D.; Burrow, P. D. *Acc. Chem. Res.* **1978**, *11*, 341.
29. Lefour, M. M.; Loupy, A. *Tetrahedron* **1978**, *34*, 2597.
30. Bryon Gill, G.; Wallace, B. *J. Chem. Soc., Chem. Commun.* **1977**, 380.
31. Snider, B. B. *Acc. Chem. Res.* **1980**, *13*, 426.
32. Bryon Gill, G.; Wallace, B. *J. Chem. Soc., Chem. Commun.* **1977**, 382.
33. Oppolzer, W.; Robbiani, C.; Bättig, K. *Helv. Chim. Acta* **1980**, *63*, 2015.
34. Mehta, G.; Reddy, V. *Tetrahedron Lett.* **1979**, 2625.
35. Snider, B. B.; Rodini, D. J. *Tetrahedron Lett.* **1980**, 1815.
36. Ibid., **1978**, 1399.
37. Andersen, N. H.; Golec, F. F., Jr. *Tetrahedron Lett.* **1977**, 3783.
38. Brown, R. F. C. "Pyrolytic Methods in Organic Chemistry"; Academic: New York, 1980; p. 229–246 and references therein.
39. Corey, E. J.; Yamamoto, H.; Herron, D. K.; Achiwa, K. *J. Am. Chem. Soc.* **1970**, *92*, 6635.
40. Corey, E. J.; Yamamoto, H. *J. Am. Chem. Soc.* **1970**, *92*, 6636.
41. Bahurel, Y.; Cottier, L.; Descotes, G. *Synthesis* **1974**, 118.
42. Ohloff, G. *Chem. Ber.* **1960**, *93*, 2673.
43. Greene, A. E. *J. Am. Chem. Soc.* **1980**, *102*, 5337.
44. Corey, E. J.; Anderson, N. H.; Carlson, R. M.; Paust, J.; Vedejs, E.; Ulattas, I.; Winter, R. E. K. *J. Am. Chem. Soc.* **1968**, *90*, 3245.
45. Oppolzer, W.; Mahalanabis, K. K.; Bättig, K. *Helv. Chim. Acta* **1977**, *60*, 2388.

9

Electrocyclic Reactions

This chapter deals with reactions involving an equilibrium between a moiety with $(n + 2)\ \pi$ electrons and one with $n\pi + 2\sigma$ electrons. The overall process can occur either thermally or photochemically; in the former case, the HOMO of the substrate is involved, in the latter, the LUMO. The principle of MO control of pericyclic reactions is clearly evident in this field because the symmetry of the MO participating in the process determines the stereochemistry of the product.

The MO of a polyene, in terms of its symmetry plane (p), can be symmetric or asymmetric (Figure 9.1). The ring closure requires a 90° twisting rotation of the terminal orbitals to allow a bonding overlap; the new σ-bond is thus formed. If the MO is symmetric, one rotation must be clockwise and another counterclockwise; the overall process requires a *disrotation.* If the MO is asymmetric, both rotations must be clockwise; the process demands a conrotation. The inverse process (that is, the electrocyclic ring opening that transforms a σ-bond into two π orbitals) requires the same operations. A somewhat different approach to these and other pericyclic reactions can be realized through the Möbius systems as reported by Dewar (*1*) and Zimmerman (*2*).

A disrotatory or conrotatory process obviously dramatically influences the configuration of the cycloalkene or of the polyene (in the inverse reaction). The presence of substituents on the twisting atoms causes a *cis* or *trans* relationship in the cycloalkene and a (*Z*) or (*E*) configuration of the polyene. Hence, to predict the configuration of the product, it is sufficient to detect the symmetry of the MO involved in the process. For neutral all-carbon polyenes, an even number of atomic orbitals is involved; for charged substrates or in the presence of heteroatoms, an odd number of atomic orbitals must be taken into account. As an example, 1,5-electrocyclization is an important process in heterocyclic chemistry (*3, 4*).

Electrocyclic reactions are important tools in the synthesis of natural products. Several photochemical processes of the 1960s have been reviewed by Sammes (*5*). The cyclobutene ring opening as a means of a diene generation for Diels–Alder reaction was reviewed by Oppolzer

0065-7719/83/0180-0361$13.25/1

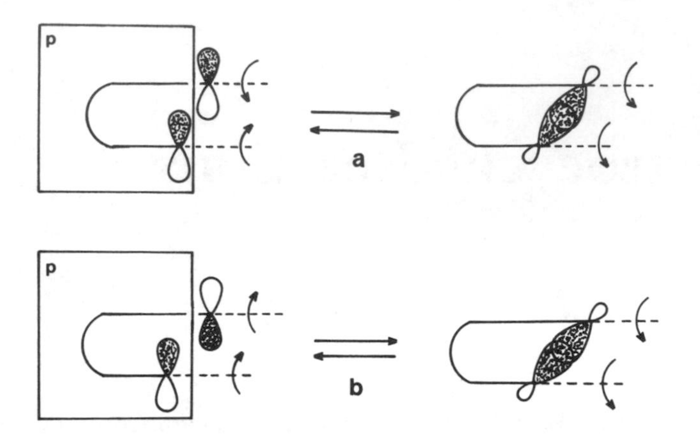

Figure 9.1. Disrotatory and conrotatory ring closure for symmetric (a) and asymmetric (b) polyenes.

(*6*) and Kametani (*7*). However, the importance of the new materials and the usefulness of the process justify an addendum in the field of natural products.

Electrocyclic Reactions of All-Carbon Polyenes

Here again we use examples from the field of vitamin D to introduce the subject (*8–10*). As discussed in Chapter 7 (in the section on [1,*m*] sigmatropic processes), ergosterol (**1204**) isomerizes photolytically to precalciferol (**1205**). This is an electrocyclic process that converts a cyclohexadiene to a hexatriene. The conrotatory character is utilized by submitting **1205** to a further irradiation to give a new electrocyclic ring closure to lumisterol (**1656**). These processes must occur at temperatures low enough (< 0 °C) to exclude formation of calciferol (**1138**) (by [1,7] sigmatropic shift). By heating **1205** at temperatures above 100 °C, a third electrocyclic ring closure takes place. Pyrocalciferol (**1657**) and isopyrocalciferol (**1658**) are thus formed. The process is disrotatory, as is immediately derived from the configuration of these products (Scheme 9.1).

There are good reasons to believe that these photochemical conversions induced by direct light absorption ($\pi \rightarrow \pi^*$ transition with no sensitizer) are singlet reactions, with the LUMO of the triene as well as its HOMO involved in the thermal conversion to **1657** and **1658.** Both processes are represented in Figure 9.2.

The products from thermal treatment of precalciferol undergo further electrocyclic modification under photochemical conditions. However, because the formation of the hexatriene derivative is a conrotatory process, one of the two cyclohexene rings would assume the *trans* configuration. Hence a four-electron process occurs forming two bicyclo[2.2.0]hexenes—photopyrocalciferol (**1659**) from **1657** and photoisopyrocalciferol (**1660**) from **1658.** The LUMO of the 4π system obviously

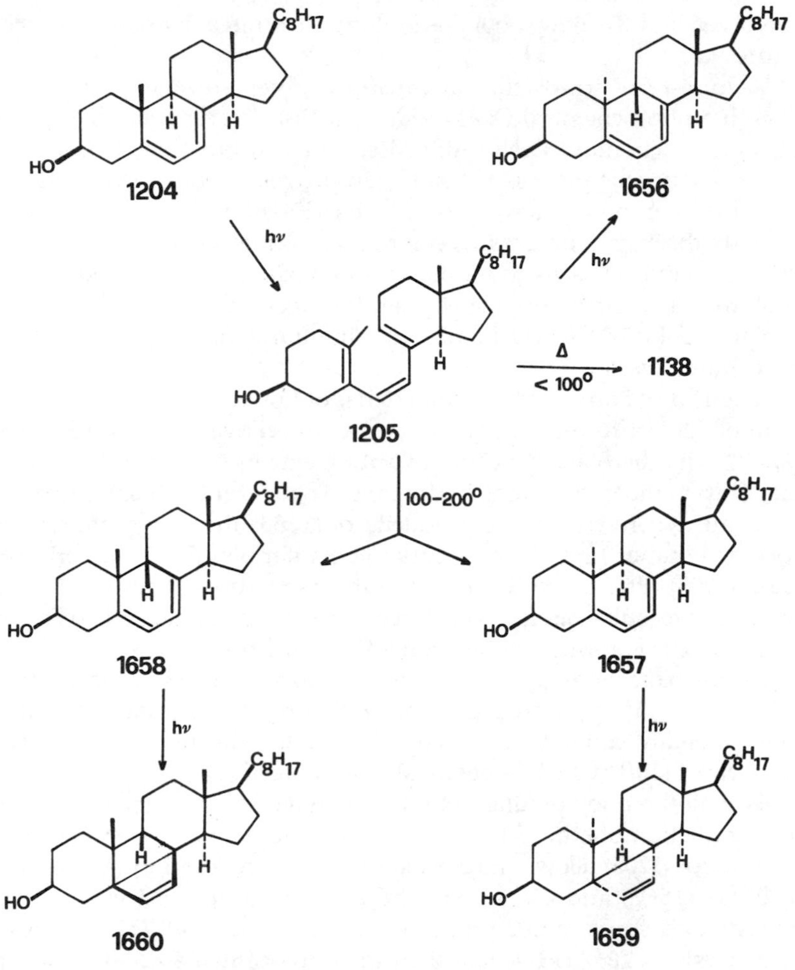

Scheme 9.1

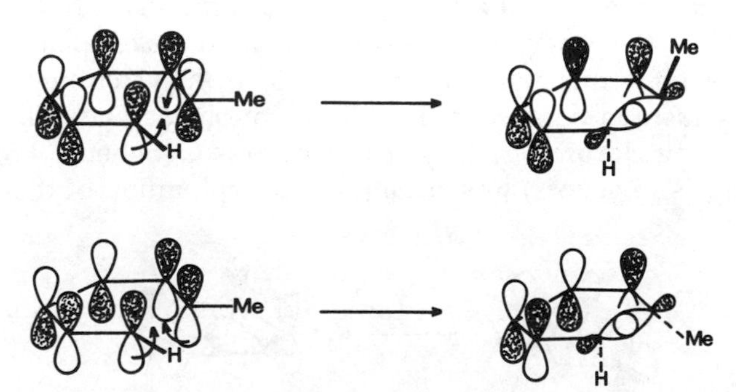

*Figure 9.2. MO representation of photochemical and thermal ring closure of precalciferol (**1205**).*

is involved and the new σ bond is built up through a disrotatory process (Figure 9.3).

A long-term irradiation of calciferol gives suprasterols I and II through a photochemical Diels–Alder reaction (*see* Chapter 5, "The So-Called Photochemical Diels–Alder Reactions"), but the necessity that this be a pericyclic process has not been proved. Photoisomerization of the (*Z*)-isomer precalciferol (**1205**) to its (*E*)-isomer tachysterol (**1661**) proceeds through a process whose nature is not yet fully clarified, but this compound helps us to introduce a new class of photoproducts, the toxisterols. This name originates from the toxic properties of long-term irradiated solutions of calciferol with simultaneous destruction of any antirachitic potency.

The irradiation of tachysterol (**1661**), obtained from an ethereal solution of **1204**, affords a vinylcyclobutene derivative, toxisterol E_1 (**1662**) (*11, 12*). The disrotatory closure of a 4π-electron system gives a *cis* relationship to the cyclobutene hydrogens. Toxisterol E_1 (**1662**) is unstable and heating converts it into precalciferol (**1205**) through a conrotatory process. What at first glance seems to be a simple (*E*)(*Z*) isomerization (**1661**→**1205**) differs greatly from the inverse process because it is the result of two subsequent electrocyclic processes. A second toxisterol, the D_1 (**1663**), is formed photochemically from precalciferol (**1205**) after long-term irradiation (*11, 12*). The reaction proceeds antarafacially through a [1,5] sigmatropic shift. Toxisterol D_1 is unstable and on heating readily converts into tachysterol (**1661**) through a second [1,5] sigmatropic shift, which is obviously suprafacial (Scheme 9.2).

Not all the photoproducts of calciferol have a pericyclic origin. Long-term irradiation of ethereal or ethanolic solutions of ergosterol (**1204**) or precalciferol (**1205**) allows the isolation of two bicyclo[3.1.0]hexenes: toxisterol C_1 (**1667**) and C_2 (**1668**). The former compound is the major product, and some minor products are formed as well (*11, 12*). An x-ray analysis of **1667** (*13*) established unequivocally a structure that can arise from a [$4\pi a + 2\pi a$] process only. Therefore, **1667** and **1668** are formed via a nonconcerted pathway (*14*), possibly involving the relaxation of the π→π* state (**1664**) through zwitterionic states **1665** and **1666.** Incidentally, at this stage, an electrocyclic reaction takes place; the negative end of **1665** (the allylic anion: a 4π-electron system) cyclizes in a conrotatory fashion because its HOMO is involved. The following cyclopentene ring closure of **1666** gives toxisterols C (Scheme 9.3). Jacobs and Havinga's review (8) was an important exploration of this field.

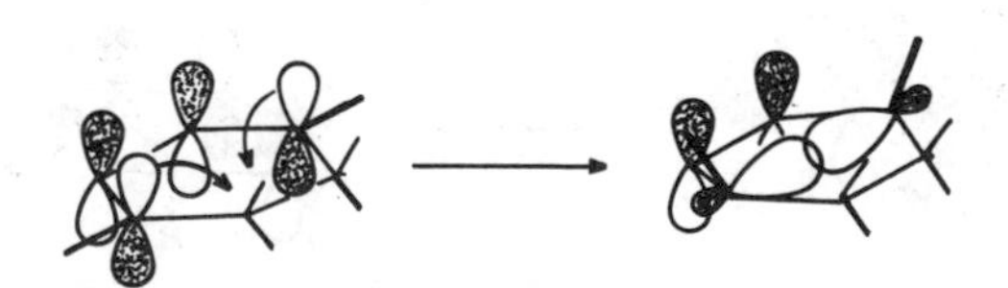

*Figure 9.3. MO representation of photochemical conversion of pyrocalciferols (**1657** and **1658**) to photopyrocalciferols (**1659** and **1660**).*

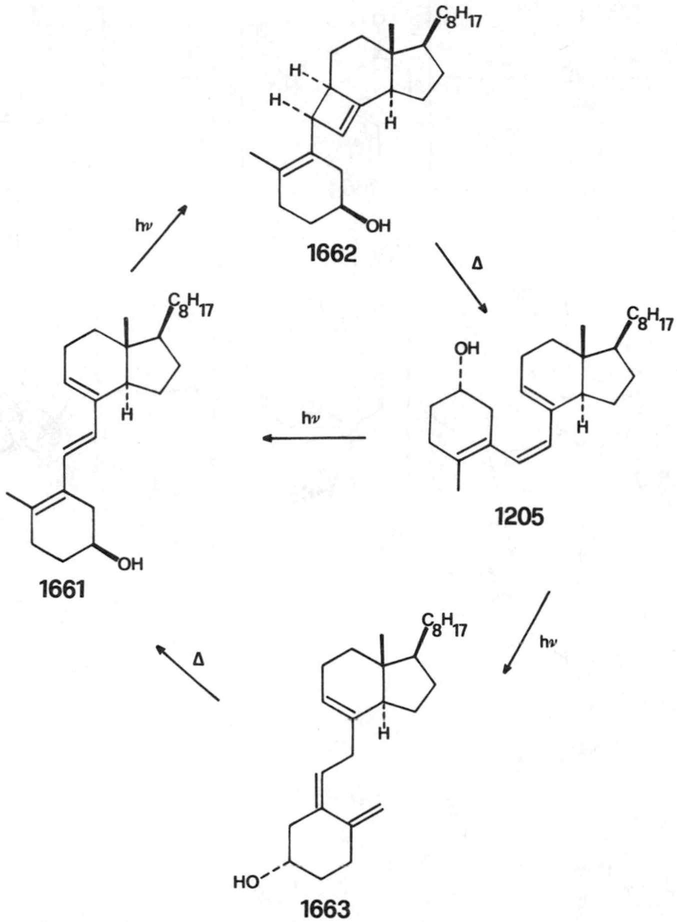

Scheme 9.2

4π Electrocyclic Processes. In Chapter 3 we mentioned the photoconversion of colchicine (**139**) to β- and γ-lumicolchicines (**140** and **141**, respectively) (*15*), as a good example of generation of a cyclobutene from a diene fragment in a seven-membered ring. The same process was used as the key step in the synthesis of the pheromone grandisol (**21**) from eucarvone (**1669**) in 20% overall yield via a disrotatory ring closure to **1670** (*16*) (Scheme 9.4).

Another pheromone, (±)-ipsenol (**953**), was generated by starting from **1671**. Although the electrocyclic cyclobutene ring opening of **1673** is the last step of the synthesis, the key step seems to be the double bond generation by anchimeric assistance of the oxyanion **1672** (*17*) (Scheme 9.5).

In closing this section we note the similarity of this topic with the discussions in Chapter 5 dealing with the generation of dienes for Diels–Alder reaction by electrocyclic cyclobutene ring openings.

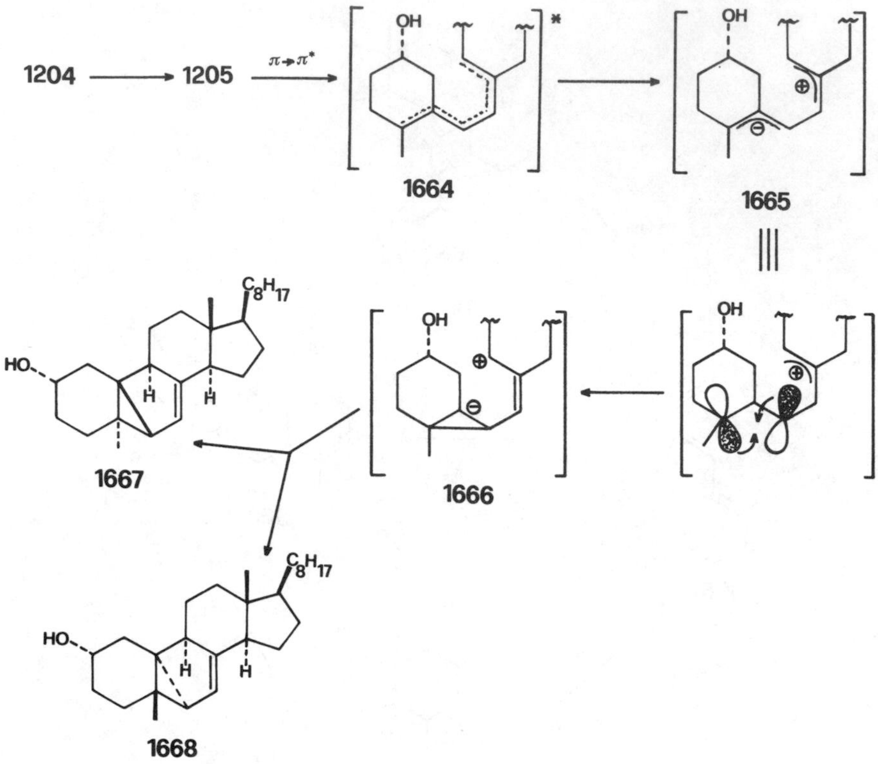

1204 → 1205 $\xrightarrow{\pi \rightarrow \pi^*}$ [1664]* → [1665] ≡ [ ] → [1666] → 1667, 1668

Scheme 9.3

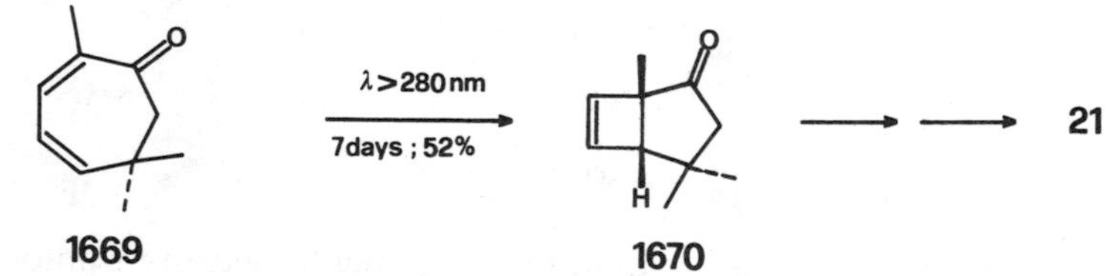

1669 $\xrightarrow[\text{7days ; 52\%}]{\lambda > 280\text{nm}}$ 1670 → → 21

Scheme 9.4

6π Electrocyclic Processes. The most familiar of these processes is the photocyclization reaction of stilbenes (*18*), recently rationalized in terms of change in bond orders upon excitation (*19*). The reaction, in accordance with the generally accepted mechanism, starts from the stilbene singlet S_1 excited state (originally from the triplet T_1). Upon conrotatory ring closure, stilbene gives a *trans*-dihydrophenanthrene that subsequently evolves to the products. In the presence of oxygen, the usual product is phenanthrene (Scheme 9.6).

Based on this approach, the dehydroaporphane ring system **1675** was synthesized by starting from tetrahydroisoquinoline derivative **1674**

(*20*). When the starting material is a mixture of (*Z*)- and (*E*)-isomers, a photoisomerization to the reacting (*E*) species is the initial step (Scheme 9.7).

Analogously, tylophorine (**1093**) (*21*) was obtained from septicine (**436**), and (±)-nuciferine (**1678**) from **1676**, after oxidation/reduction of the dihydroderivative **1677** (22) (Scheme 9.8).

Difficulties arise in the loss of two hydrogen atoms from the dihydrophenanthrene derivative, and by-products are sometimes formed.

A cleaner nonoxidative variant consists of the irradiation of 2′-chloro, bromo, or iodo derivatives. The following mechanism can be proposed: (*E*)/(Z) reversible irradiation; reversible electrocyclic ring closure to dihydrophenanthrene derivative; and, as a last step, the irreversible loss of hydrogen chloride (or other hydrogen halide acid) to yield the phenanthrenic compound (Scheme 9.9).

An alternative, homolysis of the carbon–halogen bond to give a radical intermediate, seems less probable.

Thus aristolochic acid (**1680**) was obtained from **1679** (*23*), nuciferine

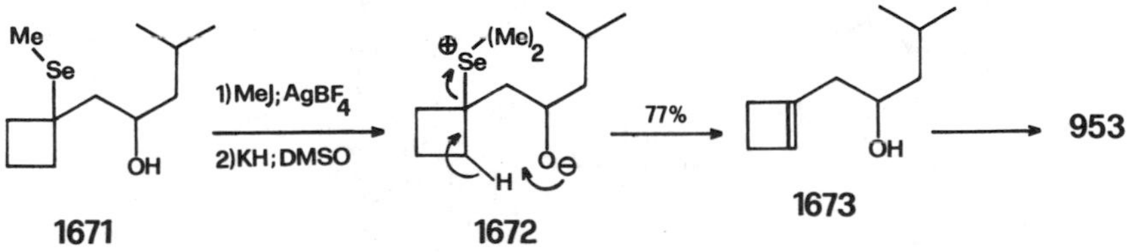

Scheme 9.5

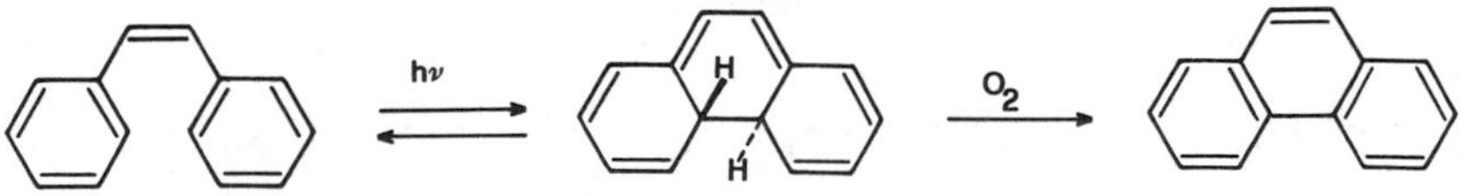

Scheme 9.6

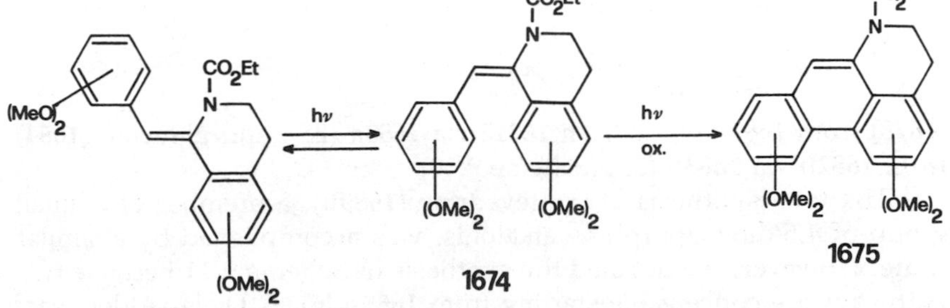

Scheme 9.7

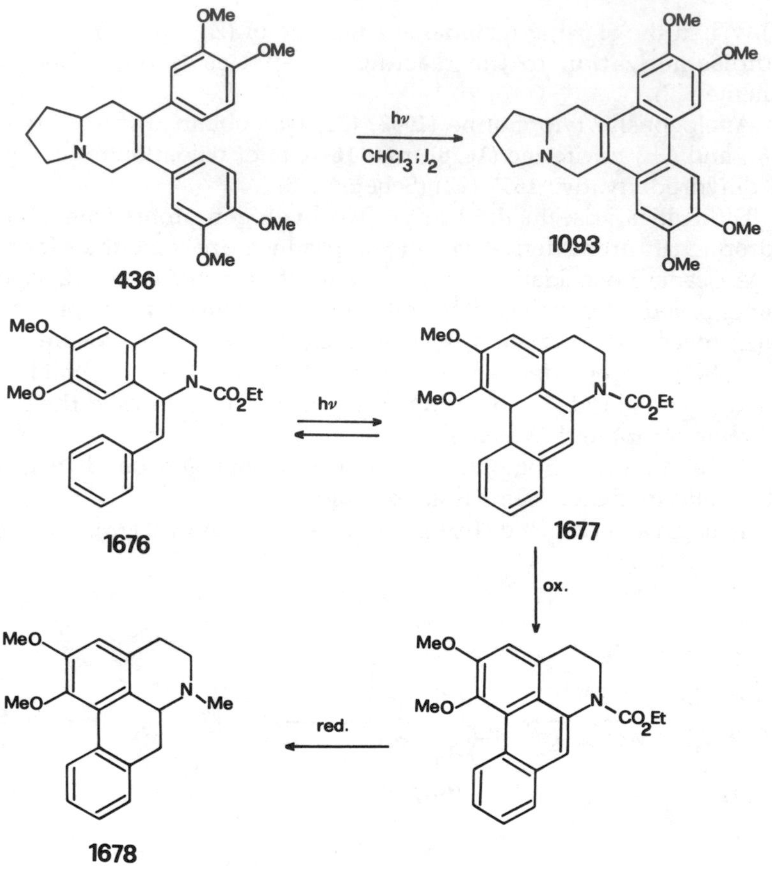

Scheme 9.8

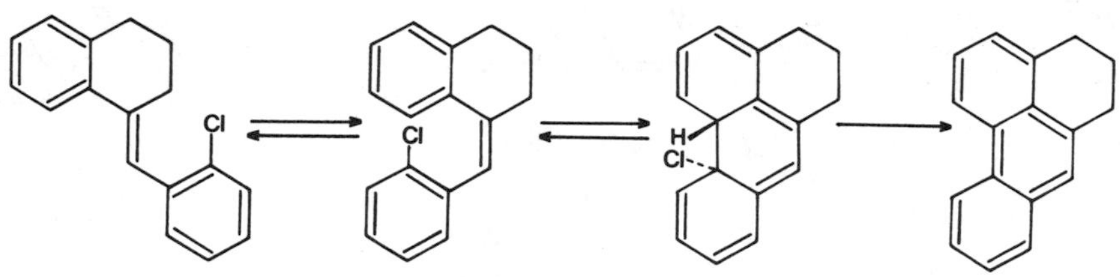

Scheme 9.9

(**1678**) from **1681** (*24*) or from **1682a** via **1683a** (*25*), and glaucine (**1684**) from **1682b** via **1683b** (*25*) (Scheme 9.10).

The total synthesis of pontevedrine (**1689b**), a member of a small group of 4,5-dioxoaporphine alkaloids, was accomplished by a similar route. However, we detailed the synthesis in Scheme 9.11 because two pathways are conceivable starting from **1685** (*26*). A Diels–Alder reaction with singlet oxygen gives dioxo compound **1687** via endoperoxide

1686. Loss of methanol affords a mixture of (*Z*)- and (*E*)-**1688**; the mixture was irradiated to produce demethylpontevedrine (**1689a**), which is easily transformed into the target compound **1689b**. The latter compound can also be obtained through a "one-pot" synthesis in 23% yield from **1685** under photooxidative conditions in basic medium.

A somewhat similar process entails the photocycloaddition of 2-vinylbiphenyl derivatives (*27*), as were utilized in the synthesis of juncusol (**1692**), a vinyl phenanthrene derivative isolated from the marsh grass *Juncus roemerianus*. The key step is the ring closure of **1690** to dihydrophenanthrene derivative **1691**, which was converted within two steps to **1692** (*28*) (Scheme 9.12). Four of the 6 π electrons may belong to two exocyclic double bonds of an indole ring. Thus two simple

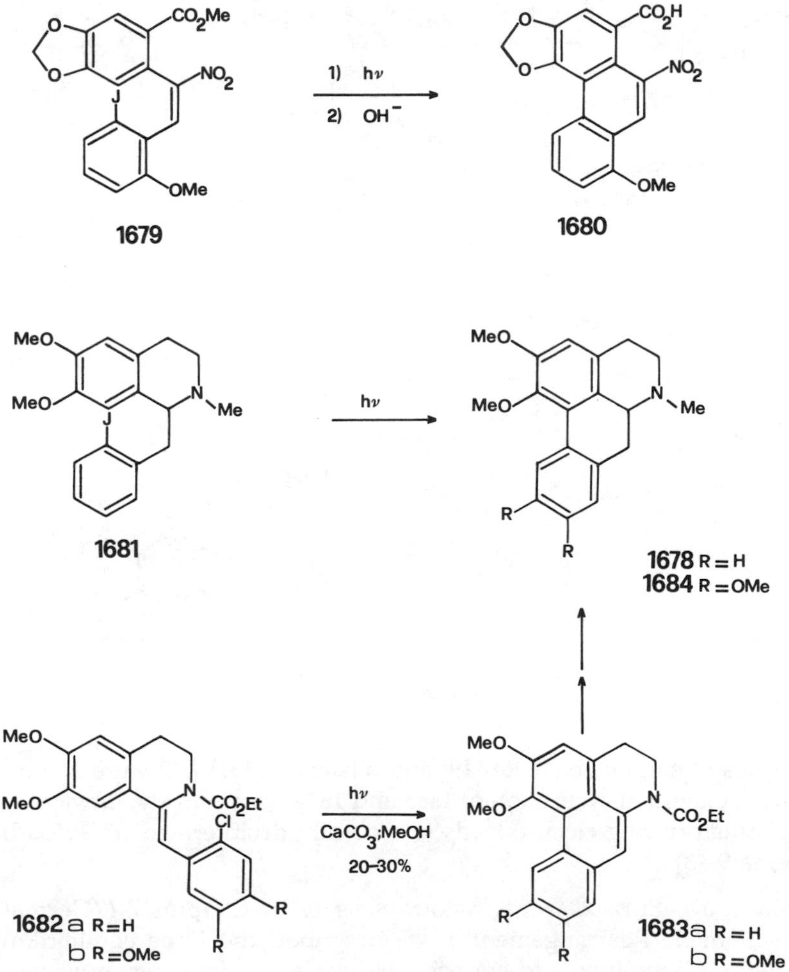

Scheme 9.10

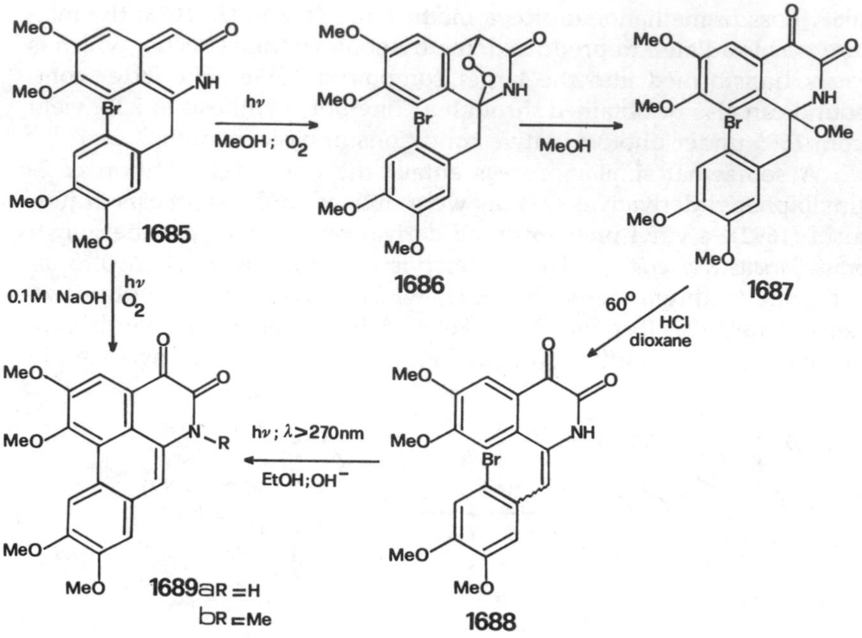

Scheme 9.11

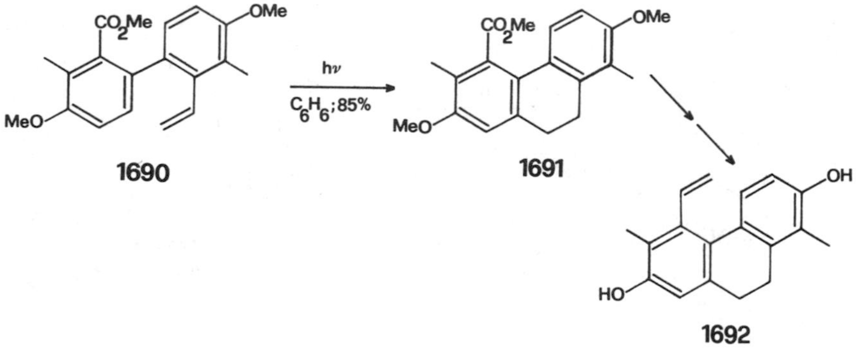

Scheme 9.12

syntheses of ellipticine (**1696**) (*29*) and olivacine (**731**) (*30*) were accomplished by thermal cyclization of **1693** and **1694**, respectively. Obviously the disrotatory ring closure leads to a *cis*-dihydrointermediate **1965a,b** (Scheme 9.13).

GERMACRANE–EUDESMANE REARRANGEMENT. In Chapter 7 ("Germacrane–Elemene Rearrangement"), we described the Cope equilibrium between elemene-type and germacrane-type sesquiterpenes; however, a further equilibrium may be achieved, namely an electrocyclic conversion to eudesmanolides. Eudesmanolides are a large family of naturally occurring sesquiterpenes based on the eudesmane skeleton (*31*). The rings are generally *trans*-fused; hence the equilibrium with germacranes must occur photochemically (conrotatory process). Formation of the less common *cis*-fused skeleton requires thermal conditions (disrotatory process) (Scheme 9.14).

In general, the photochemical conditions shift the equilibrium in favor of the germacranes. Accordingly, the first total synthesis of dihydrocostunolide (**1298**) was realized by Corey and Hortmann (*32, 33*), who started from the readily available α-santonin (**1296**). The same substrate was used later by Grieco and Nishizawa (*34*) in the synthesis of costunolide (**1270**) via elemene. α-Santonin (**1296**) was converted in five

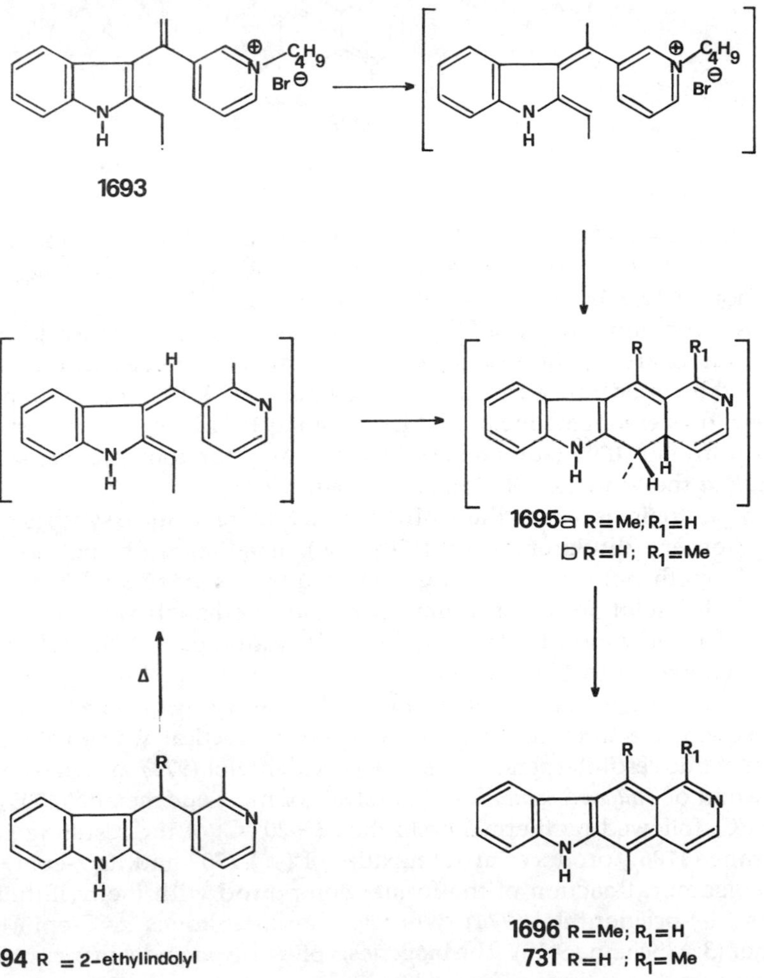

1693

1695a R=Me; R_1=H
b R=H; R_1=Me

Δ

1694 R = 2-ethylindolyl

1696 R=Me; R_1=H
731 R=H ; R_1= Me

Scheme 9.13

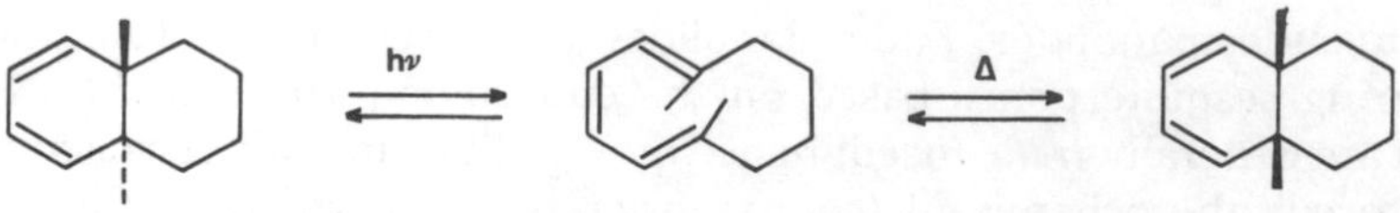

Scheme 9.14

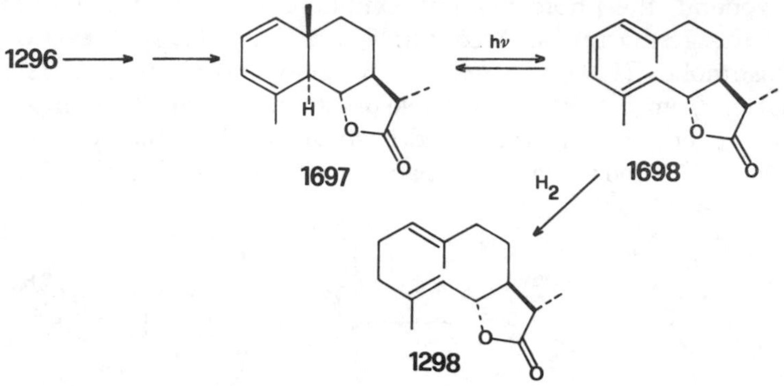

Scheme 9.15

steps to eudesm-1,3-dien-6,14-olide (**1697**), whose photochemical equilibrium with **1698** was quenched by hydrogenation, thus allowing the isolation of **1298** in an overall 10% yield (Scheme 9.15).

A recent modification of the Corey synthesis, one which avoids the photostationary equilibrium between eudesmane and germacrane, was realized by prior transformation of α-santonin into **1699**. Irradiation provides a diastereomeric mixture of the dienone **1701** because the initially formed trienol **1700** tautomerizes irreversibly. Four conventional steps complete the synthesis of **1298** (*35*) (Scheme 9.16).

α-Santonin is again the starting material for a nice synthesis of sesquiterpene dihydronovanin (**1704**) (*36*). Irradiation of enol acetate **1702** in methanol at −20 – 40 °C, followed by intermediate hydrolysis in a chilled solution of potassium hydroxide–methanol–water to avoid thermal recyclization to the *cis*-fused eudesmane, gave **1703,** which is easily converted to **1704** (Scheme 9.17).

A *trans–cis* isomerization of the eudesmane ring realized by two subsequent electrocyclic processes opened a practical way to the *cis*-eudesmanic sesquiterpene alcohol (+)-occidentalol (**974**), isolated from the wood of *Thuja occidentalis* L. Photolysis of *trans*-eudesmane (**1705**) at −78 °C, followed by thermal cyclization (−20 °C) of the resulting germacrane (**1706**), produces at 1:2 mixture of (+)-**1707** and its 7-epi-(−)-stereoisomer. Reaction of the former compound with methyllithium gives (+)-occidentalol (**974**), whereas the latter gives its 7-epi-(−)-isomer (*37*) (Scheme 9.18). The biogenesis of *cis*-fused eudesmanes could follow a route via germacranes (*38*).

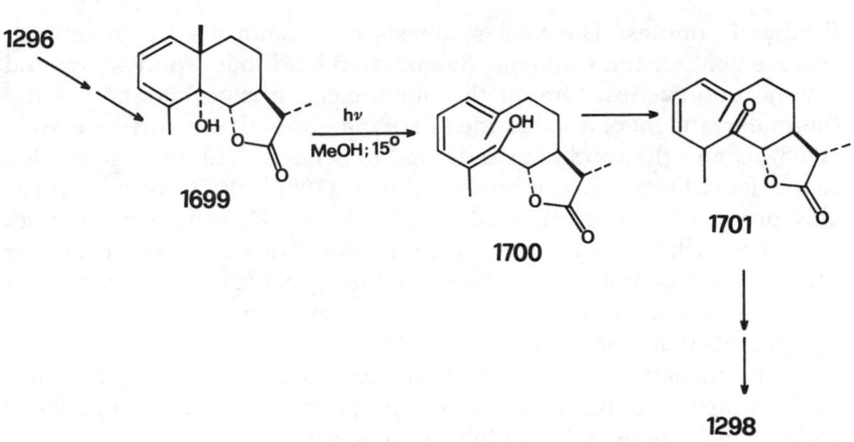

Scheme 9.16

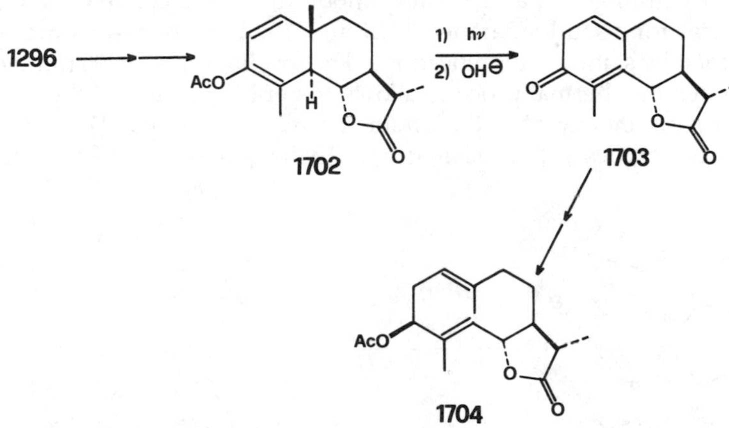

Scheme 9.17

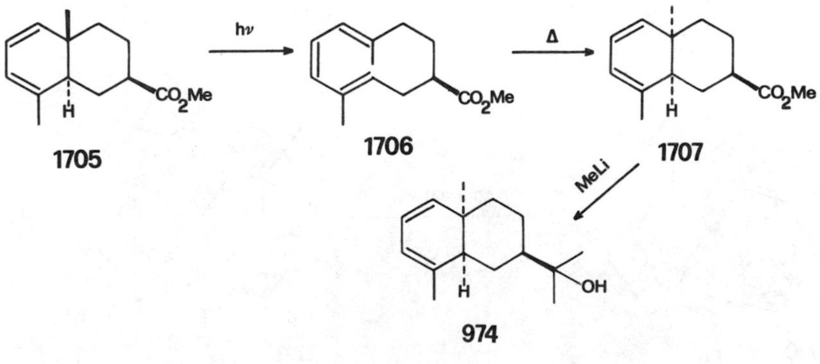

Scheme 9.18

Further Examples. The total synthesis of vitamin B_{12} was a difficult achievement for the synthetic organic chemist. Model studies explored several approaches. One of the most exciting experiments was the thermal treatment of the chloride salt of the nickel(II) secocorrinate oxide (**1708**). Under the conditions indicated in Scheme 9.19, this process led to the neutral nickel(II) D-pyrrolocorrinate (**1709**) (*39*). The reaction probably proceeds via two intermediates, **1710** and **1711**, and consists of an electrocyclization involving in the ground state 16 π electrons. Therefore the process is conrotatory, as illustrated in Figure 9.4. A similar process was proposed to occur in the biosynthesis of vitamin B_{12} (*40*).

A particular case of electrocyclization involves six electrons, 4 π and 2 σ, with the latter belonging to a cyclopropane. Thus the driving force of the transformation of divinylcyclopropane into 1,3-cycloheptadiene (Scheme 9.20) is the relief of the strain energy.

This approach was used in the total synthesis of (±)-damsinic acid (**1715**) and (±)-confertine (**19**) by thermal treatment of **1713**. This compound rearranges in a disrotatory mode to **1714**, a convenient starting material for pseudoguaianes **1715** and **19**, both of which are easily obtained by standard modifications. The irradiation ($\lambda > 290$ nm), which precedes the thermal process, allows the epimerization of the *trans* cyclopropane moiety of **1712**. Under thermal conditions, **1712** preferentially undergoes a [1,5] sigmatropic hydrogen shift to the undesired

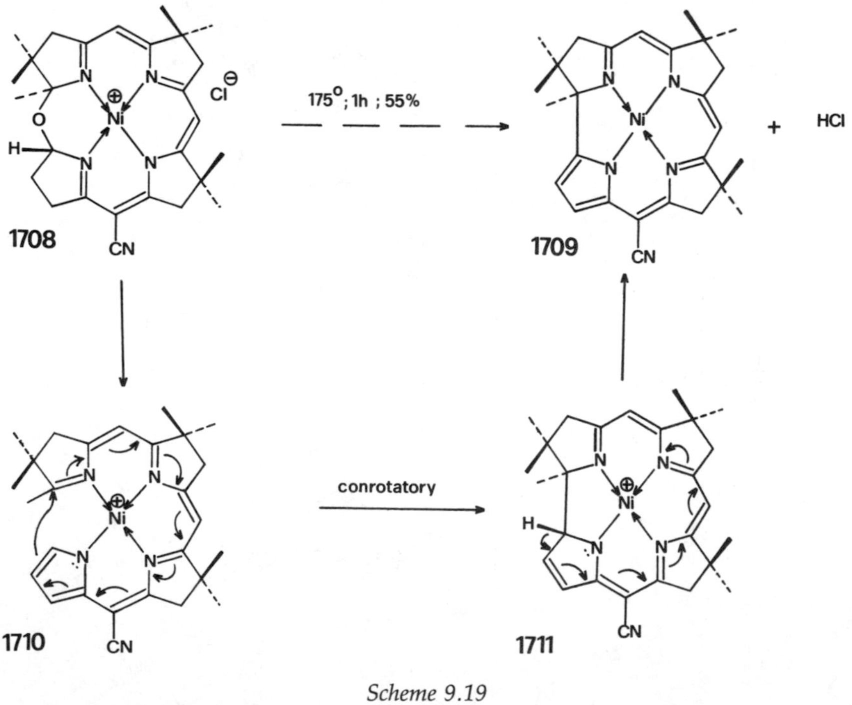

Scheme 9.19

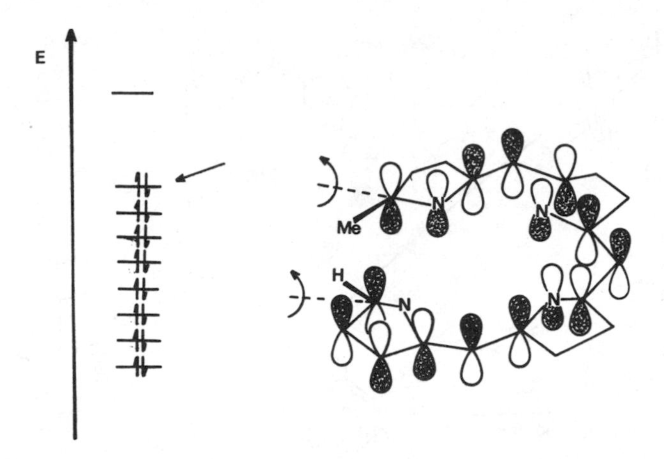

*Figure 9.4. Thermal conrotatory ring closure of **1710** (with HOMO involvement).*

product **1716** (*41*) (Scheme 9.21). This example exemplifies the importance of good overlap among the interacting orbitals.

Similarly (±)-dihydrospiniferin-1 (**1722**), a furanosesquiterpene from the Mediterranean sponge *Pleraplysilla spinifera,* was synthesized by starting from the octalone **1717**. This compound was transformed through a Simmons–Smith reaction into **1718**, which was then converted in few steps to **1719**. Treatment of **1719** with a trace of acid promotes rearrangement to **1721**, which is easily transformed into the target compound. The ring opening can be regarded as a thermally allowed disrotatory electrocyclic process with **1720** as an intermediate (*42*) (Scheme 9.22).

Electrocyclic Reactions of Heteropolyenes

This section will be divided in two parts: the first dealing with the reaction of stable heteropolyenes, and the second focussing on their generation by photochemical processes, a topic primarily related to enamides.

Heteropolyenes. Several examples of conversion of α,β-γ,δ-diunsaturated carbonyl compounds to pyrans are known.

Of the more than one hundred aporphine alkaloids, (+)-thalphenine (**1725**) is the only one with a condensed pyran ring. Its total synthesis was achieved by irradiating **1723** to give **1724**, which is easily methylated to **1725**. It is difficult to establish the mechanism of the reaction; however, after the photosubstitution step, oxyanion **1727** could

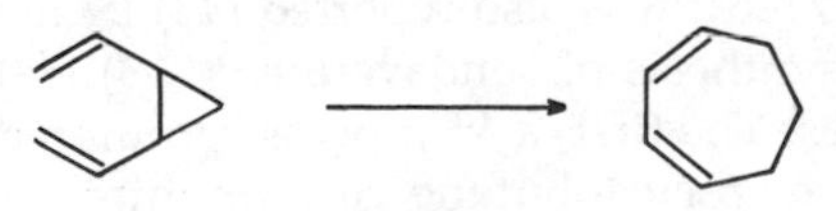

Scheme 9.20

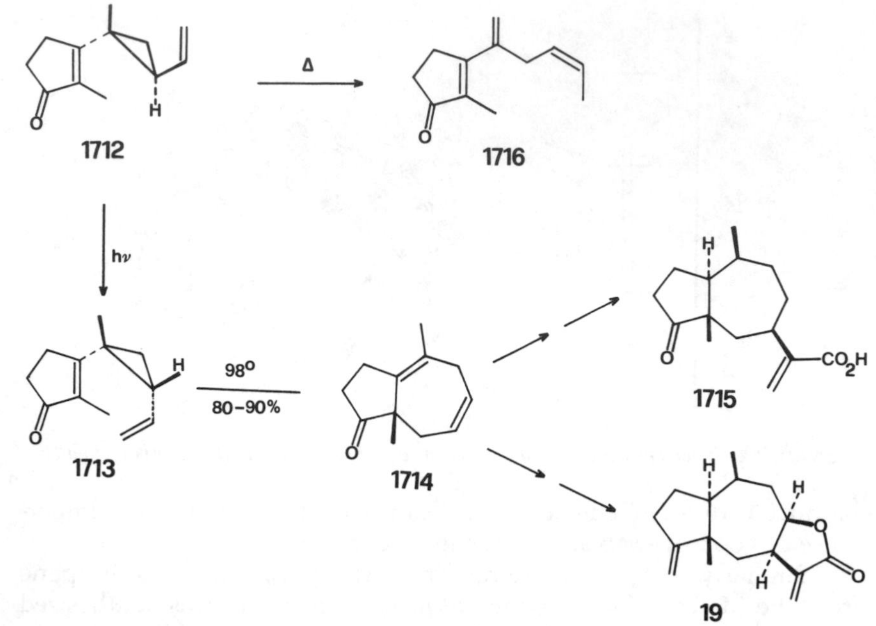

Scheme 9.21

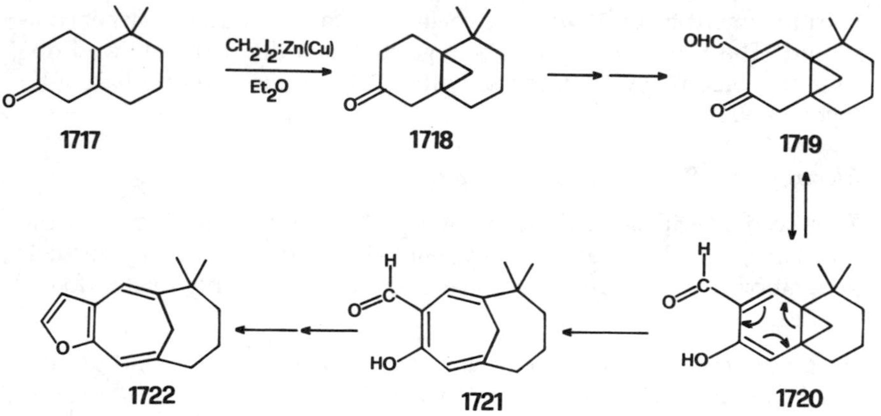

Scheme 9.22

lose methoxylate to generate **1728**, suitable for a 6π electrocyclic process (thermal or photochemical). Conversion of (±)-**1725** into two other alkaloids found in the same thalictrum species [thaliglucine (**1726a**) and thaliglucinone (**1726b**)] was also reported (*43*) (Scheme 9.23).

Kametani's synthesis of sendaverine (**1733**) (*44*) features two electrocyclic processes. The first, a 4π process common in Kametani's work, consists of the benzocyclobutene ring opening of **1729** to **1730**; the second, a 6π process, gives the isobenzopyran (**1731**). The resulting

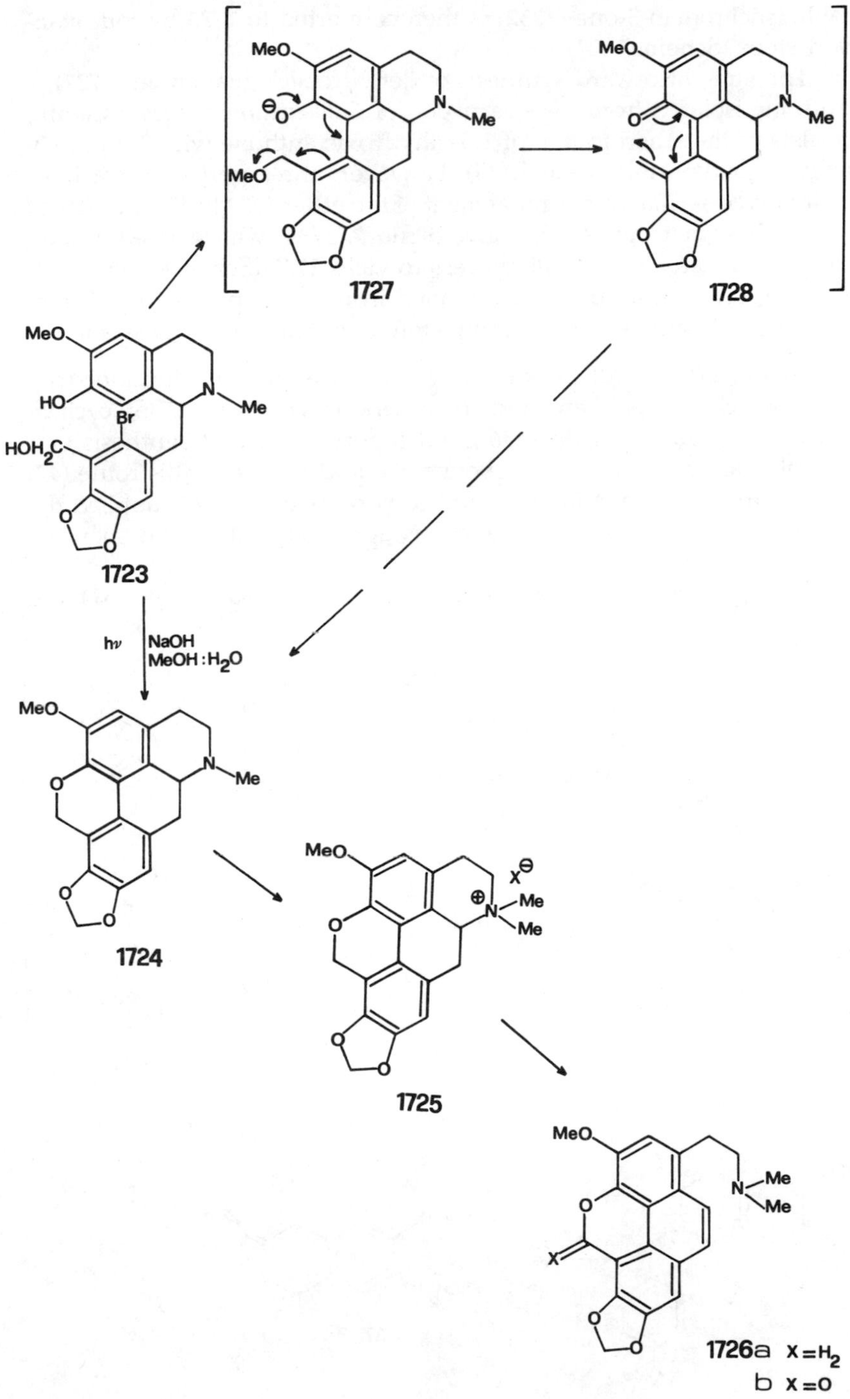

MeO HO Br HOH2C N Me O O 1723

hν NaOH MeOH : H2O

1727 1728 1724 1725

1726a X = H2
b X = O

Scheme 9.23

stable isochroman-3-one (**1732**) is then converted to **1733** by four standard steps (Scheme 9.24).

The straightforward synthesis of dehydrocycloguanandin (**1737**), a xanthone isolated from the heartwood of *Calophyllum brailiense* Camb., consists of the etherification of 1,5-dihydroxyxanthone with 3-bromo-3-methyl-1-butyne in 9% yield (*45*). However, three pericyclic reactions are involved—Claisen rearrangement of the ether **1734** to **1735**, followed by a [1,5] sigmatropic shift to give dienone **1736**, which finally undergoes a 6π heteroelectrocyclic process to yield **1737** (Scheme 9.25).

Azapolyenes are briefly mentioned in this first part of the section only; the following section is completely devoted to their cyclization.

DIENAMIDE PHOTOCYCLIZATION. This topic was mainly developed by two research groups. Lenz and coworkers described the photocyclization of a dienamide in 1966 (*46*), and reported the first synthesis of a natural product of the protoberberine alkaloid family by this route (*47*) in 1967. In the 1970s, Ninomiya and coworkers extensively applied the reaction to alkaloid synthesis, and this approach to the field soon became very useful.

Because several reviews have covered the topic (*48–51*) the

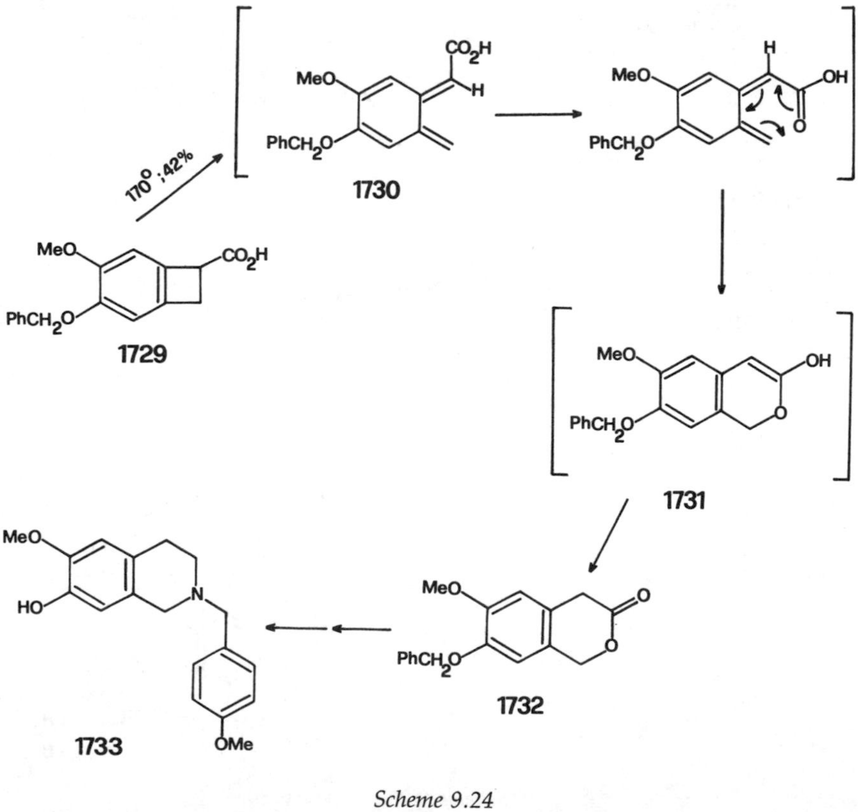

MeO PhCH$_2$O CO$_2$H 1729

170°; 42%

MeO PhCH$_2$O CO$_2$H H 1730

MeO PhCH$_2$O H OH O

MeO PhCH$_2$O OH O 1731

MeO PhCH$_2$O O O 1732

MeO HO N OMe 1733

Scheme 9.24

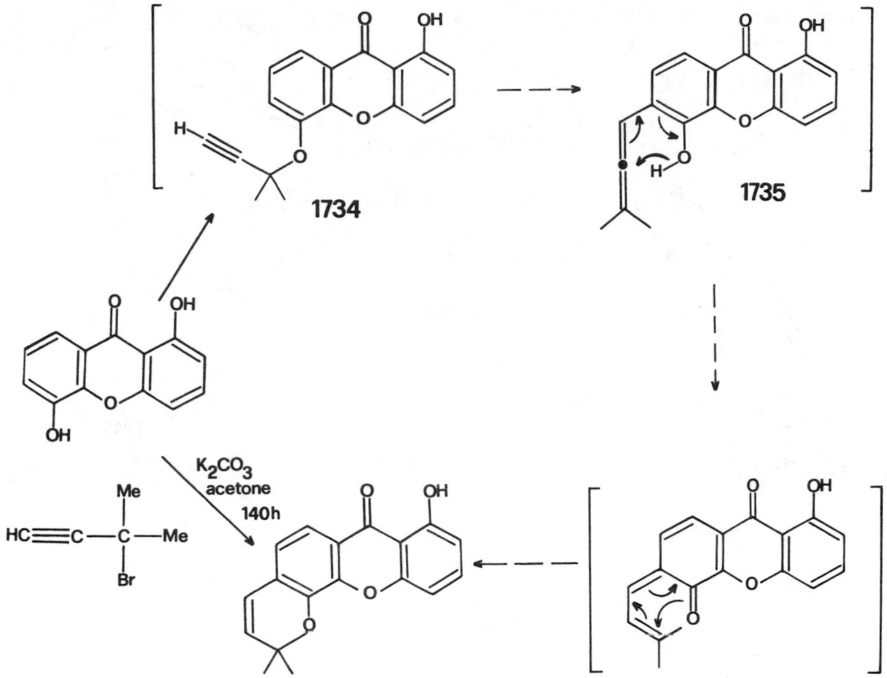

HC≡C—C(Me)(Me)Br, K_2CO_3, acetone, 140h

1734 1735 1736 1737

Scheme 9.25

syntheses will be reviewed schematically after an introduction to the mechanism.

The irradiation of a dienamide of general formula **1738** with a $\lambda >$ 280 nm gives rise to a pyridine derivative **1741** or **1742**. The mechanism of this process is summarized in Scheme 9.26 and consists of three steps: a valence-bond isomerization (**1738**→**1739**), a heteroelectrocyclic ring closure to zwitterion **1740**, and a group shift.

The first step for all amides is a ground state process. Application of the Mulliken population analysis to the occupied orbitals of the formamide shows that the nitrogen atom loses a part of its π electron density to oxygen. Through this resonance, the dipolar form contributes more than 20% to the ground state. Thus the extensive conjugation of dienamide increases the contribution of the dipolar form **1739**.

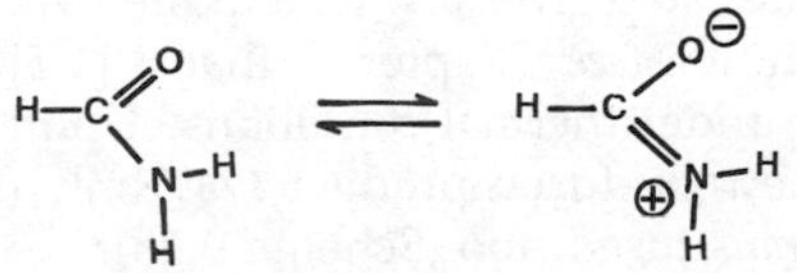

The presence of the nitrogen lone pair is critical for the electrocyclic step because diethyl 2-biphenylmalonate (**1743**) reacted upon irradiation to give **1744** under basic conditions only (*52*) (Scheme 9.27).

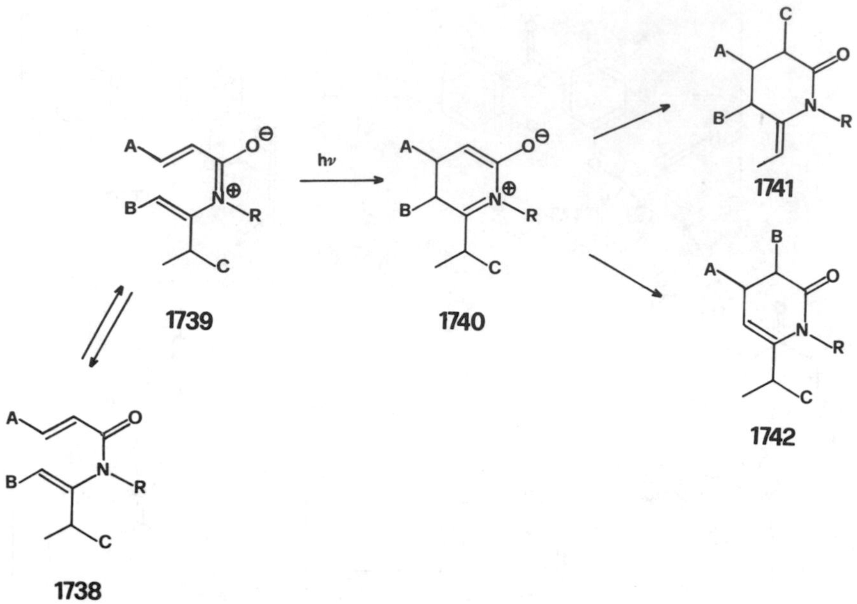

Scheme 9.26

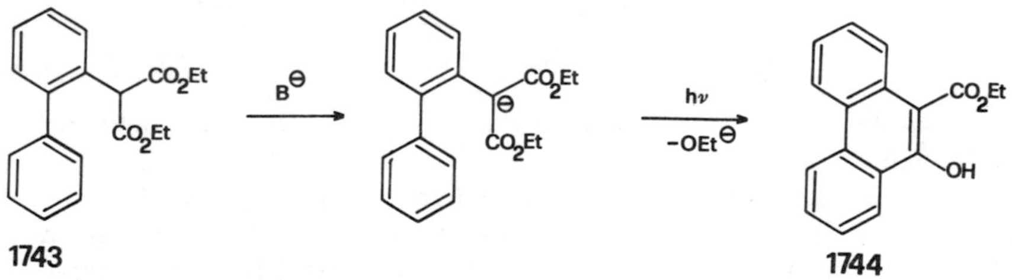

Scheme 9.27

Electrocyclization is a 6π process occurring under photochemical conditions, hence involving the LUMO of **1739**. It is difficult to provide direct proof of the conrotatory character of the process because the dipolar species **1740**, which for the configuration of **1738** should have A and B groups in a *trans* relationship, undergoes group shift. (The most common such shift is a [1,5] sigmatropic rearrangement to give **1742**.) However, only this last step gives any clue to the nature of the transient species; it is well known (*see* Chapter 7) that a [1,5] sigmatropic shift occurs suprafacially under thermal conditions. Upon irradiation, dienamide **1745** gives the *trans*-fused product **1747** only (*53*); therefore, intermediate **1746** is *trans*-fused, too (Scheme 9.28).

The MO sequence, starting from LUMO of **1745** and taking into account the HOMO of the π system and the LUMO of the σ fragment of **1746**, rationalizes the obtained configuration of **1747** (Figure 9.5).

A nice demonstration of the [1,5] sigmatropic shift can be accomplished by irradiating the pentadeutero-substituted amide **1748**, which gives exclusive migration of an *ortho* deuteron to the expected position in **1749** (*54*) (Scheme 9.29).

Finally, the solvent can be dramatically important for the success of the reaction. Generally better yields are obtained in nonpolar aprotic solvents (*55*).

To demonstrate the usefulness of this cyclization, we detail the syntheses of two alkaloids taken from older examples. The protoberberine alkaloid berberbine (**1753**) was obtained in 1967 by irradiation of *N*-acetyldienamide **1750** under acidic conditions. Intermediate **1751** undergoes protonation and loss of water to give 8-methylprotoberberine iodide (**1752**), which can be reduced with borohydride to the desired product **1753** (*47*) (Scheme 9.30).

(±)-Crinan (**1758**), the basic ring system of crinine and the related alkaloids occurring in Amaryllidaceae plants, was synthesized from a mixture of two dienamides **1754** and **1755**. These dienamides were obtained from piperonyl chloride and benzylimino-2-allylcyclohexanone.

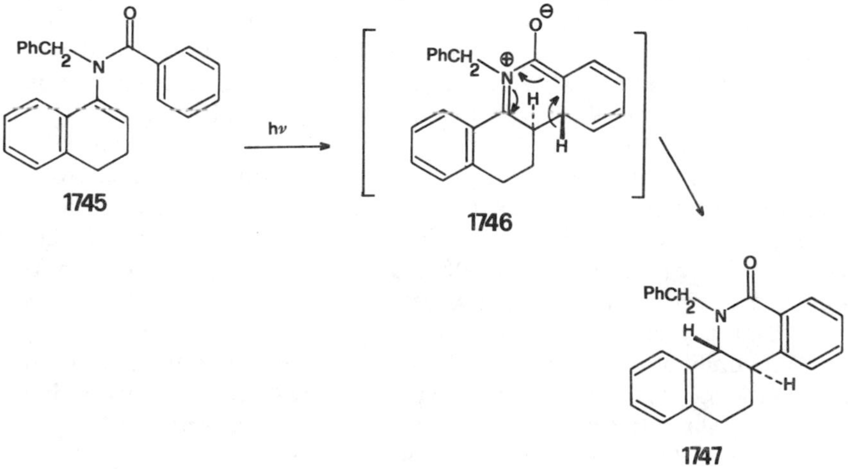

Scheme 9.28

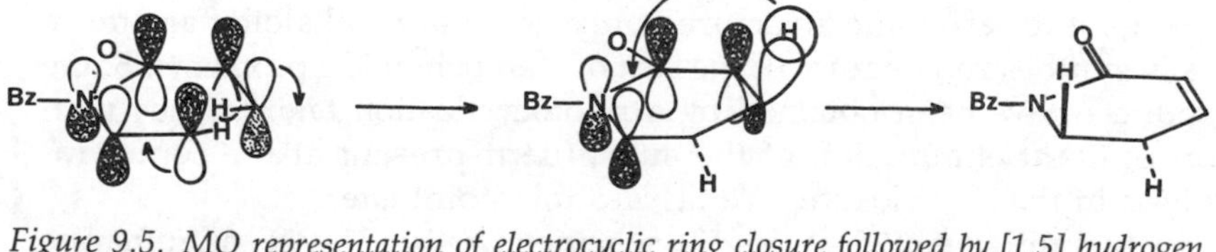

*Figure 9.5. MO representation of electrocyclic ring closure followed by [1,5] hydrogen shift from **1745** to **1747**.*

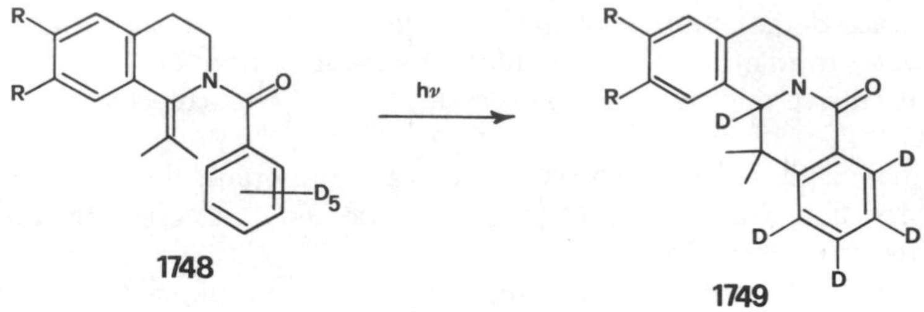

Scheme 9.29

Irradiation with long-wavelength light (λ > 300 nm) causes isomerization of the undesired isomer **1759** to **1755**. Irradiation with high energy light (λ = 254 nm) gave cyclization to **1757**, converted in a few steps to **1758** (*56, 57*) (Scheme 9.31). The allyl group of **1755** is important because it later becomes the crinan bridge; the conrotatory ring closure to give the *trans* dipolar intermediate **1756** is critical because the suprafacial [1,5] migration of hydrogen generates the configuration of **1757** that is retained in the final product.

The extensive application of this reaction is evident from the host of alkaloids, or structurally related compounds, that have been synthesized by this approach. The ring realized by photocyclization is represented by heavy lines in the formulae. The following indole alkaloids are listed in Scheme 9.32: flavopereirine (**1759**) (*58*), yohimbane (the basic skeleton) **1760** (*58*) and its partially aromatic derivative demethoxycarbonylhydrogambirtannine (the yohimbine skeleton) (**1761**) (*59, 60*), nauclefine (**1762a**) (*61, 62*), angustidine (**1762b**) (*59, 60*), angustoline (**1762c**) (*60, 63*), naucletine (**1762d**) (*60*), angustine (**1762e**) (*60*), and dihydroangustine (**1762f**) (*60*). The majority of these compounds are corynanthe alkaloids isolated, for example, from *Strychnos angustiflora*, a medicinal plant from southern China.

The family of protoberberine alkaloids is diverse and their skeleton is a suitable target for the dienamide photocyclization. Thus xylopinine (**1086b**) (*64–66*), tetrahydropalmatine (**1763a**) (*65*), sinactine (**1763b**) (*65, 67*), cavidine (**1763c**) (*67, 68*), and corytenchirine (**1763d**) (*69*) (isolated from the Taiwan plant *Corydalis ochotensis*), were approached in this manner. Their structures are listed in Scheme 9.33.

Benzo[*c*]phenanthridine alkaloids can be divided into two groups. The first group includes the fully aromatized structures with quaternary nitrogen; the second includes the hexahydro compounds having A and D rings with aromatic structures only. Aromatic alkaloids are more easily synthesized because of the lack of stereochemical problems. Some products have been obtained by a photocyclization taking place with concomitant elimination of the substituent present at the center involved in the ring closure. We discuss this point later.

Thus nitidine (**1764a**) (*70–72*), avicine (**1764b**) (*71, 72*) sanguinarine (**1764c**) (*73*), chelirubine (**1764d**) (*74*), chelerythrine (**1764e**) (*75*), and fa-

Scheme 9.30

Scheme 9.31

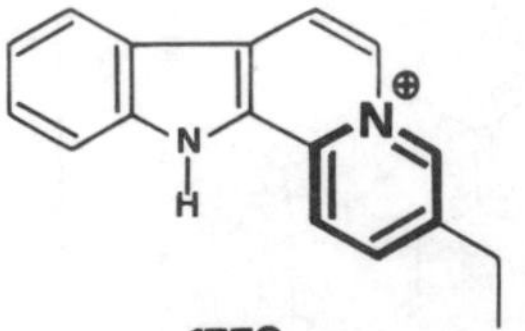

1759 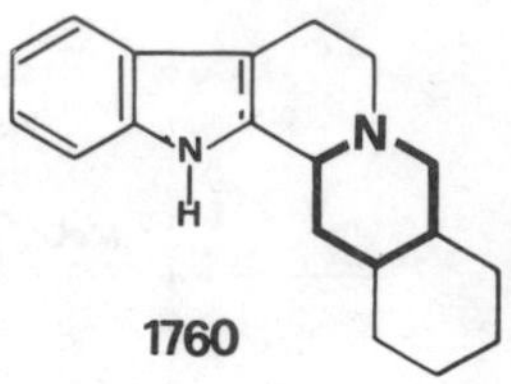 **1760** 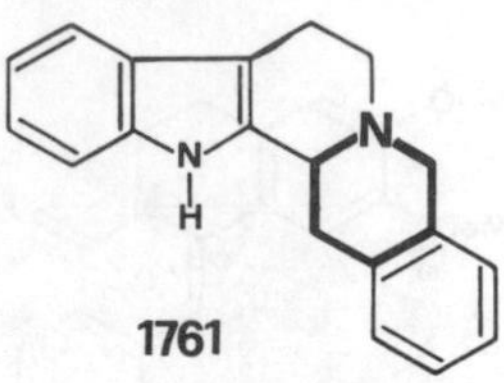 **1761**

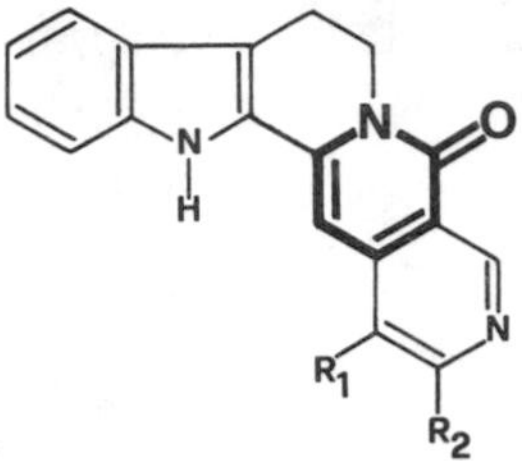

1762 a $R_1=R_2=H$
b $R_1=H$; $R_2=Me$
c $R_1=CH(OH)Me$; $R_2=H$
d $R_1=COMe$; $R_2=H$
e $R_1=-CH{=}CH_2$; $R_2=H$
f $R_1=Et$; $R_2=H$

Scheme 9.32

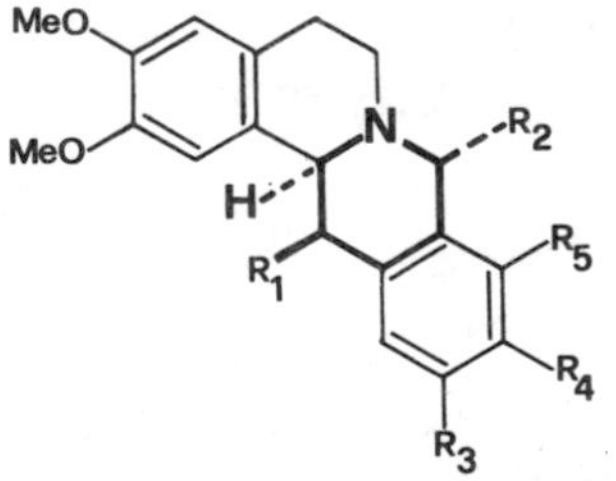

1086 b $R_1=R_2=R_5=H$; $R_3=R_4=OMe$
1763 a $R_1=R_2=R_3=H$; $R_4=R_4=OMe$
b $R_1=R_2=R_3=H$; $R_4,R_5=-OCH_2O-$
c $R_1=Me$; $R_2=R_3=H$; $R_4,R_5=-OCH_2O-$
d $R_1=R_5=H$; $R_2=Me$; $R_3=OH$; $R_4=OMe$

Scheme 9.33

garonine (**1764f**) (*76*), or strictly related structures already converted into the alkaloid, were synthesized through this approach. Their formulae are reported in Scheme 9.34.

Difficulties arise in the synthesis of nonaromatic structures because several stereochemical problems have to be solved, for example the *cis* junction of B/C rings. Nevertheless, homochelidonine (**1765a**) (*75, 77*), corynoline (**1765b**) (*78*), 12-hydroxycorynoline (**1765c**) (*78*), and 11-epicorynoline (**1765d**) (*79*) have been obtained through multiple-step pathways where the dienamide photocyclization is the first step. The structures of these compounds are depicted in Scheme 9.35.

Of the significant complex molecules obtained by photocyclization, we discuss here the lycorine-type compounds anhydrolycorine (**1766**) (*80*) and α-anhydrodihydrocaranine (**1767**) (*81*). The syntheses of γ-licorane (**1768**) (*81*) and of the key intermediates **1769** (*82*) and **1770** (*83*) for the synthesis of alkaloids emetine and haemanthidine, respectively, are also interesting. The structures for these compounds are shown in Scheme 9.36.

A promising extension of this reaction is the reductive photocyclization of dienamides in the presence of a hydride (*84*). When **1771** is irradiated in the presence of sodium borohydride, the cyclic intermediate **1772** undergoes reduction by a concomitant incorporation of hydride ion at the 4a position. Simultaneous incorporation of solvent at 6a or 8 positions gives two hydrogenated lactams **1773a** and **1773b**,

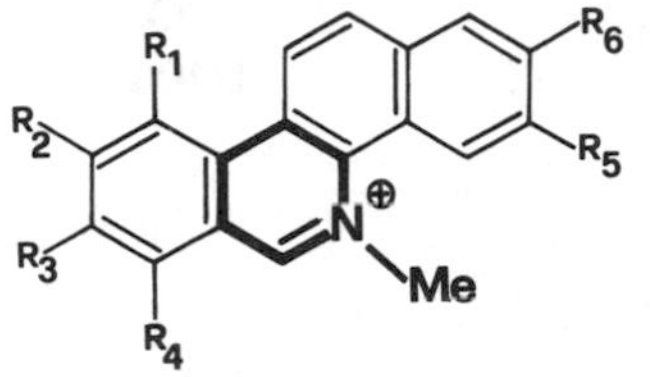

1764 a $R_1=R_4=H$; $R_2=R_3=OMe$; $R_5,R_6=-OCH_2O-$;
b $R_1=R_4=H$; $R_2,R_3=R_5,R_6=-OCH_2O-$;
c $R_1=R_2=H$; $R_3,R_4=R_5,R_6=-OCH_2O-$;
d $R_1=OMe$; $R_2=H$; $R_3,R_4=R_5,R_6=-OCH_2O-$;
e $R_1=R_2=H$; $R_3=R_4=OMe$; $R_5,R_6=-OCH_2O-$;
f $R_1=R_4=H$; $R_2=R_3=R_5=OMe$; $R_6=OH$;

Scheme 9.34

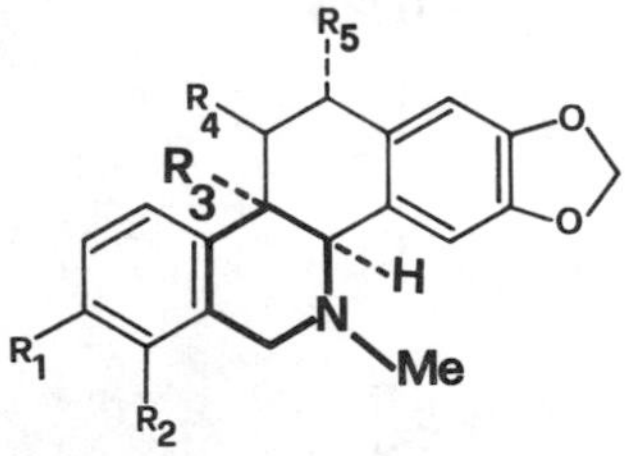

1765 a $R_1=R_2=OMe$; $R_3=R_5=H$; $R_4=\beta$-OH ;
b $R_1,R_2=-OCH_2O-$; $R_3=Me$; $R_4=\beta$-OH ; $R_5=H$
c $R_1,R_2=-OCH_2O-$; $R_3=Me$; $R_4=\beta$-OH ; $R_5=OH$
d $R_1,R_2=-OCH_2O-$; $R_3=Me$; $R_4=\beta$-OH ; $R_5=H$

Scheme 9.35

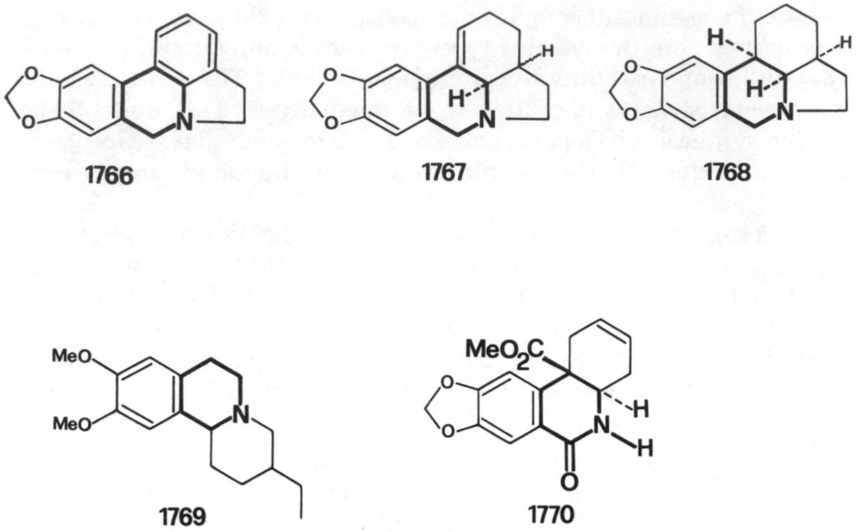
1766
1767
1768
MeO
MeO
1769
MeO_2C
1770

Scheme 9.36

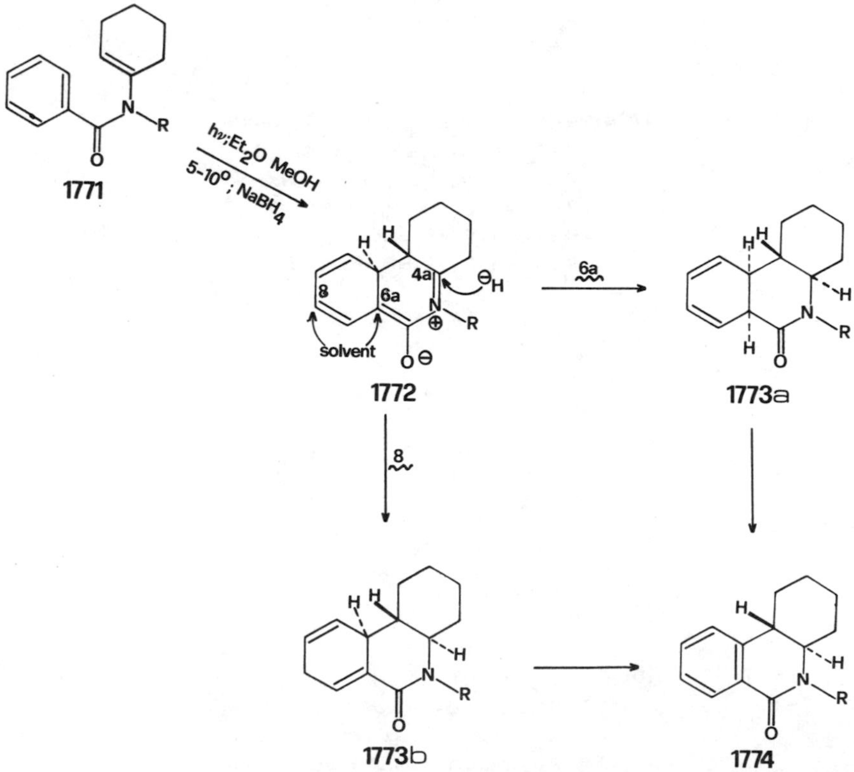
1771
hν; Et_2O MeOH
5-10°; $NaBH_4$
1772
solvent
6a
1773a
8
1773b
1774

Scheme 9.37

respectively (Scheme 9.37). These compounds are readily dehydrogenated to **1774**, which retains the configuration of the C4a chiral center; therefore, this approach shows potential for the synthesis of natural products.

Moreover, the use of chiral hydride complex (1:1 lithium aluminum hydride/quinine) in the reductive photocyclization of **1775** induces chirality at C13a of **1776** (*85*). Reduction to the optically active xylopinine (**1086b**) results in photochemical asymmetric synthesis of this alkaloid through an approach that could have future developments (Scheme 9.38).

We end this significant section by outlining the synthesis of narciprimine (**1778b**), a derivative of the alkaloid narciclasine. In this synthesis, the ring closure occurs with elimination of a substituent as a neutral molecule. Thus irradiation of **1777** proceeds with elimination of hydrogen bromide to give low yields (8%) of phenanthridone derivative **1778a**, which can be further debenzoylated to **1778b** (*86*) (Scheme 9.39).

Incidentally, the conrotatory ring closure places hydrogen and bromine in an *anti* relationship, suitable for elimination.

Polar Electrocyclic Reactions

The reaction of imine **1779** with methacryloyl chloride afforded a mixture of the open amide **1780** and cyclized product **1781**. This latter compound was further converted to costaclavine (**1782a**), a clavine alkaloid isolated from the saprophytic culture of the agropyrum-type ergot fungus; its 8-epi-isomer **1782b**, and festuclavine (**1782c**), the *trans*-fused isomer (*87*) (Scheme 9.40).

These results can be interpreted in terms of a possible thermal cyclization of **1780** to **1781** under experimental conditions. The thermal cyclization could be used in competition with the photochemical process; thus the synthesis of clavine skeleton can also be accomplished by irradiation of *N*-methacrylamides (*88*) similar to **1780**.

This assumption was supported by further synthesis of some alkaloid-related substances. Thus the thermal cyclization of **1783** proceeds regiospecifically to give a good yield of **1784**, which is easily reduced to

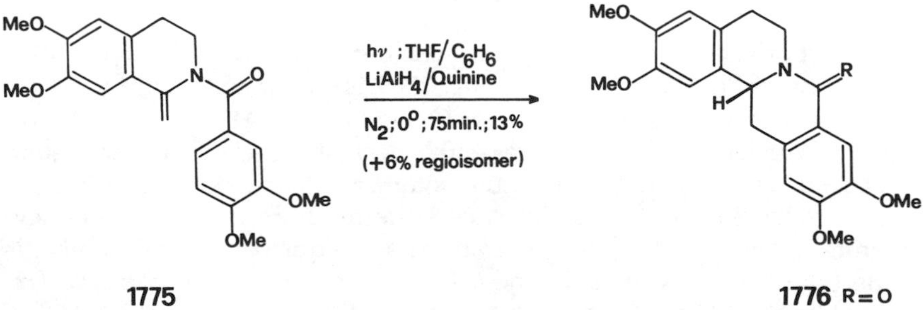

MeO MeO N O OMe OMe

1775

hν ; THF/C_6H_6
$LiAlH_4$/Quinine
N_2; 0°; 75min.; 13%
(+6% regioisomer)

MeO MeO N R H OMe OMe

1776 R=O
1086b R=H_2

Scheme 9.38

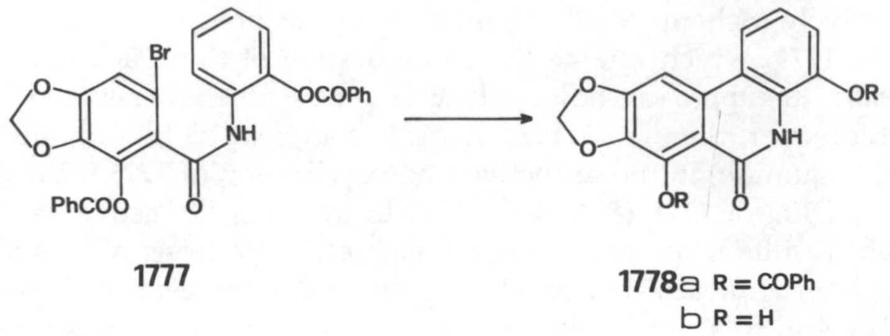

1777

1778a R = COPh
b R = H

Scheme 9.39

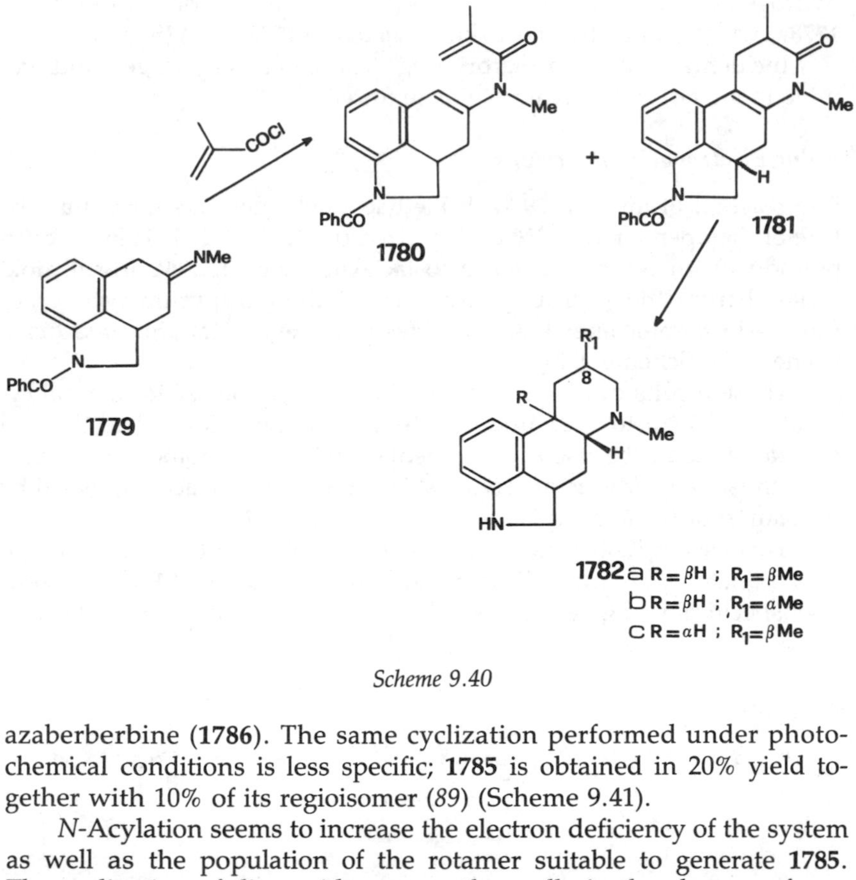

1779

1780

1781

1782a R = βH ; R_1 = βMe
b R = βH ; R_1 = αMe
c R = αH ; R_1 = βMe

Scheme 9.40

azaberberbine (**1786**). The same cyclization performed under photochemical conditions is less specific; **1785** is obtained in 20% yield together with 10% of its regioisomer (*89*) (Scheme 9.41).

N-Acylation seems to increase the electron deficiency of the system as well as the population of the rotamer suitable to generate **1785**. The cyclization of dienamides occurs thermally in the absence of acylating agents, but higher temperatures are required. A nice example is the total synthesis of alamarine (**1789**), an alkaloid isolated from *Alangium lanarckii* Thw. Heating **1787** produces **1788** in a 43% yield; **1788** can be easily converted by acid debenzylation and sodium borohydride reduction to **1789**. A comparison between the photochemical and the thermal route in the presence of acylating agents favors the thermal approach (*90*) (Scheme 9.42).

The mechanism of the thermal cyclization of dienamides is still open for discussion; the role of an electron-attracting group on the system is not yet clarified. The process of the electrocyclic ring closure, which should occur in a disrotatory way, is not conclusively defined. Certainly the overall system has a polar character. In this section, we explore the system and the process.

When undergoing an electrocyclic process, azapolyenes usually require protonation of the nitrogen atom, possibly as a method to limit the transition involved to the $\pi \rightarrow \pi^*$ only. Thus the anil **1790** cyclizes upon irradiation under strongly acidic conditions in the presence of oxygen to give good yields of calycanine (**1791**) (*91*), a degradation product of the alkaloid calycanthine (Scheme 9.43).

The Kametani electrocyclic ring opening of a benzocyclobutene was used to generate two of the three double bonds required for a 6π electrocyclic ring closure. The third double bond is a protonated azamethylene group. Thus thermolysis converted the hydrochloride **1792** to **1794** through dehydrogenation of the intermediate **1793**. This intermediate is easily reduced to the protoberberine alkaloid (±)-xylopinine (**1086b**) (*92*) (Scheme 9.44).

Similarly, thermolysis of 1-benzocyclobutenyl-3,4-dihydro-β-carboline (**1795**) gave decadehydro-17-methoxyyohimbane (**1796**). This compound was easily reduced to hexadehydro-17-methoxyyohimbane, the 17-methoxy derivative of **1761** (*93*) (Scheme 9.45).

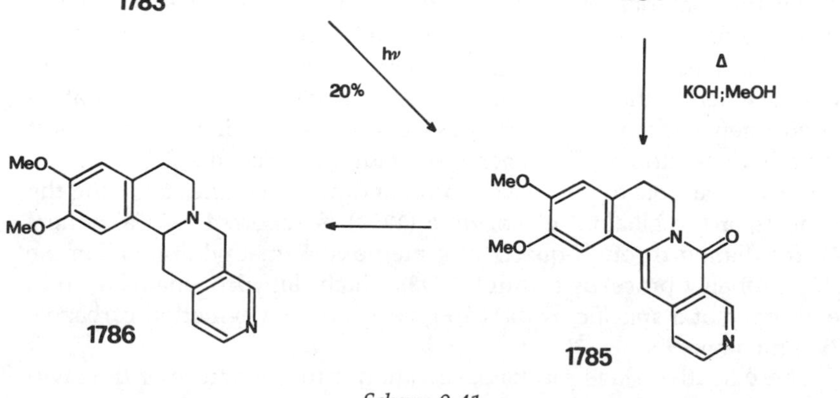

Scheme 9.41

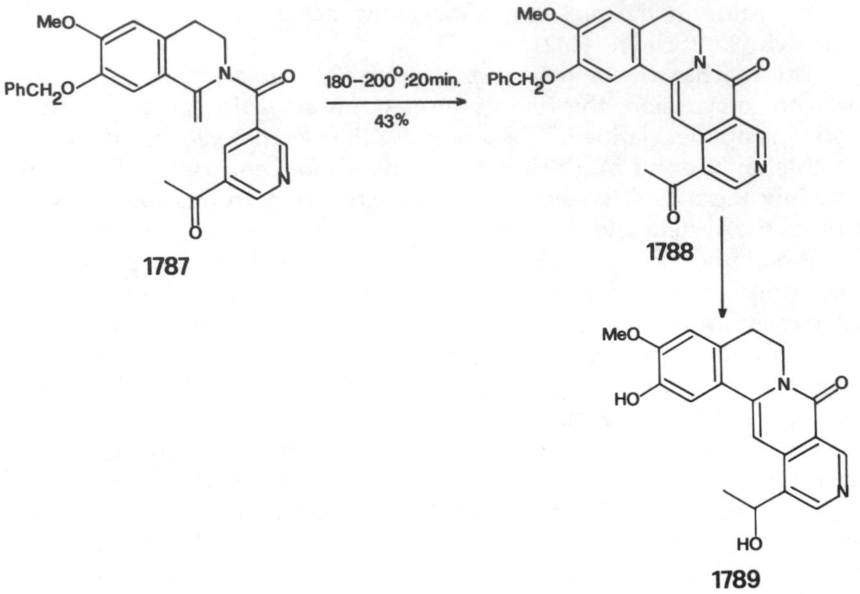

1787 → 1788 ($180-200°$; 20 min; 43%) → 1789

Scheme 9.42

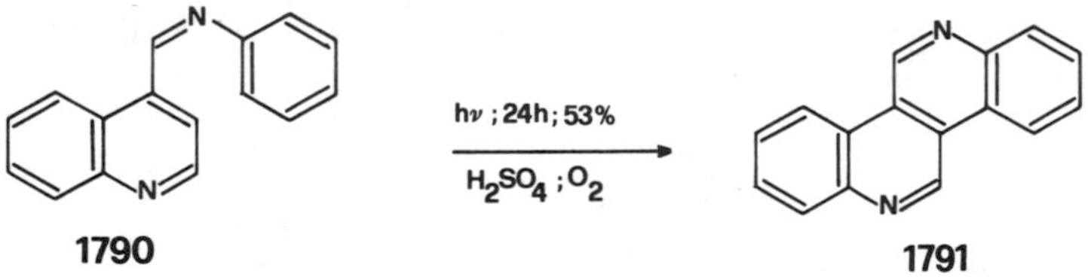

1790 → 1791 ($h\nu$; 24h; 53%; H_2SO_4; O_2)

Scheme 9.43

The polar nature of a 6π electrocyclic process is either intrinsic in the starting material or developed from a neutral species. The latter alternative occurs if five atoms are involved and if the third is a heteroatom with at least a lone pair. In this case, the electrocyclic ring closure gives a zwitterion that can undergo further reaction, generally oxidation or hydrogen migration. The process is conrotatory if it occurs, as it generally does, under photochemical conditions (Scheme 9.46).

A nice example of photocyclization of diphenylamines allowing the synthesis of the alkaloid glycozoline (**1799**) is reported by Carruthers (*94*). Irradiation through quartz of 4-methoxy-4′-methyldiphenylamine (**1797**) probably proceeds through **1798**, which dehydrogenates even in the absence of a specific oxidant to give 3-methoxy-6-methyl carbazole (**1799**) (Scheme 9.47)

The oxidative stage masks observation into the nature of the zwitterion. A heterocycle with a lower resonance energy would have a lower

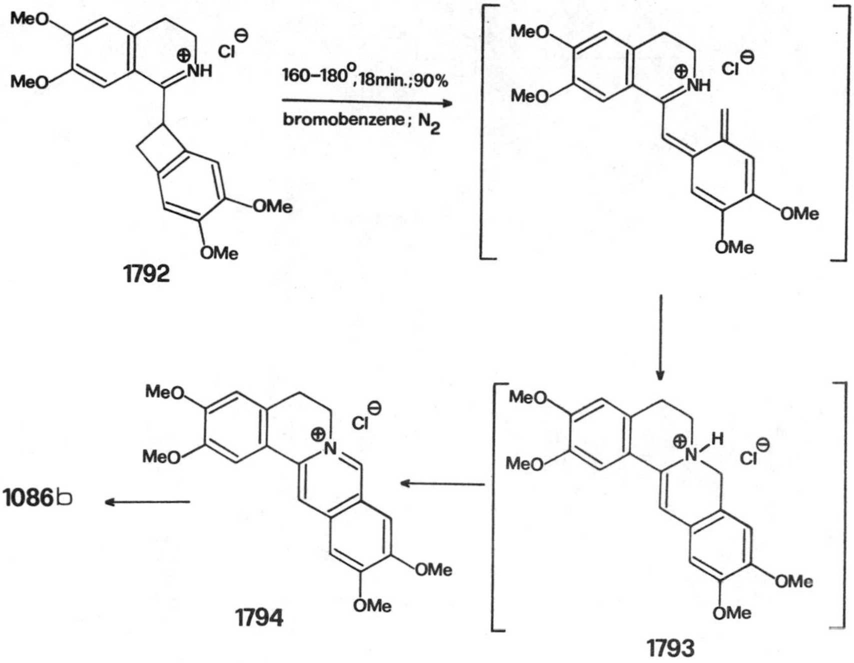

160–180°, 18min.; 90%
bromobenzene; N_2

1792

1793

1794

1086b

Scheme 9.44

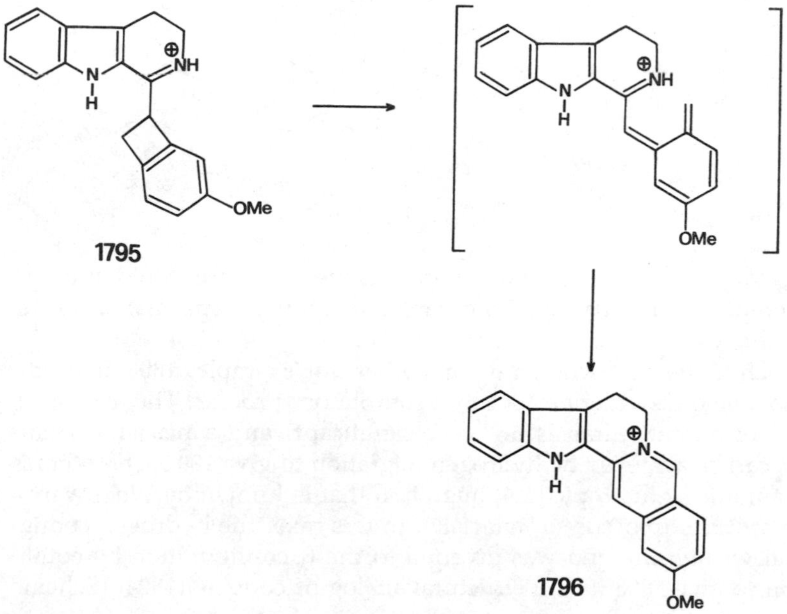

1795

1796

Scheme 9.45

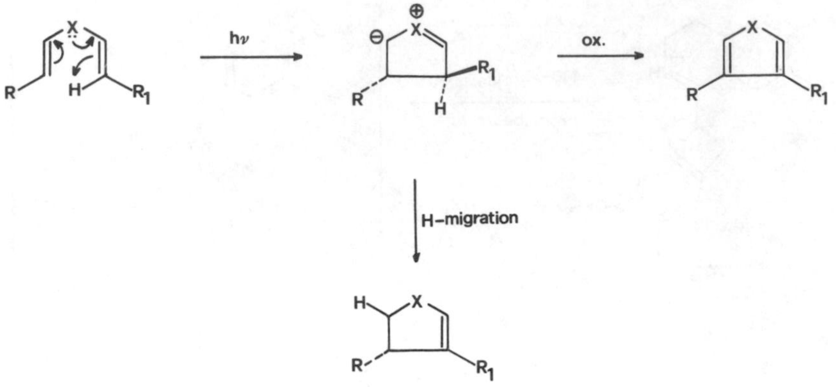

Scheme 9.46

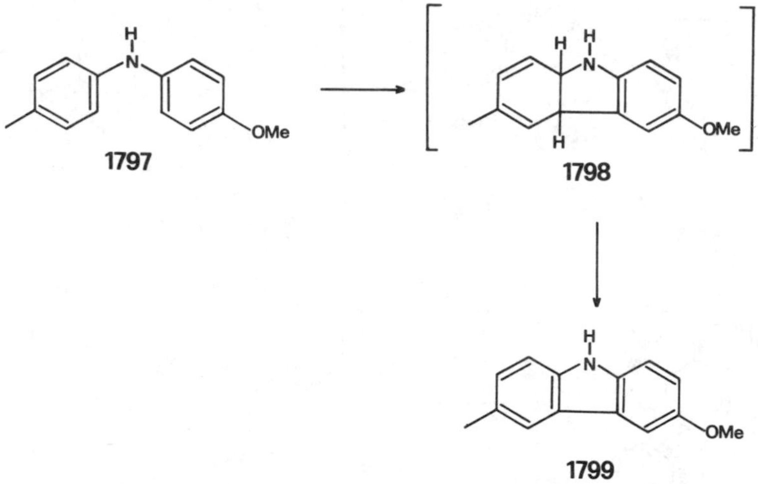

Scheme 9.47

energy gain from aromatization; therefore, the zwitterion could undergo a different reaction pathway, perhaps one allowing speculation on its structure.

Such is the case when a divinyl ether, for example, **1800**, is irradiated to give the zwitterion **1801** by a conrotatory process. The resonance energy of a benzofuran is not very significant, and a partial aromatization can be achieved by hydrogen migration to give **1802**. The process is an anionic sigmatropic [1,4] migration that is known by Woodward–Hoffman rules to occur suprafacially. In this case, the hydrogen configuration was wrong and was inverted to the β configuration by equilibration to give **1803** (*95*), a structural analog of codeine (**1804**) (Scheme 9.48).

The same approach was applied in the synthesis of (±)-lycoramine

(**1809**), a galanthamine-type alkaloid found in Amaryllidaceae. Irradiation of **1805** gives dihydrobenzofuran **1807** if the ring closure is conrotatory via zwitterion **1806** and the [1,4] anionic hydrogen shift is suprafacial. A *cis*-fused dihydrofuran **1808** is actually obtained; the inversion of the chiral center to give the more stable isomer occurs after ketalization with methanol-trimethylorthoformate-sulfuric acid. A few further steps involving the azacycloheptane ring closure give the target compound **1809** (*96*) (Scheme 9.49).

If the substrate is polar, the adduct is also polar before undergoing loss or gain to neutrality. The number of electrons involved in the process depends on the anionic or cationic character of the substrate. For five-membered substrates, HOMO is ψ_2 or ψ_3 for cations and anions,

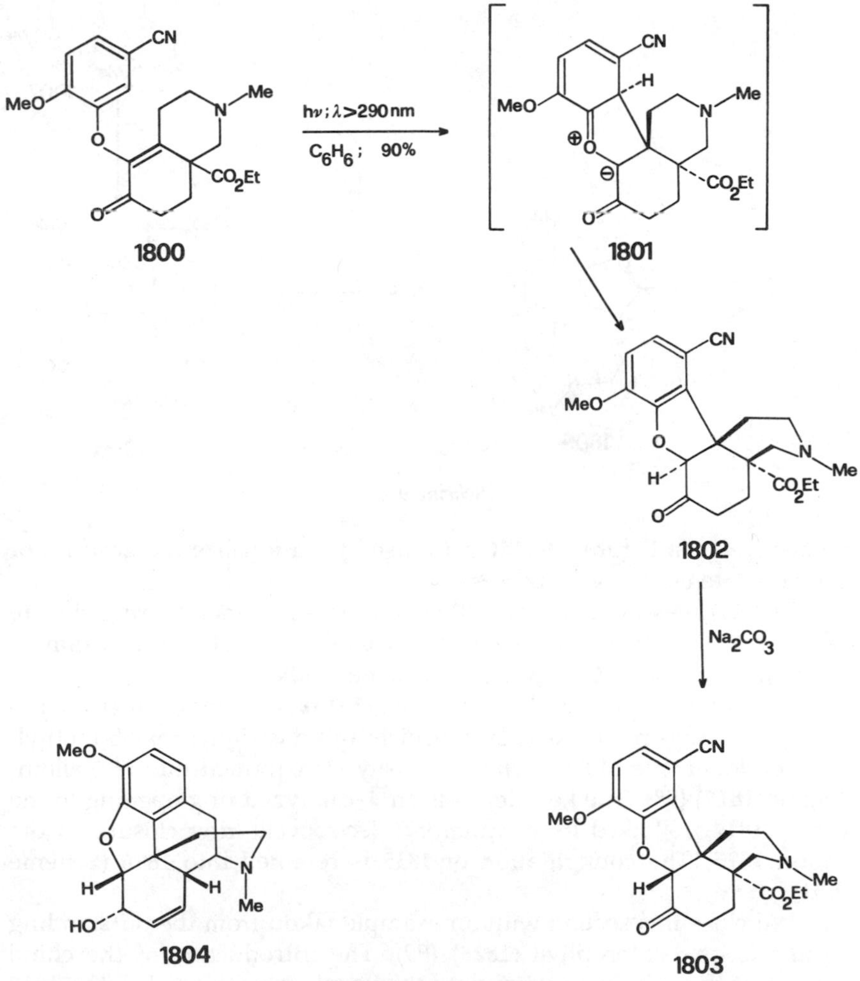

Scheme 9.48

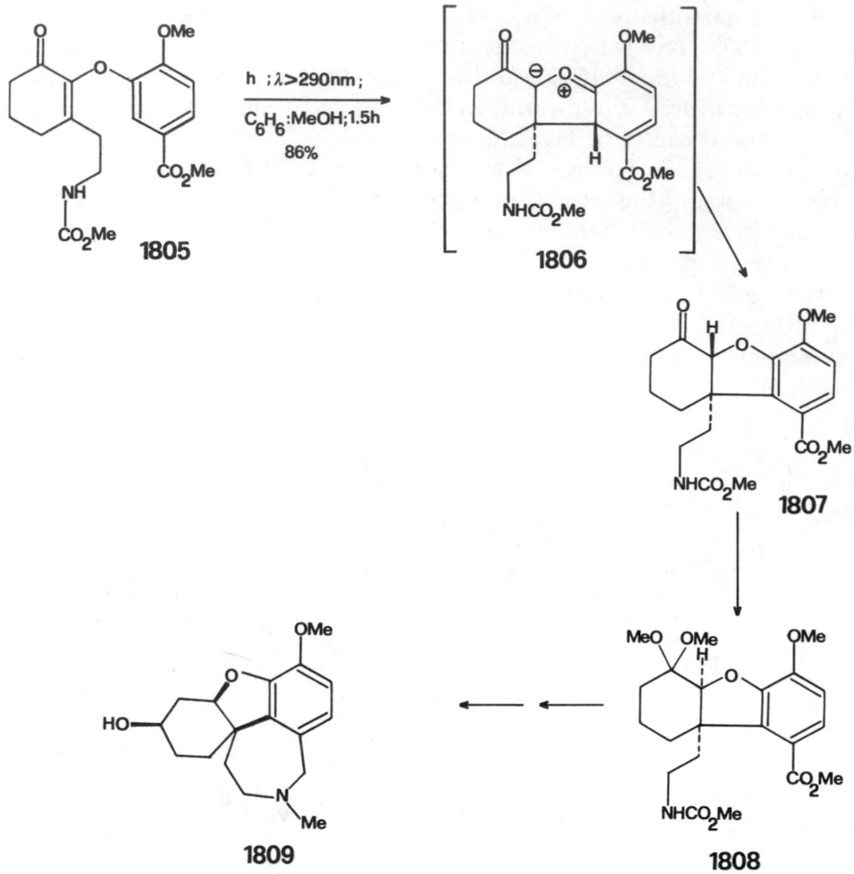

Scheme 9.49

respectively. In Figure 9.6, MOs are used to rationalize the ground- or excited-state electrocyclic process.

The synthesis of amarine (**1812**) by disrotatory thermal ring closure of anion **1811** from hydrobenzamide **1810** (*97*) is a well-known example of an anionic six-electron process (Scheme 9.50).

When 5-methyl-2-furylallyl carbinol (**1813**) is treated with zinc chloride, a rearrangement takes place furnishing 4-allyl-5-hydroxy-5-methyl-3-oxocyclopentene (**1816**), which is easily transformed into (±)-allethrolone (**1817**) (*98*). The key step is a Zn^{2+}-catalyzed ring opening to the cation **1814**, followed by conrotatory electrocyclic ring closure to the cation **1815**. The configuration of **1815** is retained into **1816** (Scheme 9.51).

We close this section with an example taken from the outstanding synthesis of α-chlorophyll (**1823**) (*99*). The introduction of the chiral centers of the chlorin structure into the porphyrin achiral skeleton **1818** was accomplished by simple warming in acetic acid at 110 °C. The pro-

Figure 9.6. ψ_2, ψ_3, and ψ_4 MOs of five-membered cations and anions, and their influence on the electrocyclic process.

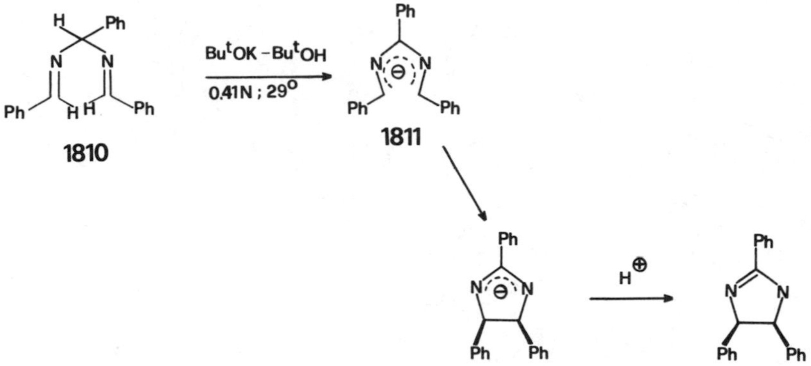

$Bu^tOK - Bu^tOH$

0.41N ; 29°

1810 1811 1812

Scheme 9.50

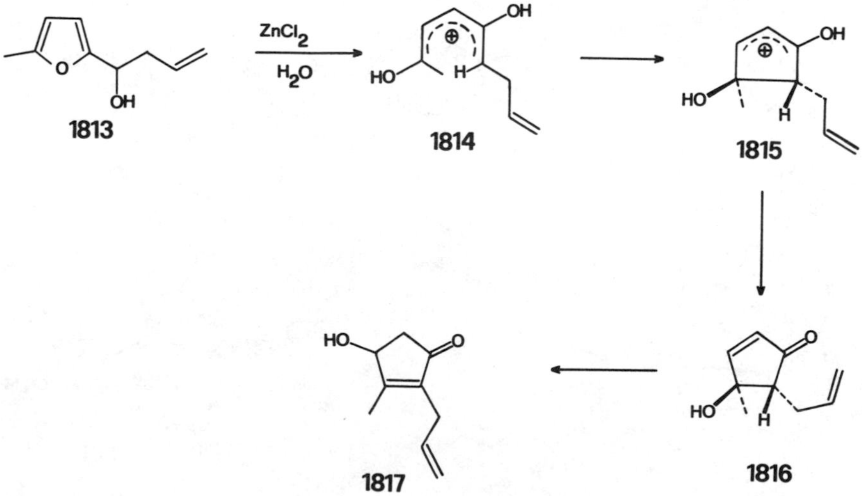

$ZnCl_2$

H_2O

1813 1814 1815 1816 1817

Scheme 9.51

cess can be explained (*100*) by formation of dienyl cation **1819** that undergoes a conrotatory ring closure to cyclopentenyl cation **1820**, which tautomerizes to **1821**. In the additional 11 steps toward chlorophyll, the cyclopentadiene ring of **1821** is broken by photooxidation, a process that involves a [2 + 2] cycloaddition of singlet oxygen to the peripheral double bond. The triplet (ground state) →singlet (excited) process is promoted by chlorine, which acts as sensitizer. The 1,2-dioxetane undergoes a [2 + 2] cycloreversion to give the two carbonyl groups of **1822**. Thus a series of trivial steps can be appreciated in light of the overall synthetic strategy (Scheme 9.52).

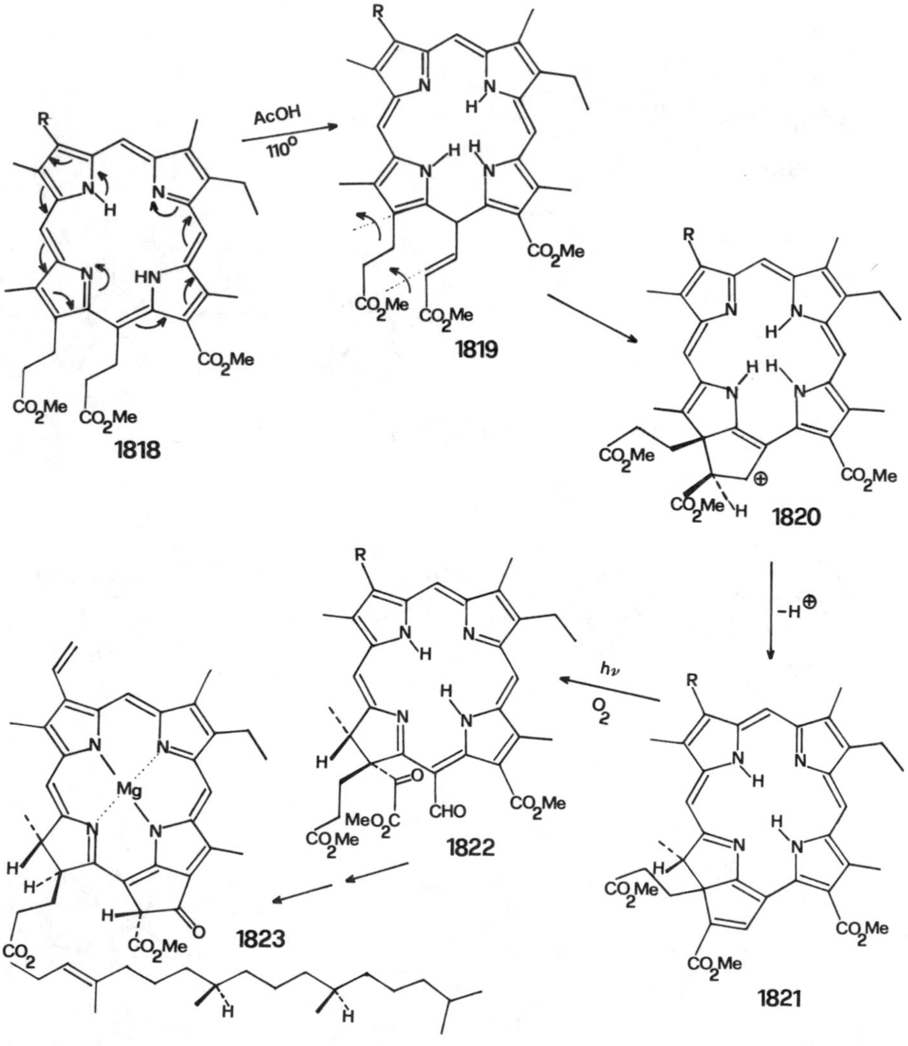

Scheme 9.52

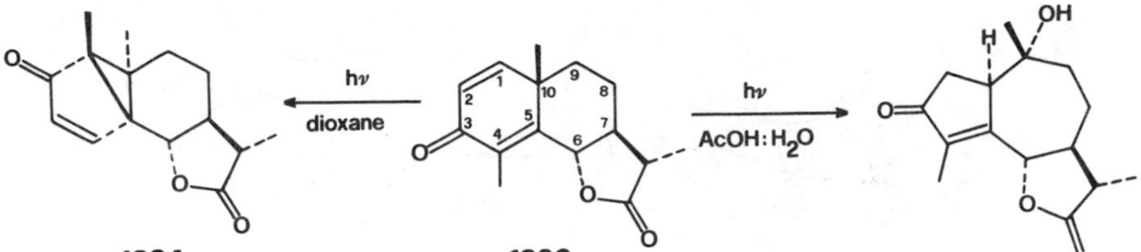

Scheme 9.53

Santonin-Type Photorearrangements

In 1957 Arigoni and coworkers (*101*) found that irradiation of santonin (**1296**) in dioxane (or alcohol) solution yields lumisantonin (**1824**). In 1958 Barton (*102*) showed that irradiation in aqueous acetic acid gives "isophotosantonic lactone" (**1825**) (Scheme 9.53). These pioneering works were further supported by extensive studies, and the photorearrangements of cross-conjugated cyclohexadienones became a favorite topic in photochemistry. Several reviews have been published (*103, 104*) [the first by Barton (*105*) in 1959], and much effort was devoted to clarification of the mechanism.

The cyclohexadienone–bicyclohexenone rearrangement formally can be considered as a concerted $[\sigma 2a + \pi 2a]$ cycloaddition (*106*). If the LUMO/LUMO interaction is considered, the process represented in Figure 9.7 rationalizes nicely the formation of lumisantonin from santonin.

This fascinating mechanism allows a justification of the configuration of the adduct, but it does not take into account the triplet $n \rightarrow \pi^*$ excited species detected along the reaction pathway. In 1961–62 Zimmerman and Schuster (*107, 108*) proposed a mechanism that satisfactorily explains the photochemical process, although several points are still open for discussion (*109*). Also, the existence of some species is still uncertain (*110, 111*).

The rearrangement, according to the Zimmerman and Schuster mechanism, is not a concerted process. The process can be broken down to a series of events, some of which involve pericyclic reactions.

1. Formation of the $n \rightarrow \pi^*$ singlet excited state (S*).
2. Singlet (S*)→triplet (T*) intersystem crossing.
3. Disrotatory electrocyclic ring closure to a bicyclohexene species, still as an excited species of the $n \rightarrow \pi^*$ type.
4. Decay to a ground state zwitterion.
5. A [1,4] suprafacial cationic shift occurring with inversion of configuration (Scheme 9.54).

The points under discussion are the interposition of the zwitterion as a ground state species and the singlet→triplet intersystem crossing.

We describe these rearrangements in this section because the crucial

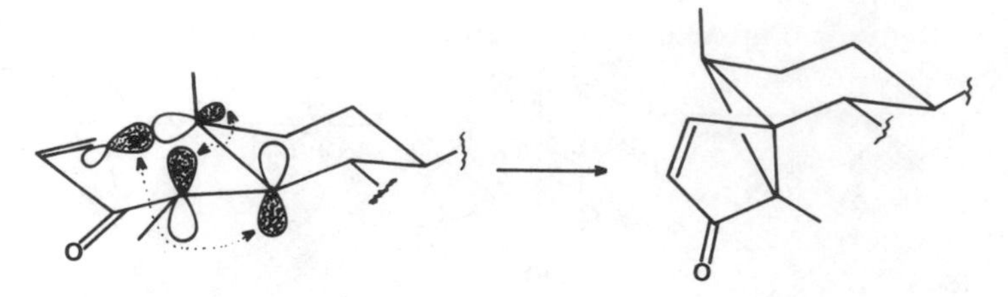

Figure 9.7. LUMO/LUMO interactions for the santonin–lumisantonin rearrangement.

Scheme 9.54

electrocyclic step determines the nature of all the reaction products. Two points are of specific interest for their MO implications. The first is the electrocyclic process that must occur in a disrotatory fashion. MOs of cross-conjugated cyclohexadienes resemble those of 3-methylene-1,4-pentadiene (*112*). Its Ψ_4 (π^*—a singly occupied MO in the process depicted in Scheme 9.54) and Ψ_3 are symmetric, hence the ring closure requires disrotation (Figure 9.8).

The second step involving MO control is the [1,4] cationic shift. The LUMO of the allylic cation interacts with the HOMO of the σ bond, and the arrangement can occur suprafacially with inversion of configuration at the migrating center only (Figure 9.9).

The rearrangement was found to be a useful tool for several syntheses of natural products. Thus dehydro-α-vetivone (**1826**) was irradiated to give **1827**, which was converted in few steps to β-vetivone (**7**) (*113*). Compound **1828** gave **1829**, which was easily transformed into 11,12-dihydronootkatone (**1830**) (*114*). This product was also obtained by selective hydrogenation of the natural nootkatone (**507**). Sometimes the selectivity of the reaction is low; for example, the irradiation of 1-dehydrotestosterone acetate (**1831**) gives nine photoproducts; two of them are primary products, one of which is **1832** (*115*) (Scheme 9.55).

The irradiation of **1833a** in dioxane afforded **1834a**. Acid-promoted ring opening of **1834a** occurred with cleavage of the C1–C10 bond to give spiro ketone **1835a**, which was further converted to β-vetivone (**7**) (*116*).

Similarly irradiation of **1833b** produced **1834b** which was cleaved to spiroketone **1835b** a convenient intermediate for the construction of a second member of the spirovetivane family of sesquiterpenes: (±)-α-vetispirene (**1836**) (*117*) (Scheme 9.56).

All these photorearrangements occurred in dioxane, and a lumisantonin-type product was obtained. When acetic acid (glacial or aqueous) is substituted for dioxane, the behavior is quite different. If **1837** is irradiated 5 h in aqueous acetic acid, **1838a** is obtained in 28% yield; irradiation for 3 h in glacial acetic acid gives **1838b** in 89% yield. Both products can be converted to a spiro ketone structurally related to **1835b** and easily converted to α-vetispirene (**1836**) (*118*).

Analogously, irradiation of 1-dehydrotestosterone acetate (**1831**) in aqueous acetic acid gave 11 products, 7 of which were identical with those obtained by irradiation of the same substrate in dioxane. However, **1832** was not isolated. One of the major products (15% yield) of the reaction was the spiroenone **1840**, which can also be obtained by acid and hydroxyspiroenones in aqueous acetic acid.

A reasonable explanation of these results is that cyclopropyl ketones related to lumisantonin are the primary photoproducts. They are cleaved in the acidic media giving acetoxyspiroenones in glacial acetic acid and hydroxy spiroenones in aqueous acetic acid.

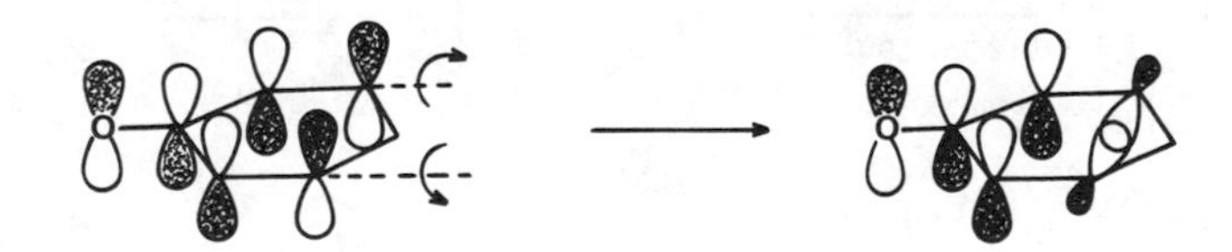

Figure 9.8. Disrotatory ring closure of the cyclohexadienone singly occupied MO (π).*

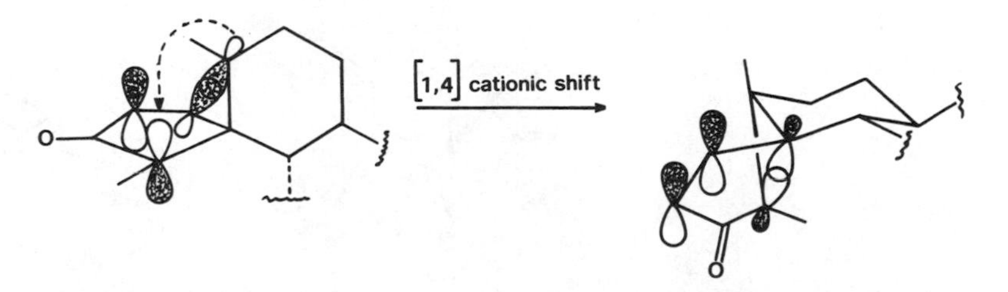

Figure 9.9. MO interaction in [1,4] cationic shift, the last step to lumisatonin.

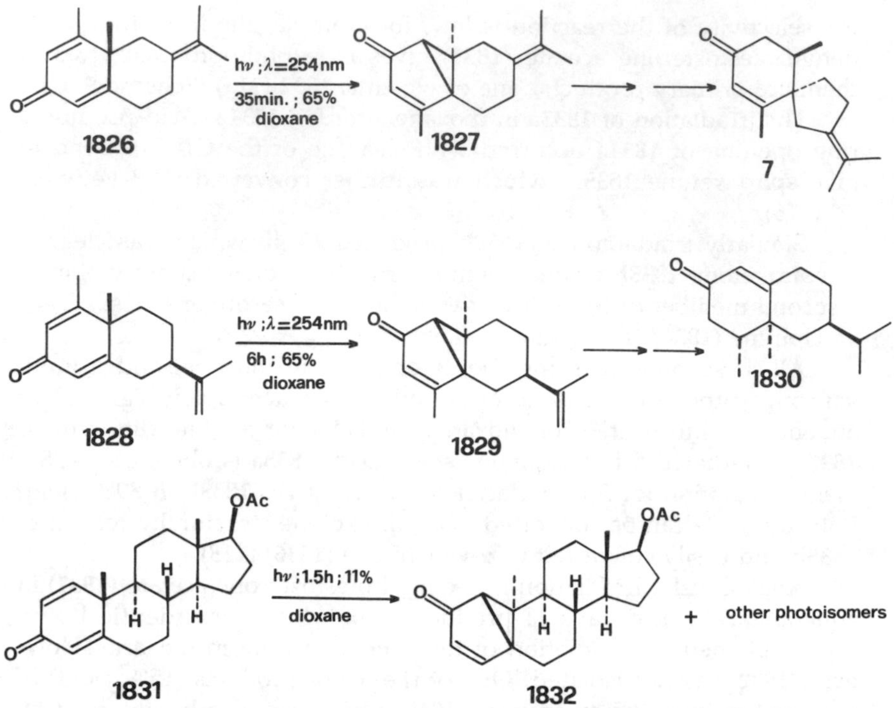

1826 → hν ; λ=254nm, 35min. ; 65%, dioxane → 1827 → → 7

1828 → hν ; λ=254nm, 6h ; 65%, dioxane → 1829 → → 1830

1831 (OAc, H, H, H) → hν ; 1.5h ; 11%, dioxane → 1832 (OAc, H, H, H) + other photoisomers

Scheme 9.55

1833a R = αMe
b R = βMe

→ hν ; λ = 254nm, dioxane, a: 2.5h ; 59%, b: 6h ; 70% → 1834a,b → H_2SO_4; AcOH; Ac_2O → 1835a,b

(a) → 7

(b) → 1836

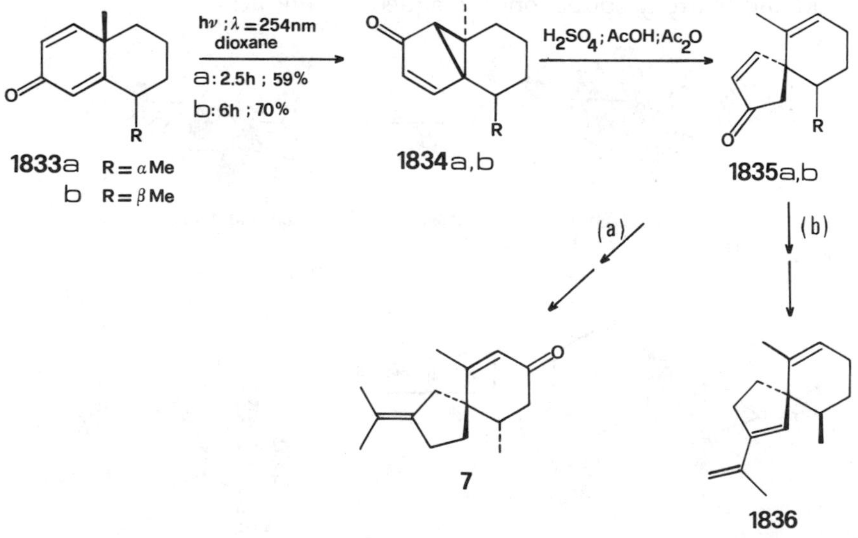

Scheme 9.56

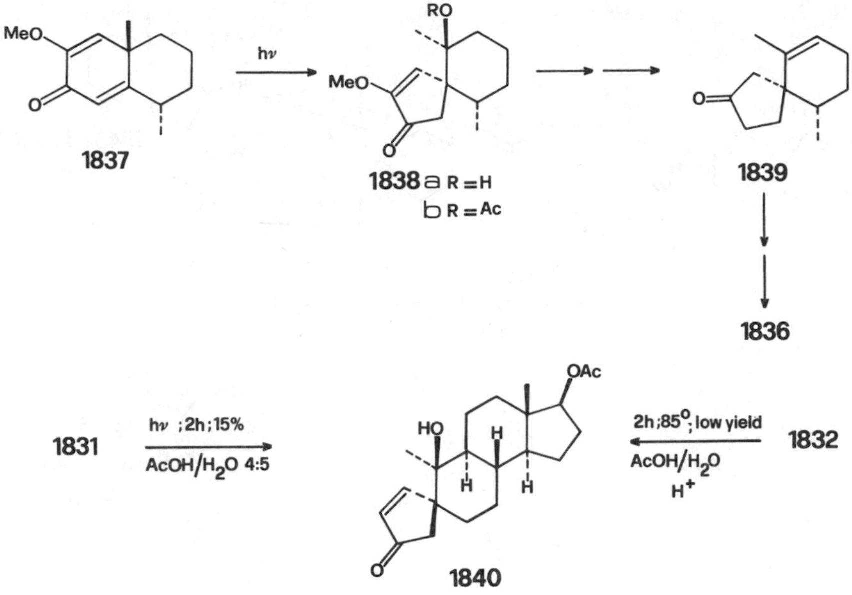

Scheme 9.57

But one point needs to be clarified. None of these products possesses a structure related to isophotosantonic lactone.

In 1964–65 Kropp (*120, 121*), investigating the effect of substituents on the cyclohexadienone ring, pointed out the dramatic effect of a methyl group α to the carbonyl in the photoarrangement of 8a-methyl-6, 7, 8, 8a-tetrahydro-1(5*H*) naphthalenones (**1841a–c**). Under acidic conditions and by using a dioxane as solvent, 2-methyl **1841a**, 4-methyl **1841b**, and unsubstituted derivative **1841c** gave rise to lumisantonine-type products **1842a–c.** In aqueous acetic acid, **1841a** gives hydroxyspiroeneone **1843a**, **1841b** gives the azulenic hydroxyenone **1844b**, and **1841c** yields a mixture of products **1843c** and **1844c** (Scheme 9.58).

All previously reported examples fit into a general rule, the basic points of which can be deduced from Scheme 9.58. Substrates in dioxane always give a lumisantonin-type product, whereas the acidic photoisomerization of 2-methyl substituted substrates produces spiroenones analogous with **1841a**.

When the substrate is 4-methyl substituted, the behavior under acetic conditions should resemble that of santonin (**1296**) and of **1841b**. α-Santonin, its 6-epimer, β-santonin and its 6-epimer are cleaved to the azulene derivatives, as are artemisin acetate, its 8-epimer, and its 6-epi-8-epimer (*105*).

The acidic photorearrangement of 4-methyl substituted derivatives is a useful tool in the synthesis of natural products possessing the azulene skeleton (Scheme 9.59). Thus *O*-acetyl isophotosantonic lactone (**1845**), obtained from α-santonin, is the starting reagent for the

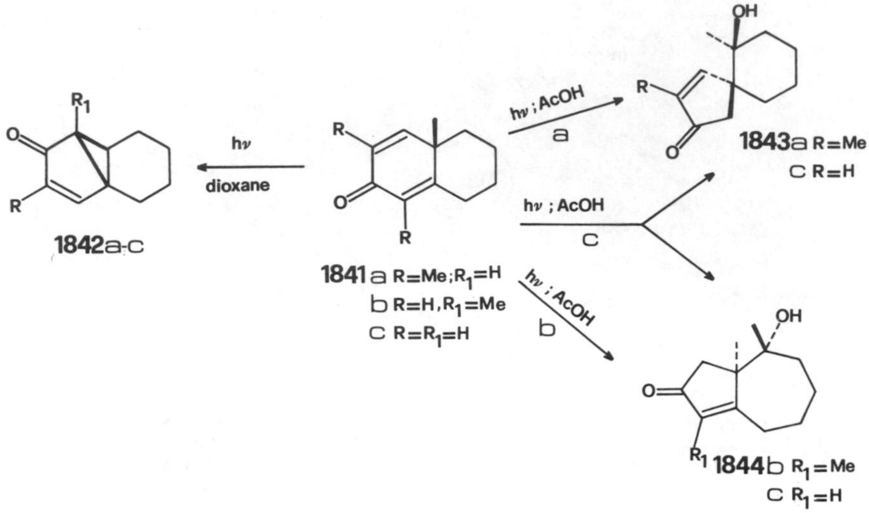

Scheme 9.58

syntheses of (+)-pachydictyol-A (**1641**) (*122*), a previously mentioned diterpene alcohol, and of desacetoxymatricarin (**1846**) (*123*), a sesquiterpene lactone found in the genus *Artemisia*. Similarly, the photochemical rearrangement of **1847** provided the tricyclic azulene derivative **1848**, which was then converted to (−)-4-epiglobulol (**1849**) and (+)-4-epiaromadendrene (**1850**) (*124*).

The first total synthesis of a grayanoid was realized by starting from the phenanthrenic dienone **1851**. This reaction gave a very good yield of **1852**, which was converted in several steps to grayanotoxin II (**1853**) (*125*) (Scheme 9.60).

The process can also be extended to the synthesis of non-azulenic derivatives by using a suitable substrate. Thus the irradiation of **1854** derived the enone **1855**, which has the three asymmetric centers in the configuration required for its conversion into (±)-oplanone (**1856**) (*126*) (Scheme 9.61).

As we have seen, all 4-substituted dienones showed a santonin-like behavior in glacial or aqueous acetic acid. A plausible mechanism would involve the attack of the nucleophile (OH in aqueous acetic acid, AcO in the absence of water) on the primary photoproduct (lumisantonin from santonin). This proposed mechanism is strongly supported by the acid-catalyzed conversion of **1824** to **1825** (*121*). The role of the substituent is illustrated in Scheme 9.62 (route a). Because this route favors the breaking of the C5–C10 bond in the protonated substrate, the transition state is more stable. Alternatively, the breaking of the same bond, in light of the Zimmerman mechanism, occurs in a protonated zwitterion (Scheme 9.62, route b). In both cases the nucleophile must enter *cis* to the tertiary hydrogen.

In Scheme 9.57 we saw that irradiation of 1-dehydrotestosterone acetate (**1831**), under acidic conditions, gives spiroenone **1840**; however, the spiro-compound **1843c** is obtained together with **1844c** by irradiating the unsubstituted derivative **1841c** (Scheme 9.58).

The dichotomy related to unsubstituted enones is further confirmed by irradiation of bicyclic dienone **1857**. This compound behaves like santonin does to give **1859**, which is converted in a few steps to (±)-3-oxo-α-cadinol (**1860**) and (±)-α-cadinol (**1861**) (*127*). A reasonable explanation can be advanced in terms of competitive cyclopropyl ring opening of the primary photoproduct **1858**. If there are no substituents

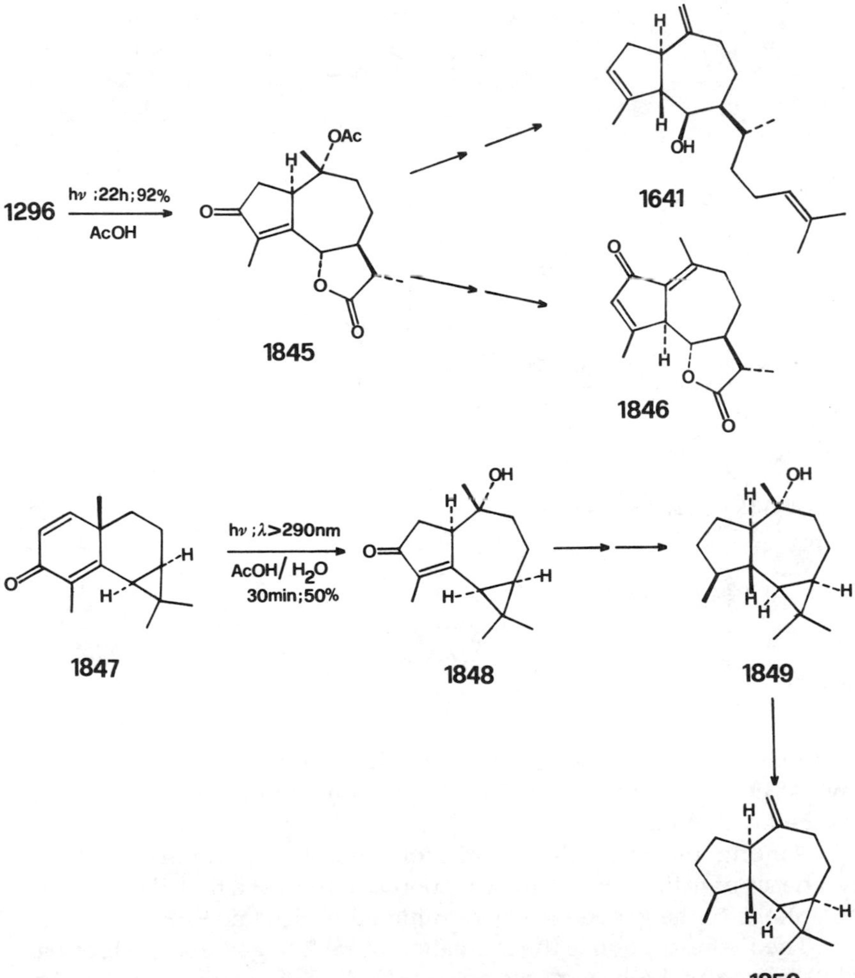

Scheme 9.59

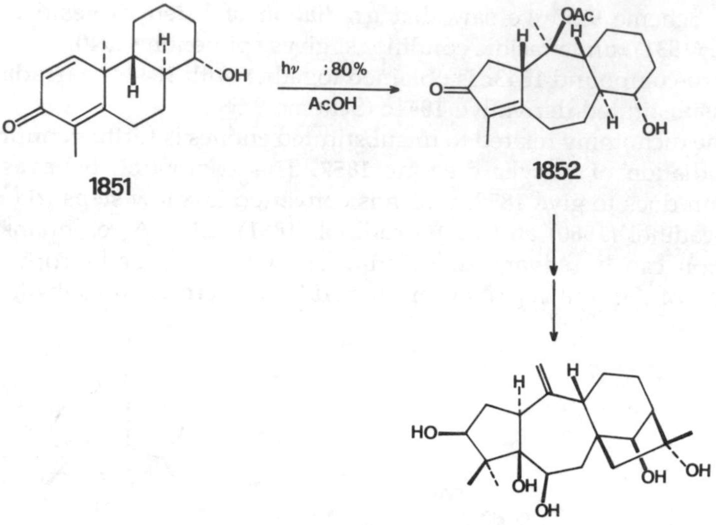

hν ; 80%
AcOH

1851 1852 1853

Scheme 9.60

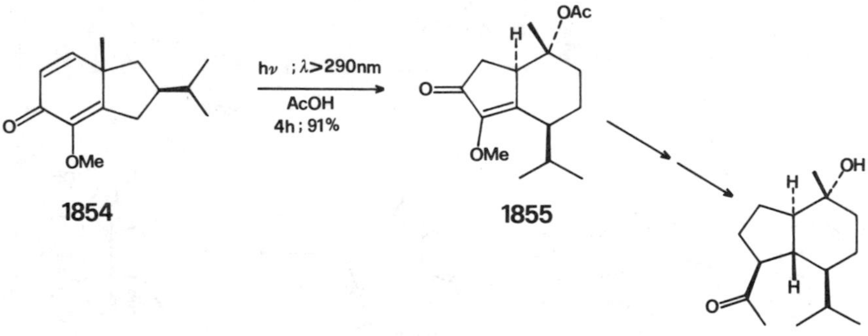

hν ; λ>290nm
AcOH
4h ; 91%

1854 1855 1856

Scheme 9.61

to stabilize a single transition state, the cleavage of the C5–C10 bond to give **1859** or the cleavage of the C1–C10 bond to give a spiro product can be favored (Scheme 9.63).

Among the examples of photoisomerization involved in the syntheses of natural products, we report two that, at a first glance, seem exceptions to the general behavior outlined in Scheme 9.58.

The first exception is the irradiation of **1862** to give **1863**, which has the skeleton of 4,4-bisnorgrayanotoxin (*128*). This compound thus behaves as a santonin-like product despite the substituent in position 2 (Scheme 9.64). If we consider the nature of the substituent, certainly an aldehyde group cannot offer good stabilization to an incoming positive

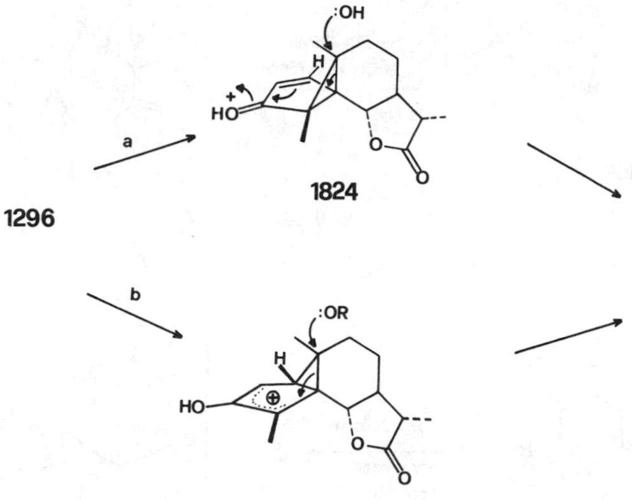

1296 —a→ 1824 → 1825

1296 —b→ → 1825

:OH
H
HO
1824
:OR
H
HO

Scheme 9.62

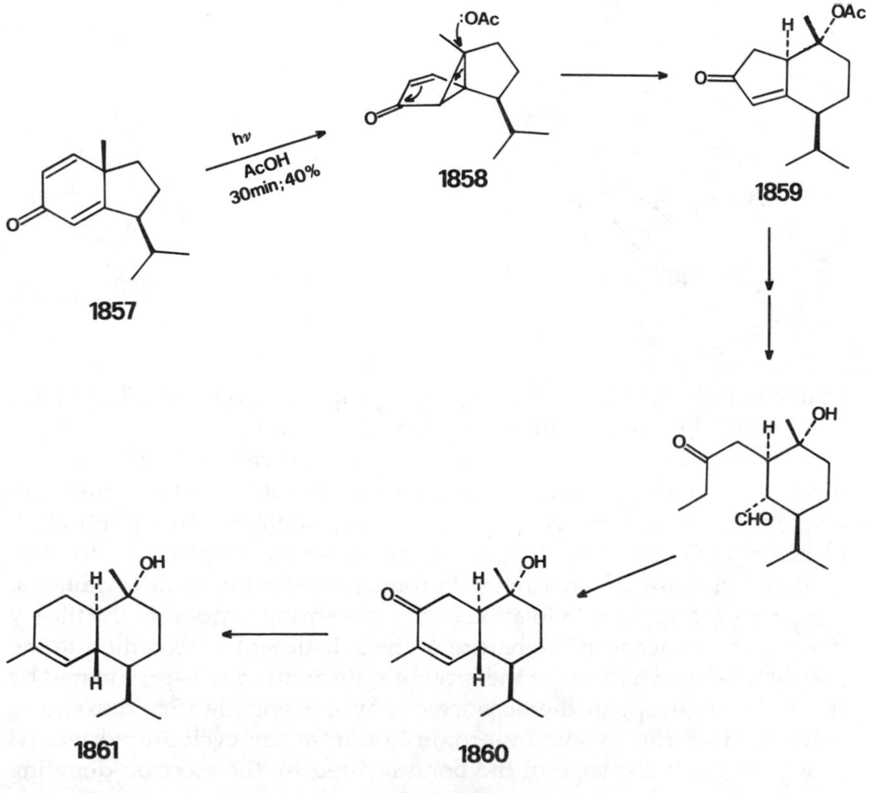

1857 —hν, AcOH, 30min; 40%→ 1858 → 1859 → → 1860 → 1861

:OAc
O
1858
H
OAc
O
1859
O
OH
H
CHO
H
OH
O
H
1860
H
OH
H
1861

Scheme 9.63

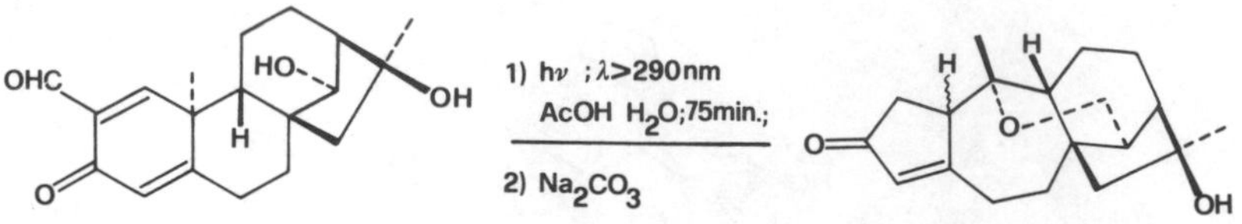

Scheme 9.64

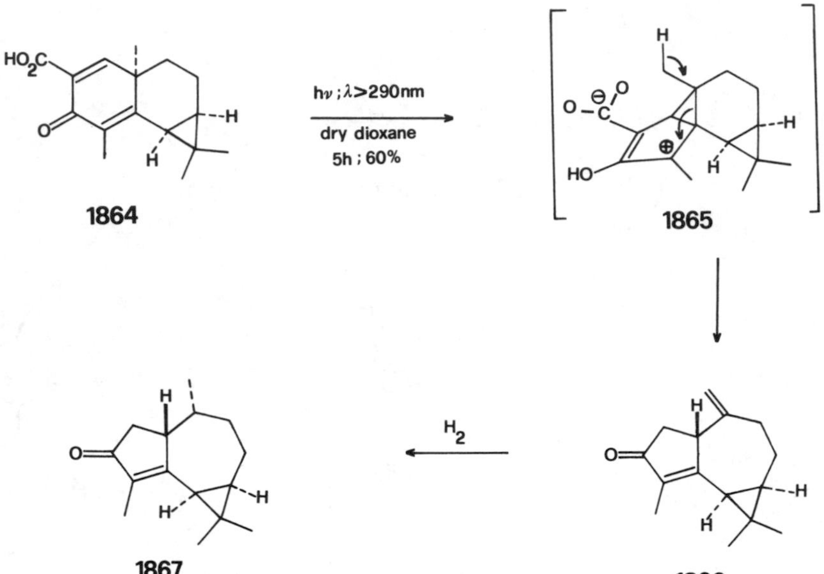

Scheme 9.65

charge in the transition state of the cyclopropyl ring opening. Therefore, what seems to violate actually supports the theory.

The second exception is the photoisomerization of **1864** in dry dioxane to produce the azulenone derivative **1866**, which is then hydrogenated to the sesquiterpene (−)-cyclocolorenone (**1867**) (*129*) (Scheme 9.65). The hypothetical violation is demonstrated by the formation in dioxane of a rearranged product, typical for acidic conditions. But again, the apparent violation becomes a strong support to the theory if we take into account the nature of the substituents. According to the Zimmerman mechanism, if the ground state zwitterion is protonated by the carboxy group, in the absence of any nucleophile (dry dioxane as solvent), **1865** can evolve by proton loss from the cyclopropyl methyl group, through cleavage of the bond favored by the electron-donating substituent.

Literature Cited

1. Dewar, M. J. S. *Angew. Chem., Int. Ed. Engl.* **1971**, *10*, 761.
2. Zimmerman, H. E. *Acc. Chem. Res.* **1971**, *4*, 272.
3. Huisgen, R. *Angew. Chem., Int. Ed. Engl.* **1980**, *19*, 947.
4. George, M.V.; Mitra, A.; Sukumaran, K. B. *Angew. Chem., Int. Ed. Engl.* **1980**, *19*, 973.
5. Sammes, P. G. *Quat. Rev.* **1970,** *24*, 37.
6. Oppolzer, W. *Synthesis* **1978,** 793.
7. Kametani, T.; Nemoto, H. *Tetrahedron* **1981**, *32*, 3.
8. Jacobs, H. J. C.; Havinga, E. *Adv. Photochem.* **1979**, *11*, 305.
9. Ho, T. L. L. *Synth. Commun.* **1977**, *7*, 351.
10. Knudsen, C. G.; Carey, S. C.; Okamura, W. H. *J. Am. Chem. Soc.* **1980,** *102*, 6355.
11. Boomsma, F.; Jacobs, H. J. C.; Havinga, E.; Van der Gen, A. *Tetrahedron Lett.* **1975**, 427.
12. Boomsma, F.; Jacobs, H. J. C.; Havinga, E.; Van der Gen, A. *Rec. Trav. Chim. Pays-Bas* **1977**, *96*, 104.
13. De Kok, A. J.; Boomsma, F.; Romers, C. *Acta Crystallogr.* **1976**, *B32*, 2492.
14. Jacobs, H. J. C.; Boomsma, F.; Havinga, E.; Van der Gen, A. *Rec. Trav. Chim. Pays-Bas* **1977**, *96*, 113.
15. Chaplan, O. L.; Smith, H. G.; King, P. W. *J. Am. Chem. Soc.* **1963**, *85*, 806.
16. Ayer, W. A.; Browne, L. M. *Can J. Chem.* **1974**, *52*, 1352.
17. Halazy, S.; Krief, A. *Tetrahedron Lett.* **1980**, 1997.
18. Floyd, A. J.; Dyke, S. F.; Ward, S. E. *Chem. Rev.* **1976**, *76*, 509, and references therein.
19. Minot, C.; Roland-Gosselin, P.; Thal, C. *Tetrahedron* **1980**, *36*, 1209.
20. Yang, N. C.; Lenz, G. R.; Shani, A. *Tetrahedron Lett.* **1966**, 2941.
21. Russel, J. H. *Naturwissenschaften* **1963**, *50*, 443.
22. Cava, M. P.; Havlicek, S. C.; Lindert, A. *Tetrahedron Lett.* **1966**, 2937.
23. Kupchan, S. M.; Wormser, H. C. *Tetrahedron Lett.* **1965**, 359.
24. Kupchan, S. M.; Kanojia, R. M. *Tetrahedron Lett.* **1966**, 5353.
25. Cava, M. P.; Mitchell, M. J.; Havlicek, S. C.; Lindert, A.; Spangler, R. J. *J. Org. Chem.* **1970**, *35*, 175.
26. Castedo, L.; Estevez, R.; Saa, J. M.; Suau, R. *Tetrahedron Lett.* **1978**, 2179.
27. Padwa, A.; Doubleday, C.; Mazzu, A. *J. Org. Chem.* **1977**, *42*, 3271.
28. McDonald, E.; Martin, R. T. *Tetrahedron Lett.* **1978**, 4723.
29. Bergman, J.; Carlsson, R. *Tetrahedron Lett.* **1977**, 4663.
30. Ibid., **1978**, 4055.
31. Fischer, N. H.; Olivier, E. J.; Fischer, H. D. *Fortschr. Chem. Org. Naturst.* **1979**, *38*, 134.
32. Corey, E. J.; Hortmann, A. G. *J. Am. Chem. Soc.* **1963**, *85*, 4033.
33. Ibid., **1965**, *87*, 5736.
34. Grieco, P. A.; Nishizawa, M. *J. Org. Chem.* **1977**, *42*, 1717.
35. Fujimoto, Y.; Shimizu, T.; Tatsuno, T. *Tetrahedron Lett.* **1976**, 2041.
36. Watanabe, M.; Yoshikoshi, A. *Tohoku Daigeku Hisui. Kagaku Kenk. Hokoku* **1973**, 23, 53; *Chem. Abstr.* **1974**, *81*, 169643v.
37. Hortmann, A. G.; Daniel, D. S.; Martinelli, J. E. *J. Org. Chem.* **1973**, *38*, 728.
38. Hortmann, A. G. *Tetrahedron Lett.* **1968**, 5785.
39. Eschenmoser, A. *Chem. Soc. Rev.* **1976**, *5*, 377.
40. Scott, A. I. *Tetrahedron* **1975**, *31*, 2639.
41. Wender, P. A.; Eissenstat, M. A.; Filosa, M. P. *J. Am. Chem. Soc.* **1979**, *101*, 2196.
42. Marshall, J. A.; Conrow, R. E. *J. Am. Chem. Soc.* **1980**, *102*, 4274.
43. Shamma, M.; Hwang, D. *Tetrahedron* **1974**, *30*, 2279.
44. Kametani, T.; Enomoto, Y.; Takahashi, K.; Fukumoto, K. *J. Chem. Soc., Perkin Trans. 1* **1979**, 2836.
45. Quillinan, A. F.; Schinmann, F. *J. Chem. Soc., Perkin Trans. 1* **1972**, 1382.
46. Yang, N. C.; Shani, A.; Lenz, G. R. *J. Am. Chem. Soc.* **1966**, *88*, 5369.
47. Lenz, G. R.; Yang, N. C. *J. Chem. Soc., Chem. Commun.* **1967**, 1136.
48. Ninomiya, I. *Heterocycles* **1974,** *2*, 105.

49. Lenz, G. R. *Synthesis*, **1978**, 489.
50. Ninomiya, I. *Heterocycles* **1980**, *14*, 1567.
51. Ninomiya, I.; Naito, T. *Heterocycles* **1981**, *15*, 1433.
52. Yang, N. C.; Lin, L. C.; Shani, A.; Yang, S. S. *J. Org. Chem.* **1969**, *34*, 1845.
53. Ninomiya, I.; Naito, T.; Mori, T. *Tetrahedron Lett.* **1969**, 3643.
54. Lenz, G. R. *J. Org. Chem.* **1976**, *41*, 2201.
55. Ogata, Y.; Takagi, K.; Ishino, I. *J. Org. Chem.* **1971**, *36*, 3975.
56. Ninomiya, I.; Naito, T.; Kiguchi, T. *J. Chem. Soc., Chem. Commun.* **1970**, 1669.
57. Ninomiya, I.; Naito, T.; Kiguchi, T. *J. Chem. Soc., Perkin Trans. 1* **1973**, 2261.
58. Ninomiya, I.; Tada, Y.; Kiguchi, T.; Yamamoto, O.; Naito, T. *Heterocycles* **1978**, *9*, 1527.
59. Ninomiya, I.; Takasugi H.; Naito, T. *J. Chem. Soc., Chem. Commun.* **1973**, 732.
60. Ninomiya, Y.; Naito, T.; Takasugi, H. *J. Chem. Soc., Perkin Trans. 1* **1976**, 1865.
61. Sainsbury, M.; Uttley, N. L. *Chem. Soc., Chem. Commun.* **1977**, 319.
62. Sainsbury, M.; Uttley, N. L. *J. Chem. Soc., Perkin Trans. 1* **1977**, 2109.
63. Ninomiya, I.; Naito, T. *Heterocycles* **1974**, *2*, 607.
64. Ninomiya, I.; Naoto, T. *J. Chem. Soc., Chem. Commun.* **1973**, 137.
65. Ninomiya, I.; Naito, T.; Takasugi, H. *J. Chem. Soc., Perkin Trans. 1* **1975**, 1720.
66. Kametani, T.; Honda, T.; Sugai, T.; Fukumoto, K. *Heterocycles* **1976**, *4*, 927.
67. Ninomiya, I.; Takasugi, H.; Naito, T. *Heterocycles* **1973**, *1*, 17.
68. Ninomiya, I.; Naito, T.; Tagasugi, H. *J. Chem. Soc., Perkin Trans. 1* **1975**, 1791.
69. Kametani, T.; Ujiie, A.; Ihara, M.; Fukumoto, K.; Lu, S. T. *J. Chem. Soc., Perkin Trans. 1* **1976**, 1218.
70. Begley, W. J.; Grimshaw, J. *J. Chem. Soc., Perkin Trans. 1* **1977**, 2324.
71. Kessar, S. V.; Singh, G.; Palakrishnan, P. *Tetrahedron Lett.* **1974**, 2269.
72. Ninomiya, I.; Naito, T.; Ishii, H.; Ishida, T.; Ueda, M.; Harada, K. *J. Chem. Soc., Perkin Trans. 1* **1975**, 762.
73. Ninomiya, I.; Kiguchi, T. Unpublished results.
74. Ishii, H.; Ueda, E.; Nakajima, K.; Ishida, T.; Ishikawa, T.; Harada, K.; Ninomiya, I.; Naito, T.; Kiguchi, T. *Chem. Pharm. Bull.* **1978**, *26*, 864.
75. Ninomiya, I.; Yamamoto, O.; Naito, T. *Heterocycles* **1977**, *7*, 137.
76. Ninomiya, I.; Naito, T.; Ishii, H. *Heterocycles* **1975**, *3*, 307.
77. Ninomiya, I.; Yamamoto, O.; Naito, T. *Heterocycles* **1977**, *7*, 131.
78. Ibid., **1976**, *4*, 743.
79. Ninomiya, I.; Yamamoto, O.; Naito, T. *J. Chem. Soc., Chem. Commun.* **1976**, 437.
80. Hara, H.; Hoshino, O.; Umezawa, B. *Tetrahedron Lett.* **1972**, 5031.
81. Iida, H.; Aoyagi, S.; Kibayashi, C. *J. Chem. Soc., Chem. Commun.* **1974**, 499.
82. Ninomiya, I.; Kiguchi, T.; Tada, Y. *Heterocycles* **1977**, *6*, 1799.
83. Ninomiya, I.; Kaito, T. Paper presented to the Kinki Regional Meeting of the Pharmaceutical Society of Japan, Osaka, **1972**, reported in Ref. 51.
84. Naito, T.; Tada, Y.; Nishiguchi, Y.; Ninomiya, I. *Heterocycles* **1981**, *16*, 1137.
85. Naito, T.; Tada,Y.; Ninomiya, I. *Heterocycles* **1981**, *16*, 1141.
86. Mondon, A.; Krohn, K. *Chem. Ber.* **1972**, *105*, 3726.
87. Ninomiya, I.; Kiguchi, T. *J. Chem. Soc., Chem. Commun.* **1976**, 624.
88. Ninomiya, I.; Kiguchi, T.; Naito, T. *Heterocycles* **1976**, *4*, 973.
89. Naito, T.; Ninomiya, I. *Heterocycles* **1980**, *14*, 959.
90. Naito, T.; Miyata, O.; Ninomiya, I.; Pakrashi, S. C. *Heterocycles* **1981**, *16*, 725.
91. Clark, V. M.; Cox, A. *Tetrahedron* **1966**, *22*, 3421.
92. Kametani, T.; Ogasawara, K.; Takahashi, T. *Tetrahedron* **1973**, *29*, 73.
93. Kametani, T.; Kajiawara, M.; Fukumoto, K. *Chem. Ind. (London)* 1973, 1165.
94. Carruthers, W. *J. Chem. Soc., Chem. Commun.* **1966**, 272.
95. Schultz, A. G.; Lucci, R. D. *J. Chem. Soc., Chem. Commun.* **1976**, 925.
96. Schultz, A. G.; Yee, Y. K.; Berger, M. H. *J. Am. Chem. Soc.* **1977**, *99*, 8065.

97. Hunter, D. H.; Sim, S. K. *J. Am. Chem. Soc.* **1969**, *91*, 6202.
98. Piancatelli, G.; Scettri, A.; David, G.; D'Auria, M. *Tetrahedron* **1978**, *34*, 2775.
99. Woodward, R. B.; Closs, G. L.; Le Goff, E.; Ayer, W. A.; Dutler, H.; Leimgruber, W.; Beaton, J. M.; Hannah, J.; Lwowski, W.; Bickelhaupt, F.; Hauck, F. P.; Sauer, J.; Bonnett, R.; Ito, S.; Valenta, Z.; Buchschacher, P.; Langemann, A.; Volz, H. *J. Am. Chem. Soc.* **1960**, *82*, 3800.
100. Fleming, I. "Selected Organic Syntheses"; Wiley: London, 1978; pp. 119–20.
101. Arigoni, D.; Bosshard, H.; Bruderer, H.; Büchi, G.; Jeger, O.; Krebaum, L. J. *Helv. Chim. Acta* **1957**, *40*, 1732.
102. Barton, D. H. R.; DeMayo, P.; Shafiq, M. *J. Chem. Soc.* **1958**, 140.
103. Schaffner, K.; Demuth, M. In "Rearrangements in Ground and Excited States"; DeMayo, P., Ed.; Academic: New York, **1980**; Vol. 3, pp. 281–348.
104. Wagner, P. J.; Hammond, G. S. *Advances in Photochem.* **1968**, *5*, 111.
105. Barton, D. H. R. *Helv. Chim. Acta* **1959**, *42*, 2604.
106. Woodward, R. B.; Hoffmann, R. *Angew. Chem., Int. Ed. Engl.* **1969**, *8*, 781.
107. Zimmerman, H. E.; Schuster, D. I. *J. Am. Chem. Soc.* **1961**, *83*, 4486.
108. Zimmerman, H. E.; Schuster, D. I. *J. Am. Chem. Soc.* **1962**, *84*, 4527.
109. Schuster, D. I. *Acc. Chem. Res.* **1978**, *11*, 65.
110. Zimmerman, H. E.; Pasteris, R. J. *J. Org. Chem.* **1980**, *45*, 4864.
111. Ibid., 4876.
112. Streitwieser, A. "Molecular Orbital Theory of Organic Chemists"; Wiley: New York, **1961**, p. 60, 61.
113. Caine, D.; Chu, C. Y. *Tetrahedron Lett.* **1974**, 703.
114. Caine, D.; Graham, S. L. *Tetrahedron Lett.* **1976**, 2521.
115. Dutler, H.; Ganter, C.; Ryf, H.; Utzinger, E. C.; Weinberg, K.; Schaffner, K.; Arigoni, D.; Jeger, O. *Helv. Chim. Acta* **1962**, *45*, 2346.
116. Marshall, J. A.; Johnson, P. C. *J. Org. Chem.* **1970**, *35*, 192.
117. Caine, D.; Boucugnani, A. A.; Pennington, W. R. *J. Org. Chem.* **1976**, *41*, 3632.
118. Caine, D.; Boucugnani, A. A.; Chao, S. T.; Dawson, J. B.; Ingwalson, P. F. *J. Org. Chem.* **1976**, *41*, 1539.
119. Ganter, C.; Utzinger, E. C.; Schaffner, K.; Arigoni, D.; Jeger, O. *Helv. Chim. Acta* **1962**, *45*, 2403.
120. Kropp, P. J. *J. Am. Chem. Soc.* **1964**, *86*, 4053.
121. Ibid. **1965**, *87*, 3914.
122. Greene, A. E. *J. Am. Chem. Soc.* **1980**, *102*, 5337.
123. White, E. H.; Eguchi, S.; Marx, J. N. *Tetrahedron* **1969**, *25*, 2099.
124. Caine, D.; Gupton, J. T., III. *J. Org. Chem.* **1975**, *40*, 809.
125. Gasa, S.; Hamanaka, N.; Matsunaga, S.; Okuno, T.; Takeda, N.; Matsumoto, T. *Tetrahedron Lett.* **1976**, 553.
126. Caine, D.; Tuller, F. N. *J. Org. Chem.* **1973**, *38*, 3663.
127. Caine, D.; Frobese, A. S. *Tetrahedron Lett.* **1977**, 3107.
128. Shiozaki, M.; Mori, K.; Hiraoka, T.; Matsui, M. *Tetrahedron*, **1974**, *30*, 2647.
129. Caine, D.; Ingwalson, P. F. *J. Org. Chem.* **1972**, *37*, 3751.

Appendix

Summary Table of Natural Products

Natural Product	*Class*	*Formula Number*	*Chapter*	*References*
Acetylemodin	Pigment	**1195**	6	15
2-Acetyl-5-hydroxy emodin	Pigment	**649**	5	104
β-Acoradiene	Terpene	**1655**	8	45
γ-Acoradiene	Terpene	**918**	5	240, 241
δ-Acoradiene	Terpene	**919**	5	240, 241
Acoragermacrone	Terpene	**329**	3	163
			7	50
			8	26
β-Acoratriene	Terpene	**1575**	8	16
β-Acorenol	Terpene	**1654**	8	45
Acorenone	Terpene	**1651**	8	45
Acorenone B	Terpene	**36**	2	33, 34
			8	45
Adaline	Alkaloid	**1039**	5	323
α-Agarofuran	Terpene	**1113**	5	371
Ajugarin I	Terpene	**527**	5	40
Alamarine	Alkaloid	**1789**	9	90
Alantolactone	Terpene	**1568**	8	10
Albene	Terpene	**518**	5	30, 31
Allethrolone	Terpenoid	**1817**	9	98
α-Allokainic acid	Alkaloid	**1585**	8	18, 20, 33
Allosedamine	Alkaloid	**429**	4	48
Alnusenone	Terpene	**704**	5	121, 122
Altersolanol B	—	**929**	5	246
Amarine	—	**1812**	9	97
Amorphane	Terpene	**575a**	5	68
α-Amorphene	Terpene	**1259**	7	32
Andranginine	Alkaloid	**835**	5	184
Angustidine	Alkaloid	**1762b**	9	59, 60

Natural Product	*Class*	*Formula Number*	*Chapter*	*References*
Angustine	Alkaloid	**1762e**	9	60
Angustoline	Alkaloid	**1762c**	9	60, 61
α-Anhydrodi-hydrocaranine	Alkaloid	**1767**	9	81
Anhydro-lycorine	Alkaloid	**1766**	9	80
Annotinine	Alkaloid	**219**	3	99, 100
Apiose	Sugar	**294**	3	137
Aristolene	Terpene	**1544**	8	1
Aristolochic acid	Terpenoid	**1680**	9	23
Aristolone	Terpene	**49**	2	49
			3	172
			4	27
Aromadendrene, 4-epi	Terpene	**1850**	9	124
Ascaridole	Terpene	**1111**	5	370
Aspidofrac-tinine	Alkaloid	**564**	5	64
Aspidosper-mine	Alkaloid	**892**	5	219
AT-125	—	**372**	4	21, 22
Atisine	Alkaloid	**231**	3	106
α-Atlantone	Terpene	**505**	5	14, 242
			8	34
β-Atlantone	Terpene	**922**	5	242
			8	34
Atractylon	Terpene	**331**	3	164
Aureolic acid	—	**700**	5	120
Avicine	Alkaloid	**1764b**	9	71, 72
Bakkenolide A	Terpene	**1538**	7	148, 149
Bakuchiol, methyl ether	Terpene	**1338**	7	75
Berberine	Alkaloid	**1753**	9	47
α-Bergamotene	Terpene	**246a**	3	117
β-Bergamotene	Terpene	**246b**	3	117
Biotin	Vitamin	**128**	3	23
			4	54–56
			7	95
β-Bisabolene	Terpene	**1619**	8	34
2,β-Bisabolene	Terpene	**1318**	7	63
γ-Bisabolene	Terpene	**1339**	7	77
α-Bisabolol	Terpene	**469**	4	67
Botryodiplodin	Antibiotic	**1451**	7	119
α-Bourbonene	Terpene	**149**	3	55, 121

Natural Product	*Class*	*Formula Number*	*Chapter*	*References*
β-Bourbonene	Terpene	**150**	3	55, 113
Brefeldin A	—	**25**	2	23, 24
Brevicomin	Pheromone	**1037**	5	323, 325, 326
Bulgarane	Terpene	**576a**	5	68
β-Bulnesene	Terpene	**270**	3	125
Burchellin	Lignan	**1187**	6	12
Cacalol	Terpene	**1335**	7	72
Cadinane	Terpene	**576b**	5	68
α-Cadinol	Terpene	**1861**	9	127
α-Cadinol, 3-oxo	Terpene	**1860**	9	127
Calamenene	Terpene	**1303**	7	51
			8	25, 26
Calameon	Terpene	**333**	3	165
Calciferol	Vitamin	**1138**	*See* Vitamin D2	
Calycanine	Alkaloid	**1791**	9	91
Campherenone	Terpene	**1155**	6	6
Camphor	Terpene	**1603**	8	24
Camptothecin	Alkaloid	**1413**	7	105–107
Cantharidin	Alkaloid	**499**	5	9
Carpanone	Lignan	**1065**	5	341
Caryophyllene	Terpene	**154**	3	56
α-Caryophyllene alcohol	Terpene	**172**	3	70
Catharanthine	Alkaloid	**828**	5	248, 306
			7	83
Cavidine	Alkaloid	**1763c**	9	67, 68
Cecropia juvenile hormone	Pheromone	**1355**	7	80, 121, 147
			8	39, 40
α-Cedrene	Terpene	**797**	5	153, 154
β-Cedrene	Terpene	**264**	3	123
			5	153, 154
Cedrol	Terpene	**796**	5	153, 154
Celebixanthone, methyl ether	Xanthone	**1332**	7	70
Cembrene	Terpene	**1402**	7	100
Chalcogran	Pheromone	**1042**	5	328
Chamaecynone	Terpene	**638**	5	99, 100
Chamazulene	Terpene	**1144a**	6	1

Natural Product	*Class*	*Formula Number*	*Chapter*	*References*
α-Chamigrene	Terpene	**34**	2	32, 34
β-Chamigrene	Terpene	**579**	5	69
Chanoclavine I	Alkaloid	**447**	4	58, 66
Chartreusin	Antibiotic	**535**	5	48
Chasmanine	Alkaloid	**1024a**	5	313, 314
Chelerythrine	Alkaloid	**1764e**	9	75
Chelidonine	Alkaloid	**846**	5	191
Chelirubine	Alkaloid	**1764d**	9	74
Chinensinaphthol	Lignan	**736f**	5	130
Chinensinaphthol, methyl ether	Lignan	**736g**	5	130
Chlorophyll	Pigment	**1823**	9	99
Cholesterol	Steroid	**480**	5	2
			7	89
Chrysanthemic acid	—	**2**	2	6, 11, 12
Chrysanthemic acid, methyl ester	—	**2**	3	131
Chrysophanol	Pigment	**687**	5	116, 272
Civetone (*cis*)	—	**1203**	6	17
α-Clausenane	Terpene	**1128**	5	377
Clerodane (intermediate to)	Terpene	**519**	5	32
Cocaine	Alkaloid	**421**	4	47, 61, 62
Colchicine	Alkaloid	**139**	5	8
Confertine	Terpene	**19**	2	20
			9	41
δ-Coniceine	Alkaloid	**1090**	5	360
Copacamphene	Terpene	**184**	3	75
α-Copaene	Terpene	**975**	5	271
β-Copaene	Terpene	**256**	3	120
			5	271
Corey aldehyde	Intermediate	**32**	2	31
			7	35
Coriolin	Terpene	**669**	5	109, 110
Coriolin B	Terpene	**670**	5	110
Coronafacic acid	Phytotoxin	**514**	5	24, 195
			7	31

Natural Product	*Class*	*Formula Number*	*Chapter*	*References*
Corrin	—	**1232**	7	15, 16
Cortisone	Steroid	**479**	5	2
Corynoline	Alkaloid	**1765b**	9	78
Corynoline, 11-epi	Alkaloid	**1765d**	9	79
Corytenchirine	Alkaloid	**1763d**	9	69
Costaclavine	Alkaloid	**1782a**	9	87
Costaclavine, 8-epi	Alkaloid	**1782b**	9	87
Costunolide	Germacranolide	**1270**	7	48
			9	34
Crinan	Alkaloid	**1758**	9	56, 57
Crotepoxide	—	**960**	5	263, 264, 374, 375
Cryptomerion	Terpene	**1622**	8	34
Cryptotanshinone	Pigment	**606**	5	84
α-Cuparenone	Terpene	**4**	2	8
			3	13
			6	5
β-Cuparenone	Terpene	**115**	3	13
Cybullol	—	**1115**	5	372
Cyclocolorenone	—	**1867**	9	129
Cyclocopacamphene	Terpene	**342**	3	175
			4	59
α,β-Cycloeudesmol	Terpene	**15**	2	14
β,β-Cycloeudesmol	Terpene	**12**	2	14
Cyclonerodiol	Terpene	**1572**	8	13
Cyclopiperstachine	Alkaloid	**837a**	5	185
Cyclosativene	Terpene	**450**	4	60
Cyclostachine A	Alkaloid	**837**	5	186
α-Cyperone	Terpene	**345**	3	177
β-Cyperone	Terpene	**346**	3	177
Cytochalasin B	Alkaloid	**1025**	5	318

Natural Product	*Class*	*Formula Number*	*Chapter*	*References*
α-Damascenone	Terpene	**904**	5	232, 233
β-Damascenone	Terpene	**923**	5	243, 244
δ-Damascenone	Terpene	**508**	5	17
Damsinic acid	—	**1715**	9	41
Daunomycinone	Antibiotic	**972**	5	269, 275, 283, 290
Dehydroanhydropicropodophyllin	Lignan	**511c**	5	21
Dehydrocycloguanandin	Xanthone	**1737**	9	45
Dehydropodophyllotoxin	Lignan	**736a**	5	130, 132
4-Demethoxydaunomycinone	Antibiotic	**987**	5	276–278, 286, 288–290, 298, 299
Dendrobine	Alkaloid	**786**	5	149, 150
Dendrolasin	Terpene	**1126a**	5	376
Dentatin	—	**1328**	7	68
Denudatine	Alkaloid	**1022**	5	310, 311
13-Deoxydelphonine	Alkaloid	**1024b**	5	313, 314
Deoxyerythrolaccin	Pigment	**1196**	6	16
Deoxyerythrolaccin, trimethyl ether	Pigment	**654**	5	105
3-Deoxy-1-hydroxyvitamin D_3	Vitamin	**1220**	7	12
15-Deoxyprostaglandin E_1	—	**1109**	5	369
Deoxytaylorione	—	**283**	3	132
			7	77
Desethylibogamine	Alkaloid	**86**	2	73
Desmosterol	Steroid	**1269**	7	36
Dictyolene	—	**1644**	8	43
Dictyopterene B	—	**387**	4	30
Dictyopterene C′	—	**1241**	7	23

Natural Product	*Class*	*Formula Number*	*Chapter*	*References*
3,6-Dideoxy-arabinohexo-pyranose	Sugar	**1045**	5	330
3,6-Dideoxy-ribohexo-pyranose	Sugar	**1046**	5	330
Digitopurpone	Anthraquinone	**711**	5	123, 126
Diketo-coriolin B	—	**671**	5	110
Dihydro-angustine	Alkaloid	**1762f**	9	60
Dihydro-antirhine	Alkaloid	**112**	3	7
4α-Dihydro-cleavamine	Alkaloid	**1476a**	7	126
Dihydro-costunolide	Germacranolide	**1298**	9	32–35
13,14-Dihydro-11-deoxyprostaglandin	—	**163**	3	66, 67
Dihydrojasmone	—	**1213**	7	9, 117
Dihydro-lycorine	Alkaloid	**520**	5	33
Dihydronovanin	—	**1704**	9	36
Dihydro-spiniferin-1	Terpene	**1722**	9	42
Dihydroxy-heliotridane	Alkaloid	**66**	2	57
(*E*,*E*)-3,7-Dimethyl-2,6-decadiene-1,10-diol	Pheromone	**1454**	7	120
(*E*)-2,4-Dimethyl-2-hexenoic acid	—	**1483**	7	129
3,7-Dimethyl-pentadec-2-yl acetate	Pheromone	**1248**	7	27
Diphyllin	Lignan	**736d**	5	130
(*E*,*Z*)-7,9-Dodecadien-1-yl acetate	Pheromone	**1409**	7	103
(*E*)-7-Dodecen-1-ol	Pheromone	**1486a**	7	130
β-Dolabrin	Terpenoid	**391**	4	31
Eburnamonine	Alkaloid	**3**	2	7

Natural Product	*Class*	*Formula Number*	*Chapter*	*References*
Ectocarpene	—	**389**	4	30
Elaeocarpine	Alkaloid	**418**	4	47
Elaeocarpus alkaloid dienone	Alkaloid	**93**	2	76
Elaeokanine A	Alkaloid	**420**	4	47
			5	359
Elemol	Terpene	**1292**	7	46
Elenolic acid, methyl-ester	—	**523**	5	35
Ellipticine	Alkaloid	**1696**	9	29
Elliptone	Rotenoid	**1334**	7	71
Emodin	Pigment	**686**	5	116
Emodinanthron	Pigment	**969**	5	268
Epizonarene	Terpene	**811**	5	160
Epoxydon	—	**959**	5	262
Eremophilane	Terpene	**814**	5	162
Eremophilone	Terpene	**1321**	7	64
Eriolangin	Terpenoid	**124b**	3	19
Eriolanin	Terpenoid	**124a**	3	18, 19
Ervinceine	Alkaloid	**832c**	5	182, 183
Estradiol	Steroid	**864b**	5	202, 203, 218
Estrone	Steroid	**864a**	5	199, 200, 209, 210, 212, 213
			7	111
Evodiamine	Alkaloid	**1096a**	5	361
Fagaronine	Alkaloid	**1764f**	9	76
Festuclavine	Alkaloid	**1782c**	9	87
Fichtelite	Terpene	**812**	5	161
Ficisterol	Steroid	**1393**	7	97
Flavopereirine	Alkaloid	**1759**	9	58
Flavoskyrin	Pigment	**1053**	5	332
1-Formyl-6,7-dihydro-5*H*-pyrrolizine	Pheromone	**404**	4	40
Fragrantol	Terpene	**146**	3	52
Friedelin	Terpene	**705**	5	121, 122
Frontalin	Pheromone	**1036**	5	323, 325
Frullanolide	Terpene	**1465**	7	124
Fuerstione	—	**1058a**	5	333
Fujenoic acid (intermediate to)	—	**510**	5	19, 20

Natural Product	*Class*	*Formula Number*	*Chapter*	*References*
Fulvoplumierin	—	**509**	5	18
Fumagillin	—	**561**	5	63
Fusidic acid	Steroid	**1050**	5	331
Futoenone	Lignan	**1188**	6	12
Garriyne	—	**1555**	8	6
Gascardic acid	Terpene	**1313**	7	59
Geijerone	Terpene	**1289**	7	45
			8	11
Genepic acid	Antibiotic	**169**	3	68
Germacradiene	Terpene	**318**	3	156
Germanicol	—	**1564**	8	9
Gibberellic acid	Terpenoid	**589**	5	75, 76, 168
Gibberone	Terpenoid	**70**	2	59
Glaucine	—	**1684**	9	25
Globulol	Terpene	**23**	2	22
Globulol, 4-epi	Terpene	**1849**	9	124
Glycozoline	Alkaloid	**1799**	9	94
Grandisol	Pheromone	**21**	2	19
			3	46–51, 148
			9	16
Grayanotoxin II	—	**1853**	9	125
Griseofulvin	Antibiotic	**643**	5	103
Griseofulvin, epi	—	**641**	5	102
Guaiazulene	Terpene	**1144b**	6	1
Guanandin	Xanthone	**1329**	7	69
Guianin	Lignan	**1185**	6	12
Gymnomitrol	Terpene	**1180**	6	11
Haemanthamine	Alkaloid	**517**	5	28, 29
Haemanthidine	Alkaloid	**521**	5	34
Hastanecine	Alkaloid	**67**	2	57
Heliocide H2	—	**591**	5	80
Heliotridine	Alkaloid	**1098**	5	363
Helminthosporal	—	**186**	3	75
Helminthosporin	Pigment	**688**	5	116, 272
Hentriacontane-14,16-dione	—	**374**	4	23
Hexahydro-17-methoxy-yohimbane	—	**1084b**	5	356
Hibane	Terpene	**206d**	3	90
			5	192
Hibaol	Terpene	**206b**	3	90
			5	192
Hibaone	Terpene	**206c**	3	90

Natural Product	*Class*	*Formula Number*	*Chapter*	*References*
Hibayl acetate	Terpene	**206a**	3	90
α-Himachalene	Terpene	**942**	5	253
			8	8
β-Himachalene	Terpene	**190**	3	76, 77
			5	253
			7	24
			8	8
Hinesol	Terpene	**38**	2	35, 36
Hinesol, epi	Terpene	**39**	2	35, 36
Hirsutane	Terpenoid	**279**	3	128, 129
Hirsutene	Terpene	**192**	3	78, 151, 152
Homochelidonine	Alkaloid	**1765a**	9	75, 77
Humbertiol	Terpene	**1337**	7	74
Hydroxycotinine	Alkaloid	**422**	4	47
Hydroxyloganin	Terpene-glucose	**197b**	3	83
Hydroxyvitamin D_4	Vitamin	**1227**	7	13
Ibogaine	Alkaloid	**1007**	5	303
Ibogamine	Alkaloid	**933**	5	247, 302, 303, 305,
Ibogamine, epi	Alkaloid	**1006**	5	302, 304, 305
Illudol	Terpene	**147**	3	53
			5	112
Ipomeamarone	Terpene	**1129a**	5	377
Ipomeamarone, epi	Terpene	**1129b**	5	377
Ipsdienol	Pheromone	**1404**	7	101
Ipsenol	Pheromone	**953**	5	260
			7	101
			8	35
			9	17
Isabelin	Terpene	**325**	3	161
Ishwarane	Terpenoid	**222**	3	101
Ishwarone	Terpenoid	**27**	2	25, 26
Islandicin	Pigment	**710**	5	123, 126
Isoacoragermacrone	Terpene	**328**	3	163
			7	50
Isoalantolactone	Terpene	**332**	3	164
Isocalamendiol	Terpene	**334**	3	166
Isocaryophyllene	Terpene	**118**	3	14, 56, 57
Isochanoclavine I	Alkaloid	**465**	4	66

Natural Product	*Class*	*Formula Number*	*Chapter*	*References*
Isocomene	Terpene	**261**	3	122
			8	22
Isocrypto-tanshinone	Pigment	**604**	5	84
9-Isocyano-pupukeanane	Terpene	**772**	5	145
Isodehydro-abietenolide	—	**1522**	7	145
Isodeoxypodo-phyllotoxin	Lignan	**513**	5	23
Isoelaeocarpine	Alkaloid	**419**	4	47
Isoguanandin	Xanthone	**1330**	7	69
Isolinderalactone	Terpene	**1277**	7	47
Isomarasmic acid	Terpene	**339**	3	173, 174
			4	28, 29
			5	37
Isomarasmic acid, methyl ester	Terpene	**174**	3	71
Isophyllocladene	Terpene	**225**	3	103, 104
Isoretronecanol	Alkaloid	**63**	2	56
			4	38, 39
Isoretro-necanolate	—	**402**	4	38
Isotanshinone II	Pigment	**603**	5	84
Jasmone	Terpenoid	**8**	2	10
			3	17
			8	41
Jasmonic acid, methyl ester	—	**949**	5	258
Juncusol	—	**1692**	9	28
Junenol, 10-epi	Terpene	**181**	3	74
Justicidin A	Lignan	**736e**	5	130
Justicidin B	Lignan	**511a**	5	21
Justicidin E	Lignan	**740**	5	132
Juvabione (*erythro*)	Terpene	**1265**	7	34
Juvabione (*threo*)	Terpene	**567**	5	65, 66

Natural Product	*Class*	*Formula Number*	*Chapter*	*References*
α-Kainic acid	—	**1582**	8	19
Karahanaenone	Terpene	**1172**	6	10
			7	61, 62
Karahana ether	Terpenoid	**945**	5	255
Kaurene	Terpene	**73**	2	61, 62
Kermesic acid	Pigment	**679**	5	114
Kessane, 5-epi	Terpene	**178**	3	73
Kessanol	Terpene	**1631**	8	37
Kessanol, 8-epi	Terpene	**1630**	8	37
Khusimone	Terpene	**790**	5	151
			8	7
Klaineanone (intermediate to)	—	**854**	5	194
Laburnine	Alkaloid	*See* Trachelanthamidine		
Laccaic acid, tetramethyl ether	Pigment	**657**	5	105
Lachnanthocarpone	Pigment	**805**	5	157, 158
Lanceol	Terpene	**1480**	7	128
Lanceolic acid, ethyl ester	Terpene	**1479**	7	127, 128
Lasalocid A	Antibiotic	**982**	5	273
			7	118
Lasiodiplodin	—	**639**	5	101
Ligularone	Terpenoid	**611**	5	85
Linderalactone	Terpene	**1276**	7	47
Lineatin	Pheromone	**148**	3	54
Loganin	Terpene-glucose	**197a**	3	82, 84
Loganin aglycone	Terpene	**130**	3	26
Longicyclene	Terpene	**61**	2	54, 55
Longifolene	Terpene	**276**	3	127
α-Longipinene	Terpene	**250a**	3	118
β-Longipinene	Terpene	**250b**	3	118
Luciduline	Alkaloid	**461**	4	63, 64
α-Lumicolchicine	Alkaloid	**142**	3	39
Lupinine	Alkaloid	**415**	4	47
Lupinine, epi	Alkaloid	**416**	4	47
Lycopodine, 12-epi	Alkaloid	**291**	3	134
Lycoramine	Alkaloid	**1809**	9	96

Natural Product	Class	Formula Number	Chapter	References
α-Lycorane	Alkaloid	**552**	5	58, 156
β-Lycorane	Alkaloid	**551**	5	58
γ-Lycorane	Alkaloid	**1768**	9	81
Lycoricidine	Alkaloid	**1076**	5	352
Lysergic acid	Alkaloid	**1105**	5	365
Maalienone, epi	—	**344b**	3	177
Majurone	Terpene	**57**	2	52
Manicone	Pheromone	**1484**	7	129
Marasmic acid	Terpenoid	**525**	5	38, 148
Matricarin, desacetoxy	Terpene	**1845**	9	123
Menthol	Terpene	**516**	5	27
15β-Methoxy-yohimban-17-ol	—	**1505**	7	137
Methylenomycin A	Antibiotic	**193**	3	80, 81
O-Methyl-joubertiamine	—	**1416**	7	109
7-Methyl-juglone	Pigment	**683**	5	116
Minovine	Alkaloid	**832b**	5	183, 307, 308
Modhephene	Terpene	**1600**	8	23
Monilformin	Mycotoxin	**121**	3	16
Morphine	Alkaloid	**486**	5	4
Multifiden	—	**956**	5	261
α-Multistriadin	Pheromone	**1040**	5	325, 327
Murolane	Terpene	**575b**	5	68
Muscone	Terpene	**10**	2	13
Myrtine	Alkaloid	**417**	4	47
Napelline	Alkaloid	**1023**	5	312
Narciprimine	Alkaloid	**1778b**	9	86
Nauclefine	Alkaloid	**1762a**	9	61, 62
Naucletine	Alkaloid	**1762d**	9	60
Neolindera-lactone	Terpene	**1295**	7	47
Neotorreyol	Trepene	**1126b**	5	376
Nilgherron A	Pigment	**1060a**	5	333
Nilgherron B	Pigment	**1060b**	5	333
Nitidine	Alkaloid	**1764a**	9	70–72
Nonactic acid	—	**1437**	7	116

Natural Product	*Class*	*Formula Number*	*Chapter*	*References*
(*E*)-6-Nonen-1-ol	Pheromone	**302**	3	139
			7	130
Nootkatone	Terpene	**507**	5	16
Nootkatone-11,12-dihydro	Terpene	**1829**	9	114
Norepi-maalienone	—	**344a**	3	176
Norketotrich-odiene	Terpenoid	**171**	3	69
Norpatchoulenol	Terpenoid	**761**	5	143
(*E*)-Nuciferol	Terpene	**1518**	7	144
Nuciferine	Alkaloid	**1678**	9	22, 24, 25
Occidentalol	Terpene	**974**	5	270
			9	37
Ochotensimine	Alkaloid	**1608**	8	27
Olivacine	Alkaloid	**731**	5	128
			9	30
Oogoniol	Hormone	**1390**	7	96
Oplanone	Terpene	**1856**	9	126
Ormosanine	Alkaloid	**201**	3	87, 88
7-Oxodesethyl-catharanthine	Alkaloid	**1015**	5	306
Pachybasin	Pigment	**978**	5	272
Pachydictyol A	Terpene	**1641**	8	43
			9	122
Palitantin	Antibiotic	**524**	5	36
Panamine	Alkaloid	**203**	3	87, 88
α-Panasinsene	Terpene	**253a**	3	119
β-Panasinsene	Terpene	**253b**	3	119
Patchouli alcohol	Terpene	**760**	5	142
Penicillin (derivatives)	Antibiotic	**77**	2	66
Pentalenolactone	Terpene	**621**	5	93, 94
Perhydrogephyro-toxin	Alkaloid	**558**	5	62
			7	138
Perillaketone	Terpene	**1127**	5	377
Perillene	Terpene	**1125**	5	376
Periplanone B	Pheromone	**1262**	7	33

Natural Product	*Class*	*Formula Number*	*Chapter*	*References*
Peroxy-costunolide	Germacranolide	**1546**	8	4
Peroxyparthe-nolide	Germacranolide	**1548**	8	4
Phomarin	Pigment	**684**	5	116
Phoracantholide I	Macrolide	**1498**	7	133
Phoracantholide J	Macrolide	**1367**	7	85, 133
Phycocyanobilins	Pigment	**96**	2	77
Phyllocladene	Terpene	**72**	2	61, 62
			3	103, 104
Phyllostine	—	**958**	5	262
Physcion	Pigment	**685**	5	116
β-Pinene	Terpene	**240**	3	114, 115
α-Pipitzol	—	**1190**	6	13
β-Pipitzol	—	**1191**	6	13
Piptanthine	Alkaloid	**202**	3	88
Pontevedrine	Alkaloid	**1689b**	9	26
Porantheridin	Alkaloid	**433**	4	49
Precapnelladiene, 11-epi	Terpene	**267**	3	124
Prefenic acid (sodium salt)	—	**617**	5	89–92
Preisocala-mendiol	Terpene	**1302**	7	50, 51
			8	25, 26
Presecamine	Alkaloid	**824**	5	176
Presqualene alcohol	Terpene	**51**	2	50
Progesterone	Steroid	**368**	4	20
			7	92
Propylure	Pheromone	**1472**	7	125
Prostaglandin A_2	—	**1383**	7	93
Prostaglandin E_1	—	**582**	5	71
			8	44
Prostaglandin $F_{2\alpha}$	—	**53**	2	51
			5	10, 11
			7	94
Pseudocytidine	Nucleoside	**1169**	6	9
Pseudoionone	—	**1251**	7	28, 29

Natural Product	*Class*	*Formula Number*	*Chapter*	*References*
Pseudouridine	Nucleoside	**1168**	6	9
Pumiliotoxin C	Alkaloid	**555**	5	60, 113, 147, 164, 165
9-Pupukeanone	—	**773**	5	146
Pyridoxine	Vitamin	**1062**	5	335–337
Pyrovellerolactone	Terpene	**1218**	7	10
Quasimarin	Terpenoid	**599**	5	83
Quassin	Terpenoid	**939**	5	249–251
Quebrachamine	Alkaloid	**1476b**	7	126
Reserpine	Alkaloid	**288b**	3	133
			5	5, 7
Reserpine, 3-epi	Alkaloid	**288a**	3	133
Retronecine	Alkaloid	**406**	4	41, 47
			5	363
Rubrocomatulin, pentamethyl ether	Pigment	**646**	5	104
Rutecarpine	Alkaloid	**1096b**	5	361
Ryanodine (intermediate to)	Insecticide	**572**	5	67
Sabina ketone	Terpene	**29**	2	28–30
Sabinene	Terpene	**30**	2	28–30
Sanguinarine	Alkaloid	**1764c**	9	73
Santolinic acid, methyl ester	Terpene	**1462**	7	123
Sarracenin	Terpene	**1432**	7	114
Sativene	Terpene	**183**	3	75
			4	60
Sativendiol (*cis*)	Terpene	**185**	3	75
			7	8
Scabequinone	—	**1336**	7	73
Scopine	Alkaloid	**1165**	6	8
Sedamine	Alkaloid	**430**	4	48
Sedridine	Alkaloid	**428**	4	48
Selina-3,7(11)-diene	Terpene	**810**	5	159
Semicorrin	—	**355**	4	10–14
Sendaverine	Alkaloid	**1733**	9	44

Natural Product	Class	Formula Number	Chapter	References
Seneol	—	**1118**	5	373
Senepoxyde	—	**961**	5	264, 265, 373
Septicine	Alkaloid	**436**	4	50
			9	21
Serratinine	Alkaloid	**89**	2	74
			5	45, 46
Sesibiricin	—	**1325**	7	66
Sesquicarene	Terpene	**46**	2	37–41
Seychellene	Terpene	**664**	5	107, 108, 140, 141
Showdomycin	Nucleoside	**1170**	6	9
Shyobunone	Terpene	**327**	3	162, 163
			7	51, 122
Siccanin	Antibiotic	**526**	5	39
Silybin	Alkaloid	**1056**	5	170–173
Silydianin	Alkaloid	**1057**	5	170–173
Sinactine	Alkaloid	**1763b**	9	65, 67
α-Sinensal	Terpene	**1310**	7	58
Sirenin	Terpene	**47**	2	42–48
Solanoquinone	—	**531**	5	47
Songorine (intermediate to)	Alkaloid	**84**	2	71, 72
			3	85, 86
Squalene	Terpene	**1351**	7	80, 90
Squaric acid	—	**120**	3	15
Stachenone	Terpene	**506**	5	15
Stemarin	Terpene	**228**	3	105
Steviol, methyl ester	Terpenoid	**236**	3	110, 111
			7	82
Stipitatic acid	—	**411**	4	44
Stipitatonic acid	—	**176**	3	72, 77
Streptonigrin	Alkaloid	**1082**	5	353–355
Supinidine	Alkaloid	**403**	4	39, 47
Suprasterol I	—	**1139**	5	382, 383
Suprasterol II	—	**1140**	5	382, 383
Tabersonine	Alkaloid	**827**	7	83, 104
Taiwanin C	Lignan	**511b**	5	21, 132

Natural Product	*Class*	*Formula Number*	*Chapter*	*References*
Taiwanin E	Lignan	**736b**	5	130, 132
Taiwanin E, methyl ether	Lignan	**736c**	5	130
Talatisamine	Alkaloid	**233**	3	108, 109
Tanshinone II	Pigment	**607**	5	84
Taylorione	Terpenoid	**284**	3	132
Tazettine	Alkaloid	**522**	5	34, 95
Telekin	Terpene	**1567**	8	10
Teloidine	Alkaloid	**1166**	6	8
δ-Terpineol	Terpene	**717**	5	124
n-Tetradeca-2(*E*)-4,5-trienoic acid, methyl ester	Pheromone	**1406**	7	102
(*E*)-7-Tetradecen-1-ol	Pheromone	**1486b**	7	130
Tetrahydropodophyllotoxin	Lignan	**736a**	5	130, 132
Tetrahydrocannabinol	Alkaloid	**1069**	5	343
Tetrahydropalmatine	Alkaloid	**1763a**	9	65
Tetrodotoxin	—	**528**	5	41–44
Thaliglucine	Alkaloid	**1726a**	9	43
Thaliglucinone	Alkaloid	**1726b**	9	43
Thalphenine	Alkaloid	**1725**	9	43
Thienamycin	Antibiotic	**79**	2	67–69
			3	21, 22
			4	25, 51
Thromboxane B_2	—	**1420**	7	110
α-Thujaplicin	Terpene	**1160**	6	7
β-Thujaplicin	Terpene	**390**	4	31, 44
			6	7
Thujopsadiene	Terpene	**59**	2	52
Thujopsene	Terpene	**58**	2	52, 53
Tilidine	—	**557**	5	61
Toddaculin	—	**1326**	7	66
Torreyal	Terpene	**1126c**	5	376
Torreyol	Terpene	**822**	5	167
Trachelanthamidine	Alkaloid	**64**	2	56
			4	39
Trachylobane	Terpene	**224**	3	102
Trichodermol	Terpenoid	**681**	5	115
Tropanediol	Terpene	**1164**	6	8

Natural Product	*Class*	*Formula Number*	*Chapter*	*References*
Tropine	Alkaloid	**1163**	6	8
Turmerone	Terpene	**951**	5	259
Tylophorine	Alkaloid	**1093**	5	360
			9	21
Uracil, 5-(4′,5′-dihydroxy-pentyl)	—	**160**	3	64
Valencane	Terpene	**815**	5	162
Valeranone	Terpene	**17**	2	18, 19
Valerianine	Alkaloid	**1038**	5	323
Veatchine	Alkaloid	**232**	3	107
Velleral	Terpenoid	**104**	3	5, 6
Vellerolactone	Terpenoid	**108**	3	6
Vermiculine	Antibiotic	**378**	4	24
Vernolepin	Terpene	**629**	5	96
Vernomenin	Terpene	**630**	5	96
Verrucarin E	Alkaloid	**967**	5	267
Verrucarol (intermediate to)	Terpene	**515**	5	25, 26
			7	143
α-Vetispirene	Terpene	**1836**	9	117, 118
β-Vetivone	Terpene	**7**	2	8
			9	113, 116
Vincadifformine	Alkaloid	**832a**	5	181–183
Vitamin B_6	Vitamin		*See* Pyridoxine	
Vitamin D_2	Vitamin	**1138**	7	3, 4
Vitamin E, acetate	Vitamin	**1396**	7	98
Warburganal	Terpenoid	**512**	5	22
Widdrol	Terpene	**1399**	7	99
Xanthorin, trimethyl ether	Pigment	**651**	5	104
Xylopinine	Alkaloid	**1086b**	5	356, 357
			9	64–66, 85, 92
α-Ylangene	Terpene	**976**	5	271
β-Ylangene	Terpene	**977**	5	271
Yohimbane	Alkaloid	**1760**	9	58
Zizaane (ring)	Terpenoid	**273**	3	126

Index

Index

Copy editor and indexer: Robin Giroux
Production editor: Frances F. Reed
Jacket designer: Anne G. Bigler
Managing editor: Janet S. Dodd

Typesetting: The Sheridan Press, Hanover, Pa.,
and Hot Type Ltd., Washington, D.C.
Printing: Port City Press, Inc., Washington, D.C.